HANDBOOK OF RESEARCH ON MEDICINAL CHEMISTRY

Innovations and Methodologies

HANDBOOK OF RESEARCH ON MEDICINAL CHEMISTRY

Innovations and Methodologies

Edited by

Debarshi Kar Mahapatra, PhD

*Assistant Professor, Department of Pharmaceutical Chemistry,
Dadasaheb Balpande College of Pharmacy, Rashtrasant Tukadoji
Maharaj Nagpur University, Nagpur, Maharashtra, India*

Sanjay Kumar Bharti, PhD

*Assistant Professor, Division of Pharmaceutical Chemistry,
Institute of Pharmaceutical Sciences, Guru Ghasidas Vishwavidyalaya
(A Central University), Bilaspur – 495009, Chhattisgarh, India*

APPLE
ACADEMIC
PRESS

Apple Academic Press Inc.
3333 Mistwell Crescent
Oakville, ON L6L 0A2 Canada

Apple Academic Press Inc.
9 Spinnaker Way
Waretown, NJ 08758 USA

© 2018 by Apple Academic Press, Inc.

First issued in paperback 2021

Exclusive worldwide distribution by CRC Press, a member of Taylor & Francis Group
No claim to original U.S. Government works

ISBN 13: 978-1-77-463662-6 (pbk)
ISBN 13: 978-1-77-188544-7 (hbk)

Library and Archives Canada Cataloguing in Publication

Handbook of research on medicinal chemistry : innovations and methodologies / edited by Debarshi Kar Mahapatra, PhD (Assistant Professor, Department of Pharmaceutical Chemistry, Kamla Nehru College of Pharmacy, RTM Nagpur University, Nagpur, India), Sanjay Kumar Bharti, PhD (Assistant Professor, Divi-sion of Pharmaceutical Chemistry, Institute of Pharmaceutical Sciences, Guru Ghasidas Vishwavidyalaya (A Central University), Bilaspur – 495009, Chhattisgarh, India).
Includes bibliographical references and index.
Issued in print and electronic formats.
ISBN 978-1-77188-544-7 (hardcover).--ISBN 978-1-315-20741-4 (PDF)
1. Pharmaceutical chemistry. 2. Drugs--Research--Methodology. I. Mahapatra, Debarshi Kar, editor II. Bharti, Sanjay Kumar, editor
RS403.H36 2017   615'.19   C2017-904170-3   C2017-904171-1

Library of Congress Cataloging-in-Publication Data

Names: Mahapatra, Debarshi Kar, editor. | Bharti, Sanjay Kumar, editor.
Title: Handbook of research on medicinal chemistry : innovations and methodologies / editors, Debarshi Kar Mahapatra, PhD Sanjay Kumar Bharti, PhD.
Description: Toronto ; New Jersey : Apple Academic Press, 2017. | Includes bibliographical references and index.
Identifiers: LCCN 2017026981 (print) | LCCN 2017027643 (ebook) | ISBN 9781315207414 (ebook) | ISBN 9781771885447 (hardcover : alk. paper)
Subjects: LCSH: Pharmaceutical chemistry. | Drugs--Research--Methodology. Classification: LCC RS403 (ebook) | LCC RS403 .H34 2017 (print) | DDC 615.1/9--dc23
LC record available at https://lccn.loc.gov/2017026981

Apple Academic Press also publishes its books in a variety of electronic formats. Some content that appears in print may not be available in electronic format. For information about Apple Academic Press products, visit our website at **www.appleacademicpress.com** and the CRC Press website at **www.crcpress.com**

CONTENTS

LIST OF CONTRIBUTORS

Premlata K. Ambre
Department of Pharmaceutical Chemistry, Bombay College of Pharmacy, Kalina, Santacruz (E), Mumbai – 400098, India

Anirudh V. Belubbi
Molecular Simulations Groups, Department of Pharmaceutical Chemistry, Bombay College of Pharmacy, Kalina, Santacruz (E), 400098, India

Evans C. Coutinho
Department of Pharmaceutical Chemistry, Bombay College of Pharmacy, Kalina, Santacruz (E), Mumbai – 400098, India

Aakash Deep
Department of Pharmaceutical Sciences, Ch. Bansi Lal University, Bhiwani – 127021, India

Anish N. Gomatam
Department of Pharmaceutical Chemistry, Bombay College of Pharmacy, Kalina, Santacruz (E), Mumbai – 400098, India

Nirzari Gupta
Department of Chemistry, University of Alabama at Birmingham, Birmingham, AL 35294, USA

Shweta Jain
ADINA College of Pharmacy, Sagar, M.P., 470003, India

Gunjan Joshi
Center for Biomedical Engineering and Technology, School of Medicine, University of Maryland, Baltimore, USA

Dipali Kamble
Ashokrao Mane College of Pharmacy, Peth-Vadagaon, District Kolhapur, Maharashtra, India

Sidhartha S. Kar
Department of Pharmaceutical Chemistry, IGIPS, Bhuabneswar – 751015, India

Sudha Kharade
Ashokrao Mane College of Pharmacy, Peth-Vadagaon, District Kolhapur, Maharashtra, India

T. N. V. Ganesh Kumar
Chebrolu Hanumaiah Institute of Pharmaceutical Sciences, Chandramoulipuram, Chowdavaram, Guntur, Andhra Pradesh – 522019, India

Manav Malhotra
M. K. Drugs, F-10 Industrial Focal Point, Derabassi – 140507, India

Elvis A. F. Martis
Molecular Simulations Groups, Department of Pharmaceutical Chemistry, Bombay College of Pharmacy, Kalina, Santacruz (E), 400098, India

Bibhudatta Mishra
Wilmer Eye Institute, John Hopkins University School of Medicine, 400 N Broadway, Baltimore, MD 20287, USA

Santosh Nandan
Chemworx Pvt. Ltd., Kalina, Santacruz (E), Mumbai – 400098, India

Sachinkumar Patil
Ashokrao Mane College of Pharmacy, Peth-Vadagaon, District Kolhapur, Maharashtra, India

Shitalkumar Patil
Ashokrao Mane College of Pharmacy, Peth-Vadagaon, District Kolhapur, Maharashtra, India

Amit Roy
Department of Pharmaceutics, Columbia College of Pharmacy, Raipur, Chhattisgarh, India

Ram Kumar Sahu
Department of Pharmaceutics, Columbia College of Pharmacy, Raipur, Chhattisgarh, India

Archana Sharma
Institute of Pharmaceutical Sciences, Kurukshetra University, Kurukshetra – 136119, India

Prabodh Chander Sharma
Institute of Pharmaceutical Sciences, Kurukshetra University, Kurukshetra – 136119, India

Shivani Sharma
Department of Pharmaceutical Chemistry, Indo-Soviet Friendship (ISF) College of Pharmacy, Ferozepur Road, Moga – 142001, India

Ankur Vaidya
Pharmacy College, Uttar Pradesh University of Medical Sciences, Saifai, Etawah, 206130, Uttar Pradesh, India

Amit Yadav
Institute of Pharmaceutical Sciences, Kurukshetra University, Kurukshetra – 136119, India

LIST OF ABBREVIATIONS

API	active pharmaceutical ingredients
ACTH	adrenocorticotrophic hormone
ADH	alcohol dehydrogenase
AP	alkaline phosphatase
APP	amyloid precursor protein
BPH	benign prostatic hyperplasia
BZF	benzoflavone moiety
BBB	blood brain barrier
BP	blood pressure
BHA	butylated hydroxyanisole
CIP	Cahn-Ingold-Perlog
CEA	carboxyethyl-L-argininesynthase
CVD	cardiovascular disease
CBER	Center for Biologics Evaluation and Research
CDER	Center for Drug Evaluation and Research
CNS	central nervous system
CDS	chemical delivery systems
CI	chemical ionization
$CRTH_2$	chemoattractant receptor-homologous molecule
CG	coarse-grained
CV	collective variables
CBPA	competitive binding protein technique
CUDA	compute unified device architecture
CG	conjugate gradient
DP	D-prostanoid receptor
DOR	delta opioid receptor
DCC	dicyclohexylcarbodiimide
DPPH	diphenyl-1-picrylhydrazyl
DUR	drug utilization review
EPA	eicosapentaenoic acid
ESI	electrospray ionization
EBT	embryonic bovine tracheal
EAE	encephalomyelitis

ELS	endosomes and lysosomes
EE	energy expenditure
EC	enterochromaffin cells
ELISA	enzyme linked immunosorbent assay
EC	epicatechin
EGC	epigallocatechin
EGCG	epigallocatechin gallate
EET	epoxyeicosatrienoic acid
ENL	erythema nodosum leprosum
EMDT	ethyl-5-methoxy-N,N-dimethyltryptamine
EI	eudismic index
ER	eudismic ratio
ES	extended-spectrum
ESBLs	extended-spectrum β-lactamases
ERK	extracellular signal-regulated kinases
FAERS	FDA adverse event reporting system
FRAP	ferric reducing-antioxidant power
FRET	fluorescence resonance energy transfer
FSH	follicle-stimulating hormone
FDA	Food and Drug Administration
GPCR	G-protein-coupled receptor
GC	gallocatechin
GI	gastrointestinal tract
GA	general anesthetics
GTE	green tea extract
GTP	green tea powder
GH	growth hormone
HF	Hartree-Fock
HCC	hepatocellular carcinoma
HDL	high density lipoprotein-cholesterol
HPLC	high performance liquid chromatography
HTRF	homogeneous time resolved fluorescence
HLE	human lens epithelial
HMBA	hydroxy methyl benzoic acid
HETE	hydroxyeicosatetraenoic acid
IBD	inflammatory bowel disease
IR	infrared
IMDH	inosine monophosphate dehydrogenase

IRS-1	insulin receptor substrate-1
IL-6	interleukin 6
IPAT	intraperitoneal adipose tissues
IBS	irritable bowel syndrome
KE	kinetic energy
LOSCs	laser optical synthesis chips
LTB_4	leukotriene B_4
LTs	leukotrienes
LCAO	linear combination of atomic orbitals
LE	loteprednol etabonate
LDL	low density lipoprotein
LH	luteinizing hormone
LSD	lyzergic acid diethylamide
MHT	malignant hyperthermia
MBHA	Marshall benzhydryl amine linker
MS	mass spectrometry
MALDI	matrix assisted laser desorption/ionization
MTD	maximum tolerated dose
MPI	message parsing interfaces
MRSA	methicillin-resistant *Staphylococcus aureus*
MSSA	methicillin-susceptible *Staphylococcus aureus*
MMC	migrating motor complex
MED	minimal erythema dose
MAC	minimum alveolar concentration
MICs	minimum inhibitory concentrations
MAPK	mitogen activated protein kinase
MD	molecular dynamics
MO	molecular orbital
MAO	monoamine oxidase
MAOA	monoamine oxidase A
MAOI	monoamine oxidase inhibitors
OMe-PEG	monomethyl polyethylene glycol
MCs	mononuclear cells
MPA	mycophenolic acid
NMTT	*N*-methylthiotetrazole
TCP	*N*-tetrachlorophthaloyl
NDDO	neglect of diatomic differential overlap
NCEs	new chemical moieties

NO	nitric oxide
NVOC	nitro veratryl oxycarbonyl
NBH	nitrobenzhydryl linker
NBHA	nitrobenzhydrylamine linker
NSAID	nonsteroidal antiinflammatory drugs
NA	noradrenaline
NET	noradrenaline transporter
NMR	nuclear magnetic resonance
ODV	O-desmethylvenlafaxine
OR	odds ratio
PT	parallel tempering
PBP	penicillin binding proteins
PFOS	perfluorooctane sulfonate
PFOA	perfluorooctanoic acid
PBC	periodic boundary conditions
PAS	peripheral anionic site
PAM	phenyl acetamido methyl
PDE-5	phosphodiesterase-5
PRA	plasma renin activity
PDGF	platelet-derived growth factor
PARP	poly-adenosine diphosphate ribose polymerase
PEG	polyethylene glycol
PE	potential energy
PES	potential energy surface
PI	propidium iodide
PGI	prostacyclin
PGHS-1	prostaglandin endoperoxide H synthase-1
PG	prostaglandins
PSA	prostate-specific antigen
PKA	protein kinase A
PKC	protein kinase C
PPI	proton pump inhibitor
PPTS	pryidinium p-toluenesulfonate
QSAR	quantitative structure-activity relationship
QSMR	quantitative structure–metabolism relationship
QM	quantum mechanics
RIA	radioimmunoassay
RF	ranking factors

RBL-1	rat basophilic leukemia
ROS	reactive oxygen species
RTK	receptor tyrosine kinase
REM	replica exchange method
REMD	replica exchange molecular dynamics
RPE	retinal pigment epithelial
RMDD	retrometabolic drug design
RMSD	root mean squared deviation
RMSF	root mean squared fluctuation
SEVI	semen-derived enhancer of virus infection
SERT	serotonin reuptake transporter
SMC	smooth muscle cell
SD	soft drugs
PBP	penicillin-binding proteins
SRS	spontaneous reporting systems
SD	steepest descent
STZ	streptozotocin
SHRSP	stroke-prone spontaneously hypertensive rats
SAR	structure activity relationship
SBP	systolic blood pressure
TBAF	tetrabutyl ammonium fluoride
TCP	tetrachlorophthaloyl
TMA	tetramethyl ammonium
TF	theaflavins
TD	thiamine diphosphate
TX	thromboxanes
TSH	thyroid-stimulating hormone
TRH	thyrotropin-releasing hormone
TC	total cholesterol
TBI	traumatic brain injury
TFMSA	trifluoromethane sulfonic acid
TG	triglycerides
TMSE	trimethylsilylethyl
TPH	tryptophan hydroxylase
TNF	tumor necrosis factor a
IV	intravenous
XIAP	X-linked inhibitor of apoptosis protein

FOREWORD

It is my pleasure to provide this foreword and to recommend this valuable book to all who have been involved in research in medicinal chemistry. I am absolutely delighted to learn that Debarshi Kar Mahapatra and Sanjay Kumar Bharti, both distinguished researchers in the area of medicinal chemistry, have presented a very useful reference book that encompasses various principles and applications associated with medicinal chemistry. As the title implies, *Handbook of Research on Medicinal Chemistry: Innovations and Methodologies*, this book integrates insights regarding drug discovery, pharmacological aspects of natural products, druggable targets of chemical mediators, and several classes of medicinal agents. The book is comprised of 12 well-written chapters by various reputed authors across the globe. The insightfulness of this book relies on the verity that the authors have made use of their own knowledge in this area of research, and also captures contemporary relevant literature. From Chapter 1 to 12, a good skillful blend of medicinal chemistry, chemical biology, chemical synthesis, pharmacology, formulation, and miscellaneous subjects has been achieved. The book demonstrates a prime focus toward drug discovery and development using conventional or unconventional approaches. This book is illustrated with several examples, diagrams, and figures, and is authored in a students friendly manner. Although the book mainly focuses on medicinal chemistry, it also incorporates an interesting chapter on pharmaceutical interactions, which is a part of pharmacy practice. This book is meant to serve as a reference for students, chemists, biochemists, and pharmacologists interested in learning about various interdisciplinary techniques of pharmaceutical relevance. I am very confident that this book will deliver newer aspects and applications in the field of medicinal chemistry.

—*Vivekananda Mandal, PhD*
Former WERC Researcher, Government of Japan;
Institute of Pharmaceutical Sciences
Guru Ghasidas University (Central University)
Bilaspur, Chhattisgarh, India

PREFACE

The unprecedented pace over recent years in the field of drugs has emphasized the relevance of pharmaceutical chemistry. With the development of scientific thought and practices, this subject is no longer restricted to any boundaries. This book is prepared with the objective to deliver the latest advancements in the various fields of combinatorial chemistry, drug discovery, biochemical aspects, pharmacology of medicinal agents, current practical problems, and nutraceuticals. The book aims to keep the drug molecule as the central component and to explain the associated features essential to exhibit pharmacological activity. Adopting a user-friendly format, like introduction, prototype, mechanisms, techniques, procedures, kinetics, formulations, toxicology, illustrations, etc., makes this book unique. It is a unique combination of biology, clinical aspects, biochemistry, synthetic chemistry, medicine and technology. The sole motive of the text is to provide a broad exposure to the essential aspect of pharmaceuticals. Figures, pictures, flowcharts, and illustrations make for an easy understanding of contents.

All the chapters are up to date, complete, and compiled in easy to reproduce format and availability of worthy content in only few pages. The book includes the base of USP along with all the latest therapeutic guidelines put forward by WHO & USFDA. Drug discovery remains an important part of medicinal chemistry. In Chapter 1, accelerating the drug discovery process by finding new chemical moieties (NCEs) faster using combinatorial chemistry approach is discussed. In Chapter 2, molecular dynamics simulation and methods to improve the sampling to overcome the convergence problems of classical molecular dynamics have been discussed. The chapter also focuses on ligand-protein interaction and protein folding carried out of using molecular dynamics based methods. In Chapter 3, a retrometabolic drug design approach is discussed with a focus on developing compounds by predicting the metabolic pathways of the current lead compound and to design novel analogs having predicted metabolic route, better therapeutic index, and fewer side effects.

Natural products have been popular among the growing population due to their varied advantages, and belief of being free from any side effects and are a part of many traditions across the globe. Since the awareness among the population has increased, nutraceuticals have gained adequate importance in daily life. Tea and its products are the most popular of them. Chapter 4 has focused on pharmacological targets, pharmacotherapeutics approaches, and clinical aspects of *Camellia sinensis*. This chapter focuses on the rational and science behind the use of natural supplements (or nutraceuticals) in daily routine.

Men and women have since ages been curious to improve, revive and maintain their sexual efficiency, and aphrodisiacs substances are employed for this purpose. Sexual dysfunction among any of the sexes often results in psychological stress, and affects quality of life. Chapter 5 describes various aphrodisiac molecules obtained from both plant and animal sources for treating sexual dysfunction. These molecules, in the form of whole plant, are used in many nations and are considered safe for long-term use. The aim of Chapter 6 is to describe the biosynthesis, metabolism, pathophysiological roles, agonists and antagonists of serotonin. Chapter 7 deals with the biochemistry, synthesis, pathophysiological functions, biochemical aspects, and mechanism of action(s) of eicosanoids.

Chapter 8 highlights the emergence of drug resistance parasites that have led to evolution of new approaches, such as modification of existing agents, discovery of new natural compounds, and identification of new targets. This chapter reviews the potential of newly synthesized compounds against plasmodial resistance for the prevention and treatment of malaria. Chapter 9 describes the chemistry, essential structural features, pharmacology, synthesis, structure activity relationships and applications of recently discovered and newer generations of β-lactam analogs.

A research note on anesthetics has been presented in Chapter 10, where pharmacodynamics, pharmacokinetic, drugs, their synthesis, mechanism of action, toxicity aspects are described exhaustively. Chapter 11 describes a detailed note on stereochemical aspects in medicinal chemistry. This chapter comprehensively describes fundamentals, importance, influence, role and advantages of stereochemistry in drug chemistry and pharmacology. Chapter 12 depicted the common drug interactions in daily practices.

ACKNOWLEDGMENTS

The editors would like to thank all the authors across the globe for their generous contributions. Without their support the book would not have been compiled. We respect their interest and sincere commitment for contributing and working with us even they are have such busy schedules. Thanks are also due to Apple Academic Press for their priceless support of this book.

—*Debarshi Kar Mahapatra*
Sanjay Kumar Bharti

ABOUT THE EDITORS

Debarshi Kar Mahapatra, PhD
Assistant Professor, Department of Pharmaceutical Chemistry, Dadasaheb Balpande College of Pharmacy, RTM Nagpur University, Nagpur, India

Debarshi Kar Mahapatra, PhD, is currently Assistant Professor in the Department of Pharmaceutical Chemistry, Dadasaheb Balpande College of Pharmacy, RTM Nagpur University, Nagpur, India. He taught medicinal and computational chemistry at both the undergraduate and postgraduate levels and has mentored students in their various research projects. His area of interest includes computer-assisted rational designing and synthesis of low molecular weight ligands against druggable targets, drug delivery systems, and optimization of unconventional formulations. He has published research, book chapters, reviews, and case studies in various reputed journals and has presented his works at several international platforms, for which he received several awards from a number of bodies. He has also authored a book titled *Drug Design*. Presently, he is serving as a reviewer and editorial board member for several journals of international repute. He is a member of a number of professional and scientific societies, such as the International Society for Infectious Diseases (ISID), the International Science Congress Association (ISCA), and ISEI.

Sanjay Kumar Bharti, PhD
Assistant Professor, Institute of Pharmaceutical Sciences, Guru Ghasidas Vishwavidyalaya (A Central University), Bilaspur, India

Sanjay Kumar Bharti, PhD, is Assistant Professor at the Institute of Pharmaceutical Sciences, Guru Ghasidas Vishwavidyalaya (A Central University), Bilaspur, India. He has working experience in several organizations, such as Win-Medicare Pvt. Ltd, Meerut, India (as a chemist) and the National Institute of Pharmaceutical Education and Research (NIPER), Hajipur (as a lecturer). He has published several research papers, one book, several book chapters and review articles in various reputed journals. His research interests include synthesis of Schiff's base, heterocyclic compounds, and metallopharmaceuticals for therapeutics. He is an active member of various scientific and pharmaceutical organizations, including IPA, IPGA, ISCA, etc. Dr. Bharti has completed a BPharm from IT-BHU, Varanasi, India, in 2003, an MPharm from RGPV, Bhopal, India, in 2004, and a PhD from IIT-BHU, Varanasi, India, in 2011.

PART I

TECHNIQUES IN DRUG DISCOVERY

CHAPTER 1

COMBINATORIAL CHEMISTRY: ROLE IN LEAD DISCOVERY

PREMLATA K. AMBRE,[1] ANISH N. GOMATAM,[1]
SANTOSH NANDAN,[2] and EVANS C. COUTINHO[1]

[1]*Department of Pharmaceutical Chemistry, Bombay College of Pharmacy, Kalina, Santacruz (E), Mumbai – 400098, India, Tel: +91-22-26671871; E-mail: Premlata.ambre@gmail.com*

[2]*Chemworx Pvt. Ltd., Kalina, Santacruz (E), Mumbai – 400098, India*

CONTENTS

1.1   INTRODUCTION TO COMBINATORIAL CHEMISTRY

Accelerating the drug discovery process by finding new chemical moieties (NCEs) faster is a major objective in the field of medical research. Hunting for a drug among the vast numbers of chemotypes, is a daunting task as is evidenced by the small number of molecules that are approved as drugs even though significant resources are expended on this activity. The drug discovery process generally involves identification of a validated biological target (enzyme, receptor, DNA & RNA) and developing a biological assay to hunt for novel chemotypes as a starting point for drug design. After finding a novel chemotype with the required biological activity multiple rounds of optimization with respect to pharmacokinetics and drug metabolism lead to clinical candidates. Drug hunters have evolved various strategies in order to perform this task efficiently, among these combinatorial chemistry is a strategy that has been used to efficiently sample the large chemical space that is available to them. The starting point or the initial discovery of a novel chemotype with the required biological activity is often called a "lead compound" and this phase of drug discovery is often called the "lead discovery" phase. A suitable lead compound is a small molecule with measurable and reproducible activity in the primary assay(s). In case the target is unknown then random screening may be need to test many thousands or millions of compounds in order to discover a lead. Combinatorial chemistry is designed to support the lead discovery process. It is a technique by which a large number of structurally distinct molecules can be synthesized together at a time and submitted for focused or varied pharmacological assay. Combinatorial chemistry has enabled the identification of many leads in drug discovery programs.

1.2   PRINCIPLES OF COMBINATORIAL CHEMISTRY

"Combinatorial synthesis involves synthesis of diverse analogs under the same reaction conditions, in the same reaction vessel followed by identification of the biologically most active compound for further development using high-throughput screening" (Furka, 2007; Jung, 2008).
The main principle behind this combinatorial chemistry is the synthesis of products all at a time from all possible combinations of a given set of building blocks (starting material). The collection of these synthesized compounds is referred to as a combinatorial library. This title is normally used

when synthesis is done producing compound mixtures; however, if the step-wise synthesis of individual compounds is made then they are designated as 'arrays'. The combinatorial libraries are generally structurally related by a central core structure, a common backbone, termed the scaffold. The total number of compounds within the combinatorial library is determined by the number of building blocks used per reaction, and the number of reaction steps, in which a new building block is incorporated. The number of all synthesized compounds (N) is given by the equation $N = b^n$, where b is an equal number of building blocks used in each reaction (1, 2, 3, ...) and n is the number of reaction steps, in which a new building block is introduced (Gallop, 1994; Houghten, 1991; Swartz, 2000).

1.3 COMBINATORIAL LIBRARIES

The creation of libraries composed of hundreds or thousands or millions of molecules by combining and mixing solid supports and various building blocks or by parallel addition of substrates at a time during the organic reaction is the main goal of combinatorial chemistry. A library of compounds is an ensemble of compounds that have originated from the same assembly strategy and building blocks. The members of a library exhibit a common core structure with different appended substituents. This assembly strategy defines the position and sequence of addition of the building blocks. The assembly strategy can be illustrated by the examples shown in Figure 1.1.

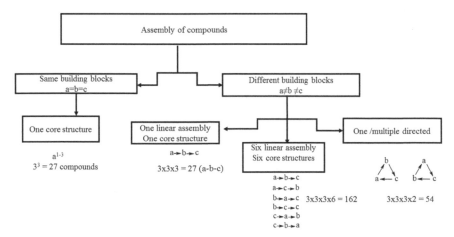

FIGURE 1.1   Strategy for preparation of assembly of compounds or libraries.

If a, b, and c are the three different building blocks assembled in a linear fashion, then the expected library would have either 27 products or in a nonlinear fashion the library would have 162 products with the basic core structure. If the building blocks are assembled in a cyclic fashion then the same building block assembly gives rise to 54 products in a library. In addition to this assembly strategy of reactive building blocks, the focus remains on the solid-phase chemistry that uses suitable resins and linkers on which the synthesis of the structurally diverse set of small-molecular weight compound libraries is achieved (Obrecht, 1998).

A library made by the process described above, suffers from low hit rates (low actives), partly because the library members possess poor structural diversity and have poor physicochemical properties, and partly because they are produced with the aim for quantity rather than quality. The key to synthesizing a diverse combinatorial library is a thorough knowledge about the relationship between the structure of a molecule and its physiological function.

An ideal combinatorial library should be relatively small and should contain chemically and functionally diverse compounds; each having a distinct biological activity, oral bioavailability and less toxicity (Welsch et al., 2010).

A good library, thus, has to take into account the requirements of both the products that will be made and the reagents that will be used to synthesize them. A general scheme for the design of a combinatorial library is demonstrated in Figures 1.2 and 1.3.

Following are the general considerations while designing a combinatorial library:

a. overall similarity or diversity to a target;
b. drug-like properties;
c. predicted activity;
d. deconvolution/decoding strategy;
e. availability of reagents;
f. cost of reagents;
g. combinatorial constraints (Brown et al., 2000).

To achieve good success while synthesizing such a library, careful thought must be given in the following five steps:

i) use of computational techniques to enumerate all possible compounds, creating a so called 'virtual library', keeping in mind factors such as reagent availability and synthetic feasibility;
ii) choosing a subset out of these compounds which is representative of the virtual library;

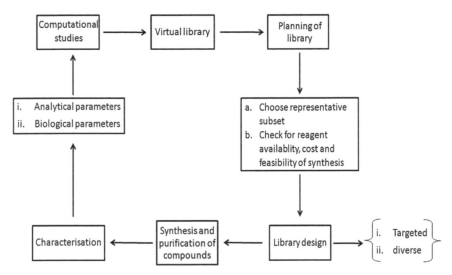

FIGURE 1.2 A schematic for the design, synthesis and characterization of a combinatorial library.

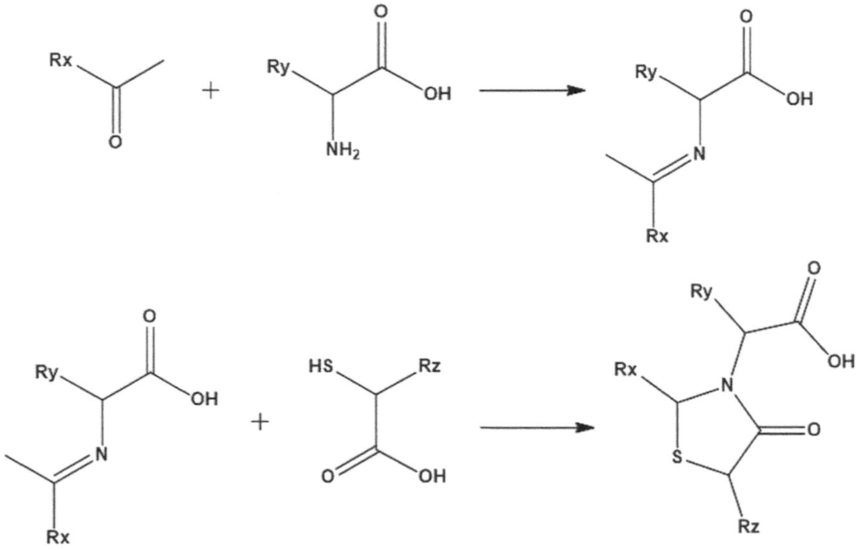

FIGURE 1.3 A general scheme to design a combinatorial library having three sites of variation.

iii) library design;

iv) purification followed by screening of the components in the library;

v) determining the chemical structure of the active molecules (Lewis, 2002).

There are various methods for the preparation of a combinatorial library, the choice of the method being determined by the type of library desired. Some of these types will be discussed briefly.

1.3.1   SCAFFOLD/BACKBONE BASED LIBRARY

The basic structural component that is the starting point for the production of a chemical library is termed the 'scaffold'. It is a fixed part of the library on which functional groups may be attached or substituted, and it is common to all the compounds in the particular series. The scaffolds employed generally are 'privileged structures'; molecules that are capable of acting as ligands for a wide array of receptors.

For a molecule to be successful as a drug, it has to be both selective and potent. These attributes are determined by the compound's stereochemistry and rigidity. The stereochemistry depends on the number of chiral centers present in the molecule, which make the molecule selective through stereospecific binding on the target. Rigidity improves binding affinity as a consequence of the lower entropic loss on binding. Natural products, because of their functional role, are found with a large number of chiral centers and show greater rigidity compared to their synthetic counterparts, for these reasons, majority of scaffolds in use today are based on natural products.

Ideal properties of a scaffold:
- The scaffold should be easily synthesizable by simple reactions and inexpensive reagents.
- It should be readily available if the product is to be obtained from natural sources.
- It should be amenable to chemical modification and introduction of different functional groups.
- It should not have any highly reactive groups.
- It should be free from structural elements that could engender toxicity, such as alkylating agents or planar three ring aromatic systems.
- It should abide by Lipinski's rule of five, that is,
  - i)   the molecular weight should be less than 500.
  - ii) log P value should be less than 5.
  - iii) there should be less than 5 H-bond donors (calculated as the sum of O-H's and N-H's).

iv) there should be less than 10 H-bond acceptors (calculated as the sum of O's and N's).

Scaffolds can be broadly classified into two types:

(a) Functional scaffold: A compound having a fixed molecular function (e.g., antagonism at a particular receptor, inhibition of an enzyme) or a specific indication (e.g., antifungal or proapoptotic). A functional scaffold is used when there is a need to optimize properties of the lead (e.g., potency or bioavailability) while retaining the basic activity of the molecule. Their major use is for the generation of focused libraries which are target class-specific. However, their main drawback is their limited utility in the discovery of lead compounds for a wide range of targets, as they have been biased towards a target class-specific activity.

(b) Structural scaffold: As the name suggests, structural scaffolds are molecules with specific structural features. The structural feature in question may be a particular ring system, chiral centers, functional groups and so on. These scaffolds may help in lead identification for a broad range of targets and are useful to increase the available chemical space. Since they do not have an activity bias, they are more universally applicable than libraries based on functional scaffolds (Brown et al., 2000; Eckard et al., 2010; Lewis2002; Welsch et al., 2010).

1.3.2 RANDOM/DIVERSE LIBRARIES

As mentioned previously, a diverse chemical library is a collection of compounds which aims to act at multiple targets. Since these compound collections are much larger as opposed to focused libraries, they have the obvious disadvantage of cost of reagents and screening. However, they may generate leads which may not be amenable by other methods because of the added chemical space they encompass. The logic is that greater the molecular diversity higher is the chance for a ligand to be recognized by a given receptor during screening.

1.3.3 VIRTUAL LIBRARIES

A virtual library contains all possible molecules that may be synthesized, taking into consideration the constraints of the chemistry being used and reagents that are commercially available and compatible with the chemistry being employed. Care should be taken in the selection of a scaffold during the initial stages. Small, flexible scaffolds such as guanidine or lysine are preferred over hydantoins or benzodiazepines owing to the limitations in conformational space spanned by these ring structures. Commonly applied filters for scaffolds and substituents include hydrophobicity, molecular weight, hazardous or toxic features, price and biological relevance. Another commonly employed filter is 'drug-likeness' in that if they are to be found active against the target in question, then they should less likely show problems in their absorption, distribution, metabolism, excretion or toxicology, so as to shorten the drug discovery and development cycle (Fauchère et al., 1998).

1.3.4 TARGETED OR FOCUSED LIBRARIES

The common practice in drug discovery over the past few years is the creation of 'diverse' chemical libraries; compounds collections which aim to act at multiple targets or proteins (Harris et al., 2011) The logic behind such a design is that if more number of compounds are synthesized, greater will be the chemical space that is covered and hence greater the chances of generating a lead compound (Drewry and Young, 1999). But unfortunately, the synthesis and screening of such a large collection of compounds has proved to be a costly affair. As a result, the focus has now shifted towards building 'targeted' or focused' libraries. These are libraries that are smaller, containing higher quality compounds that have been designed keeping in mind a specific target or protein. The obvious advantage of designing such a library is that they eliminate compounds which are unlikely to bind to the target; as a result, fewer number of compounds have to be screened to identify leads. Also, the hit rates observed with such a library are greater compared to diverse libraries, thus reducing the hit-to-lead timescale.

An understanding of the target is an essential prerequisite for designing a focused library. The design can be based on the structural information available about the target (e.g., kinase receptors, where an abundance of crystallographic data is available) or sequence or mutagenesis data (e.g., ion channels). An alternative approach can be based on known ligands for the target in question provided high quality structural information or prior knowledge of the binding site interactions is available. Targeted libraries are usually obtained as a subset of much larger collections using computational techniques. These libraries commonly employ a single core entity with attachment points to which specific substituents, or side chains can be appended to arrive at the desired compounds. When selecting the scaffold and the substituents, synthetic feasibility must be given careful consideration. Generally, all possible combinations of compounds are considered and a subset is chosen for synthesis. Thus the process of selection of compounds out of all possibilities is very important (Harris et al., 2011). This selection process is governed by the chemical space of interest for the target in question, followed by finalization of the set of molecules which will represent that space. A filter may be the introduction of a constraint that the molecules must be 'drug-like', that is, they must have properties which give them a reasonably good chance of having good oral bioavailability and a decent pharmacokinetic profile (Drewry and Young, 1999).

Methods for generating targeted/focused libraries are:

(a) *In silico based design*

This involves use of available structural information such as is available in the Protein Databank to help dig out drug receptor interactions. In the absence of experimental structures, homology models can be built for many target proteins, since there is an abundance of available template proteins. The first step in *in silico* design involves choosing the right template or the scaffold. This is achieved by docking various core molecules into the binding site of the target. Once the scaffold is finalized, substituents are attached to each possible position on the template, for assessing substituents that provide the best fit for the target. These combinations of scaffold and substituents are then 'scored', and the best ones are chosen as the subset for further synthesis, based on electrostatic and steric complementarity with the target. The correct docking of the template along with the substituents is crucial for the success of this approach (Beavers and Chen, 2002).

(b) *X-ray crystal structure based design*

This method is used in the design of fragment based libraries when high resolution X-ray crystal structures of the target are available. This helps to generate small molecules of molecular weight lower than normal drug molecules. Fragments which bind well to the target are identified and subsequently linked to each other through a scaffold. The structure based chemical library that is generated is bioassayed later. However, a limitation of using X-ray crystallography to design a library is the static nature of the crystal. Ligand-binding pockets are flexible in nature, and a protein-ligand structure solved by this method does not reflect this aspect, which might result in errors in the library design.

(c) *Library design based on NMR*

The technique, abbreviated as 'SAR by NMR', uses a high-field NMR spectrometer and large quantities of a pure, labeled protein. Isotopes such as ^{15}N, ^{19}F or ^{13}C are used to monitor protein-ligand interactions, provided there is adequate structural information on the binding site. Very weak binders (having activity in the millimolar range) can be detected by this method (Orry et al., 2006).

Assessment of the virtual library/subsequent synthesis: Virtual libraries are invariably too large to be synthesized in their entirety, and are frequently redundant in terms of the chemical space they cover. A process termed as 'library subsetting' is employed to obtain a representative subset of this virtual library, and this subset is termed as a 'design library'. Library subsetting involves assessment of the virtual library in order to quantify its diversity (Pandeya and Thakkar, 2005). The objective of these investigations is to obtain maximum diversity while keeping the library as small as practically possible. Structural descriptors are used as a means of assessing the molecular diversity, which provide information about hydrophobic, electronic, steric parameters etc., which can separate active moieties from inactive ones. The information gleaned from such a study depends to a large extent upon the choice of descriptors used. Such a library should display 'neighborhood behavior', that is, minor changes in diversity should produce minor changes in biological activity, which can be reflected in a plot of the changes in the biological activity vs. the difference in descriptor values. This process generates a set of compounds that is maximally diverse, out of which any number can be synthesized depending on the desired size of the design library (Fauchère et al., 1998).

The subset of molecules chosen for synthesis should be representative of the chemical space covered by the entire library. An important factor that governs the decision to synthesize a set of compounds is the diversity it adds to the existing compound collection. The collection may be diverse with respect to the chemical space it covers, but the information it contains should not overlap with that present in the preexisting collection.

While classifying a library as a diverse one, care should be taken to mention whether the diversity refers to the substituents attached to the scaffold or the product molecule as a whole. Generally, it is easier to analyze diversity in substituents. For example, a diversity analysis carried out in a scaffold with three points of diversity, using ten different substituents would require screening of a total of 30 compounds. However, analysis of the fully enumerated product set would involve 1,000 compounds. Analysis of diversity in substituents, although convenient may not be a reliable method as the fragments may interact with each other. For this reason diversity calculations are generally performed on the final products (Drewry and Young, 1999).

1.4 CHARACTERIZATION OF LIBRARIES OBTAINED FROM COMBINATORIAL SYNTHESIS

In the last few years, combinatorial technology has rapidly evolved from the production of peptide libraries to synthesis of small organic compounds. Synthesis as well as screening has become largely an automated process. Because of this, the bottleneck has shifted from production of compound libraries to analytical characterization. Analytical characterization involves use of various available analytical techniques to comprehensively dig out information about the designed library. Ideally, analytical characterization should not only provide the medicinal chemist with information about the final quality of the synthesized library, but also help him during the synthesis of the said library with vital parameters such as optimization of reaction conditions, selection of building blocks, etc. Thus the key to a successful characterization of a compound library is carrying out chemical and analytical work in parallel with compound library development.

During analysis of any compound obtained as the product of a chemical reaction, a synthetic chemist seeks to address three key issues: identity (have we been successful in synthesizing what we wanted?), quality (is our compound a pure one?, have we also generated some undesirable side products?), and quantity (how much have we made?, what is the yield?). Traditionally, analysis involves structure elucidation using both Nuclear Magnetic Resonance (NMR) and high resolution Mass Spectrometry (MS) with additional confirmation provided by IR spectroscopy. Purity estimation is carried out by chromatographic techniques, either HPLC or TLC with UV/VIS detection.

By definition, a combinatorial library is a collection of compounds that is highly diverse, synthesized in small amounts, and which may be a mixture of several components. This obviously means that the analytical chemist will have to work with a set of compounds that have varying properties, develop unbiased methods which apply to the entire collection, provide results in a short span of time, cost effective, all these while working with a small quantity of the material. It goes without saying that this will put a strain on the analytical method to be used. The technique involved should be fast, robust, informative, sensitive, and should be conveniently combined with separation techniques (for ease of handling mixtures of compounds). Furthermore, analytical demands are different at various stages of library development. In the early stages, information content is crucial when it becomes important to elucidate the structure of the desired compound and any other side product. During evaluation of building blocks and optimization of reaction conditions, the priority is to combine the analytical method with a separation technique. At the final stages, when thousands of compounds are to be analyzed, the primary requirement is high throughput. We shall first consider the strategies to be used in different stages of library development, followed by an application of the methods to various types of libraries.

1.4.1 ANALYTICAL METHODS USED FOR FINAL CHARACTERIZATION OF COMPOUND LIBRARIES

After library development has passed the stage of optimization and development, the focus is more on high-throughput rather than detailed analysis. Thousands of samples have to be analyzed for characterization of the production library. At this stage, analysis is carried out to provide confirmation that the intended compound has been successfully synthesized. Other goals are quantification of the product and purity assessment.

    a. Confirmation of structure using mass spectrometry: It is the method of choice for high throughput structure confirmation as the technique is based on the measurement of a very basic parameter of a compound – the mass to charge ratio. The technique does not depend on the presence of any functional groups or chromophores in a molecule. It is a highly sensitive method, femto moles of a sample can be measured easily. It is also fast, with analysis times of only a few seconds. The ionization techniques commonly employed are electrospray ionization (ESI), chemical ionization (CI) and matrix assisted laser desorption/ionization (MALDI).

b. NMR: Although NMR is a highly informative technique for compound characterization, NMR has limited application in this field because of the following drawbacks. It is slow, insensitive, consumes expensive deuterated solvents and requires homogenous samples.

c. Infrared (IR) Spectroscopy: IR spectroscopy is mostly used for functional group identification. The applications include determining the extent to which the reactant gets converted to the product, and observing the time taken for completion of the reaction, by observing the changes in the intensity of bands which correspond to specific functional groups.

d. Purity assessment: The technique of choice is high performance liquid chromatography (HPLC). HPLC is ideal for high throughput screening, since it is easy to automate, robust, has high resolving power and is not biased towards certain features of a molecule, such as charge.

e. Quantification: Quantification along with the requirement of high throughput is very challenging. Neither UV nor MS detectors are capable of carrying out quantitative analysis without a reference standard. The key to this is to use a proper reference standard and arrive at conclusions that are based on a combination of results from various analytical techniques.

1.5 GENERAL METHODS OF COMBINATORIAL SYNTHESIS

1.5.1 BIOLOGICAL LIBRARY APPROACH

A biological library is made up of a pool of microorganisms, which express various polypeptides on their surfaces. Each microorganism used in the library display shows one kind of peptide sequence, and represents a clone. Every clone which is part of the library can be replicated multiple times, and the resulting progeny will express the same polypeptide on its surface. A specific sequence of DNA coding for the polypeptide is inserted into the microorganism for library construction (either into a plasmid or into its genome). Library construction usually starts with synthesis of degenerated oligonucleotides. The nucleotides are single stranded; they are made double stranded and ligated into appropriate vectors. This is followed by insertion into host cells where they undergo replication and are displayed (Figure 1.4) (Miertus and Fassina, 2014).

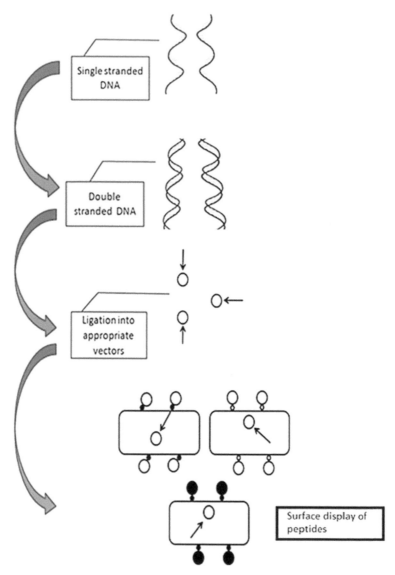

FIGURE 1.4 Construction of a biological display library.

1.5.1.1 Plasmid Display

The major applications of plasmid display lie in the field of protein engineering. Phage display and polysome have been used, but in most cases, their use is limited to the display of peptide libraries.

Plasmid display is much more efficient in displaying target proteins because:

1. As compared to the limited space on the surface, larger number of proteins can be produced in the cytoplasm
2. Following gene expression, subsequent translocation into the cell surface is not required
3. The DNA structure employed in this display has high stability and so different conditions (salt concentration, temperature, pH) can be used for screening.

The methodology employed is as follows: plasmid display involves directly linking proteins to plasmid encoding target genes in the cytoplasm. It is the process of synthesizing a protein in vivo, followed by linking it to the plasmid in the host cell's cytoplasm as shown in Figure 1.5. The process involves fusing the target proteins to an anchor, which is mainly the DNA binding proteins, which then binds to the appropriate DNA sequence in the plasmid. Thus the target protein is linked to the plasmid via the DNA binding protein, forming a plasmid protein complex. This complex is purified using an affinity resin and eluted. The plasmid is repurified and transformed into competent *E. coli* cells on an agar plate. The transformed cells can be used for the next round of screening. Formation and maintenance of a strong protein

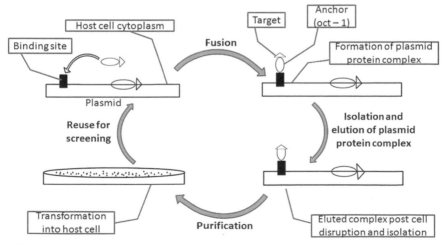

FIGURE 1.5 Plasmid display.

plasmid complex during screening are the most important issues that need to be addressed. Thus it becomes important to choose the anchor very carefully.

1.5.1.2   Polysome Approach

Although libraries displayed on phages or plasmids are a rich source of ligands for numerous targets, both display systems depend on in vivo gene expression, and the diversity and size of the library is eventually determined by the biological constraints and transformation capacity of the host microorganism. Polysome libraries are in vitro display systems, which, as the name suggests, displays the library on polysomes. They have the advantage of not being dependent on bacterial transformation, and they can also generate libraries which are much larger and much more diverse compared to other cell based systems. It is very useful for peptide expression. A discussion on the construction of a peptide library using polysome display follows (Figure 1.6).

1.5.1.2.1   Library Construction

The method begins with the construction of a DNA library which is made up of random peptide coding sequences. This library is incubated in a DNA dependent, *E. coli* coupled in vitro translation/transcription system. After terminating protein synthesis with an appropriate reagent such as

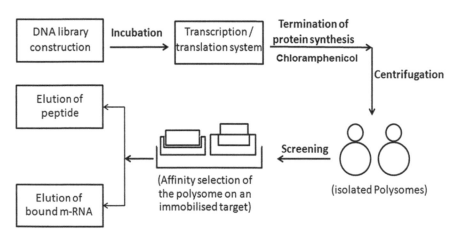

FIGURE 1.6   Polysome display of peptide libraries.

chloramphenicol, the polysomes are isolated by high speed centrifugation. The polysomes consist of peptides, which are linked to their corresponding encoding mRNAs. These are screened using affinity selection of the peptides on an immobilized target. The bound mRNA is dissociated using EDTA and by polymerase chain reaction, can be used to produce a template for new rounds of in vitro selection and synthesis. Identification of the enriched peptides is carried out by cloning the template and sequencing a portion of it. Their binding specificities may be determined using appropriate assays.

1.5.1.3   Phage Display Libraries

Phage display is a convenient method to generate and screen combinatorial libraries for various purposes. It is mostly used for the design of peptide libraries. Using concepts of molecular biology, it is possible to design phage libraries which display 10^{10} different peptides, and subsequently screen them for peptide ligands to metals, proteins, substrates of proteases or cell surfaces within a few weeks.

Vectors: Most phage display systems involve a bacteriophage that infects *E. coli*. The most popular bacteriophage is M13. Their genomes are small and can therefore be easily used for the construction of large libraries. They do not cause cell lysis unlike other bacteriophages. The M13 bacteriophage particle consists of one DNA molecule and five coat proteins. As displayed in Figure 1.7, these coat proteins are protein products of 5 genes, namely genes 3, 6, 7, 8, and 9, and are abbreviated as p3, p6, p7, p8 and p9, respectively.

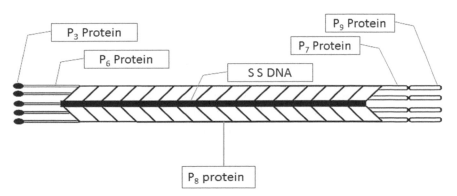

FIGURE 1.7   The M13 bacteriophage particle.

The coat proteins package the DNA into a stable molecule that is infectious for *E. coli* (Huang et al., 2011).

Design/Infection: The phage particle is first attached to the *F. pilus* of a male *E. coli* cell. This is mediated by the pIII coat protein. Following entry of the bacteriophage into the cell, its coat is removed and DNA replication is initiated using the host. The DNA is packaged into a new phage particle, which is extruded via the cell wall into the medium (Miertus and Fassina, 2014).

Advantages of phage display combinatorial peptide libraries:

i) Phage display is perfect for the construction of peptide libraries.

ii) The generated libraries are highly stable, and can remain so for an indefinite period of time if stored under proper conditions.

iii) Phage libraries can be conveniently renewed by infection of bacteria and harvesting the particles from cultures.

iv) Ligands of interest can be identified easily, in a short span of time and with modest effort.

v) Using DNA sequencing, the identity of selected peptides can be deduced very simply and at a low cost.

vi) Large libraries can be represented in a very small volume. This makes screening of the libraries a very simple process (Huang et al., 2011).

1.5.2 CHEMICAL APPROACH LIBRARY SYNTHESIS USING PARALLEL SOLID PHASE METHOD

As the name implies, parallel synthesis involves synthesis of a series of compounds carried out simultaneously. Each reaction is carried out in a different vessel and all the operations are carried out in parallel. At the end of the reaction, the products are individually cleaved from the support and collected in designated vessels. The advantage of parallel synthesis is that the number of operations carried out is the same; the only change is that it requires serial transport of the solvents and reagents into each reaction vessel. The collection of compounds prepared by this method is called compound libraries.

1. *Geyzen's Multipin method:* The Multipin method was developed by Geyzen, and was originally designed for synthesis of peptides. The apparatus involved a series of microtiter plates as reaction vessels, and cover plates with polyethylene pins that fitted into the vessels. The tip of the polyethylene rods (also referred to as 'pins') were

coated with polyacrylic acid. The wells contained a solution of the coupling reagents and the amino acids used to build the peptides. The coated tips of the pins were immersed into the wells, and kept in solution until the reactions were complete. The peptide sequence depends on the order of amino acids added to the well. The assay can be carried out while the peptides are still attached to the rods. (Furka, 2007). Screening is carried out using enzyme linked immunosorbent assay (ELISA), where the binding affinity of the peptides to antibodies is determined. (Pandeya and Thakkar, 2005). Recent applications of the multipin method include studies on T-cell proliferation, Substance P-receptor binding and SAR studies on ligands for endothelin receptors (Zhao and Lam, 1997).

2. *Houghton's tea-bag approach:* Developed by Houghten, the tea bag method enables preparation of over 200 fully characterizable peptides at a time, on a much larger scale as compared to the Multipin method. The technique involves placing solid phase resins in solvent permeable polypropylene bags, which are called 'tea-bags' (Rinnová and Lebl, 1996). Amino acids are then coupled to the solid phase resins by immersing the bag in a solution of individual activated monomers (Pandeya and Thakkar, 2005). These are then grouped, according to sequence of amino acids that will be attached in a particular synthetic step. Washing and deprotection are carried out by mixing the tea bags in a vessel, followed by their separation and characterization (Rinnová and Lebl, 1996). The advantages here are smaller number of operations and smaller number of reaction vessels (Furka, 2007). Applications include epitope mapping and studying antigen antibody interactions (Rinnová and Lebl, 1996).

3. *SPOT technique of Frank:* The SPOT technique is similar to the Multipin method, but more economical and easier to perform. The only difference is that cellulose is used as the solid support in place of polyethylene pins. In the presence of certain activating reagents, the solution of protected amino acids are 'spotted' on to the functionalized cellulose paper (Zhao and Lam, 1997). These spots can be considered as reaction vessels where the reactions take place, and as many as 2000 peptides can be synthesized at a time (Furka, 2007) (Figure 1.8). By treating the cellulose membrane with appropriate reagents, the entire synthetic procedure can be carried out simultaneously. The peptides that are formed can be either analyzed directly

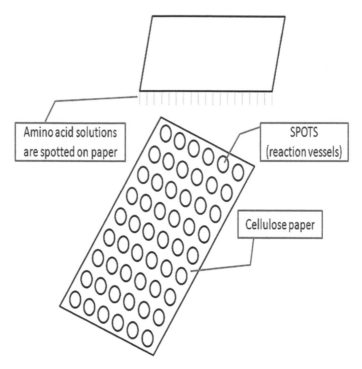

FIGURE 1.8   SPOT synthesis.

using ELISA based studies or cleaved and applied in solution (Zhao and Lam, 1997).

4. *Light directed peptide synthesis:* This technique is based on the principle of spatially addressable synthesis, in which the identity of a compound depends on its location on a substrate. Here, the addition of a reagent is carried out on predetermined sites on the solid support. The technique is a combination of photolithiography and solid phase peptide synthesis. It is demonstrated in Figure 1.9. The substrate is attached to a photolabile protected covalent linker, the protecting group usually being a nitroveratryloxycarbonyl (NVOC) group. With use of light, the photolabile linker is removed. Removal of these groups, also called deprotection, causes activation of selected areas. The activation is followed by amino acid coupling which occurs selectively in the activated areas. The amino acid solution is removed, and a different region on the substrate is activated, followed by coupling. The product is determined by the deactivation pattern, and the sequence of reactants

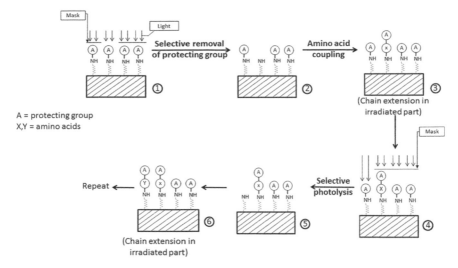

FIGURE 1.9 Spatially addressable light directed peptide synthesis.

used. Since photolithiography is used, the total number of compounds that can be made depends on the number of sites that can be addressed with proper resolution. Direct assessment of the compound is possible since the position is precisely known (Pandeya and Thakkar, 2005).

1.6   APPLICABILITY OF COMBINATORIAL TECHNOLOGY

Over the years scientists have explored the applications of combinatorial chemistry in peptide, oligonucleotide as well as small molecule synthesis. Solid phase synthesis demands highly optimized repetitive reactions that are suitable for automation. Early work on small molecule combinatorial chemistry was reported by Leznoff and co-workers (Wong et al., 1974) and Frechet and co-workers (Fréchet and Seymour, 1978). It was realized that different solid supports are required for nonpeptide and peptide synthesis. Furthermore, unless each reaction step in a solid-phase synthesis is optimized, poor quality products are often the result. Strong interest in the synthesis of small molecules was finally kindled by Ellman and Bunin in 1992, when they illustrated the synthesis of benzodiazepine analogs prepared by rapid parallel synthesis, yielding compounds of pharmaceutical interest (Bunin and Ellman, 1992). After the success story of benzodiazepines synthesis, combinatorial synthesis became a well-established tool in the pharmaceutical industry for generation of large sets of small molecules for lead finding and optimization (Dolle, 2002; Golebiowski et al., 2001). Combinatorial chemistry is mainly divided into solution phase and solid phase synthesis.

1.6.1 SOLUTION PHASE SYNTHESIS

Organic synthesis has traditionally been done in solution phase in which starting materials and reagents are dissolved in a solvent to generate a homogeneous mixture for the reaction to proceed. Solution phase organic reactions are "worked up" after every step in a multisynthesis through a series of extractions whereby the mixture is partitioned between an organic phase and an aqueous phase of known pH. Reaction progress as well as product identity and purity may be checked by well-established chromatographic and spectroscopic methods and the time needed for chemistry development in the preparation of libraries in solution phase is much less than for solid-phase approaches. In contrast to the large number of advantages of solution-phase parallel synthesis, there is one major disadvantage, namely the "purification problem". Solution and solid-phase synthesis are both convenient and easily automated, the limitation to solution-phase parallel synthesis is the isolation of the desired compounds. Thus, the throughput ratio of libraries in automated solution-phase synthesis is directly proportional to the work-up procedures and to the purification process. Therefore, easy and efficient purification methodologies are required for high speed solution-phase synthesis (Golebiowski et al., 2001). The following methods are used in the solution phase synthesis of combinatorial chemistry.

1. Solution phase combinatorial synthesis using monomethyl polyethylene glycol (OMe-PEG) solid support. It is amphiphilic in nature and soluble in water as well as in many organic solvents (e.g., methylene chloride, ethanol, toluene, acetone, and chloroform) except in diethyl ether. The synthesis is started by reacting the acid group of a building block to the hydroxy group of OMe-PEG (Figure 1.10). The product is precipitated by adding diethyl ether and the excess reagent and other impurities are removed by washing. The solid product is redissolved in fresh solvent and the second stage of synthesis is carried out using a similar reaction and washing procedure. At the end of the synthesis the product may be cleaved from

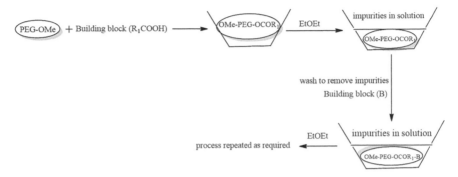

FIGURE 1.10   Solution phase libraries using monomethyl PEG.

the OMe-PEG, purified and assayed. In some cases the product is assayed when it is still attached to the OMe-PEG. This approach may be carried out using either the parallel synthesis or split and mix methods, the latter being carried out while the products of a stage are in solution (Han and Janda, 1996).

Dendrimers have been used as soluble supports in solution phase combinatorial chemistry in a similar fashion to OMe-PEG. Chemically dendrimers are branched oligomers (small polymers) with regular structures (Figure 1.11). This method helps the user to attain a high yield of synthesized compounds as multiple copies of each molecule are synthesized per dendrimer. The final product is released from the dendrimer and purified, and its structure is determined using the history of the process and/or standard analytical techniques. An important advantage of this technique is that it usually gives high yields (Kim et al., 1996).

2. Polyfluorinated organic compounds are frequently used as soluble support in solution phase combinatorial synthesis even though it is insoluble in water and organic solvents but soluble in liquid

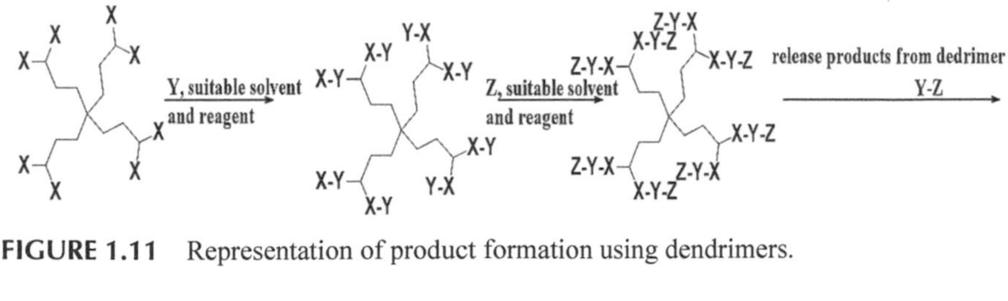

FIGURE 1.11   Representation of product formation using dendrimers.

perfluoroalkanes. The polyfluoro compounds usually have structures with long chains of CF_2 units attached via a short hydrocarbon chainto a silicon or tin atom (represented in a reaction of isothioazole by Studer et al. (1997) schematically in Figure 1.12) (Studer et al., 1997). The library is produced by reacting the initial building block with a fluorous reagent. The product of this reaction may be separated from the reaction mixture by extraction into a perfluoroalkane solvent that is not miscible with water or the organic solvent used in the reaction. This process of purification by extraction is repeated at each stage of the synthesis. However, at the end of the synthesis the product is detached from the fluorocarbon reagent prior to purification by extraction into a suitable organic solvent.

3. Solution phase libraries are produced using resin-bound scavenging agents when the excess of reagents are required to be eliminated along with by-products from the reaction (schematics represented in Figure 1.13). These resins or solid phases are known as scavenging or sequestering agents. They consist of resin beads to which is permanently attached a suitable residue with an organic functional group that will readily react with the relevant reactant or by-product. This reaction results in the formation of a solid complex containing the excess reagent/by product, which can be removed from the reaction mixture by filtration through a cartridge containing an appropriate resin. The by-products of some of these scavenger reactions, such as the calcium

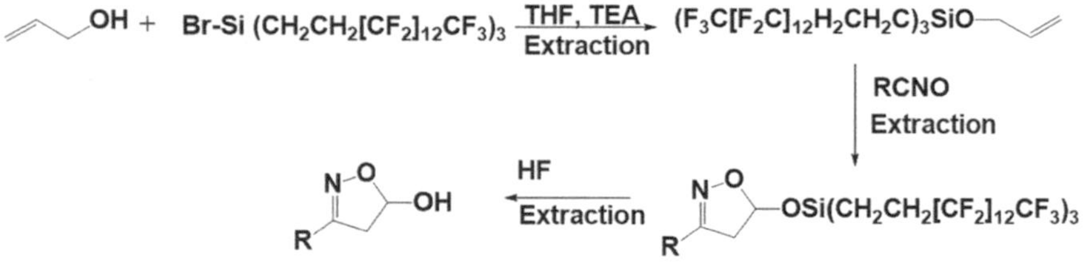

FIGURE 1.12   Reported reaction by Studer et al. of preparation of isoxazole library using fluorous resin.

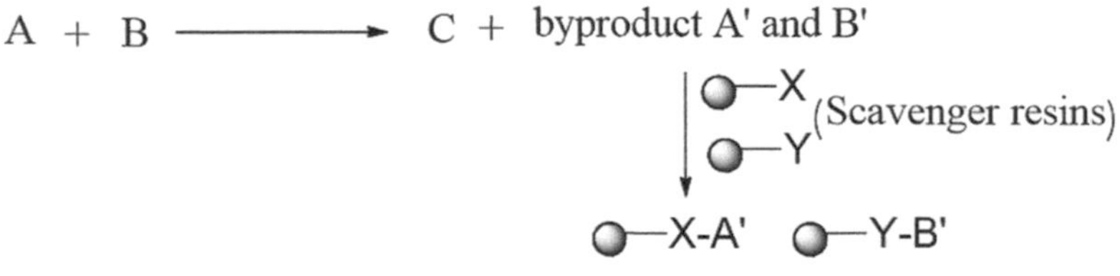

FIGURE 1.13   Schematics of resin bound scavengers to capture products.

sulphonate resin used to remove excess of tetrabutyl ammonium fluoride (TBAF) from a number of desilylation reactions, are also solid and may be removed by filtration. Resin-bound scavengers for the removal of by-products of both the reaction and the reagents operate in the same way as those used for the removal of excess reagents, except they are normally used during the reaction. For example, carboxylic acids have been sequestered by the use of the anionic Amberlite-68 resin and 4-nitrophenol has been removed by ion exchange with a quaternary ammonium hydroxide resin. The use of these sequestering agents during the reaction usually helps to drive the reaction to completion.

4. Resin capture of products: In this technique the resin has a functional group that can sequester the product. However, there is also the possibility of breaking the bond linking the product to the resinto form the original functional group of the product. At the end of the reaction the product is captured on the resin and the excess reagents, and reagent by-products are washed away with suitable solvents. The product is released from the resin, dissolved in a suitable solvent and the resin removed by filtration. For example, Blackburn et al. synthesized a library of 3-aminoimidazo[1,2-a]pyridines and pyrazines by this technique. They used a cation exchange resinto capture the products (Jung, 2008).

1.6.1.1 Solid Phase Synthesis

Solid-phase synthesis was invented by Merrifield in 1963; he used polystyrene resin beads to aid the synthesis of peptides. This strategy was expanded and further investigations on solid-phase synthesis towards organic compounds was carried by many other groups including Leznoff, Camps, Frechet, Rapaport in the 1970s (Früchtel and Jung, 1996; Hermkens et al., 1996; Thompson and Ellman, 1996). The ideal properties of a solid support are that it should be insoluble in the solvents used in the solid phase synthesis and should not react with the reagents used in the synthesis. The organic reactions take place mostly at the surface and sometimes inside the solid support. These supports are mostly used in the form of small resin beads that swell in the solvents used in the synthesis. The reactions in other kinds of supports take place only at the surface. These supports are polymers or glass beads, rods, sheets etc. and with the exception of their surface layer do not swell in the solvents. Solid

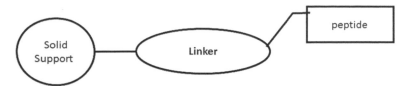

FIGURE 1.14A Attachment of Linkers, protective groups in peptide synthesis.

supports are usually made up of a core and a linker. The starting molecule of the synthesis is attached to the support *via* the linker (Figure 1.14A). The core provides insolubility to the support, determines the swelling properties, while the linker directs the functional group for attachment of the starting molecule. Reactions in solid-phase chemistry are selective and efficient and consequently the synthesis must be planned carefully. Otherwise, the purification of the final products can be a challenge. The scale of solid-phase synthesis is limited and generally restricted by the amount of the solid support and its loading capacity; the preparation of multimilligram quantities can be cumbersome and expensive for large combinatorial libraries (Jung, 2008). Following is a brief on the different types of supports and linkers used in solid phase synthesis.

1. *Polymeric Supports:* Cross-linked polystyrene resins are the most commonly used supports for solid phase synthesis. These resins are synthesized from styrene and divinylbenzene by suspension polymerization. The ratio of divinylbenzene to styrene determines the density of cross-links. Higher crosslink density increases the mechanical stability of the beads. Lowering the crosslink density, on the other hand, increases swelling and increases the accessibility of the functional groups buried inside the beads. The bead size of the resin is an important factor for the combinatorial synthesis. Smaller the resin, faster the reactions, but use of very small beads may cause problems during filtration. Normally 1–2% divinylbenzene is used with styrene. Cross-linked polystyrene is very hydrophobic so it swells only in an apolar solvent like THF, toluene, DCM, dioxane, CAN, DMF and methanol but not in water (Merrifield, 1963; Merrifield et al., 1966).

    Polyethylene glycol grafted polystyrene support (PEG) has a 1–2% polystyrene core with ethylene glycol chains covalently attached. The PEG chains render hydrophilic character to the resin and it swells in water and methanol but poorly in ethanol and ether (Jung, 2008).

Glass beads with controlled pore size can also be considered as supports. These supports are mechanically stable, they can be functionalized by attaching linkers but they do not swell in solvents. Sometimes macroscopic objects can be used as solid supports are known as SynPhase crowns or SynPhase lanterns. In this technique polymer chains are grafted into the surface of an appropriate monomer and then the terminal chains are connected with the linker group, for example, styrene with polyolefin when grafted by radiation acts as a solid support (Geyzen et al., 1984; Morales and Bunin, 2003).

Mimotopes are devices with molded polypropylene surface on which is grafted either a hydrophilic copolymer of methacrylic acid/dimethyl acrylamide or the relatively hydrophobic polystyrene. The polymer is then suitably derivatized to allow the incorporation of a linker system. Another solid support lantern is made up of uniformly spaced flat rings. They are of two types D-series and I-series. D-series have larger surface area or volume ratio and have a higher loading capacity of 35 μmol/unit. Lanterns are mainly used for the synthesis of libraries of small molecules (Vaino and Janda, 2000; Wu et al., 2003).

2. *Linker:* A linker is abifunctional molecule, that binds irreversibly to the solid support (resin) and exhibits a reversible binding site for the coupling of desired molecules so that further chemical reactions may be carried out. The linker remains unchanged during the synthesis and on cleavage of the product remains attached to the solid support. The solid support attached to the linker is reusable. Cleavage of the bond between the linker and the synthesized library depends on the strength of its attachment to the resin. The synthesized library can be cleaved from the linker at the end of the synthesis either with acid, base or nucleophilic reagents, hydrogenolysis, enzymatic, palladium-catalyzed, photochemical, or by oxidative and reductive cleavage methods.

A large number of linkers have been prepared and discussed in a number of reviews. Linkers allowing the cleavage of a certain functional group have been named as mono-functional linkers. However, if an attachment is made cleavable to generate more than one functional group, it is named as multifunctional linkers. Linkers that are copolymerized into resin beads can be either of the integral or nonintegral type. Those which are not part of the polymer core can

TABLE 1.1 List of Linkers, Anchoring Groups and Cleavage Reagents and Deprotecting Reagents

Type	Linker	Anchoring group	Carrier	Cleavage reagent	Ref.
Linkers cleaved using electrophilic reagents					
Strong Acid Cleavable Linker	Free carboxylic acid of benzyloxycarbonyl (Cbz) or Boc *N*-protected amino acids	Alcohols		HF or Trifluoromethane sulfonic acid (TFMSA)	[40–42]
	Phenylacetamido-methyl (PAM) linker	Peptide ester	R= p-Nitrophenol/N-Oxasuccinimide, tertiary-amine	50% TFA in CH$_2$Cl$_2$	[43]
Strong Acid Cleavable Linker	Marshall benzhydryl amine linker (MBHA)	Carboxamide		TFMSA and HBF$_4$/ thioanisole in TFA	[44–47]

Type	Linker	Anchoring group	Carrier	Cleavage reagent	Ref.
			Me, NH_2	HBr in TFA, or HF/anisole to cleave the MBHA	
	Benzhydrol		OR	1–2% TFA in CH_2Cl_2	
	OMPPA (4-(3-Hydroxy-4-methoxypentyl)) phenylacetic acid linker	Boc protected amino acid	Me, Me, BocHN(Me)COCO, O	On cleavage with HF it undergoes intramolecular cyclization	[48]
Mild acid cleavable linker	Wang Linker p-Alkoxybenzyl alcohol linker	α-Aminophosphonates and Phosphonic acid derivatives	BocN(Me)COH$_2$CO, O	1–10% TFA in CH_2Cl_2	[49, 50]

TABLE 1.1 (Continued)

Type	Linker	Anchoring group	Carrier	Cleavage reagent	Ref.
	SASRIN linker (superacid sensitive resin)	Acids		1% TFA	[51, 52]
Mild acid cleavable linker	Rink linker alkoxy groups onto the benzhydryl system	Hydroxamic acids, alcohols, phenols, amines, anilines, thiols, and thiophenols.		i) TFA from 10% to 95% in CH$_2$Cl$_2$ ii) Higher concentrations of TFA are required for peptide amide cleavage (generally 95% in CH$_2$Cl$_2$).	[53]

Type	Linker	Anchoring group	Carrier	Cleavage reagent	Ref.
	Sieber linker	Carboxamide or peptide-amides		1% TFA in CH_2Cl_2 after 1 min	[54]
	Indole linker	Chloroformates, isocyanates, sulfonyl chlorides, or acids	OHC	TFA in CH_2Cl_2 (2–50%)	[55]
Mild acid cleavable linker	Trityl linkers	Anchor selectively one alcohol group of a diol, preferably primary alcohols.	R= H/ 2'Cl/4'-Me/4'-OMe	anhydrous 2% TFA or dry HBr	[56–58]

TABLE 1.1 (Continued)

Type	Linker	Anchoring group	Carrier	Cleavage reagent	Ref.
	Ketal linkers	a carbonyl group onto a diol-based linker or diols onto a carbonyl linker		10% TFA in CH$_2$Cl$_2$ with a trace amount of water or methanol	[59, 60]
	THP and Ketal linkers for Alcohol Immobilization	Purine derivatives to generate hindered secondary alcohols (stable to strong bases and nucleophiles)		TFA in CH$_2$Cl$_2$ with a trace amount of water or methanol for the dioxane linker	[61]

Type	Linker	Anchoring group	Carrier	Cleavage reagent	Ref.
	Semicarbazone linker	Ketone (trifluoromethyl ketone)		Aq. HCl and acetic acid	[62]
Mild acid cleavable linker	Imine, enol, ether and enamine linkers	Ketones or C-terminal peptide		3% TFA in CH_2Cl_2	[63]
	t-Alkyloxycarbonyl based linker	Immobilize amine to release 1° and 2° amines or N-terminal peptide		TFA	[64]
	Aryltriazene linker using diazonium salt	1° and 2° amines		10% TFA	[64]

TABLE 1.1 (Continued)

Type	Linker	Anchoring group	Carrier	Cleavage reagent	Ref.
Linkers cleaving using nucleophilic reagents					
Oxygen nucleophile	Saponification	Acid or alcohol		Hydrazine ammonia, ethanolamine, methylamine	[65, 66]
Enzyme cleavable	4-acyloxybenzyloxy linker	Acid		Lipase or peptidase; Exception protease gives only 1–2% yield	[67, 68]
Nucleophilic transesterification	4-Hydroxymethyl-benzoic acid (HMBA)	Ester		MeOH/ MeONa/ THF, MeOH/Et$_3$N, K$_2$CO$_3$, CH$_2$N$_2$	[69]

Type	Linker	Anchoring group	Carrier	Cleavage reagent	Ref.
Nitrogen nucleophile	Sulfonate-based polystyrene linker	1° Amine, imidazole or thiolates		1° Amines, imidazoles, thiolate	[70, 71]
	Diketopiperazine cyclo release	C-terminal proline and glycine anchored peptides		0.1 M Acetic acid DCM	[72] [73]
	Hydantoin	Peptide linkage		Et$_3$N, MeOH	[74]
Nitrogen nucleophile	Pyrazolones	β-ketoester		ArNHNH$_2$	[75]

TABLE 1.1 (Continued)

Type	Linker	Anchoring group	Carrier	Cleavage reagent	Ref.
Safety Catch Linkers sensitive to nucleophiles upon activation	Sulfide or sulfone safety catch	2-Aminopyrimidines		mCPBA, $R_1R_2NH_2$	[76, 77]
	Kenner Safety catch linker	1° Amines		1° Amines	[78]
	Boc benzamide activation	Alcohol		LiOH, 5% H_2O_2	[79]
	Wieland Safety-Catch Linker	Amine		$BnNH_2$	[80]

Type	Linker	Anchoring group	Carrier	Cleavage reagent	Ref.
Carbon Nucleophile	Thioester linker	Ketone		Grignard reagent	[81]
Halogen nucleophile	Wang linker	Secondary amines or N-benzyl tertiary amines	$R_1, R_2 = $ alkyl	α-chloroethylchloroformate	[82]
	Trialkylsilyl linkers	diketopiperzaines	R = Cl/ OR/ R' = Me/ Ph/i Pr/nBu R'' = Me/Ph/iPr/nBu	TBAF, AcOH, and THF	[83]

TABLE 1.1 (Continued)

Type	Linker	Anchoring group	Carrier	Cleavage reagent	Ref.
Photocleavable linker	o-Nitrobenzyl-Based Linkers, Nitrobenzhydryl linker (NBH)	Boc-peptide		Sonication and a mixture of $CH_2Cl_2/$ trifluoroethanol	[84]
	Nitrobenzhydrylamine linker (NBHA)		R=OH (NBH) R =NH2 (NBHA)		
	Nitroveratryl linkers	Peptides or oligonucleotides		Irradiation at wavelengths above 320 nm	[85]

Type	Linker	Anchoring group	Carrier	Cleavage reagent	Ref.
Multidetachable linkers	Boc-aminoacyl-4-[4-(oxymethyl) phenylacetoxymethyl]-3-nitrobenzamidomethyl resin	Different families of compounds		Acidolysis or hydrogenolysis for benzyl ester and photolysis for other ester	[86]

be considered as nonintegral (or grafted) in nature. Although the core structure of the linker may remain unchanged, the group placed between the linker and the support can modify the cleavage conditions and also alter the degree of linker cleavage. These groups are called linker attachments (Jung et al., 2007). Table 1.1 shows the classes of frequently used linkers, anchoring groups and cleavable reagents.

3. *Protecting Groups:* Generally in combinatorial chemistry protecting groups are used when a chemical reaction is to be carried out selectively at one reactive site in a multifunctional compound, and other reactive sites must be temporarily blocked. The protecting group must react selectively in good yield to give a protected substrate. The protective group must be selectively removed in good yield by readily available, preferably nontoxic reagents that do not attack the regenerated functional group. The protective group should form a derivative (without the generation of new stereogenic centers) that can easily be separated from side products associated with its formation or cleavage. Protecting groups are subcategorized as "Orthogonal protection" if two functional groups are protected using two different protecting reagents and later in the synthesis the multiply protected groups are removed one at a time (Figure 1.14B). Applications of protection and deprotection steps in organic synthesis will help to control the chemistry. The protecting groups used for different functional groups and the deprotecting reagents are tabulated in the Table 1.2.

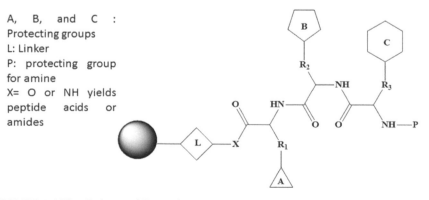

A, B, and C :
Protecting groups
L: Linker
P: protecting group
for amine
X= O or NH yields
peptide acids or
amides

FIGURE 1.14B   Orthogonal Protecting groups.

TABLE 1.2 List of Protecting Reagents and Deprotecting Reagents

Protecting groups	Protection reaction	Cleavage reagent	Reference
Amino functional group (R-NH$_2$)			
Benzylcarbonyl (Z) group	Amino-protection in peptide synthesis	HBr/AcOH, HBr/TFA	[87]
t-Butoxycarbonyl (Boc) (Boc$_2$O)		TFA	[88, 89]
9-Fluorenylmethoxycarbonyl (Fmoc) group reagent used is FmocCl		Basic condition 20% piperidine in DMF	[90–93]

TABLE 1.2 (Continued)

Protecting groups	Protection reaction	Cleavage reagent	Reference
Tetrachlorophthaloyl (TCP)		15% hydrazine DMF	[94]
Allyloxycarbonyl (Alloc) group		Pd(PPh$_3$)$_4$ and morpholine	[95–97]
2-(4-biphenyl) isopropoxycarbonyl (Bpoc)		t-Butyl group	[98–101]
Trityl (Trt),		t-Butyl group	[100]

Protecting groups	Protection reaction	Cleavage reagent	Reference
α, α-Dimethyl-3,5-dimethoxybenzyloxycarbonyl (Ddz)		t-Butyl group	[100]
2-(4-Nitrophenylsulfonyl)ethoxycarbonyl (Nsc)		20% piperidine and DMF, Dioxane in DMF(1:1)	[102–104]
Guainidino group			
Benzyl ester		Saponification, HBr/AcOH, HF, catalytic hydrogenation but not by TFA	[105]
t-Butyl ester		TFA	[106, 107]

TABLE 1.2 (Continued)

Protecting groups	Protection reaction	Cleavage reagent	Reference
Allyl(Al)		Pd(Ph$_3$)$_4$ (0.1eq) and scavengers (PhSiH$_3$ 10eq), DCM	[108, 109, 110]
Phenacyl (pac)		Sodium thiophenoxide or Zn in AcOH	[111, 112]
Trimethylsilylethyl (TMSE)		TBAF	[113]
p-Hydroxyphenacyl (pHP)		Photolysis	[114]
4,5-Dimethoxy-2-nitrobenzyl (Dmnb)		Photolysis	[115, 116]

Protecting groups	Protection reaction	Cleavage reagent	Reference
Alcoholic and phenolic hydroxyl group			
Benzyl ether	Ph—CH$_2$—O—	HF, HBr/AcOH, catalytic hydrogenolysis	[117]
Trityl ether	Ph$_3$C-O-	CF$_3$CO$_2$H t-BuOH	[118]
Silylethers	Me$_3$SiO— (Me can be replaced with Et/ iPr)	HCl or TFA	[119]
Tetrahydropyranyl ether	(tetrahydropyranyl structure)	Pryidinium p-toluenesulfonate (PPTS) in EtOH Or TsOH in MeOH	[120, 121]
Benzoate	(benzoate ester structure)	K$_2$CO$_3$/KOH	[122]
t-Butyl carbonate(Boc)	(t-Butyl carbonate structure)	TFA	[88, 89]
Trichloroethylcarbonate	(trichloroethylcarbonate structure)	Zn, AcOH	[123]

TABLE 1.2 (Continued)

Protecting groups	Protection reaction	Cleavage reagent	Reference
Allyl carbonate		Pd$_2$(dba)$_3$, dppe, Et$_3$NH in THF	[124]
Perfluoroalkylsulfonate		KOH or palladium-Pd(OAc)$_2$ mediated reductive cleavage	[125]
Carbonyl Protecting Groups			
Acetal/ketal methanol in dry HCl or MeOH (MeO)$_3$CH.		TFA, CHCl$_3$, H$_2$O in 1,3-dithiane and a dioxolane acetal or TsOH, acetone.	[126, 127]
S,S'-dialkylacetal (RSH in HCl, or RSSi(CH$_3$)$_3$, ZnI$_2$, Et$_2$O.)		m-CPBA; Et$_3$N Ac$_2$O, H$_2$OorHg(ClO$_4$)$_2$, MeOH, CHCl$_3$ "	[128, 129, 130]
Guainidino group			
Nitroguanidine	O$_2$N-NHCONH-	Resists HBr/ AcOH but cleaved by liq. HF	[131]

Protecting groups	Protection reaction	Cleavage reagent	Reference
2,2,5,7,8-Pentamethylchroman-6-sulphonyl (Pmc) group	SO$_2$NHCONH—	TFA	[132]
4-Methoxy-2,3,6-trimethylbenzenesulphonyl (Mtr)	SO$_2$NHCONH—	TFA (less sensitive and takes few hours to cleave)	[133]
Amide groups (in side chains of asparagine and glutamine) Tritylation		TFA-EDT-H$_2$O(EDT-ethanedithiol)	[134, 135]

TABLE 1.2 (Continued)

Protecting groups	Protection reaction	Cleavage reagent	Reference
Cyclopropyldimethylcarbinyl (Cpd)		TFA-thioanisole-EDT-anisole (90:5:3:2)	[136, 137]
9-Xanthenyl (Xan)		TFA scavengers	[138, 139, 140]
4,4′-Dimethoxybenzhydryl (Mbh)		TMSBr (trimethylsilylbromide) thioanisole EDT-m-cresol in TFA	[141, 142]

1.7 SPLIT AND MIX SYNTHESIS

Arpad Furka and co-workers invented the mix and split method in 1988. This method is used to generate large number of libraries as it produces one type of compound on each bead. Mix and split synthesis is executed in following three simple steps: first the solid support is divided into equal portions, each portion is coupled individually with one building block as shown in Figure 1.15. The final product is then mixed, homogenized and divided into equal portions for the next step of reaction with different building blocks.

Mix and split technique forms all products in any reaction vessel and the products are evenly distributed among the reaction vessels for the next reaction step. This method was mainly designed for peptide libraries. However, it has been used widely for organic libraries. As organic molecules are prepared by multistep synthesis, one can easily implement the split and mix synthesis for preparation of organic libraries. However, if every step needs to be screened for activity, then chemical tags are attached to the solid supports during the synthetic steps and they must be different from those that are applied for coupling of the building blocks (the two reactions need to be orthogonal). This process is called encoding or chemical tagging of the resin. The chemical tag must be cleavable at the end of the synthesis from the beads separately from

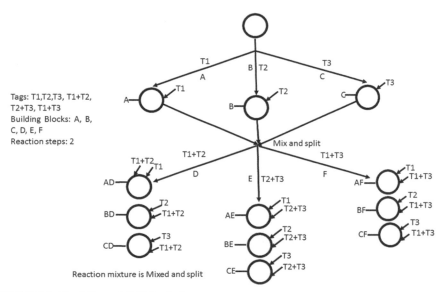

FIGURE 1.15 Mix and Split Technique.

the products. It must reveal information of the route of the beads by a simple analysis using spectroscopic methods rather than determining the structure of the compound. Chemical tagging can be done using peptide or oligonucleotide sequences as chemical tags (Brenner and Lerner, 1992; Needels et al., 1993; Nielsen et al., 1993; Nikolaiev et al., 1992) or by binary encoding method where the coding units are halobenzenes carrying a hydrocarbon chain with varying length attached to the solid support with a cleavable spacer (Kerr et al., 1993). Generally, separate linker functionality is used for the tagging sequence to prevent interference with the target chemistry. If the compound libraries are prepared in small concentrations in μmol to nmol yield, the solid support is then enclosed with electronic transponders into permeable capsules at the time of the reaction. Electronic transponders are used to encode the reagents and to encode the order of their addition. Radiofrequency signals can then be used to decode upon completion of the reaction. A capsule/micro-reactor contains 0.3 mL polypropylene with mesh sides capable of holding (in addition to the electronic chip) 30 mg of resin. About 15–30 μmol of compound can be prepared in each capsule. However, this technique is generally limited in many laboratories as it requires computers and readers, and because of the limited space present in a microreactor to carry out the reactions (Moran et al., 1995; Nicolaou et al., 1995). Another system of optical coding encompasses laser optical synthesis chips (LOSCs). The supports are 1×1 cm^2 polystyrene grafted square plates. The medium carrying the code is a 3×3 mm^2 ceramic plate in the center of the support. The code is imprinted into the ceramic support by a CO_2 laser in the form of a two-dimensional bar code. Before each synthetic step, the pooled chips are scanned and sorted according to the principles of the combinatorial synthesis as directed by a computer. This modified technique was demonstrated in the synthesis of a 27-membered oligonucleotide library (Xiao et al., 1997). The visual tagging technique came into picture recently to overcome the drawbacks observed in radiofrequency tagging. It is a color-coding strategy. This method helps to distinguish 96 different microreactors when 12 different colored caps and 8 different colored beads are placed among the resin beads. An equal portion of resin is divided into 8 reaction vessels for attachment of 8 different building blocks. Upon completion of the reaction the resin in the vessel is separated into 12 equal portions and placed in a microreactor with a few colored beads (color beads used for this purpose are same for each pool but different for each reactor). One capsule from each pool of 12 is removed and capped with the same colored cap. A second capsule from each pool is removed and capped with a different colored cap and so on (step B). After color tagging the capsules are combined and mixed and divided into 12 reactors as per their cap color (Guiles et al., 1998).

1.8 SCREENING

Combinatorial libraries generated from genomics, proteins, and peptides libraries are screened by HTS technology. This method can screen large compound libraries at a rate of few thousand compounds per day or per week. The main goal of the technique is to accelerate the drug discovery process. It is of vital importance, because parallel and combinatorial chemical synthesis generates a vast number of novel compounds. HTS implements fluorescence resonance energy transfer (FRET) and homogeneous time resolved fluorescence (HTRF) techniques for identification of novel hits. Initially combinatorial libraries are screened for primary assay. This assay is less quantitative than biological assays. If a compound from an examined library gives a positive result or "HIT" then a precise secondary screening is conducted and the IC_{50} is calculated. Secondary screening is performed by adopting biological and biochemical tests (Figure 1.16) (Evans et al., 1988; Gallop, 1994; Gordon et al., 1994; Szymański et al., 2011). Two main strategies are practiced for screening the combinatorial libraries:

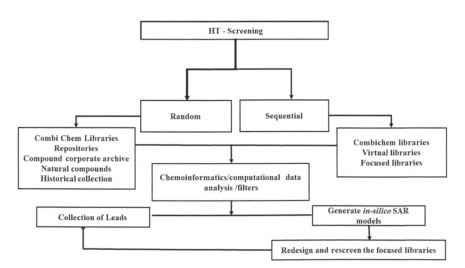

FIGURE 1.16 High throughput screening protocol.

1.8.1   RANDOM SCREENING

It is also known as primary library or mass random screening. The compound libraries obtained from repositories, corporate compound archives, and historical collections or from natural products are screened for primary assay. The assay is used as a filter to identify promising hits. A major goal of this method of screening is to increase the chance of finding biologically active compounds from an unbiased and diverse library with a new and unexpected structural scaffold as a lead candidate. For identifying a hit from the diverse compounds using random screening method, there is no requirement for pharmacophore knowledge or information on enzyme –ligand interactions.

1.8.2   SYSTEMATIC SCREENING

This method works in association with chemoinformatic technology. The main feature of this approach is that only a small set of the library is assayed and results are statistically analyzed. From this a model for structure activity relationship is prepared and the next set of libraries are planned to improve the activity. This step is repeated several times to identify the most promising compounds for lead optimization. Chemoinformatics methods play an important role while designing libraries of combinatorial synthesis. The methods applied include pharmacophore based virtual screening, diversity selection, recursive partitioning, *in silico* prediction methods for drug likeness or lead likeness.

1.9   STRATEGIES FOR SYNTHESIS OF SMALL MOLECULE LIBRARIES

Small molecules libraries are prepared using various strategies of which the more popular ones are listed here: (Mannhold et al., 2006)
   1. *Unbiased/Random Libraries:* These libraries are generated when insufficient knowledge is available on the molecular target and it mainly follows a method that is driven by synthesis. They are envisioned to generate a diverse collection of chemical structures, for the identification of "hits" for any number of targets. Normally, the screening hit results are further investigated, with the goal of generating validated new "lead" structures. These libraries are commonly

referred to as "lead identification", "lead discovery" or "lead finding" libraries. Typically, an unbiased library will consist of structures having a common chemical core or scaffold (or template) and in some cases the compound scaffold is described by a "privileged structure" (DeSimone et al., 2004; Guo and Hobbs, 2003; Horton et al., 2003; Horton et al., 2002; Nicolaou et al., 2000). The unbiased libraries are further divided into two categories: compound libraries and natural product libraries. Compound libraries consist of molecules which are prepared in house or purchased from external vendors. Natural libraries consists of molecules isolated from various sources (Jia, 2003; Ortholand and Ganesan, 2004).

2. *Directed Libraries:* Also known as focused libraries. Many compounds in this type of library are defined by the pharmacophoric feature of the target. This library may be designed on the basis of privileged structures. Privileged structures are the dataset of active molecules available against multiple targets (Bondensgaard et al., 2004; Müller, 2003).

1.10   DECONVOLUTION/IDENTIFYING THE HIT FROM THE SCREENING

Deconvolution is used to identify a HIT/bioactive/potent molecule from the multicomponent combinatorial libraries. It is mainly divided into three types, iteration, positional scanning, and omission.

1. *Iteration Method:* In this method the bioactive molecule is identified by iterative synthesis. It can be demonstrated by the example of tripeptide synthesis. The original library is initially released from the solid support for biological assays. Normally along with the generated library the product in the last step of coupling is separated and given for screening. If any molecule in the library turns out to be bioactive then the terminal amino acid in the library is identified from the secondary library. To identify the second position amino acid the second last step in the tripeptide before mixing and coupling are separated and then coupled with the bioactive terminal amino acid previously screened for bioactivity. The bioactive hit is known as the last two sequences of bioactive amino acids. To identify the first amino acid in the hit one needs to separate the first step coupled

amino acids products and further combine then with the next two bioactive sequences of amino acids which is then screened. The most bioactive molecule reveals the sequence of the bioactive tripeptide (Geyzen et al., 1984; Houghten, 1991).

2. *Positional Scanning:* In case of the iterative method chemists are required to prepare the sublibraries to identify the hit molecule. This is accomplished by prepreparing the set of libraries that identifies the active hit in the positional scanning method. This concept was first explained in a patent filed by Furka et al. (Furka, 1995) in May, 1992 and subsequently put into practice by Pinilla et al. (Houghten et al., 1994). In case of positional scanning a sublibrary is prepared at every step of the combichem reaction along with the original library. Then the biological testing is carried out to identify the N-terminal amino acids at the every step of sublibrary. In principle, along with the original library, sublibraries are also prepared and screened. If the result is positive then all components of the set are required to be tested. From the result, the amino acid sequence of the bioactive peptide can be deduced. Furka explained the model of 20 building blocks of a pentapeptide. Along with the synthesis of the main library few sublibraries aliquots were prepared. While synthesizing the combichem libraries at every step after coupling the amino acids, one fifth of the total resin is removed before the mixing and recombining step and that library is called step 1 sublibrary. Similarly in steps 2, 3, 4 and 5 one fifth of the resin is separated before the coupling and mixing step and labeled as sublibrary 2, 3, 4 and 5, respectively. All of these sublibraries are released from the solid support and screened along with the original library to identify the hit.

3. *Omission Libraries:* Omission libraries can be prepared using full libraries by omitting one amino acid in all coupling positions. One must remember that the same amino acid is omitted irrespective of its number or position occupied in the sequences in all coupling positions (Câmpian et al., 1998). If the prepared bioactive peptide contains the omitted amino acid that means the omitted amino acid is responsible for the bioactivity on the other hand if the prepared peptide gives negative results in the presence of the omitted amino acid it means that the omitted amino acid is not responsible for the biological activity. This technique is simpler and less laborious than iterative screening and positional scanning methods.

1.11   ROLE IN DRUG DISCOVERY

Combinatorial chemistry can contribute to drug discovery by helping to identify biologically active compounds in conjunction with high-throughput screening thus accelerating the discovery of lead compounds with a desired pharmacological profile. Combinatorial chemistry has the power to generate highly diverse compound collections for random screening. It is one of the most efficient tool to identify lead molecules in a short span of time with minimum cost. It is heartening to note that some compounds are undergoing clinical trials that were discovered by combinatorial chemistry (Figure 1.17). The impact of combinatorial chemistry on lead identification is dictated by the size and composition of the library.

As of date we have witnessed two generations of methods in combinatorial chemistry. In the first-generation the main focus was on peptide and oligonucleotide synthesis and in the second generation researchers directed their interest on small-molecule chemistry. Many pharma companies have been an integral part of the development of technologies in the second-generation combinatorial chemistry venture. One among many is Pfizer; which from the beginning used methods in combinatorial chemistry that were developed in-house. Another pharma company, Sepracor has spawned a combinatorial chemistry subsidiary of Dainippon Sumitomo Pharma America Holdings Inc., Eli Lilly got a foot-hold in this technology by acquiring Sphinx and

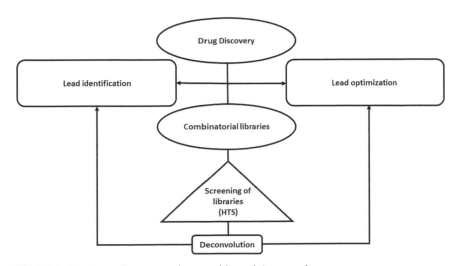

FIGURE 1.17   Drug discovery using combinatorial approach.

Glaxo took over Affymax to kick-start their combinatorial process. Presently the current trend in drug discovery is to combine combinatorial chemistry with structure-based drug design, molecular biology and biological target identification. Arris Pharmaceutical, Ariad Pharmaceuticals, Chiron, 3-Dimensional Pharmaceuticals, and Neurogen are some pharma companies that have taken this approach. Combinatorial chemistry requires huge investment in terms of and infrastructure and other facilities.

Pfizer has a compound in early clinical trials that researchers there wouldn't have discovered without combinatorial chemistry. Gunther Jung (Jung, 2008) has tabled a the list of industries that are progressing in the field of combinatorial chemistry synthesis, analysis and screening and has also reviewed molecules that are in clinical trials which are the products of combinatorial chemistry (Burgess and Lim, 1997). Several reviews have covered the companies (at present more than 180) that are using combinatorial chemistry in early drug discovery phase and have also listed a large number of leads are now in preclinical stage of development. In spite of the vast efforts by several companies, it is disheartening to note that there is no drug till date on the market that has been exclusively discovered by the application of combinatorial chemistry (Thayer, 1996). The following paragraphs gives a glimpse of some efforts by pharma companies to use combinatorial chemistry in the drug discovery process. The nineteenth century saw many scientists aggressively applying combinatorial chemistry tools to

SCHEME 1   Synthesis of dipeptides by Merck Lab using the Kaiser resin.

obtain leads. In 1999, Merck lab used a target based approach (the serine proteinase enzyme thrombin) and combinatorial tools to improve the bio-availability of a lead molecule L-371912 with K_i of 5 nM and prepared a library of 200 compounds. From the primary library they successfully developed L-372460 with K_i of 1.5 nM with improved bioavailability (Lumma et al., 1998). The group used a Kaiser oxime resin which was coupled to Boc protected proline using dicyclohexylcarbodiimide (DCC) (Scheme 1). In the next step deprotection was carried out with trifluoroacetic acid followed by HOBt/EDC coupling to Boc-D-diphenylalanine to get dipeptide attached to the resin. The peptide was cleaved from the resin with a collection of amines RCH_2NH_2 and triethylamine at room temperature. The released amides were deprotected with TFA to give the desired inhibitors.

(i) DCC, CH_2Cl_2, (ii) CF_3COOH, CH_2Cl_2, (iii) Boc (D)diPhe, DCC, CH_2Cl_2, and (iv) RCH_2NH_2, 3–6 days, 25°C, TFA.

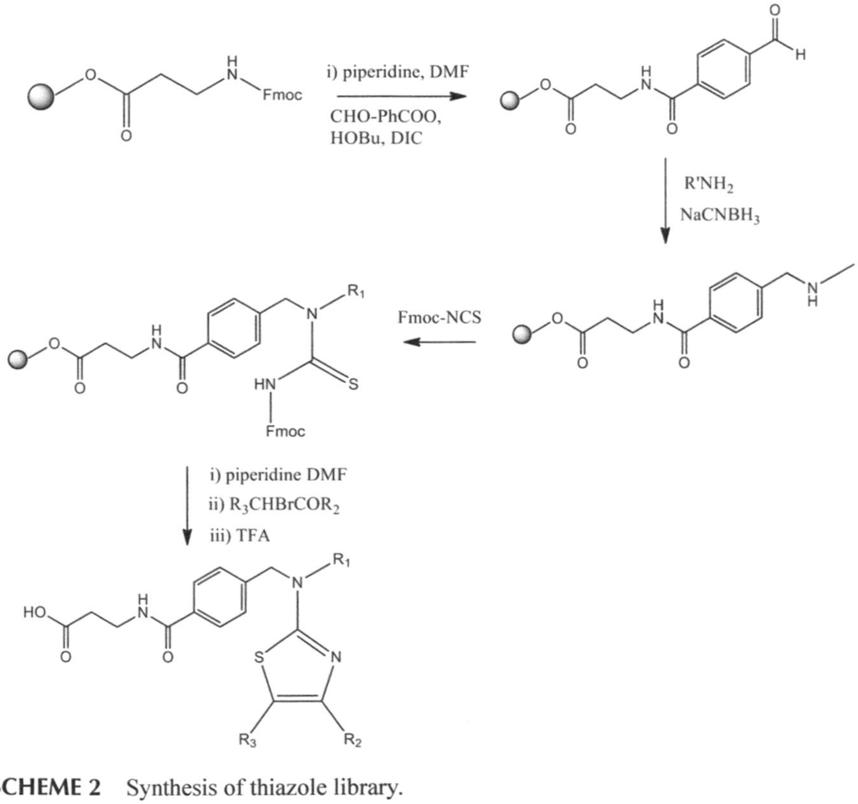

SCHEME 2   Synthesis of thiazole library.

Roland Dolle has successfully reviewed the literature covering the period from 1999 to 2010 and delivered to readers a vast dataset of chemical libraries that have been designed against various targets. We quote here a significant example in this section:

hGluR: Glucagon maintains glucose homeostasis during the fasting state by promoting hepatic gluconeogenesis and glycogenolysis. Antagonizing the glucagon receptor is expected to result in reduced hepatic glucose overproduction, leading to overall glycemic control and a possible treatment for type 2 diabetes. Over 800 thiazole analogs were prepared using Fmoc-β-Ala-Wang resin which was deprotected and coupled to 4-formylbenzoic acid which on reductive animation with primary amines followed by treatment with isocyanate gave the thiourea. Condensation of the Fmoc protected thiourea with α-bromoketones lead to the thiazole library (Scheme 2) (Madsen et al., 2009).

ACKNOWLEDGEMENT

The authors would like to acknowledge Mr. Joginder Singh Paneysar and Mr. Rajendra R. Rane for their invaluable contributions in the preparation of manuscript.

KEYWORDS

- chemistry
- combinatorial
- drug discovery
- library
- synthesis
- technology

REFERENCES

Akaji, K., et al. (1990). Tetrafluoroboric acid as a useful deprotecting reagent in Fmoc-based solid-phase peptide syntheses (Fmoc = fluoren-9-ylmethoxycarbonyl). *J. Chem. Soc., Chem. Commun., 4*, 288–290.

Albericio, F. (2000). Orthogonal protecting groups for Nα-amino and C-terminal carboxyl functions in solid-phase peptide synthesis. *Peptide Science, 55*(2), 123–139.

Albericio, F., et al. (1984). Solid phase synthesis and HPLC purification of the protected *1–12* sequence of apamin for rapid synthesis of apamin analogs differing in the C-terminal region. *Tetrahedron, 40*(21), 4313–4326.

Alcaro, M. C., et al. (2004). On-resin head-to-tail cyclization of cyclotetrapeptides: optimization of crucial parameters. *Journal of Peptide Science, 10*(4), 218–228.

Anderson, G. W., & Callahan, F. M. (1960). t-Butyl Esters of Amino Acids and Peptides and their Use in Peptide Synthesis. *Journal of the American Chemical Society, 82*(13), 3359–3363.

Anderson, G. W., & McGregor, A. C. (1957). t-Butyloxycarbonylamino acids and their use in peptide synthesis. *Journal of the American Chemical Society, 79*(23), 6180–6183.

Atherton, E., et al. (1979). The polyamide method of solid phase peptide and oligonucleotide synthesis. *Bioorganic Chemistry, 8*(3), 351–370.

Atherton, E., Sheppard, R. C., & Wade, J. D. (1983). Side chain protected N α-fluorenylmethoxycarbonylamino-acids for solid phase peptide synthesis. N α-Fluorenylmethoxycarbonyl-NG-4-methoxy-*2,3,6*-trimethylbenzenesulphonyl-L-arginine. *Journal of the Chemical Society, Chemical Communications, 19*, 1060–1062.

Baxter, E. W., et al. (1998). Arylsulfonate esters in solid phase organic synthesis. II. Compatibility with commonly used reaction conditions. *Tetrahedron Letters, 39*(9), 979–982.

Beavers, M. P., & Chen, X. (2002). Structure-based combinatorial library design: methodologies and applications. *Journal of Molecular Graphics and Modelling, 20*(6), 463–468.

Bergmann, M., & Zervas, L. (1932). Über ein allgemeines Verfahren der Peptid-Synthese. *Berichte der deutschen chemischen Gesellschaft (A and B Series), 65*(7), 1192–1201.

Bodanszky, M. (1984). Alkyl esters of amino acids. *International Journal of Peptide and Protein Research, 23*(1), 111–111.

Böhm, G., et al. (1998). A novel linker for the attachment of alcohols to solid supports. *Tetrahedron Letters, 39*(22), 3819–3822.

Bondensgaard, K., et al. (2004). Recognition of privileged structures by G-protein coupled receptors. *Journal of Tetrahedron Letters,, 47*(4), 888–899.

Bourgault, S., Letourneau, M., & Fournier, A. (2007). Development of photolabile caged analogs of endothelin-*1*. *Peptides, 28*(5), 1074–1082.

Bräse, S., et al. (1999). Triazenes as robust and simple linkers for amines in solid-phase organic synthesis. *Tetrahedron Letters, 40*(11), 2105–2108.

Brenner, S., & Lerner, R. A. (1992). Encoded combinatorial chemistry. *Proceedings of the National Academy of Sciences, 89*(12), 5381–5383.

Brown, R. D., Hassan, M., & Waldman, M. (2000). Combinatorial Library Design for Diversity, Cost Efficiency, and Drug-Like Character. *Journal of Molecular Graphics and Modeling, 18*(4), 427–437.

Bunin, B. A., & Ellman, J. A. (1992). A general and expedient method for the solid-phase synthesis of *1, 4*-benzodiazepine derivatives. *Journal of the American Chemical Society, 114*(27), 10997–10998.

Burgess, K., & Lim, D. (1997). Resin type can have important effects on solid phase asymmetricalkylation reactions. *Chem. Commun., 8*, 785–786.

Butwell, F. G., Haws, E. J., & Epton, R. (1988). Advances in ultra-high load polymer-supported peptide synthesis with phenolic supports: *1*. A selectively-labile c-terminal spacer group for use with a base-mediated n-terminal deprotection strategy and fmoc amino acids. in Makromolekulare Chemie. Macromolecular Symposia. Wiley Online Library.

Câmpian, E., et al. (1998). Deconvolution by omission libraries. *Bioorganic & Medicinal Chemistry Letters, 8*(17), 2357–2362.

Carey, R. I., et al. (1997). Preparation and properties of Nα-Bpoc-amino acid pentafluorophenyl esters. *The Journal of Peptide Research, 49*(6), 570–581.

Carpino, L. A. (1957). Oxidative Reactions of Hydrazines. IV. Elimination of Nitrogen from *1, 1*-Disubstituted-*2*-arenesulfonhydrazides*1–4. Journal of the American Chemical Society, 79*(16), 4427–4431.

Carpino, L. A., & Chao, H.-G., (1995). *H. Chem. Abstr, 124,* 146865.

Carpino, L. A., & Han, G. Y. (1970). 9-Fluorenylmethoxycarbonyl function, a new base-sensitive amino-protecting group. *Journal of the American Chemical Society, 92*(19), 5748–5749.

Carpino, L. A., Schroff, H. N., Chao, H.-G., Mansour, E. M. E., & Albericio, F. (1995). Peptides 1994, *Proceedings of the 23rd European Peptide Symposium,* 155–156.

Carpino, L. A., & Han, G. Y. (1972). 9-Fluorenylmethoxycarbonyl amino-protecting group. *The Journal of Organic Chemistry, 37*(22), 3404–3409.

Chucholowski, A., et al. (1996). Novel Solution-and Solid-Phase Strategies for the Parallel and Combinatorial Synthesis of Small-Molecular-Weight Compound Libraries. *CHIMIA International Journal for Chemistry, 50*(11), 525–530.

Colvin, E., Raphael, R., & Roberts, J. (1971). The total synthesis of (±)-trichodermin. *Journal of the Chemical Society D: Chemical Communications, 15,* 858–859.

Conti, P., et al. (1997). A new cleavage strategy for the solid-phase synthesis of secondary amines. *Tetrahedron Letters, 38*(16), 2915–2918.

Corey, E., et al. (1981). *Tetrahedron Letters, 22*(36), 3455–3458.

Corey, E., Niwa, H., & Knolle, J. (1978). Total synthesis of (S)-*12*-hydroxy-*5, 8, 14*-cis,-*10*-transeicosatetraenoic acid *(*Samuelsson's HETE*). Journal of the American Chemical Society, 100*(6), 1942–1943.

Cros, E., et al. (2004). N-Tetrachlorophthaloyl (TCP) Protection for Solid-Phase Peptide Synthesis. European Journal of Organic Chemistry, 17, 3633–3642.

DeSimone, R., et al. (2004). Privileged structures: applications in drug discovery. *Combinatorial Chemistry & High Throughput Screening, 7*(5), 473–493.

Dolle, R. E. (2002). Comprehensive survey of combinatorial library synthesis: 2001. *Journal of Combinatorial Chemistry, 4*(5), 369–418.

Dressman, B. A., Spangle, L. A., & Kaldor, S. W. (1996). Solid phase synthesis of hydantoins using a carbamate linker and a novel cyclization/cleavage step. *Tetrahedron Letters, 37*(7), 937–940.

Drewry, D. H., & Young, S. S. (1999). Approaches to the design of combinatorial libraries. *Chemometrics and Intelligent Laboratory Systems, 48*(1), 1–20.

Eckard, P., et al. (2010). Five Natural Product-Based, Chemically and Functionally Diverse Libraries. *Combinatorial Synthesis of Natural Product-Based Libraries,* 99.

Ellison, R. A. L., E. R., Chiu, C.-W., (1975). Tetrahedron Lett, 499.

Estep, K. G., et al. (1998). Indole resin: A versatile new support for the solid-phase synthesis of organic molecules. *The Journal of Organic Chemistry, 63*(16), 5300–5301.

Evans, B., et al. (1988). Methods for drug discovery: development of potent, selective, orally effective cholecystokinin antagonists. *Journal of Medicinal Chemistry, 31*(12), 2235–2246.

Evans, D. A., et al. (1977). Thiosilanes, a promising class of reagents for selective carbonyl protection. Journal of the American Chemical Society, 99(15), 5009–5017.

Fauchère, J.-L., et al. (1998). Combinatorial chemistry for the generation of molecular diversity and the discovery of bioactive leads. *Chemometrics and Intelligent Laboratory Systems, 43*(1), 43–68.

Fréchet, J. M., & Haque, K. E. (1975). Use of polymers as protecting groups in organic synthesis. II. Protection of primary alcohol functional groups. *Tetrahedron Letters, 16*(35), 3055–3056.

Fréchet, J. M., & Nuyens, L. J. (1976). Use of polymers as protecting groups in organic synthesis. III. Selective functionalization of polyhydroxy alcohols. *Canadian Journal of Chemistry, 54*(6), 926–934.

Fréchet, J. M., & Seymour, E. (1978). Use of Polymers as Protecting Groups in Organic Synthesis. VII. Preparation of Monobenzoates of Acyclic Triols. Israel *Journal of Chemistry, 17*(4), 253–256.

Friede, M., et al. (1991). Incomplete TFA deprotection of N-terminal trityl-asparagine residue in fmoc solid-phase peptide chemistry. *Peptide Research, 5*(3), 145–147.

Früchtel, J. S., & Jung, G. (1996). Organic chemistry on solid supports. *Angewandte Chemie International Edition in English, 35*(1), 17–42.

Funakoshi, S., et al. (1988). Combination of a new amide-precursor reagent and trimethylsilyl bromide deprotection for the Fmoc-based solid phase synthesis of human pancreastatin and one of its fragments (Fmoc= fluoren-9-ylmethoxycarbonyl). *Journal of the Chemical Society, Chemical Communications, 24*, 1588–1590.

Furka, A. (1995). History of combinatorial chemistry. *Drug Development Research, 36*(1), 1–12.

Furka, Á. (2007). *Combinatorial Chemistry Principles and Techniques.*

Fyles, T. M., & Leznoff, C. C. (1976). The use of polymer supports in organic synthesis. V. The preparation of monoacetates of symmetrical diols. *Canadian Journal of Chemistry, 54*(6), 935–942.

Gallop, M. A., et al. (1994). Applications of Combinatorial Technologies to Drug Discovery. 1. Background and Peptide Combinatorial Libraries. *Journal of Medicinal Chemistry, 37*(9), 1233–1251.

Gayo, L. M., & Suto, M. J. (1997). Traceless linker: Oxidative activation and displacement of a sulfur-based linker. *Tetrahedron Letters, 38*(2), 211–214.

Genet, J. P., Blant, E., Savignac, M., Lemeune, S., Lemaire-Audoire, S., & Bernard, J. M. (1993). *Synlett J.* 680–682.

Geyzen, H. M., Meloen, R. H., & Barteling, S. J. (1984). Use of peptide synthesis to probe viral antigens for epitopes to a resolution of a single amino acid. *Proceedings of the National Academy of Sciences, 81*(13), 3998–4002.

Givens, R. S., et al. (2000). *Journal of the American Chemical Society, 122*(12), 2687–2697.

Golebiowski, A., Klopfenstein, S. R., & Portlock, D. E. (2001). Lead compounds discovered from libraries. *Current Opinion in Chemical Biology, 5*(3), 273–284.

Gordon, E. M., et al. (1994). Applications of combinatorial technologies to drug discovery. 2. Combinatorial organic synthesis, library screening strategies, and future directions. *Journal of Medicinal Chemistry, 37*(10), 1385–1401.

Guiles, J. W., Lanter, C. L., & Rivero, R. A. (1998). A visual tagging process for mix and sort combinatorial chemistry. *Angewandte Chemie International Edition, 37*(7), 926–928.

Guo, T., & Hobbs, D. W. (2003). Privileged structure-based combinatorial libraries targeting G protein-coupled receptors. *Assay and Drug Development Technologies, 1*(4), 579–592.

Han, H., & Janda, K. D. (1996). Azatides: Solution and liquid phase syntheses of a new peptidomimetic. *Journal of the American Chemical Society, 118*(11), 2539–2544.

Han, Y. S., Sole, N. A., Tejbrant, J., & Barany, G. (1996). *Pept. Res, 9,* 166.

Hanessian, S., & Xie, F. (1998). Polymer-bound p-alkoxybenzyl trichloracetimidates: Reagents for the protection of alcohols as benzyl ethers on solid-phase. *Tetrahedron Letters, 39*(8), 733–736.

Harris, C. J., et al. (2011). The design and application of target-focused compound libraries. *Combinatorial Chemistry & High Throughput Screening, 14*(6), 521.

Heathcock, C. H., & Ratcliffe, R. (1971). Stereoselective total synthesis of the guaiazulenic sesquiterpenoids. alpha.-bulnesene and bulnesol. *Journal of the American Chemical Society, 93*(7), 1746–1757.

Hendrickson, J. B., & Kandall, C. (1970). The phenacyl protecting group for acids and phenols. *Tetrahedron Letters, 11*(5), 343–344.

Hermkens, P. H., Ottenheijm, H. C., & Rees, D. (1996). Solid-phase organic reactions: a review of the recent literature. *Tetrahedron, 52*(13), 4527–4554.

Horton, D. A., Bourne, G. T., & Smythe, M. L. (2002). Exploring privileged structures: the combinatorial synthesis of cyclic peptides. *Journal of Computer-Aided Molecular Design, 16*(5–6), 415–431.

Horton, D. A., Bourne, G. T., & Smythe, M. L. (2003). The combinatorial synthesis of bicyclic privileged structures or privileged substructures. *Chemical Reviews, 103*(3), 893–930.

Houghten, R. A., et al. (1991). *Generation and Use of Synthetic Peptide Combinatorial Libraries for Basic Research and Drug Discovery.*

Houghten, R., et al. (1994). Optimal peptide length determination using synthetic peptide combinatorial libraries. in *Peptides-American Symposium.* Escom Science Publishers.

Huang, R., Pershad, K., Kokoszka, M., & Kay, B. K. (2011). *Phage-Displayed Combinatorial Peptides. Amino Acids, Peptides and Proteins in Organic Chemistry: Protection Reactions, Medicinal Chemistry, Combinatorial Synthesis,* Volume 4, 451–471.

Hulme, C., et al. (1998). Novel safety-catch linker and its application with a Ugi/De-BOC/Cyclization (UDC) strategy to access carboxylic acids, 1,4-benzodiazepines, diketopiperazines, ketopiperazines and dihydroquinoxalinones. *Tetrahedron Letters, 39*(40), 7227–7230.

Imoto, M., et al. (1988). Synthetic approach to bacterial lipopolysaccharide, preparation of trisaccharide part structures containing KDO and 1-dephospho lipid A. *Tetrahedron Letters, 29*(18), 2227–2230.

Jia, Q. (2003). *Stud. Nat. Prod. Chem., 29,* 643–718.

Jung, G. (2008). *Combinatorial Chemistry: Synthesis, Analysis, Screening.* John Wiley & Sons.

Jung, G. (2008). *Combinatorial Peptide and Nonpeptide Libraries: A Handbook.* John Wiley & Sons.

Jung, N., Wiehn, M., & Bräse, S. (2007). Multifunctional linkers for combinatorial solid phase synthesis, in *Combinatorial Chemistry on Solid Supports.* Springer. 1–88.

Kenner, G., McDermott, J., & Sheppard, R. (1971). The safety catch principle in solid phase peptide synthesis. *Journal of the Chemical Society D: Chemical Communications, 12,* 636–637.

Kerr, J. M., Banville, S. C., & Zuckermann, R. N. (1993). Encoded combinatorial peptide libraries containing nonnatural amino acids. *Journal of the American Chemical Society, 115*(6), 2529–2531.

Kim, R. M., et al. (1996). Dendrimer-supported combinatorial chemistry. *Proceedings of the National Academy of Sciences, 93*(19), 10012–10017.

Kishi, Y., Fukuyama, T., & Nakatsuka, S. (1973). New method for the synthesis of epidithiodiketopiperazines. *Journal of the American Chemical Society, 95*(19), 6490–6492.

Koh, J. S., & Ellman, J. A. (1996). Palladium-Mediated Three-Component Coupling Strategy for the Solid-Phase Synthesis of Tropane Derivatives. *The Journal of Organic Chemistry, 61*(14), 4494–4495.

König, W., & Geiger, R. (1970). Eine neue Amid-Schutzgruppe. *Chemische Berichte, 103*(7), 2041–2051.

Kowalski, J., & Lipton, M. A. (1996). Solid phase synthesis of a diketopiperazine catalyst containing the unnatural amino acid (S)-norarginine. *Tetrahedron Letters, 37*(33), 5839–5840.

Lenard, J., & Robinson, A. B. (1967). Use of Hydrogen Fluoride in Merrifield Solid-Phase Peptide Synthesis. *Journal of the American Chemical Society, 89*(1), 181–182.

Lewis, R. A. (2002). The design of small-and medium-sized focused combinatorial libraries, in *Molecular Diversity in Drug Design*. Springer. 221–248.

Lloyd-Williams, P., et al. (1991). Solid-phase synthesis of peptides using allylic anchoring groups. An investigation of their palladium-catalyzed cleavage. *Tetrahedron Letters, 32*(33), 4207–4210.

Lodder, M., et al. (1998). Misacylated transfer RNAs having a chemically removable protecting group. *The Journal of Organic Chemistry, 63*(3), 794–803.

Loffet, A., & Zhang, H. (1993). Allyl-based groups for side-chain protection of amino-acids. *International Journal of Peptide and Protein Research, 42*(4), 346–351.

Lumma, W. C., et al. (1998). Design of novel, potent, noncovalent inhibitors of thrombin with nonbasic P-1 substructures: rapid structure-activity studies by solid-phase synthesis. *Journal of Medicinal Chemistry, 41*(7), 1011–1013.

Lyttle, M. H. H. (1992). Peptides Chemistry and Biology. *Proceedings of the 12th American Peptide Symposium*, 583–584.

McArthur, C. R., et al. (1982). Polymer supported enantioselective reactions. II. α-Methylation of cyclohexanone. *Canadian Journal of Chemistry, 60*(14), 1836–1841.

MacCoss, M., & Cameron, D. J. (1978). Facile detritylation of nucleoside derivatives by using trifluoroacetic acid. *Carbohydrate Research, 60*(1), 206–209.

Madsen, P., et al. (2009). Human glucagon receptor antagonists with thiazole cores. A novel series with superior pharmacokinetic properties. *Journal of Medicinal Chemistry, 52*(9), 2989–3000.

Mannhold, R., et al. (2006). *High-Throughput Screening in Drug Discovery*. Vol. 35. John Wiley & Sons.

Mergler, M., et al. (1988). Peptide synthesis by a combination of solid-phase and solution methods I: A new very acid-labile anchor group for the solid phase synthesis of fully protected fragments. *Tetrahedron Letters, 29*(32), 4005–4008.

Mergler, M., et al. (1988). Peptide synthesis by a combination of solid-phase and solution methods II synthesis of fully protected peptide fragments on 2-methoxy-4-alkoxy-benzyl alcohol resin. *Tetrahedron Letters, 29*(32), 4009–4012.

Mergler, M., et al. (1999). Solid phase synthesis of fully protected peptide alcohols. *Tetrahedron Letters, 40*(25), 4663–4664.

Merrifield, R. B. (1963). Solid phase peptide synthesis. I. The synthesis of a tetrapeptide. *Journal of the American Chemical Society, 85*(14), 2149–2154.

Merrifield, R. B., Stewart, J. M., & Jernberg, N. (1966). Instrument for automated synthesis of peptides. *Analytical Chemistry, 38*(13), 1905–1914.

Miertus, S., & Fassina, G. (2014). Combinatorial Chemistry and Technologies: Methods and Applications. CRC Press.

Mitchell, A. R., Erickson, B. W., Ryabtsev, M. N., Hodges, R. S., & Merrifield, R. B. (1976). tert-Butoxycarbonylaminoacyl-4-(oxymethyl) phenylacetamidomethyl-resin, a more acid-resistant support for solid-phase peptide synthesis. *Journal of the American Chemical Society, 98*(23), 7357–7362.

Miyashita, M. Y., & Grieco, P. A. (1977). *A. J. Org. Chem., 44*, p. (1438).

Mojsov, S., & Merrifield, R. (1981). Solid-phase synthesis of crystalline glucagon. *Biochemistry*, *20*(10), 2950–2956.

Montero, A., et al. (2007). Synthesis of a *24*-Membered Cyclic Peptide-Biphenyl Hybrid. *European Journal of Organic Chemistry*, *8*, 1301–1308.

Morales, G. A., & Bunin, B. A. (2003). *Combinatorial Chemistry*. Academic Press.

Moran, E. J., et al. (1995). Radio frequency tag encoded combinatorial library method for the discovery of tripeptide-substituted cinnamic acid inhibitors of the protein tyrosine phosphatase PTP*I*B. *Journal of the American Chemical Society*, *117*(43), 10787–10788.

Mullen, D. G., & Barany, G. (1988). A new fluoridolyzable anchoring linkage for orthogonal solid-phase peptide synthesis: design, preparation, and application of the N-(3 or 4)-[[*4*-(hydroxy-methyl) phenoxy]-tert-butylphenylsilyl] phenyl pentanedioic acid monoamide (Pbs) handle. *The Journal of Organic Chemistry*, *53*(22), 5240–5248.

Müller, G. (2003). Medicinal chemistry of target family directed master keys. *Drug Discovery Today*, *8*(15), 681–691.

Needels, M. C., et al. (1993). Generation and screening of an oligonucleotide-encoded synthetic peptide library. *Proceedings of the National Academy of Sciences*, *90*(22), 10700–10704.

Nefzi, A., et al. (1997). Solid phase synthesis of heterocyclic compounds from linear peptides: cyclic ureas and thioureas. *Tetrahedron Letters*, *38*(6), 931–934.

Neises, B., & Steglich, W. (1978). Simple method for the esterification of carboxylic acids. *Angewandte Chemie International Edition in English*, *17*(7), 522–524.

Nicolaou, K., et al. (1995). *Angewandte Chemie International Edition in English*, *34*(20), 2289–2291.

Nicolaou, K., et al. (2000). Natural product-like combinatorial libraries based on privileged structures. *2*. Construction of a *10,000*-membered benzopyran library by directed split-and-pool chemistry using NanoKans and optical encoding. *Journal of the American Chemical Society*, *122*(41), 9954–9967.

Nielsen, J., Brenner, S., & Janda, K. D. (1993). Synthetic methods for the implementation of encoded combinatorial chemistry. *Journal of the American Chemical Society*, *115*(21), 9812–9813.

Nikolaiev, V., et al. (1992). Peptide-encoding for structure determination of nonsequenceable polymers within libraries synthesized and tested on solid-phase supports. *Peptide Research*, *6*(3), 161–170.

Nugiel, D. A., Cornelius, L. A., & Corbett, J. W. (1997). Facile preparation of *2*, *6*-disubstituted purines using solid-phase chemistry. *The Journal of Organic Chemistry*, *62*(1), 201–203.

Obrecht, D., & Villalgordo, J. M. (1998). *Solid-Supported Combinatorial and Parallel Synthesis of Small-Molecular-Weight Compound Libraries*. *17*, Elsevier.

Orlowski, R. C., Walter, R., & Winkler, D. (1976). Study of benzhydrylamine-type polymers. Synthesis and use of p-methoxybenzhydrylamine resin in the solid-phase preparation of peptides. *The Journal of Organic Chemistry*, *41*(23), 3701–3705.

Orry, A. J., Abagyan, R. A., & Cavasotto, C. N. (2006). Structure-based development of target-specific compound libraries. *Drug Discovery Today*, *11*(5), 261–266.

Ortholand, J.-Y., & Ganesan, A. (2004). Natural products and combinatorial chemistry: back to the future. *Current Opinion in Chemical Biology*, *8*(3), 271–280.

Pandeya, S., & Thakkar, D. (2005). Combinatorial chemistry: A novel method in drug discovery and its application. *Indian Journal of Chemistry*, *44*, 335–348.

Pietta, P., & Marshall, G. R. (1970). Amide protection and amide supports in solid-phase peptide synthesis. *Journal of the Chemical Society D: Chemical Communications*, *11*, 650–651.

Pon, R. T., & Yu, S. (1997). Hydroquinone-O, O'-diacetic acid as a more labile replacement for succinic acid linkers in solid-phase oligonucleotide synthesis. *Tetrahedron Letters, 38*(19), 3327–3330.

Poupart, M.-A., et al. (1999). Solid-phase synthesis of peptidyl trifluoromethyl ketones. *The Journal of Organic Chemistry, 64*(4), 1356–1361.

Quesnel, A., & Briand, J. P. (1998). Incomplete trifluoroacetic acid deprotection of asparagine-trityl-protecting group in the vicinity of a reduced peptide bond. *The Journal of Peptide Research, 52*(2), 107–111.

Rabanal, F., et al. (1990). Study on the stability of 9-fluorenylmethoxycarbonyl in catalytic-hydrogenation conditions. In anales de quimica. Real soc espan quimica facultad de fisica quimica ciudad univ, 3 madrid, spain.

Ramage, R., et al. (1999). Comparative studies of Nsc and Fmoc as Nα-protecting groups for SPPS. *Journal of Peptide Science, 5*(4), 195–200.

Ramage, R., Green, J., & Blake, A. J. (1991). An acid labile arginine derivative for peptide synthesis: N G-2,2,5,7,8-pentamethylchroman-6-sulphonyl-L-arginine. *Tetrahedron, 47*(32), 6353–6370.

Rinnová, M., & Lebl, M. (1996). Molecular diversity and libraries of structures: Synthesis and screening. *Collection of Czechoslovak Chemical Communications, 61*(2), 171–231.

Roeske, R. (1963). Preparation of t-butyl esters of free amino acids. *The Journal of Organic Chemistry, 28*(5), 1251–1253.

Rosenthal, K., Erlandsson, M., & Undén, A. (1999). 4-(3-Hydroxy-4-methylpentyl) phenylacetic acid as a new linker for the solid phase synthesis of peptides with Boc chemistry. *Tetrahedron Letters, 40*(2), 377–380.

Rueter, J. K., et al. (1998). Arylsulfonate esters in solid phase organic synthesis. I. Cleavage with amines, thiolate, and imidazole. *Tetrahedron Letters, 39*(9), 975–978.

Sabirov, A. N., et al. (1997). FMOC-and NSC-Groups as a Base Labile N (a)-Amino Protection: A Comparative Study in the Automated SPPS. *Protein and Peptide Letters, 4*, 307–312.

Samukov, V. V., Sabirov, A. N., & Pozdnyakov, P. I. (1994). 2-(4-Nitrophenyl) sulfonylethoxycarbonyl (Nsc) group as a base-labile α-amino protection for solid phase peptide synthesis. *Tetrahedron Letters, 35*(42), 7821–7824.

Sauerbrei, B., Jungmann, V., & Waldmann, H. (1998). An Enzyme-Labile Linker Group for Organic Syntheses on Solid Supports. *Angewandte Chemie International Edition, 37*(8), 1143–1146.

Shimonishi, Y., Sakakibara, S., & Akabori, S. (1962). Studies on the Synthesis of Peptides Containing Glutamine as the C-Terminal. I. Protection of Amide-Nitrogen With Xanthyl Group During Peptide Synthesis. *Bulletin of the Chemical Society of Japan, 35*(12), 1966–1970.

Sieber, P. (1987). A new acid-labile anchor group for the solid-phase synthesis of C-terminal peptide amides by the Fmoc method. *Tetrahedron Letters, 28*(19), 2107–2110.

Sieber, P. A., Eisler, K., Kamber, B., Riniker, B., & Rink, H. (1977). 543–545.

Sieber, P., & Riniker, B. (1991). Protection of carboxamide functions by the trityl residue. Application to peptide synthesis. *Tetrahedron Letters, 32*(6), 739–742.

Simkins, R., & Williams, G. (1952). The nitration of guanidine in sulfuric acid. Part I. The reversible conversion of guanidine nitrate into nitroguanidine. *J. Chem. Soc.,* 3086–3094.

Sola, R., Méry, J., & Pascal, R. (1996). Fmoc-based solid-phase peptide synthesis using Dpr (Phoc) linker. Synthesis of a C-terminal proline peptide. *Tetrahedron Letters, 37*(51), 9195–9198.

Stelakatos, G., Paganou, A., & Zervas, L. (1966). New methods in peptide synthesis. Part III. Protection of carboxyl group. *Journal of the Chemical Society C: Organic,* 1191–1199.

Story, S. C., & Aldrich, J. V. (1992). Preparation of protected peptide amides using the Fmoc chemical protocol. *International Journal of Peptide and Protein Research, 39*(1), 87–92.

Studer, A., et al. (1997). Fluorous synthesis: A fluorous-phase strategy for improving separation efficiency in organic synthesis. *Science, 275*(5301), 823–826.

Swartz, M. E. (2000). *Analytical Techniques in Combinatorial Chemistry.* CRC Press.

Szardenings, A. K., et al. (1999). Identification of highly selective inhibitors of collagenase-*1* from combinatorial libraries of diketopiperazines. *Journal of Medicinal Chemistry, 42*(8), 1348–1357.

Szymański, P., Markowicz, M., & Mikiciuk-Olasik, E. (2011). *International Journal of Molecular Sciences, 13*(1), 427–452.

Tam, J. P. (1985). A gradative deprotection strategy for the solid-phase synthesis of peptide amide using p-*(*acyloxy*)* benzhydrylamine resin and the SN*2* deprotection method. *The Journal of Organic Chemistry, 50*(25), 5291–5298.

Tam, J. P., Heath, W. F., & Merrifield, R. (1983). An SN*2* deprotection of synthetic peptides with a low concentration of hydrofluoric acid in dimethyl sulfide: evidence and application in peptide synthesis. *Journal of the American Chemical Society, 105*(21), 6442–6455.

Tam, J. P., Heath, W. F., & Merrifield, R. (1986). Mechanisms for the removal of benzyl protecting groups in synthetic peptides by trifluoromethanesulfonic acid-trifluoroacetic acid-dimethyl sulfide. *Journal of the American Chemical Society, 108*(17), 5242–5251.

Thayer, A. M. (1996). Combinatorial chemistry becoming core technology at drug discovery companies. *Chemical & Engineering News, 74*(7), 57–57.

Thompson, L. A., & Ellman, J. A. (1994). Straightforward and general method for coupling alcohols to solid supports. *Tetrahedron Letters, 35*(50), 9333–9336.

Thompson, L. A., & Ellman, J. A. (1996). Synthesis and applications of small molecule libraries. *Chemical Reviews, 96*(1), 555–600.

Tietze, L. F., Steinmetz, A., & Balkenhohl, F. (1997). Solid-phase synthesis of polymer-bound *β*-ketoesters and their application in the synthesis of structurally diverse pyrazolones. *Bioorganic & Medicinal Chemistry Letters, 7*(10), 1303–1306.

Vaino, A. R., & Janda, K. D. (2000). Solid-phase organic synthesis: a critical understanding of the resin. *Journal of Combinatorial Chemistry, 2*(6), 579–596.

Vlattas, I., et al. (1997). The use of thioesters in solid phase organic synthesis. *Tetrahedron Letters, 38*(42), 7321–7324.

Wang, S.-S. (1973). p-Alkoxybenzyl alcohol resin and p-alkoxybenzyloxycarbonylhydrazide resin for solid phase synthesis of protected peptide fragments. *Journal of the American Chemical Society, 95*(4), 1328–1333.

Wang, S., et al. (1974). Solid phase synthesis of bovine pituitary growth hormone-*(123–131)* nonapeptide. *International Journal of Peptide and Protein Research, 6*(2), 103–109.

Welsch, M. E., Snyder, S. A., & Stockwell, B. R. (2010). Privileged Scaffolds for Library Design and Drug Discovery. *Current Opinion in Chemical Biology, 14*(3), 347–361.

Wong, J. Y., Manning, C., & Leznoff, C. C. (1974). Solid Phase Synthesis and Photochemistry of *4, 4*-Stil-benedicarbaldehyde. *Angewandte Chemie International Edition in English, 13*(10), 666–667.

Woodward, R., et al. (1966). The total synthesis of cephalosporin C1. *Journal of the American Chemical Society, 88*(4), 852–853.

Wu, Z., et al. (2003). Synthesis of tetrahydro-*1, 4*-benzodiazepine-*2*-ones on hydrophilic polyamide synphase lanterns. *Journal of Combinatorial Chemistry, 5*(2), 166–171.

Xiao, X. Y., et al. (1997). Combinatorial chemistry with laser optical encoding. *Angewandte Chemie International Edition in English 36*(7), 780–782.

Zehavi, U., & Patchornik, A. (1973). Oligosaccharide synthesis on a light-sensitive solid support. I. Polymer and synthesis of isomaltose *(6*-O-. alpha.-D-glucopyranosyl-D-glucose*). Journal of the American Chemical Society, 95*(17), 5673–5677.

Zhang, K., et al. (2013). Destruction of perfluorooctane sulfonate (PFOS) and perfluorooctanoic acid (PFOA) by ball milling. *Environmental Science & Technology, 47*(12), 6471–6477.

Zhao, Z.-G., & Lam, K. S. (1997). Synthetic peptide libraries. In *Annual Reports in Combinatorial Chemistry and Molecular Diversity.* Springer, pp. 192–209.

Zinner, H. (1950). Notiz *über* Mercaptale der d-Ribose. *Chemische Berichte, 83*(3), 275–277.

CHAPTER 2

ADVANCED TECHNIQUES IN BIOMOLECULAR SIMULATIONS

ANIRUDH V. BELUBBI and ELVIS A. F. MARTIS

Molecular Simulations Groups, Department of Pharmaceutical Chemistry, Bombay College of Pharmacy, Kalina, Santacruz (E), Maharashtra – 400098, India, E-mail: elvis_bcp@elvismartis.in, elvis@profeccoutinho.net.in

CONTENTS

ABSTRACT

Chemical biology is the area of science which lies at the interface of chemistry and biology, and path breaking discoveries in this area have given insights on interesting biological events. These insights have opened new avenues for drug discovery science, and thus, paved way for identifying new targets for developing new therapeutic agents. It has also sought answers to the biology behind old targets and has helped medicinal chemists design better chemical entities and target specific molecules. The great challenge in studying biological molecules, viz proteins, RNA and DNA, is related to their structure and functions. Experimental methods like X-ray crystallography, neutron scattering and electron microcopy help elucidate the 3D structure of a molecule and provide a fixed/rigid view of the structure of the biomolecule without much comprehension about their motion. NMR, a spectroscopic tool, is capable of showing a dynamic picture of biomolecules, however, suffers from many other drawbacks. In order to study the motion of these biomolecules in physiological medium (water), computer simulations are very useful. The fixed structure elucidated with the help of spectroscopic and crystallographic tools forms a good starting point for any theoretical study, as experimentally solved structures are less prone to artifacts than theoretical models (homology modeling, threading or *ab initio*). Hence, the crux of this chapter is molecular dynamics simulation, and methods that improve the sampling in order to overcome the convergence problems of classical molecular dynamics. We will also discuss studies on ligand-protein interaction, and protein folding carried out of using molecular dynamics based methods.

2.1   INTRODUCTIONS TO COMPUTER SIMULATIONS

It is a well-known fact that all the biological macromolecules are never static in the body, but always in a state of dynamic motion (de la Torre and Bloomfield, 1981). Owing to this dynamic behavior of macromolecules many interesting events occur and biological processes are carried out, that are essential to sustain life. Owing to highly complex and multidimensional configuration phase space approachable to macromolecules, they are very flexible in nature (Carlson, 2002; Taverna and Goldstein, 2002). A traditional perception of proteins as rigid led to the development and acceptance of "lock-and-key" theory of binding in which the protein exists in a single well-defined rigid state with only one optimal complementary ligand. However, soon it was realized that many ligands that do not binding to the active, instead bind to site away and many structural dissimilar ligands bind to the same site the lock and key theory failed to give satisfactory justifications. A simple example might throw more light on this concept and give a clear understanding, let us look at the functioning of G-Protein Coupled Receptors (GPCRs), when the ligand binds to the appropriate binding site (which is outside the bilayer membrane) there is large conformational change in the receptor, which leads to cascade of events with the final outcome of the desired biological response seen in the cell nucleus (Gouldson et al., 2000; Violin et al., 2008). Not only do receptors undergo conformational changes upon binding of the ligand, but they may also exhibit a conformational change while the ligand is in its vicinity to accommodate it in the binding pocket, which is called *induced fit* binding (Anderson et al., 1990). Many enzymes also exhibit a similar behavior, giving a satisfactory explanation to how allosteric modulators (Barak et al., 1995) and most noncompetitive inhibitors might function (Koshland, 1998). The implications for drug discovery are clear; a single protein structure is only useful to identify ligands for that particular narrow state. To obtain new leads and properly predict activity of existing inhibitors, multiple structures are the best option. The available computational methods range in the degree of protein flexibility that they accommodate. The most challenging task for computational drug design is predicting large domain motions (Carlson and McCammon, 2000).

Small molecules also possess dynamic motion in solution (Stratt and Maroncelli, 1996), giving rise to many conformations with different energy levels related to their conformations. Moreover, ligands are also found to undergo changes in their conformations, in order to bind to their biological target (Chow and Bogdan, 1997; Zavodszky and Kuhn, 2005). This is said to be the bio-active conformation and prediction of this should be the prime focus. Molecular modeling studies thrive to locate this bio-active conformations, which may or may not be the global minimum. This would reduce computational time that would have been wasted searching only for the global minimum which may not be the biological relevant conformation.

Molecular interactions of ligand and receptor are of majorly attractive and repulsive nature. The whole goal is to optimize the attractive interactions and minimize repulsive ones, if one cannot completely omit them in order to design ligands that can tightly bind to it receptor (Böhm and

Klebe, 1996; Whitesides and Krishnamurthy, 2005). Molecular recognitions are typically noncovalent events: the interacting species form a structured aggregate without forming a covalent bond (Homans, 2007).

We begin by understanding the types of motion and their respective time scales that biological macromolecules exhibits, examples are enlisted in Table 2.1. Now, in order to study these motions and their relevance in any biological event, we must run computer simulations for total time period mentioned in column 3 of Table 2.1. The local motions, whose time scales may range from few nanosecond to few microseconds are comparatively less expensive computations and now-a-days these are routinely carried out to compute free energy of binding of ligands and studies alike. Rigid body motions and large scale motions with time scales ranging from few hundreds of nanoseconds to few milliseconds to even few seconds (protein folding) require long sampling to observe events that are inherently slow.

In the following sections we discuss about molecular dynamics, its pros and cons and various other methods alternative to classical molecular dynamics simulations that are used to study complex biological phenomenon. Before we start, we would digress from the topic molecular dynamics, and spend some time explaining quantum mechanics, semiempirical methods, and introduction to molecular mechanics forcefields. Forcefields would lay foundation to understand the concept of computing free energies using molecular dynamics, and quantum mechanics and semiempirical methods would throw light on how forcefield parameters are developed for any

TABLE 2.1　Biological Motions and Their Time Scales

Type of Motion	Specific Example	Time Scale (seconds)	Distance covered
Local Motions	Atomic fluctuations	10^{-15} to 10^{-1}	0.01 to 5 Å
	Side chain Motions		
	Loop Motions		
Rigid Body Motions	Helix Motions	10^{-9} to 1	1 to 10 Å
	Domain Motions (Hinge bending, etc.)		
	Subunit motions		
Large-Scale Motions	Helix coil transitions	10^{-7} to 10^{4}	> 5 Å
	Ligand Dissociation/ Association		
	Folding and Unfolding		

chemical system. We will also discuss statistical mechanics that would lay foundation in making attempts relating microscopic state (used for computations) to macroscopic states (real world scenario), and explaining various ensembles in computer simulations and importance of Ergodic hypothesis.

2.2 QUANTUM MECHANICS

Historically, Quantum Mechanics (QM) was seen as a competitive technique to molecular mechanics as both the methods were developed by different group-seeking answer to different questions revolving around molecular modeling (Boyd, 1990). With time more avenues of both techniques were explored and currently it is widely accepted that both techniques have their own applications in molecular modeling. With the help of current technology, MM methods can be applied to thousands of atoms, semiempirical quantum chemistry to hundreds, and ab initio quantum chemistry to tens of atoms (Stewart, 1990).

To understand the mathematics of QM, we first present some basic mathematical formulations. Partial derivatives is an important concept in QM as well in MD simulations, the following equations depicts a simple partial derivative,

$$\frac{\partial}{\partial x}(e^{ax}) = a\,(e^{ax}) \tag{1}$$

where '$\delta/\delta x$' is the operator, '(e^{ax})' is the function and 'a' is the scalar value. An eigenvalue equation is an equation that has finite solutions. These solutions are called eigenvectors (or eigenfuctions), and the scalar values associated with them are called eigenvalues. In quantum mechanical theory, solutions to the Schrodinger equation (Eq. 2) are sought.

$$H\Psi = E\Psi \tag{2}$$

Here the wave function Ψ gives the x, y, z spatial coordinates of the particles in the system. The eigenvalue E is the energy of the system in that state. H is the Hamiltonian, an operator to derive the kinetic energy (KE) and potential energy (PE) of a system of electrons and the nuclei. The Hamiltonian operator "H" is based purely on the physics of the system. Unlike equations in MM, there are no chemical properties included in the Schrodinger equation; only physics of the system is considered. The chemistry that we wish to study is due to the fundamental physics (arrangement of electrons) of the system being obeyed. To solve the Schrodinger equation for all molecules, many approximations must be made.

Quantum mechanical calculations have been proven to be very accurate when used to determine the change in energy as a small molecule undergoes geometric distortions. This might lead one to believe that QM calculations are futile for large molecules (Levine and Learning, 2009). It is true that QM calculations are expensive and resource intensive when applied to large molecules as compared to MM calculations, but this does not imply that the QM calculations are useless when it applied to large molecules. Although not widely appreciated, the parameters obtained from QM calculations are used to create empirical forcefields which help in accurate modeling of large molecules using MM based techniques (Boyd, 1990). This approach brings the accuracy level of QM calculations to the large molecules although they are beyond the scope of QM. Molecular mechanics calculations are fast and fairly accurate for molecules within the domain of the forcefield parameterization but sometimes there are certain molecules (generally heterocycles) where MM calculations do not yield satisfactory results and developing forcefield parameters is highly laborious. In such cases, QM is directly applied to the complex molecule to determine amenable three-dimensional molecular structures and their relative energies.

2.3   SEMI-EMPIRICAL METHODS

Semi-empirical techniques are based on the principles of quantum chemistry thereby making them accurate compared to MM techniques. The difference between QM and Semi-empirical arises from the number of assumptions that are made in order to apply those principals to chemistry models. Although assumptions are made in both QM and Semi-empirical calculations, the number of assumptions are far greater in Semi-empirical calculations as compared to those made during QM calculations. The end result is better model obtained compared to that obtained using MM calculation and in much lesser time compared to QM calculations.

Semi-empirical techniques basically employ the Hartree-Fock (HF) equation to solve the Schrodinger equation with a few assumptions. Modern semiempirical model called "Neglect of Diatomic Differential Overlap (NDDO)" simplifies the calculation by replacing overlap matrix S, by a unit matrix, as suggested by Pople (Pople et al., 1965). Hence, the HF secular equation Eq. (3a) reduces to Eq. (3b).

$$|H-ES| = 0 \tag{3a}$$

$$|H-E| = 0 \tag{3b}$$

Existing semiempirical models differ by adding approximations while evaluating one-and two-electron integrals and by the parameterization philosophy. Currently, semiempirical models MNDO (Dewar and Thiel, 1977), AM1 (Dewar et al., 1985), and PM6 (Rezác et al., 2009) are widely used as they are parameterized using a large dataset and hence are applicable to a wide range of molecules.

The main application of semiempirical methods is to gain requirement qualitative insight about electronic structure and properties. These techniques have been used to study the thermochemistry of reactions with the viewpoint of application to design synthetic or scale up procedures. These techniques have also been employed in quantitative structure activity relationship studies (QSAR) to gain insight about reactivity or property trends for a group of similar compounds (Zerner, 1990). Initial geometry optimization of unusually large molecules or transition states that are not possible using MM methods, can be carried out using Semi-empirical methods which in turn helps with the further QM calculations. Semi-empirical calculations are the only method of choice when there are no suitable forcefields available to handle large molecular system.

2.4 HUCKEL'S MOLECULAR ORBITAL THEORY

Molecular Orbital (MO) theory is an alternate theory to the laborious Quantum Mechanical theory that gives close approximations of the electronic structure. The basic principle of this theory is based on the fact that electrons occupy the molecular orbitals that are basically atomic orbitals that are protracted over the entire molecule (Kier, 2012). This assumption takes into account not only the nuclei-electron interactions but also the potentials due to the charge distributions of all other electrons. Electrons occupy the molecular orbital in pairs and with opposite spins. A reasonable assumption made in the MO theory is that a Linear Combination of Atomic Orbitals (LCAO approximation) can be used to express the MO (Lewars, 2010). The fundamental assumption of the LCAO approximation is that an MO can be derived by combining simple functions (basis functions), which comprise a basis set. Pauling (1928) and Lennard–Jones (1929) (Cramer, 2013) suggested these methods for calculating MOs. The LCAO approximation can mathematically be represented by the following general equation.

$$\psi = C_a \psi_a + C_b \psi_b + \cdots + C_n \psi_n \tag{4a}$$

where the coefficients C weight the contributions of the atomic orbitals denoted by ψ.

The LCAO approximation can only be applied to systems where the energies of atomic orbitals are of same magnitude. This approximation is now used to solve the Schrodinger equation (Eq. 2), substituting for ψ for two atomic orbitals (a and b) and rewriting the Eq. (2), we get:

$$C_a(H - E)\psi_a + C_b(H - E)\psi_b = 0 \tag{4b}$$

Now, ψ^2 gives the probability of finding an electron in a region of space and since across all space, the probability is 1, we can write the following equation

$$\int \psi^2 d\tau = 1 \tag{4c}$$

For two atomic orbitals, Eq. (4c) is multiplied on both side by its wave function and integrated over all space that result in the following equations

$$C_a \int \psi_a(H - E)\psi_a \, d\tau + C_b \int \psi_a(H - E)\psi_b \, d\tau = 0 \tag{4d}$$

$$\psi = C_a \psi_a + C_b \psi_b + \cdots + C_n \psi_n \tag{4e}$$

Now to further simplify Eqs. (4d) and (4e), we introduce the following terms:

$$\alpha_i = \int \psi_i H \psi_i d\tau, \beta_{ij} = \int \psi_i H \psi_j d\tau, S_{ij} = \int \psi_i \psi_j d\tau \text{ and } E = \int \psi_i E \psi_i d\tau$$

Expanding the above equations and substituting the newly defined terms in Eq. (4d) and Eq. (4e), we get

$$C_a(\alpha_a - E) + C_a(\beta_{ab} - ES_{ab}) = 0 \tag{4f}$$

$$C_a(\beta_{ba} - ES_{ba}) + C_b(\alpha_b - E) = 0 \qquad (4g)$$

The aforementioned secular equations with two unknown variables C's and E are solved by simultaneous equations or by solving the determinants as shown in Eq-4h

$$\begin{vmatrix} \alpha_a - E & \beta_{ab} - ES_{ab} \\ \beta_{ba} - ES_{ba} & \alpha_b - E \end{vmatrix} = 0 \qquad (4h)$$

Two values for E are obtained by solving Eq. (4h), substituting the lower value of E in Eqs. (4f) and (4c) values for the respective MOs are obtained. These values indicate energies of the molecular orbitals which determine the electronic properties of a molecule. Hence, MO theory is widely used as a reasonably good approximation for solving the Schrödinger equation (Figure 2.1).

2.5 MOLECULAR MECHANICS

Molecular mechanics lays foundation for all the calculations related to molecular dynamics simulations and calculations of energy for any system under study. It was the pioneering work by Lifson (Bixon and Lifson, 1967),

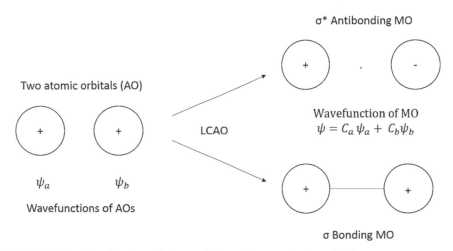

FIGURE 2.1 Two Atomic orbitals combine to give a molecular orbital.

who attempted to study the property of molecular systems using potential energy function. He developed a rather simple mathematical formulation which relates potential energy from bonds, angles torsion, and nonbonded interactions to the global energy of the whole system. This mathematical formulation is called as "forcefield", and is the heart of molecular mechanics. A forcefield equation comprises of three group of terms; one arising from bonded interactions, second from nonbonded interactions, and lastly, cross terms. Bonded interactions (covalent interactions) comprises of terms necessary to define energy changes due to bond stretching or bending (atoms directed bonded to each other; 1,2 atoms), angle bending (between 1,3 atoms), torsion angles (between 1,4 atoms) and improper torsion angles (out-of-plane bending). Non-bonded interactions comprises of terms used to include changes in the potential energy of a system due to interaction of atoms in space (noncovalent interactions), electrostatic terms includes the effect of two point charges interacting and the magnitude is governed by Coulomb's law, change in the energy due dispersive forces (attraction or repulsion) is defined by van der Waal's (vdW) term, and hydrogen-bonding term (special case of electrostatic interaction) may be included explicitly or implicitly defined in the electrostatic term (Pissurlenkar et al., 2009). The functional form of the forcefield is written as follows:

$$U(r) = \sum \frac{1}{2} k_b (b - b_0)^2 + \sum \frac{1}{2} k_\theta (\theta - \theta_0)^2 + \sum k_\Phi (1 - \cos(n\phi + \delta)) +$$
$$\sum_{j=1}^{N-1} \sum_{i=j+1}^{N} f_{ij} \{\varepsilon_{ij} [(\frac{r_{oij}}{r_{ij}})^{12} - 2(\frac{r_{oij}}{r_{ij}})^6] + \frac{q_i q_j}{4\pi\varepsilon_0 r_{ij}}\}$$

$$(5)$$

On close inspection, we see that the first two terms follow Hook's law, and hence while minimizing any system, value of b and θ are brought close to the equilibrium values denoted by subscript 0 respectively, while torsion terms vary as the function of the cosine ratio of the torsion angles (n = periodicity). Forcefields are classified into Class 1, Class 2 and Class 3 based on the presence or absence of cross terms. These cross terms are included in order to incorporate the effect of change in one term due to the change in other term. For example, we expect to see a change in the bond length due to change in the bond angle or vice-versa. Forcefields without any cross terms and restricted only to the harmonic terms are classified as **class 1** forcefields (AMBER (Wang et al., 2004), GROMOS (van Gunsteren et al., 1998), CHARMM (Brooks et al., 1983; MacKerell et al., 1998)), those which include cross terms along with the anharmonic terms are classified as **class 2** forcefields (CFF (Sun et al., 1994), MM3 (Allinger et al., 1989), MMFF94 (Halgren, 1996)), and **class 3** forcefields (MM4 (Allinger et al., 1996)) would include, in addition to the terms in the class 2 forcefields, terms that incorporate effects of electron delocalization (i.e., resonance and hyperconjugation).

2.6   ENERGY MINIMIZATION

A molecule or biomolecular system as a whole has certain energy at a given coordinate space depending on its position on the potential energy surface (PES). Every PES is known to comprise of several wells (local minima) separated most often by several hills (high energy barriers), one of wells is deeper than the rest which we call as global energy minimum. The goal of every minimization algorithm is to reach the system to this global minimum. Hence, all energy minimization algorithms are designed to locate a downhill minima on the PES. It is imperative to note that a biologically active structure may not be a global minima on the PES, and hence sometimes working with structures located at the local minimum lead to good convergence or agreement with the experimental results.

Minimization algorithms are broadly classified as nonderivative based methods, like Simplex method (Lagarias et al., 1998; Müller and Brown, 1979) and the sequential univariate method, and derivative based methods, like the steepest descent (Fletcher and Powell, 1963) and conjugate gradient method (Štich et al., 1989) (1st order derivatives) and Newton-Raphson's method (2nd order derivative). In the Simplex method, a geometric figure is constructed using the coordinates of the vertices of the molecule to form a 'simplex', that has M+1 vertices, where M denote the vertices of an existing geometric figure. Iteratively, a new vertex is included in the simplex and the energy of the system is calculated. The process is stopped when no further drop in the energy is seen on adding new vertices. This method is best suited for those structures whose energy is being calculated by MM techniques. In the Sequential univariate method, for each coordinate, two new structures are generated. The minimum is deemed to have reached when the changes in all the coordinates are sufficiently small. This technique is used when QM methods are used to calculate energies.

Energy minimization methods that use calculus to find the energy minima are further classified as first order derivative techniques and second order derivative techniques. The first derivative of energy gives the direction that indicates the location of minima and the magnitude of the first derivative indicate the steepness of the gradient of the PES. Whereas, the second derivative indicates the curvature of the function which is used to predict the function that could changes the direction along the PES.

In first order derivative techniques, the algorithms use either the steepest descent method or the conjugate gradient method for minimization. In the steepest descent method the system moves along the negative gradient, and the coordinates are changed gradually to reach minimum along the search direction, starting point for each iteration is the configuration obtained from previous iteration. The minimum point is located by performing a line search or by taking a step of arbitrary size along the direction of force. The step size is increased if the energy is lowered by the move or decreased otherwise. In the conjugate gradient method, a set of directions that are conjugate, rather than orthogonal, to the previous set of directions is used. The search begins similar to the steepest descent, that is, in the direction of gradient. For a quadratic function of M variables, the minimum will be reached in M steps using the conjugate gradient method. Newton Raphson method (Hilderbrandt, 1977) is widely used method based on the second order derivatives. This method is often used after the system is minimized using steepest descent or conjugate gradient method, as this method is well suited for system(s) close to the minimum, where the PES curvature is believed to follow quadratic function. Newton-Raphson method gives extremum values, in this context the aim is to find the minimum value. In order to achieve this minimum the first order derivative must be zero and the solution of the second order derivative must a value great than zero.

2.7   STATISTICAL MECHANICS

Computer simulations generate information at the microscopic level, which needs conversion to macroscopic observables such as pressure, energy, heat capacities, and other properties, and that requires application of statistical mechanics. The connection between microscopic simulations and macroscopic properties is established using rigorous mathematical expressions that relate macroscopic properties to the distribution and motion of the atoms and molecules of the N-body system; MD simulations provide the means to solve the equation of motion of the particles and evaluate these mathematical formulas (Chandler, 1987; McQuarrie, 1976). Hence, statistical mechanics is fundamental to the study of biological systems by molecular dynamics or Monte-Carlo simulation.

In computer simulations, the system is defined by its instantaneous mechanical state using its atomic positions and momenta. For an N-body system, the phase space will be made up of 6N dimensions, where 3N comes from positions and 3N from the momenta. We are always interested in calculating any property of the system "Z" at any particular point in the phase space "Y", which evolves with time as the function given by "Z(Y)". It is well known that experimentally observed properties (macroscopic properties), are the time-averaged properties "Z_{obs}" of Z(Y) taken over an infinitively long time interval (Allen and Tildesley, 1989).

$$Z_{obs} = <Z>_{time} = <Z(Y(t))>_{time}$$

$$<Z(Y(t))>_{time} = \lim_{t_{obs} \to \infty} \frac{1}{t_{obs}} \int_0^{t_{obs}} Z(Y(t)) dt \qquad (6a)$$

The time evolution of this equation is also solved using Newton's classical equation of motions. Unfortunately, it not suitable to integrate Eq. (6a) for infinite amount of time, instead sufficient long time period (t_{obs}) can be used. This is what is done in molecular dynamics simulations, and Eq. (6a) boils down to Eq. (6b).

$$<Z(Y(t))>_{time} = \frac{1}{t_{obs}} \sum_{t=1}^{t_{obs}} Z(Y(t)) \qquad (6b)$$

Another serious problem working with time averages is that, it is not possible to compute the thermodynamic properties, and the time evolution of a large system (i.e., Macromolecular systems) is very complex. Therefore, Gibbs (Gibbs, 1885) suggested that it was possible to replace time averages with ensemble averages, and this satisfies the Ergodic hypothesis (Malament and Zabell, 1980), which states that "any dynamic system has the same behavior averaged over time as averaged over the space of all the system's states (configurational phase space)."

Previously, the property "Z" was the function of any point "Y" in the phase space, now it will be the function of probability density "$\rho(Z)$". The function determined at fixed macroscopic parameters, like constant NPT, NVT, etc., where N stands for number of particles in the systems (including solvent atoms and ions), V stands for volume of the systems, P for pressure and T for Temperature. Thus, for NVT ensemble the probability density will be denoted as ρ_{NVT}, similarly for NPT; ρ_{NPT} or generally it will be denoted as ρ_{ens}. According to Liouville's theorem, which is essentially the law of conservation for probability density, states the no systems are created or destroyed, but at every point when one systems exits, another systems replaces it. This also explains why the properties of a system at equilibrium are time-independent. There are two ensembles commonly used in MD simulations, when we intend to compute Helmholtz free energy we must use NVT ensemble (canonical ensemble), and when we are interested in computing Gibbs free energy, NPT ensemble (isobaric-isothermal) must be used (Smit et al., 1989). Unfortunately due to space constraints, giving a complete mathematical formulation on statistical mechanics is not possible in this chapter, and hence an advanced reader can refer to an exhaustive report by Tupper et al. (Draganescu et al., 2015).

2.8   INTRODUCTION TO MOLECULAR DYNAMICS SIMULATIONS

One of the main toolkit for theoretical studies on the biological molecules is Molecular Dynamics Simulations (MD) (Binder et al., 2004; Hansson et al., 2002). This computational method calculates the time dependent behavior of any molecular system. MD simulations have provided detailed information on the fluctuations and conformational changes of proteins, nucleic acids or any other molecular system. These methods are now routinely used to investigate the structure, dynamics and thermodynamics of biological molecules and their complexes. They are also used in the determination of structures from X-ray crystallography and from NMR experiments. To simulate accurately the behavior of molecules, one must be able to account for the thermal fluctuations and the environment-mediated interactions arising in diverse and complex systems (e.g., a protein binding site or bulk solution).

The crux of MD simulations is Newton's second law of motion states that "the acceleration of an object is dependent upon two variables: the net force acting upon the object and the mass of the object."

$$F_i = m_i a_i \tag{7a}$$

MD trajectory are obtained by differentiating Eq. (7a) with respect to time,

$$\frac{d^2x}{dt^2} = \frac{F_i}{m_i} \tag{7b}$$

Equation (7b) is solved for all particles in the system (including solvent atoms and ions), and the next move is predicted using integrator algorithms (discussed in the following section).

2.8.1 SOLVING EQUATION OF MOTION (INTEGRATION ALGORITHMS) FOR N-BODY SYSTEM

All the integration algorithms assume the positions, velocities and accelerations can be approximated by a Taylor series expansion and we include the discussion most commonly used Verlet algorithms.

2.8.1.1 Verlet Algorithm

In MD, the most commonly used time integration algorithm is probably the so-called Verlet algorithm (Verlet, 1967). The Verlet algorithm uses positions and accelerations at time t and the positions from time $t - \Delta t$ to calculate new positions at time $t + \Delta t$. The Verlet algorithm uses no explicit velocities. The advantages of the Verlet algorithm are (i) it is straightforward, and (ii) the storage requirements are modest. The disadvantage is that the algorithm is of moderate precision.

The basic idea is to write two third-order Taylor expansions for the positions $r(t)$, one forward and one backward in time. Let us suppose that $\mathbf{v}$ the velocities, and a the accelerations, we have the following equations:

$$r(t + \Delta t) = r(t) + v(t) + \frac{1}{2}(a)\Delta t^2 \tag{8a}$$

$$r(t - \Delta t) = r(t) - v(t) + \frac{1}{2}(a)\Delta t^2 \tag{8b}$$

Summing the above equations and rearranging the terms gives us,

$$r(t - \Delta t) = 2r(t) - r(t - \Delta t) + (a)\Delta t^2 \tag{8c}$$

All other algorithms namely Leapfrog (Fincham, 1992), velocity-Verlet (Swope et al., 1982), Beeman (Beeman, 1976) and many more are also derived in the similar fashion.

2.8.2 SETTING UP MOLECULAR DYNAMICS SIMULATIONS

For setting up any MD simulation, one must have clear goals of what one intends to achieve from this simulations. The starting point is appropriately preparing the system under study, suitable for the MD program one intends to use (i.e., Desmond, AMBER, GROMACS, LAMMPS, etc.), and building necessary topology and parameter files. The basic information contained in the topology files would be the atoms, atom-types and their Cartesian coordinates, whereas the parameter file would contain information necessary for the forcefield, for instance bond lengths, angles, torsion, improper torsions, and atom pair information necessary for computing nonbonded interactions. Fortunately, for proteins, DNA and RNA all the necessary information for the parameters file can be obtained from the protein data bank and are supplied along with the source code of the program. The real challenge is to build parameters (except for Desmond users with Maestro as GUI) for small molecules or cofactors or ions etc., for which parameters have to be extracted from Quantum mechanical calculations. Once all the prerequisite files are ready, depending on the simulation experiment intended, type of solvent (explicit solvent; water (Jorgensen and Jenson, 1998), DMSO, methanol, etc., or implicit solvent (Onufriev, 2008) or no solvent), type of box, in case of explicit solvent model (no periodic conditions required for implicit or vacuum simulations) needs to be defined by the user, and add appropriate number ions to maintain neutrality or ionic strength.

2.8.2.1 Boundary Conditions

Surface effects in microscopic simulations can be more dangerous as compared to macroscopic experiments, and hence to reduce any artifacts due to surface effects "Periodic Boundary Conditions (PBC)" are defined. Identical images of the system are replicated infinitely, to form a lattice like system (Figure 2.2). In PBC when an atom(s) leave the central box, equivalent atom(s) from the neighboring box enter the system and hence "N" remains constant throughout the simulation. Hence, there are neither walls to the boundary nor any surface molecules. Most often PBC is switched on while run simulations with explicit solvent system. Many programs do not support PBC with implicit solvation methods.

FIGURE 2.2 Representation of a solvated protein with cubic PBC with spherical cutoff for minimum image convention (red sphere: dotted line- inner cutoff; solid line- outer cutoff).

2.8.2.2 Cutoff Schemes for Nonbonded Interactions

The intention of using cutoff schemes in molecular dynamics simulations, is to reduce the computational cost by avoiding atom pairs outside a short range of distance, dictated by the type of periodic box and its size. The atom pair for computing nonbonded interactions increases in order of N^2 for any system under study. Hence, cutoff schemes are advantageous in reducing the nonbonding atom pair list, but many artifacts arise due to its use. In case of van der Waals interactions, the potential falls to less than 1% beyond 2.5σ, and hence it will lead to fewer artifacts. Unfortunately, this is not the case while handling electrostatic interactions, wherein any cutoffs will lead to artifacts, hence a large cutoff is needed, and this would increase the computational cost. In order to define cutoffs, simple spherical cutoff (Figure 2.2) scheme can be used defining a minimum image convention such that no particle sees its own image (no interactions to be calculated with one's own image). The force/potential is set to zero beyond this distance. It was observed that, abruptly forcing the force/potential to zero leads to larger errors compared to potential/force switches. In potential/force switches, two regions are defined (dotted and solid sphere in Figure 2.2), the inner cutoff region indicates that potential/force to taper gradually until the outer cutoff region, beyond which the potential/force is set to zero (Leach, 2001).

2.8.2.3 Treating Long Range Forces

Electrostatic interactions or charge-charge interactions are most troublesome type of interactions when it comes it to placing cutoffs for cost reduction, as these interaction diminish as $1/r$. This means that the cutoff should be placed at more than half the box dimensions, this would lead to lengthy computations. In order to treat long range interactions, method like Ewald (Belhadj et al., 1991) summation, reaction field method (Lee and Warshel, 1992; Tironi et al., 1995) and cell multipole methods (Greengard and Rokhlin, 1987) are available.

2.8.2.4 Minimization

Once the system is prepared, it must be minimized thoroughly to relieve any clashes that may have developed during the addition of water molecules. This must be done in order to prevent any hotspots in the system leading to unstable simulation. It is advisable to minimize the system first with the steepest descent (SD) method followed by conjugate gradient (CG) method. It is recommended to place constrain on the solute, solvent or both and repeat the SD followed by CG for appropriate number of cycles gradually tapering the restraining force to zero.

2.8.2.5 Heating

After minimization, the systems must be heated in order to achieve the target temperature (most often 293–300K). Most commonly used thermostats for MD simulations are Langevin's (Davidchack et al., 2009), Berendsen's (Berendsen et al., 1984), and Andersen's (Andersen, 1980). The temperature for molecular dynamics simulations cannot be controlled directly, instead it is controlled by adjusting the kinetic velocity of particles. In the Berendsen's thermostat the temperature is adjusted by rescaling the velocity, while in Andersen's method the velocities are redistributed and Langevin's thermostat depends on the frictional drag of the particles in motion. Langevin dynamics, an example of stochastic dynamics, is a special case of Newtonian dynamics, wherein the equation of motion is modified to incorporate the frictional force/dampening effect of solvent molecules.

In order to begin the simulation, the velocity is first assigned randomly from the Maxwell-Boltzmann distribution at the target temperature set for simulation using random seed number generator depending the computer wall-time and date.

2.8.2.6 Equilibration

In this step the system is stabilized by exchanging kinetic and potential energy to achieve a mean total energy. Generally two parameters are checked for stability, the temperature drifts (using NVT ensemble) which ideally should not be more than ± 5K from the target temperature. These drifts may vary depending on the external bath and the coupling strength to this external bath. Second parameter is density (using NPT ensemble), wherein the density of the solvent (water for biomolecular simulations) is brought close to the experimental density of that solvent at the target temperature.

2.8.2.7 Production

In the production run all the necessary parameters, and coordinates saved for analysis depending on the property that one intends to study from molecular dynamics. It is advisable to run more than one simulation. Running many short simulations is always advantageous than running one long simulation, this ensures that system visits more configurational space owing to the random nature of initial velocity assignment.

2.8.2.8 Analysis

Analyzing the trajectories is the most difficult part of MD simulations in order to extract the desired properties. We focus on few easily computed parameters, Root Mean Squared Deviation (RMSD) gives an indication how far the system has moved from its initial configuration, and also indicates the stability of the system. Large RMSD values indicate that the system is not very stable.

For proteins, various types of RMSD can be computed depending on what atoms are included in its calculations, for instance all-atom RMSD includes all the atoms, Cα-RMSD includes only α-carbon of the protein backbone, backbone RMSD includes all atoms that make up the backbone of the protein (α-carbon, amide nitrogen, amide carbonyl carbon and oxygen), side-chain RMSD (excludes all atoms in the backbone). Another commonly used parameter is the Root Mean Squared Fluctuation (RMSF), this indicates and helps to differentiate between the flexible and rigid regions in the protein. Similar to RMSD, RMSF can also be classified based on atoms used for its calculation. Radius of gyration indicates the compactness of the system, increase in this value indicates dissociation of the drug-receptor complex or opening of the protein structure leading to larger radius of gyration. Apart from these, there various other parameters that can be calculated, like potential of mean force, order parameters, calculation of free energies, etc. Other than calculations many interesting events can be observed simply by visualizing (Programs from visualizing trajectories: VMD, UCSF Chimera) the trajectories over and over again, and focusing on a particular region one at a time.

2.8.3 PITFALLS IN CLASSICAL NEWTONIAN DYNAMICS

The major pitfall of classical molecular dynamics is, it found to get trapped in the local minimum and doesn't possess the enough energy to cross the barriers. Moreover, since there is no memory of the phase space visited, the system may sample the same space over and over again get "stuck" in the well. Once, the system is trapped it may take exorbitantly long time to surmount the barrier and sample any new space. In order to enhance the sampling, various advance techniques employ modified sampling algorithms. Few most common examples are local elevation, high temperature molecular, umbrella sampling, replica-exchange molecular dynamics, metadynamics. Due to space limitations, we would elaborate on very recent methods, namely, metadynamics, and replica-exchange molecular dynamics.

2.9 METADYNAMICS

Metadynamics (MetD) is a powerful technique for enhancing sampling in molecular dynamics simulations and reconstructing the free-energy surface as a function of few selected degrees of freedom, often referred to as collective variables (CVs). In metadynamics, sampling is accelerated by a memory-dependent bias potential, which is adaptively constructed in the space of the CVs (Barducci, 2011; Laio and Parrinello, 2002). This potential can be written as a sum of Gaussians deposited along the trajectory in the CVs space to disallow the system from revisiting same configuration space that already have been sampled. Initially, in the MetD, the bias was used to influence a coarse-grained dynamics in the CVs space that was based on a series of constrained MD simulations. During subsequent updations, the bias was applied continuously during an MD simulation either through an extended Lagrangian formalism or acting directly on the microscopic coordinates of the system (Iannuzzi, 2003F).

At time t, the metadynamics bias potential can be written as

$$V_G(S(x), t) = \sum_{V=tG,2tG,\dots} we^{-\left(\frac{(S(x)-Sv)^2}{2\delta s^2}\right)} \qquad (9)$$

where, w – height of the Gaussian; δs – width of the Gaussian; $S(x)$, t – value collective variable at time "t"; t_G – time interval for addition of new Gaussian. The sampling in MetD is very sensitive to w and δs; large values lead to faster calculations with higher probability of missing important configurations. Whereas extremely small values would lead to time consuming computations.

The advantages of metadynamics are:

1. It accelerates the sampling of rare events by pushing the system away from local free- energy minima.
2. It allows exploring new reaction pathways as the system tends to escape the minima passing through the lowest free-energy saddle point.
3. No apriori knowledge of the landscape is required, as opposed to umbrella sampling, metadynamics inherently explores the low free-energy regions first.
4. After a transient, the bias potential V_G provides an unbiased estimate of the underlying free energy.

2.9.1 COLLECTIVE VARIABLES (CV)

A CV is a function of the coordinates of the system, based on which the entire PES can be reconstructed. To ensure an effective application of metadynamics, the CVs must respect the following guidelines:

1. They should distinguish between the initial and final states and describe all the relevant intermediates.
2. They should include all the slow modes of the system.
3. They should be limited in number (limited to 2 for simplicity).

The question thus arises how to choose prior to a simulation a set of CVs for describing complex processes that may involve hundreds or thousands of particles. Unfortunately, a universal guide does not exist. However, several useful approaches have been suggested that can be grouped in two general categories:

2.9.2 SELECTING APPROPRIATE CVS

A large variety of CVs have been used and their choice usually depends on the nature of specific process studied. Examples of frequently used CVs are interatomic distances, angles, dihedrals, coordination numbers, radius of gyration, dipole moment, number of hydrogen bonds, and Steinhardt parameters. More complex coordinates have been used to describe puckering motions or the amount of alpha or beta secondary structure in polypeptide chains. Other approaches rely on identifying the most relevant degrees of freedom by means of essential dynamics. From a preliminary MD run, a principal component analysis is performed. The first few eigenvectors of the correlation matrix are then used as CVs (Iannuzzi, 2003; Sega et al., 2009; Spiwok et al., 2007) that include most of the relevant biological motions.

2.9.3 ENHANCING THE SAMPLING IN TRANSVERSE COORDINATES

In spite of the success in selecting appropriate CVs, sometimes finding a small set of CVs is extremely difficult task. An example is the case of protein folding, which takes place in a large and complex conformational space and often involves many alternative pathways. In order to study such processes, it is useful to enhance sampling along a great number of degrees of freedom besides those that can be targeted by metadynamics. In order to do this, successful strategy employed was to combine metadynamics with a replica exchange method (REM), a popular parallel tempering (PT) algorithm. In PT, multiple copies of the same system at different temperatures are independently simulated. At fixed intervals, an exchange of configurations between two adjacent replicas is attempted while respecting detailed balance. By exchanging with higher temperatures, colder replicas are prevented from being trapped in local minima (Hansmann, 1997; Sugita and Okamoto, 1999).

2.9.4 APPLICATIONS

2.9.4.1 Metadynamics Coupled with Monte Carlo Simulation

The Monte Carlo (Hart and Read, 1992; Liu and Wang, 1999; Meteopolis and Ulam, 1949) algorithm has proven useful to calculate thermodynamic quantities associated with simplified models of proteins (Hansmann and Okamoto, 1999; Li and Scheraga, 1987), and thus to gain excellent insights on the general principles underlying the mechanism of protein folding. Marini et al. (Calalb et al., 1995; Irby and Yeatman, 2000) reported that it is possible to couple metadynamics and Monte Carlo algorithms to obtain the free energy of model proteins in a way which is computationally very economical. They studied the domain of the Proto-oncogene tyrosine protein kinase Src. Src is a 536 residue protein that plays a multitude of roles in cell signaling. SH3 is a domain built out of 60 residues, displaying mainly β-strands (Grantcharova et al., 1998; Wenqing et al., 1997). Their study comprised of simulating the protein as a chain of beads centered on the C_α of the protein backbone which allows the flipping movements of the protein. Every time steps of the Monte Carlo sampling, the non Markovian energy contribution was updated by adding a Gaussian hill with appropriate height and width, centered around the current values of the collective variables. Transition probability for each step is computed using Metropolis criteria. This probability gives an estimate whether next Monte Carlo move is accepted and is calculated on the variation of the energy of the system, plus the variation of the metadynamics potential (Marini et al., 2008).

2.9.4.2 Well-Tempered Metadynamics

Metadynamics simulations are successful in most of its applications and is said to be a superior methodology then classical molecular dynamics simulations, yet not free of drawbacks. It is often difficult to decide when to terminate a metadynamics run. In fact, in a single run, the free energy does not converge to a definite value but fluctuates around the correct result, leading to an average error which is proportional to the square root of the bias potential deposition rate. Reducing this rate implies increasing the time required to fill the FES (Bussi et al., 2006; Laio et al., 2005). In order to address this issue Barducci et al. (2008) reported a method called as well-tempered metadynamics. This method is based on self-healing umbrella sampling method. They reported to have substantially improved the metadynamics method in estimating the PES that converges to more accurate results in the long time limit. As opposed to classical metadynamics, this method offers a control over the regions to sample during the construction of PES. Owing to the controllable nature authors report decrease in the computational cost (Marsili et al., 2006) .

2.9.4.3   Insights of Ligand-Protein Interactions Using Metadynamics

Metadynamics has the potential to improve the sampling of configurational space compared to classical molecular dynamics algorithm, and hence is computationally economical method. This means better choice of collective variables can ensure better convergence and meaningful results. Branduard et al. (Branduardi et al., 2005) presented the dynamical motion of a ligand Tetramethyl Ammonium (TMA) from the bulk of the solvent into a 20Å deep narrow gorge of its biological counterpart (AChE) using metadynamics simulations. This work highlighted the unique role of Peripheral Anionic Site (PAS) and its aromatic residue Trp in capturing TMA from the solution and stabilizing it close to the aromatic cage. The diffuse negative Π cloud of the indole ring is well suited to interact with quaternary cation. This led to the finding of two key events responsible for the ligand entrance are: breaking of Asp-Tyr Hbond which results in the aromatic plane reorientation of Tyr residue, and the formation of cation-Π interactions with both Trp and Phe of the internal anionic site. The center of mass of TMA and of active site of AChE was selected as one of CVs (reflecting fast motion) and a second CV was chosen to representing the Asp-Tyr Hbond breaking (reflecting slow motion). The report strongly supports the presence of positive charge on ligands binding to AChE (including the substrate acetycholine), and increase in the positive charge on the ligands have shown to increase the binding affinity of the AChE inhibitors (Martis et al., 2015).

In yet another study, Provasi et al.(Provasi et al., 2009) showed the possible entry pathways of the nonselective antagonist naloxone (NLX) from the water environment into the well-accepted alkaloid binding pocket of a delta opioid receptor (DOR). Molecular model based on the β_2-adrenergic receptor crystal structure was explored using microsecond-scale well-tempered metadynamics simulations. Collective variables used, similar to those used by Branduard et al.(Branduardi et al., 2005) distances that account for the position of NLX and the receptor extracellular loop 2 in relation to the DOR binding pocket were studied. They successfully identified the different states visited by the ligand (i.e., docked, undocked, and metastable bound intermediates), and predicted free energy of binding in close agreement with the experimental values after correcting for possible drawbacks of the sampling approach.

Gervasio et al. (Gervasio et al., 2005) reported the use of metadynamics method for docking ligands. This method is found to mimic the real dynamics of the ligand entering and leaving the enzyme binding pocket and in doing so reconstructs the PES. They showed that in case of the ligand-receptor complexes, the method successfully predicts the bound geometry without any prior knowledge of the bound conformation. Moreover, it predicts in a quantitative way the $\Delta G_{binding}$, the free energy of docking for different ligand-protein complexes, and the $\Delta\Delta G$ of binding resulting from slight modifications of the ligand. Gervasio et al. (Gervasio et al., 2005) also showed that the PES is in quantitative agreement with the one obtained by running a much longer two-dimensional umbrella sampling.

2.10   REPLICA-EXCHANGE MOLECULAR DYNAMICS

In replica exchange molecular dynamics (REMD) (Earl and Deem, 2005), a number of parallel simulations are performed at different temperatures, and the configuration are exchanged periodically using Metropolis criteria. Unlike the classical MD, in which the system may get trapped in a local minima, in REMD, with next higher temperature switch the system would gain enough kinetic energy to surmount the barrier and move out of the well.

Let us consider any system with N atoms with their respective masses m_N (where $N = 1.....N$), in MD the system is most appropriately defined by their coordinates (most often Cartesian Coordinates) represented as q_N and momenta p_N (where N has as the same notation as described earlier). Now, the Hamiltonian H(q, p) of the system which indicates the total energy of the system, that can be written as,

$$H(q, p) = \sum_{N=1}^{N} \frac{p_N^2}{2\, m_N} + U(q) \qquad (9a)$$

where U(q) represents the potential energy of the system under study. For the exchange of replicas which is governed by the Metropolis criteria, we will now understand how the Boltzmann factor and partition function (probability density function) is related to Hamiltonian of the system. In a canonical ensemble (explained in statistical mechanics) at a temperature T, each state x (q, p) of the system with H (q, p) is weighted by the Boltzmann factor:

$$\rho \ (x; T) = \frac{1}{Z} e^{-[\beta H(q,p)]} \qquad (9b)$$

where, $\beta = 1/k_b \, T$ (k_b = Boltzmann constant), Z is the partition function which may be obtained by

$$Z = \int e^{-\beta H(q,p)} \, dqdp \qquad (9c)$$

From Eqs. (9b) and (9c), we understand that every state for a system at a given temperature has the probability of existence and associated energy defined by its Hamiltonian. Let us consider a simple case of running REMD on a system using two replicas at temperature T^a and T^b such that $\delta T = (T^b - T^a)$. If this difference is small then, we could see that the replicas will have population of states overlapping, this also means that the one replica can visit the configurational space of the other replica. It is now obvious that when $\delta T = 0$, the probability of overlap is 1 and as the difference increases the probability decreases. The acceptance probability of the replicas is given by the Metropolis-Hastings criterion (Hastings, 1970; Neal, 1996),

$$Z = \int e^{-\beta H(q,p)} \, dqdp \qquad (9d)$$

where E is the energy of exchange pairs (replicas in this case) 1 and 2 and T is their respective temperatures, the pairs selected must be reversible with the same probability. From this it becomes clear that the probability of acceptance decreases as the difference increases. More elaborate discussion on Metropolis-Hastings criterion is beyond the scope of this chapter and can be found elsewhere (Ref. (Dewar and Thiel, 1977) and references therein).

Practically, to perform REMD, each replica is assigned to one temperature among the range of temperatures set by the user, the assignment is such that one and only one replica is assigned to a particular temperature. Once the user sets the maximum and the minimum temperature and the number of replicas to be simulated, the temperature gaps are defined either using quadratic or linear or manually defined temperature profiles. Alternately, to obtain an optimum temperature distribution, few trial replicas and a short MD simulation must be performed, and then select the temperature gap by monitoring the acceptance ratio between neighboring replicas (Zhou et al., 2001).

2.10.1 APPLICATIONS

There are several applications of REMD, but we focus on the use of REMD for understanding the folding pathways of proteins. A classic example which we discuss, is study on folding of Trp-cage protein.

2.10.1.1 Trp-Cage Protein

Trp-cage is a 20-mer protein (named "miniprotein) from the c-terminal part of 39-residue exendin-4 peptide. Interestingly, Neidigh's and co-workers reported, using NMR spectroscopy (Neidigh et al., 2002), that trp-cage has the capability to fold independently to its native structure even though being a truncated part of the protein. This phenomenon excited, both experimentalists (Qiu et al., 2002) and theoretical biologists to explored this protein, to validate new methods for understanding protein folding pathways, metastable states, and folding kinetics. One such theoretical study employing implicit solvation (GBSA based solvation model) REMD ($\sim$ 92 ns per replica) to elucidate the folding pathway of trp-cage protein was reported by Pitera and Swope (Pitera and Swope, 2003). It was known from several studies and also from Neidigh's that driving force for the folding of this protein is cooperativity and hydrophobicity which in an attempt to encapsulate the hydrophobic core, buries this core away from aqueous solvent.

All the folding simulations should start with extended conformation with all sidechains in "Trans" configuration to each other, in order to avoid bias that may lead unnatural folding pathway. Pitera and Swope used 23 replicas for the temperature range of 250–630 K, which is far beyond the cooperative melting transition midpoint of trp-cage protein (~315K). By simulating at such high temperatures, it can be observed that whether the melting curve of the protein can be reproduced using the current computational method used, which is essential in order to study the metastable states and molten globules.

2.11 COARSE-GRAINING

It is very unfortunate that the current computational power allows us to run a simulation which covers a time scale of the order of few hundreds of nanoseconds or microseconds that is usually insufficient to follow several important biological processes, evolving in much longer time (usually milliseconds or more). In addition, the number of degrees of freedom of biological systems is generally large, and an appropriate exploration of the phase space is possible only if a small number of appropriate reaction coordinates can be identified (Rahman and Tully, 2002; Wang and Landau, 2001). Coarse-grained (CG) models, in which small groups of atoms are treated as single particles, provide a promising approach to increasing the timescale of biomolecular simulations (Bond et al., 2007; Steve et al., 2004). The development of coarse-grained models as a fast alternative to united-atom representation, where all nonpolar hydrogens are eliminated, has a long history in the study of protein structure prediction, When coarse graining is done at a very high level of simplification such description is less reliable, but such coarse-grain models allow us to explore the early steps of many events occurring in the protein aggregation, for early steps in aggregation of amyloid forming peptides (Derreumaux and Mousseau, 2001).

2.12 CONCLUDING REMARKS

The current state of computing facilities are capable of handling far greater number of particles compared to older times. Developments in message parsing interfaces (MPI) have permitted the developers to write parallel codes for molecular dynamics programs which have enhanced the rate of computations to large extent. Recently, GPU workstations based on Compute Unified Device Architecture (CUDA) technology have revolutionized MD programs to such an extent that simulations with time scales of few nanoseconds can be computed within one day. The development of such technologies is a never-ending process, involving improvement in both hardware and software.

ACKNOWLEDGEMENTS

We are highly indebted to Prof. Dr. Evans Coutinho for his valuable suggestions for preparing this chapter. We would also like thank our colleagues. Megha, Sandhya and Devanshi for carefully proof reading the drafts of the manuscript and suggesting improvements. EAFM is also thankful to Prof. Dr. Holger Gohlke and Group (HHU), Prof. Dr. Peter Comba (University of Heidelberg) Prof. Dr. G. N. Sastry (IICT- Hyderabad) and Prof. Dr. Hendrik Zipse (LMU, Munich) for allowing a short sabbatical at Heinrich-Heine Universität (HHU), Düsseldorf, Germany in the context of MCBR4 2015. We also express our gratitude to Department of Science and Technology, Department of Biotechnology and Council for Scientific and Industrial research for funding the research in our Lab.

KEYWORDS

- computer simulations
- metadynamics
- molecular dynamics
- molecular mechanics
- protein dynamics
- quantum mechanics
- replica-exchange molecular dynamics
- semi-empirical

REFERENCES

Allen, M. P., & Tildesley, D. J., (1989). *Computer Simulation of Liquids*. Clarendon Press.

Allinger, N. L., Chen, K., & Lii, J. H. (1996). An improved force field (MM 4) for saturated hydrocarbons. *J Comput Chem. 17* (5-6), 642–668.

Allinger, N. L., Yuh, Y. H., & Lii, J. H. (1989). Molecular mechanics. The MM 3 force field for hydrocarbons. 1. *J Am Chem Soc 111* (23), 8551–8566.

Andersen, H. C. (1980). Molecular dynamics simulations at constant pressure and/or temperature. *J Chem Phys 72* (4), 2384–2393.

Anderson, H. L., Hunter, C. A., Meah, M. N., & Sanders, J. K. M. (1990). Thermodynamics of induced-fit binding inside polymacrocyclic porphyrin hosts. *J Am Chem Soc 112* (15), 5780–5789.

Barak, D., Ordentlich, A., Bromberg, A., Kronman, C., Marcus, D., Lazar, A., Ariel, N., Velan, B., & Shafferman, A. (1995). Allosteric modulation of acetylcholinesterase activity by peripheral ligands involves a conformational transition of the anionic subsite. *Biochem 34* (47), 15444–15452.

Barducci, A., Bonomi, M., & Parrinello, M. (2011). Metadynamics. *WIREs Computational Molecular Science 1* (5), 826–843.

Barducci, A., Bussi, G., & Parrinello, M. (2008). Well-tempered metadynamics: a smoothly converging and tunable free-energy method. *Phys Rev Lett 100* (2), 20603.

Beeman, D. (1976). Some multistep methods for use in molecular dynamics calculations. *J Comput Phys 20* (2), 130–139.

Belhadj, M., Alper, H. E., & Levy, R. M. (1991). Molecular dynamics simulations of water with Ewald summation for the long range electrostatic interactions. *Chem Phys Lett 179* (1), 13–20.

Berendsen, H. J., Postma, J. P. M., van Gunsteren, W. F., DiNola, A., & Haak, J. (1984). Molecular dynamics with coupling to an external bath. *J Chem Phys 81* (8), 3684–3690.

Binder, K., Horbach, J., Kob, W., Paul, W., & Varnik, F. (2004). Molecular dynamics simulations. *J Phys. Condens Matter 16* (5), S 429.

Bixon, M., & Lifson, S. (1967). Potential functions and conformations in cycloalkanes. *Tetrahedron 23* (2), 769–784.

Böhm, H. J., & Klebe, G. (1996). What Can We Learn from Molecular Recognition in Protein–Ligand Complexes for the Design of New Drugs? *Angew Chem Int Edit 35* (22), 2588–2614.

Bond, P. J., Holyake, J., Ivetac, A., Khalid, S., & Sansom, M. S. P. (2007). Coarse-grained molecular dynamics simulations of membrane proteins and peptides. *J Struct Biol 157* (3), 593–605.

Boyd, D. B., (1990). Aspects of molecular modeling. In: *Reviews in Computational Chemistry*, Boyd, D. B., (Ed.); Vol. 1, pp. 321–351.

Branduardi, D., Gervasio, F. L., Cavalli, A., Recanatini, M., & Parrinello, M. (2005). The role of the peripheral anionic site and cation-π interactions in the ligand penetration of the human ache gorge. *J Am Chem Soc 127* (25), 9147–9155.

Brooks, R. B., Bruccoleri, E. R., Olafson, D. B., States, J. D., Swaminathan, S., & Karplus, M. (1983). CHARMM: A program for macromolecular energy, minimization, and dynamics calculations. *J Comput Chem 4* (2), 187–217.

Bussi, G., Laio, A., & Parrinello, M. (2006). Equilibrium free energies from nonequilibrium metadynamics. *Phys Rev Lett 96* (9), 90601.

Calalb, M. B., Polte, T. R., & Hanks, S. K. (1995). Tyrosine phosphorylation of focal adhesion kinase at sites in the catalytic domain regulates kinase activity: a role for Src family kinases. *Mol Cell Biol 15* (2), 954–963.

Carlson, H. A. (2002). Protein flexibility and drug design: how to hit a moving target. *Curr Opin Chem Biol 6* (4), 447–452.

Carlson, H. A., & McCammon, J. A. (2000). Accommodating protein flexibility in computational drug design. *Mol Pharmacol 57* (2), 213–218

Chandler, D. (1987). *Introduction to Modern Statistical Mechanics*. Oxford University Press: New York, p. 288.

Chow, C. S., & Bogdan, F. M. (1997). A structural basis for RNA-ligand interactions. *Chem Rev 97* (5), 1489–1514.

Cramer, C. J. (2013). *Essentials of Computational Chemistry: Theories and Models*. Wiley-Blackwell.

Davidchack, R. L., Handel, R., & Tretyakov, M. (2009). Langevin thermostat for rigid body dynamics. *J Chem Phys 130* (23), 234101.

de la Torre, J. G., & Bloomfield, V. A. (1981). Hydrodynamic properties of complex, rigid, biological macromolecules: theory and applications. *Q Rev Biophys 14* (1), 81–139.

Derreumaux, P., & Mousseau, N. (2001). Coarse-grained protein molecular dynamics simulations. *J Chem Phys 126*, 025101.

Dewar, M. J., & Thiel, W. (1977). Ground states of molecules. 38. The MNDO method. Approximations and parameters. *J Am Chem Soc 99* (15), 4899–4907.

Dewar, M. J., Zoebisch, E. G., Healy, E. F., & Stewart, J. J. (1985). Development and use of quantum mechanical molecular models. 76. AM 1: a new general purpose quantum mechanical molecular model. *J Am Chem Soc 107* (13), 3902–3909.

Draganescu, A., Lehoucq, R., & Tupper, P. (2015). Hamiltonian Molecular Dynamics for Computational Mechanicians and Numerical Analysts. www.math.umbc.edu/~draga/papers/na-md-sand.pdf (accessed 05/05/2015).

Earl, D. J., & Deem, M. W. (2005). Parallel tempering: theory, applications, and new perspectives. *Phys Chem Chem Phys 7* (23), 3910–3916.

Fincham, D. (1992). Leapfrog rotational algorithms. *Mol Sim 8* (3–5), 165–178.

Fletcher, R., & Powell, M. J. (1963). A rapidly convergent descent method for minimization. *Comput J 6* (2), 163–168.

Gervasio, F. L., Laio, A., & Parrinello, M. (2005). Flexible docking in solution using metadynamics. *J Am Chem Soc 127* (8), 2600–2607.

Gibbs, J. W., (1885). *On the Fundamental Formula of Statistical Mechanics, with Applications to Astronomy and Thermodynamics*. Salem Press.

Gouldson, P. R., Higgs, C., Smith, R. E., Dean, M. K., Gkoutos, G. V., & Reynolds, C. A. (2000). Dimerization and domain swapping in G-protein-coupled receptors: a computational study. *Neuropsychopharmacol 23* (4), S60–S77.

Grantcharova, V. P., Riddle, D. S., Santiago, J. V., & Baker, D. (1998). Important role of hydrogen bonds in the structurally polarized transition state for folding of the src SH 3 domain. *Nat Struct Mol Biol 5* (8), 714–720.

Greengard, L., & Rokhlin, V. (1987). A fast algorithm for particle simulations. *J Comput Phys 73* (2), 325–348.

Halgren, T. A. (1996). Merck molecular force field. III. Molecular geometries and vibrational frequencies for MMFF 94. *J Comput Chem 17* (5–6), 553–586.

Hansmann, U. H. E. (1997). Parallel tempering algorithm for conformational studies of biological molecules. *Chem Phys Lett 281* (1–3), 140–150.

Hansmann, U. H. E., & Okamoto, Y. (1999). New Monte Carlo algorithms for protein folding. *Curr Opin Struct Biol 9* (2), 177–183.

Hansson, T., Oostenbrink, C., & van Gunsteren, W. F. (2002). Molecular dynamics simulations. *Curr Opin Struct Biol 12* (2), 190–196.

Hart, T. N., & Read, R. J. (1992). A multiple-start Monte Carlo docking method. *Proteins: Struct, Funct, Bioinf 13* (3), 206–222.

Hastings, W. K. (1970). Monte Carlo sampling methods using Markov chains and their applications. *Biometrika 57* (1), 97–109.

Hilderbrandt, R. L. (1977). Application of Newton-Raphson optimization techniques in molecular mechanics calculations. *Comput Chem 1* (3), 179–186.

Homans, S., (2007). Dynamics and thermodynamics of ligand–protein interactions. In *Bioactive Conformation I*, Peters, T., Ed. Springer Berlin Heidelberg, Vol. 272, pp. 51–82.

Iannuzzi, M., Laio, A., & Parrinello, M. (2003). Efficient exploration of reactive potential energy surfaces using Car-Parrinello molecular dynamics. *Phys Rev Lett 90* (23), 238302.

Irby, R. B., & Yeatman, T. J. (2000). Role of Src expression and activation in human cancer. *Oncogene 19* (49), 5636.

Jorgensen, W. L., & Jenson, C. (1998). Temperature dependence of TIP 3P, SPC, and TIP 4P water from NPT Monte Carlo simulations: Seeking temperatures of maximum density. *J Comput Chem 19* (10), 1179–1186.

Kier, L., *Molecular Orbital Theory in Drug Research*. Academic Press Inc., 2012; Vol. 10.

Koshland, Jr., D. E. (1998). Conformational changes: How small is big enough? *Nat Med 4* (10), 1112–1114.

Lagarias, J. C., Reeds, J. A., Wright, M. H., & Wright, P. E. (1998). Convergence properties of the Nelder--Mead simplex method in low dimensions. *SIAM J Control Optim 9* (1), 112–147.

Laio, A., & Parrinello, M. (2002). Escaping free-energy minima. *Proc Natl Acad Sci USA 99* (20), 12562–12566.

Laio, A., Rodriguez-Fortea, A., Gervasio, F. L., Ceccarelli, M., & Parrinello, M. (2005). Assessing the accuracy of metadynamics. *J Phys Chem B 109* (14), 6714–6721.

Leach, A. R., *Molecular Modeling: Principles and Applications*. Prentice Hall: 2001.

Lee, F. S., & Warshel, A. (1992). A local reaction field method for fast evaluation of long-range electrostatic interactions in molecular simulations. *J Chem Phys 97* (5), 3100–3107.

Levine, I. N., & Learning, P., (2009). *Quantum Chemistry*. Vol. 6. Pearson Prentice Hall Upper Saddle River, NJ .

Lewars, E. G., (2010). *Computational Chemistry: Introduction to the Theory and Applications of Molecular and Quantum Mechanics*. Springer Science & Business Media.

Li, Z., & Scheraga, H. A. (1987). Monte Carlo-minimization approach to the multiple-minima problem in protein folding. *Proc Natl Acad Sci USA 84* (19), 6611–6615.

Liu, M., & Wang, S. (1999). MCDOCK: a Monte Carlo simulation approach to the molecular docking problem. *J Comput Aid Mol Des 13* (5), 435–451.

MacKerell, A. D., Bashford, D., Bellott, M., Dunbrack, R., Evanseck, J., Field, M. J., Fischer, S., Gao, J., Guo, H., & Ha, S. A. (1998). All-atom empirical potential for molecular modeling and dynamics studies of proteins. *J Phys Chem B 102* (18), 3586–3616.

Malament, D. B., & Zabell, S. L., (1980). Why Gibbs Phase Averages Work—The Role of Ergodic Theory. In: *Philosophy of Science* pp. 339–349.

Marini, F., Camilloni, C., Provasi, D., Broglia, R., & Tiana, G. (2008). Metadynamic sampling of the free-energy landscapes of proteins coupled with a Monte Carlo algorithm. *Gene 422* (1), 37–40.

Marsili, S., Barducci, A., Chelli, R., Procacci, P., & Schettino, V. (2006). Self-healing umbrella sampling: a nonequilibrium approach for quantitative free energy calculations. *J Phys Chem B 110* (29), 14011–14013.

Martis, E. A., Chandarana, R. C., Shaikh, M. S., Ambre, P. K., D'Souza, J. S., Iyer, K. R., Coutinho, E. C., Nandan, S. R., & Pissurlenkar, R. R. (2015). Quantifying Ligand-Receptor Interactions for Gorge Spanning Acetylcholinesterase Inhibitors for the Treatment of Alzheimer's disease. *J Biomol Struct Dyn 33* (5), 1107–1125.

McQuarrie, D. (1976). *Statistical Mechanics*. New York: Harper and Row.

Meteopolis, N., & Ulam, S. (1949). The Monte Carlo method. *J Amer Statist Assoc 44* (247), 335–341.

Müller, K., & Brown, L. D. (1979). Location of saddle points and minimum energy paths by a constrained simplex optimization procedure. *Theor Chim Acta 53* (1), 75–93.

Neal, R. M. (1996). Sampling from multimodal distributions using tempered transitions. *Stat Comput 6* (4), 353–366.

Neidigh, J. W., Fesinmeyer, R. M., & Andersen, N. H. (2002). Designing a 20-residue protein. *Nat Struct Mol Biol 9* (6), 425–430.

Onufriev, A., Implicit solvent models in molecular dynamics simulations: A brief overview. In: *Annual Reports in Computational Chemistry*, Wheeler, A. R., Spellmeyer, D., (eds.), Vol. 4, pp. 125–137.

Pietrucci, F., & Laio, A. (2009). A collective variable for the efficient exploration of protein beta-sheet structures: application to SH 3 and GB 1. *J Chem Theory Comput 5* (9), 2197–2201.

Pissurlenkar, R. R., Shaikh, M. S., Iyer, R. P., & Coutinho, E. C. (2009). Molecular mechanics forcefields and their applications in drug design. *Curr Med Chem: Anti-Infect Agents 8* (2), 128–150.

Pitera, J. W., & Swope, W. (2003). Understanding folding and design: Replica-exchange simulations of "Trp-cage" miniproteins. *Proc Natl Acad Sci USA 100* (13), 7587–7592.

Pople, J. A., Santry, D. P., & Segal, G. A. (1965). Approximate Self-Consistent Molecular Orbital Theory. I. Invariant Procedures. *J Chem Phys 43* (10), S 129–S 135.

Provasi, D., Bortolato, A., & Filizola, M. (2009). Exploring molecular mechanisms of ligand recognition by opioid receptors with metadynamics. *Biochem 48* (42), 10020–10029.

Qiu, L., Pabit, S. A., Roitberg, A. E., & Hagen, S. J. (2002). Smaller and faster: the 20-residue Trp-cage protein folds in 4 μs. *J Am Chem Soc 124* (44), 12952–12953.

Rahman, J. A., & Tully, J. C. (2002). Puddle-skimming: An efficient sampling of multidimensional configuration space. *J Chem Phys 116*, 8750.

Rezác, J., Fanfrlík, J. I., Salahub, D., & Hobza, P. (2009). Semiempirical quantum chemical PM 6 method augmented by dispersion and H-bonding correction terms reliably describes various types of noncovalent complexes. *J Chem Theory Comput 5* (7), 1749–1760.

Sega, M., Autieri, E., & Pederiva, F. (2009). On the calculation of puckering free energy surfaces. *J Chem Phys. 130*, 225102.

Smit, B. D., De Smedt, P., & Frenkel, D. (1989). Computer simulations in the Gibbs ensemble. *Mol Phys 68* (4), 931–950.

Spiwok, V., Lipovová, P., & Králová, B. (2007). Metadynamics in essential coordinates: free energy simulation of conformational changes. *J Phys Chem B 111* (12), 3073–3076.

Steve O. Nielsen, Ivaylo Ivanov, Preston B. Moore, John C. Shelley, & Michael L. Klein (2004). Transmembrane peptide-induced lipid sorting and mechanism of L-alpha-to-inverted phase transition using coarse-grain molecular dynamics. *Biophys J 87*(4), 2107–2115.

Stewart, J. J., (1990). Semiempirical molecular orbital methods. In: *Reviews in Computational Chemistry*, Vol. 1, Boyd, D. B., (Ed.), pp. 45–81.

Štich, I., Car, R., Parrinello, M., & Baroni, S. (1989). Conjugate gradient minimization of the energy functional: A new method for electronic structure calculation. *Phys Rev B 39* (8), 4997.

Stratt, R. M., & Maroncelli, M. (1996). Nonreactive dynamics in solution: the emerging molecular view of solvation dynamics and vibrational relaxation. *J Chem Phys 100* (31), 12981–12996.

Sugita, Y., & Okamoto, Y. (1999). Replica-exchange molecular dynamics method for protein folding. *Chem Phys Lett 314* (1–2), 141–151.

Sun, H., Mumby, J. S., Maple, R. J., & Hagler, T. A. (1994). An ab Initio CFF 93 All-Atom Forcefield for Polycarbonates. *J Am Chem Soc 116* (7), 2978–2987.

Swope, W. C., Andersen, H. C., Berens, P. H., & Wilson, K. R. (1982). A computer simulation method for the calculation of equilibrium constants for the formation of physical clusters of molecules: Application to small water clusters. *J Chem Phys 76* (1), 637–649.

Taverna, D. M., & Goldstein, R. A. (2002). Why are proteins marginally stable? *Proteins: Struct, Funct, Bioinf 46* (1), 105–109.

Tironi, I. G., Sperb, R., Smith, P. E., & van Gunsteren, W. F. (1995). A generalized reaction field method for molecular dynamics simulations. *J Chem Phys 102* (13), 5451–5459.

van Gunsteren, W. F., Daura, X., & Mark, A. E. (1998). GROMOS forcefield. In *Encyclopedia of Computational Chemistry*.

Verlet, L. (1967). Computer experiments on classical fluids. I. Thermodynamical properties of Lennard-Jones molecules. *Phys Rev 159* (1), 98.

Violin, J. D., DiPilato, L. M., Yildirim, N., Elston, T. C., Zhang, J., & Lefkowitz, R. J. (2008). β2-adrenergic receptor signaling and desensitization elucidated by quantitative modeling of real time cAMP dynamics. *J Biol Chem 283* (5), 2949–2961.

Wang, F., & Landau, D. P. (2001). Efficient, multiple-range random walk algorithm to calculate the density of states. *Phys Rev Lett 86* (10), 2050–2053.

Wang, J., Wolf, R. M., Caldwell, J. W., Kollman, P. A., & Case, D. A. (2004). Development and testing of a general amber forcefield. *J Comput Chem 25* (9), 1157–1174.

Wenqing, X., Harrison, S., & Eck, M. (1997). Three-dimensional structure of the tyrosine kinase c-Src. *Nature 385*, 595–602.

Whitesides, G. M., & Krishnamurthy, V. M. (2005). Designing ligands to bind proteins. *Q Rev Biophys 38* (4), 385–396.

Zavodszky, M. I., & Kuhn, L. A. (2005). Side-chain flexibility in protein–ligand binding: The minimal rotation hypothesis. *Protein Sci 14* (4), 1104–1114.

Zerner, M. C., Semiempirical molecular orbital methods. In *Reviews in Computational Chemistry*, Boyd, D. B., Ed. Vol. 2, pp. 313–365.

Zhou, R., Berne, B. J., & Germain, R. (2001). The free energy landscape for β hairpin folding in explicit water. *Proc Natl Acad Sci USA 98* (26), 14931–14936.

CHAPTER 3

RETROMETABOLIC DRUG DESIGN

NIRZARI GUPTA

Department of Chemistry, University of Alabama at Birmingham, Birmingham, AL 35294, USA, E-mail: nirzari@uab.edu, nirzari1@gmail.com

CONTENTS

3.1 INTRODUCTION

Despite the amount of research conducted in pharmaceutical companies, the success rate for the lead molecule to reach into market is markedly low and though there are many new techniques for drug designing such as combinatorial chemistry, high throughput screening, they don't seem to have a higher significance in drug discovery. This is most likely due to our limited understanding of what turns a good lead compound into potential drug candidate. The problem is also featured by the fact that most recently launched drugs are not only derived by modifications of known drug structures or published lead structures, but are also even closely related to their original lead compounds (Bodor and Buchwald, 2008). The significance of drug design would only be fulfilled not only after increased activity, but also therapeutic index, margin of safety (TI = TD_{50}/ED_{50}, ratio between median toxic and median effective dose) and degree of selectivity.

It is known that when a drug is inserted into the body, it goes through many metabolic changes and creates various metabolites. Metabolism processes are mainly to excrete the foreign molecule from the body (Bodor, 1999). Such enzymatic transformations are dependent on the structure of the molecule. The structure of the molecule is key factor for which enzyme will modify which part of the structure (Buchwald and Bodor, 2000). These metabolites are very important for considering the drug design. The metabolites as well as active drug create various pharmacokinetic and toxic effects, which are reflected into overall effect of the drug (Bodor, 2000). Drug metabolism is undeniably important in drug design approaches. In the current drug discovery pathway, it comes into considerations too late. It is not taken into account until the drug reaches to the advanced stages of development. The basic ground of the Retrometabolic drug design is to consider metabolic effects into the early stages of the designing, and absolutely not as an after thought (Bodor and Buchwald, 2012).

3.2   PRINCIPLE OF RETROMETABOLIC DRUG DESIGN

As per the E. J. Corey's (Corey, 1988) retrosynthetic analysis, in which the synthetic pathways are designed going backwards compared to actual laboratory operations, the Retrometabolic approaches focus on the same fact that metabolic pathways can be designed in backwards direction to compare actual metabolic processes. Integration of various methodologies which consider SAR (Structural Activity Relationships) as well as SMR (Buchwald and Bodor, 1999) (Structural Metabolic Relationships) can be useful in the design of safe, locally targeted active compounds with improved therapeutic index and margin of safety. Mainly two approaches are considered into Retrometabolic Drug Design (RMDD): Design of Chemical Delivery systems (CDSs) and Design of Soft Drugs (SDs) (Bodor, 1995). CDS involves an inactive compound which is administered into the body and upon various enzymatic changes, is converted into the active drug, whereas SD involves an already active drug which is modified from lead compound, and is metabolized into inactive moiety after enzymatic changes. Both of the approaches rely on enzymatic changes, but CDS is an inactive drug which undergoes sequential distribution and enzymatic changes to ultimately release an active drug (Brouillette et al., 1996). While SD is as active as lead but designed to sequentially metabolize into inactive species (Figure 3.1). These approaches are keynotes for site targeted delivery of drugs (Buchwald and Bodor, 2014).

In both CDS and SD, drug should be present on the site of action. In CDS, only the enzymes of the site will convert inactive moieties into active drug molecule while in SD, the active molecule will be converted into inactive molecule on site of action.

For example, Retrometabolic drug design involves modern development of new corticosteroid with refined therapeutic index. Loteprednol etabonate (LE) (Buris et al., 1999) had been developed from prednisolone. It has 17α-chloromethyl ester, in lieu of a ketone group and 17β-etabonate group. LE has high lipophilic character which binds to glucocorticoid receptor with high affinity and unbound LE is metabolized to inactive metabolites (Figure 3.2) (Druzgala et al., 1991).

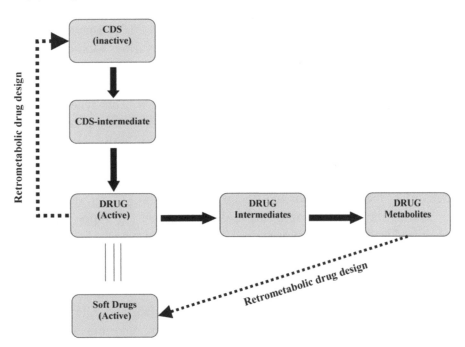

FIGURE 3.1   General approaches for Retrometabolic drug design.

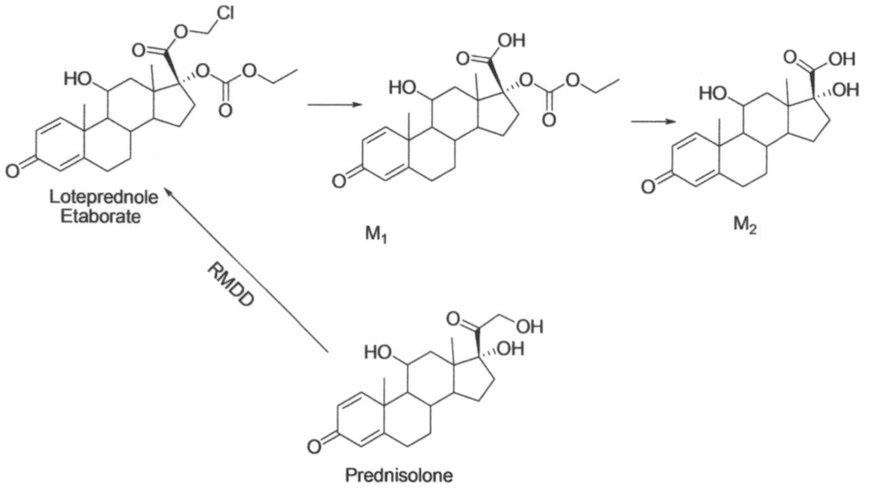

FIGURE 3.2 Structure of Prednisolone, Loteprednole Etaborate and its metabolites (M1 and M2).

3.3 CHEMICAL DELIVERY SYSTEM (CDS) APPROACH

This approach was developed when any specific exclusive enzyme is found on site of action. This concept was developed for various kind of drug targeting. It has been successful for various targeted delivery to brain, eyes and other organs. In general, CDS should involve any drug targeting system, which needs a chemical reaction to activate it. This type of system can also include a linker that make covalent link between the drug and the "carrier". Hypothetically, one bond has to be broken to activate the drug system. However, in the CDS, one or more modifications are done, which are generally monomolecular units. They provide site specific or site enhanced delivery of the drug (Bodor and Buchwald, 1997).

During the modification of the molecule, mainly two types of moieties are introduced for conversion of drug into inactive precursor form: Targetor (T) which is responsible for drug targeting delivery and site specification and Modifier functions $(F_1...F_n)$, which serve to protect particular function of the drug molecule. They also prevent premature and unwanted metabolic enzymatic conversion. We can achieve drug targeting with CDS, once site specific or exclusive or more activated enzymes are determined at the specific membrane or site (e.g., BBB Blood brain Barrier). There are several classes for CDS depending upon the modifications. Main classes are: (1)

Receptor based CDS (2) Enzymatic Physical chemical based CDS (3) Site specific enzyme activated CDS (Goskonda et al., 1999).

In the receptor based CDS approach, formation of the covalent bond between delivered CDS and some part of the active site is achieved. Here the T part of the CDS undergoes modification and acts as to anchor the drug molecule. This approach is not very well developed but it is accepted after recent discoveries. Naloxone, a pure opioid antagonist provides the best example for this approach. It was explored that opiate addiction was a very serious problem (Simpson and Sells, 1990). Therefore, the availability of long-lasting antagonists is highly desirable. The presence of an essential –SH group was suspected as necessary near opioid receptor binding site (Smith and Simon, 1980), because –SH group of N-ethylmaleimide. Chlornaltrexamine, an irreversible alkylating affinity label on the opiate receptor, destroy agonist binding capacity and maintain its antagonist effect for a period of 2–3 days. It proves that covalent binding can improve the duration of action for these receptors. A more reversible, spirothiazolidine based system was studied (Figure 3.3). Spirothiazolidine derivatives were selected because they tend to cleave carbon-sulfur bond and cause the biological cleavage of imine (Somogyi et al., 2002). They have been also evaluated before in delivery systems for controlled-release of endogenous agents for, for example, hydrocortisone or progesterone (Bodor, 1982).

The spirothiazolidine ring was attached at 6-position as the modification, because it does not affect the biological activity (Figure 3.3). Spirothiazolidine ring opening allows oxidative anchoring of the agent to –SH group via disulfide bridging. The blocking process is reversible; endogenous –SH compounds presumably will regenerate the –SH group of the receptor (Sloan and Bodor, 1982).

In the Enzymatic Physical Chemical based CDS, the drug moiety is modified to introduce protective function(s) and a T group. When it is administered, this type of CDS gets distributed throughout the body and follows various enzymatic modifications. It is converted into a complex which is still inactive in all part of the body except the active site. The efflux-influx processes at the active site act differently because the targeted specific enzyme converts CDS into active drug. Due to the presence of a specific membrane or other distributional barrier, they will allow a specific concentration to enter and finally allow the release of the active drug only at the site of action (Bhardwaj et al., 2014).

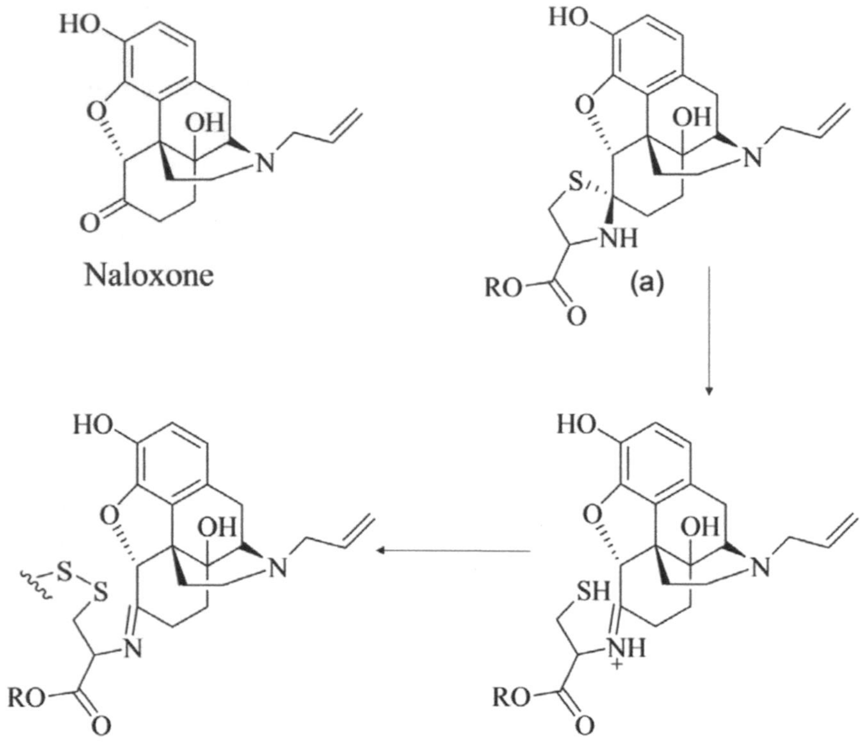

FIGURE 3.3 Spirothiozolidine based CDS (a) and Naloxone. After stepwise conversion of Spirothiozolidine based CDS, it forms disulfide linkage with nearby SH of opioid receptor.

One of the examples of this type of approach is estradiol-CDS (Figure 3.4). When estradiol-CDS is administered, it gets distributed in the whole body including Blood Brain Barrier (BBB), followed by the conversion into nonlipophilic intermediate. The intermediate can never come out due to the hydrophilic nature, so it gets 'trapped' in BBB and release the active drug estradiol (Bodor, 1987).

In the site specific enzyme activated CDS, the enzymatic conversion occurs only at the site of action as the result of the differential distribution of various enzymes throughout the body. The modification of the drug mainly involves two parts: A pharmacophore of the drug is chemically modified. In the modification part it involves incorporating the T moiety and the other protective function(s) to the pharmacophore. When the resulting CDS is administered in the body, it gets distributed throughout the body. As per the predictable enzymatic reactions, it show the pharmacological effect only at the targeted site by converting from original CDS to the active form where

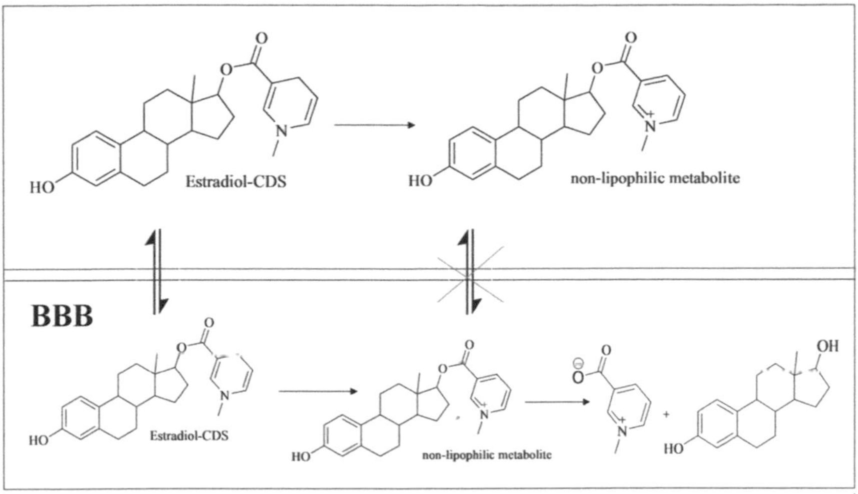

FIGURE 3.4   Target delivery of Estradiol-CDS into the BBB and its lock-in mechanism.

necessary enzymes are present; while it gets constantly eliminated from the rest of the body (Bodor, 1987; Goskonda et al., 1999).

In the beta-adrenergic antagonist-CDSs, corresponding p-amino alcohol pharmacophore part of the original molecules was replaced by a p-amino oxime or alkyloxime function. The enzymes located in the iris-ciliary body hydrolyze this oxime or alkyloxime derivative, and simultaneously, reductive enzymes located in the iris-ciliary body mount only the active S-(−) stereo-isomer of the P-blocker (Bodor and Prokai, 1990). A major drawback of the classical antiglaucoma agents are void of any cardiovascular activity because even if they show significant Intra-Ocular Pressure lowering activity, they did not produce any active p-blocker metabolically when given by intravenously.

3.3.1 MOLECULAR PACKAGING

Some neuropeptides such as Leu-enkephalin, thyrotropin-releasing hor-mone (TRH), and kyotorphin analogs (Bodor, 1997) are delivered by CDS approach for successful brain delivery. Peptides are difficult to deliver in brain, because of the rapid metabolic degradation by peptidase and lipophi-licity profile. Therefore, peptides should have mainly three characteristics for successful brain delivery. First it should enhance passive transport by having high lipophilicity. Second, it should ensure enzymatic stability for preventing premature degradation, and lastly it should provide targeting by exploiting a lock-in mechanism. These three characteristics can be achieved by a complex *molecular packaging* strategy. Here, the peptide part is attached to a bulky lipophilic functional groups (L) which leads to the Blood Brain Barrier (BBB) penetration and prevent reactions with peptidases. The efficiency of the CDS is also influenced by modifications of the Spacer group (S). Spacer is strategically selected amino acids that ensured the removal of the changed Targetor moiety (T) from peptide and lipophilic moeity. Targetor moiety (T) is a bioremovable moiety which should be attached to the spacer for targeted delivery purpose (Figure 3.5) (Prokai-Tatrai et al., 1996).

3.4 SOFT DRUG DESIGN APPROACH

The goal of the soft drug design is not to avoid metabolic pathway, but to design such active compounds which can be easily converted into inactive metabolites, and to avoid formation of toxic metabolites. The SD concept came into picture during the 1970 s, and since then it has been a great use for designing better therapeutic agents (Bodor, 1982). SDs produce targeted localized pharmacological activity but without undesirable toxic activity as they are deactivated by various enzymes when distributed in the body. SDs controls and directs metabolism as required. Prodrugs are often confused with SDs as both rely on enzymatic conversion, but where the metabolic approach activates the prodrug, SDs get deactivated at that point (Liederer and Borchardt, 2006).

SDs can be classified into following classes:
1. Soft analogs
2. Active metabolite based SDs
3. Activated SDs
4. Pro-SDs
5. Inactive metabolite based SDs.

From all the *above soft analogs* and *inactive metabolite based SDs* are proved to be the best proven approach and frequently used.

3.4.1 SOFT ANALOGS

These classes of compounds are the structural analogs of the active lead compound. They are designed to follow metabolically one step detoxification process.

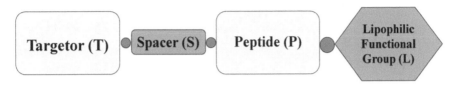

FIGURE 3.5 Molecular packaging approach used for brain-targeted delivery of neuropeptides.

Soft analogs drug design follows main points such as:

1. Newly designed soft analogs are isosteric/isoelectronic with the lead compound.
2. The inbuilt metabolically sensitive moiety is ideally the only route of metabolism.
3. Structural modification in the lead can direct to predictable metabolism.
4. The resulting metabolic products are with lower of no toxicity.
5. Pharmacological activity of the drug should not be diminished by the inbuilt metabolically sensitive moiety.
6. The metabolically sensitive moiety is located within the molecule in such a manner that the overall physical, physiochemical, steric and complementary properties of the soft analog are similar to the lead compound (Bodor and Buchwald, 2000).

The design of the quaternary ammonium salts (Figure 3.6) (Bodor et al., 1980) and some soft quaternary ammonium L-carnitine esters (Calvani et al., 1998) (Figure 3.7) are example of this type of soft drugs.

3.4.2 INACTIVE METABOLIC APPROACH

It is one of the most useful and successful approach among all classes of drug design. The main strategies for this type of design are:

1. The overall design starts with the inactive and excreted metabolite of the active lead compound.
2. One step conversion into the inactive metabolite without any metabolic conversions is maintained for new structure design.
3. By structural modifications we can control the metabolic routes as well as specific binding and transport properties of the new soft compounds.

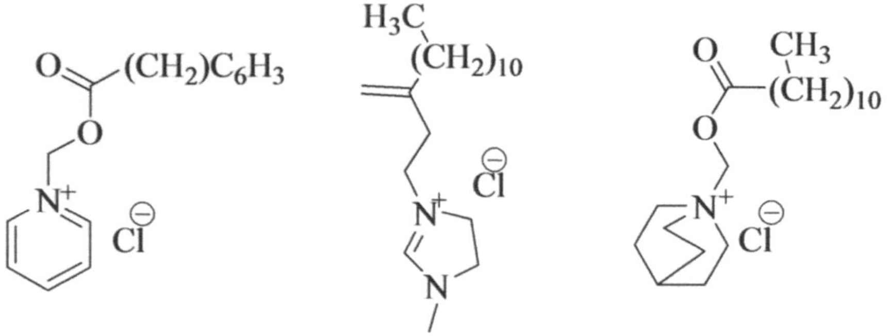

FIGURE 3.6   Quaternary ammonium salts as soft antimicrobacterials.

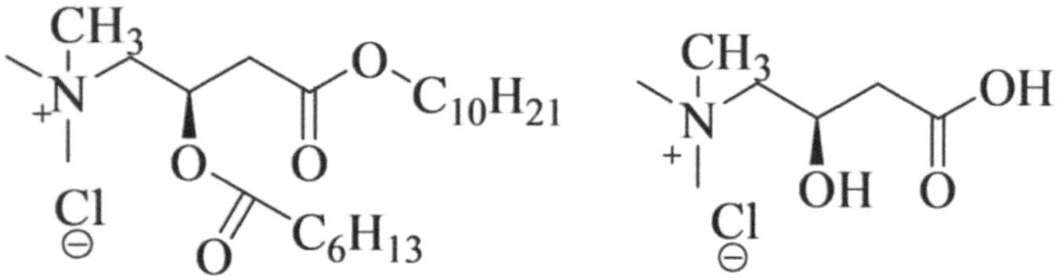

FIGURE 3.7   Quaternary ammonium L-carnitine esters as soft analogs.

4. The specific chemical modifications of the inactive metabolite which is isosteric or isoelectronic to the active metabolite of the drug or the active drug itself.

Various Anti-cholinergics (Buchwald and Bodor, 2006) and synthetic steroids (Little et al., 1999) are designed by this approach which can be shown in Figures 3.8 and 3.9.

3.4.3   CONTROLLED RELEASED ENDOGENOUS AGENTS OR NATURAL SOFT DRUGS OR PRO-SDS

A major advantage of considering biologically active agents as soft drugs is their fast and efficient metabolism rate. Some natural hormones and other biologically active neurotransmitters can be considered very active in such a design. Their metabolite intermediates do not result intoxic side products if taken within the recommended dosage. Therefore, the designing of the

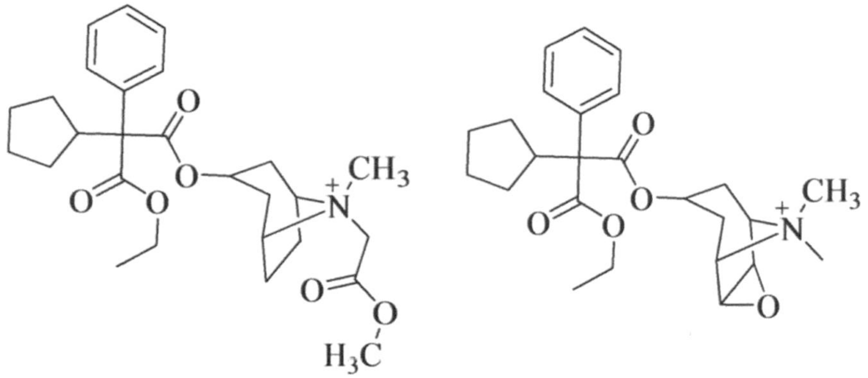

FIGURE 3.8   Soft anticholinergics.

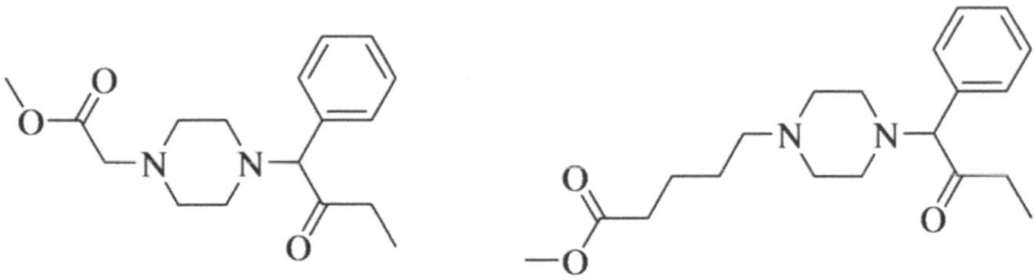

FIGURE 3.9   Highly potent synthetic steroids.

soft drugs from such natural agents can be adopted by combining prodrug as well as soft drug aspects, which is called the prosoft drug approach (Bodor, 1994).

For example, progesterone and testosterone are natural soft drugs which have high metabolism rate but it can be overcome with specific delivery approach (Sloan and Bodor, 1982). Another example is hydrocortisone, shown in Figure 3.10. The 4,5-unsaturated 3-ketone group is important for good binding and activity of hydrocortisone. Modification on that group is good target for controlled release and design for a natural soft drug.

3.4.4   ACTIVATED SOFT DRUGS

In this type of design, the compound does not include any known analogs of drugs. It is derived from the nontoxic metabolite, in which a specific pharmacophore is introduced. During the overall process, the active part

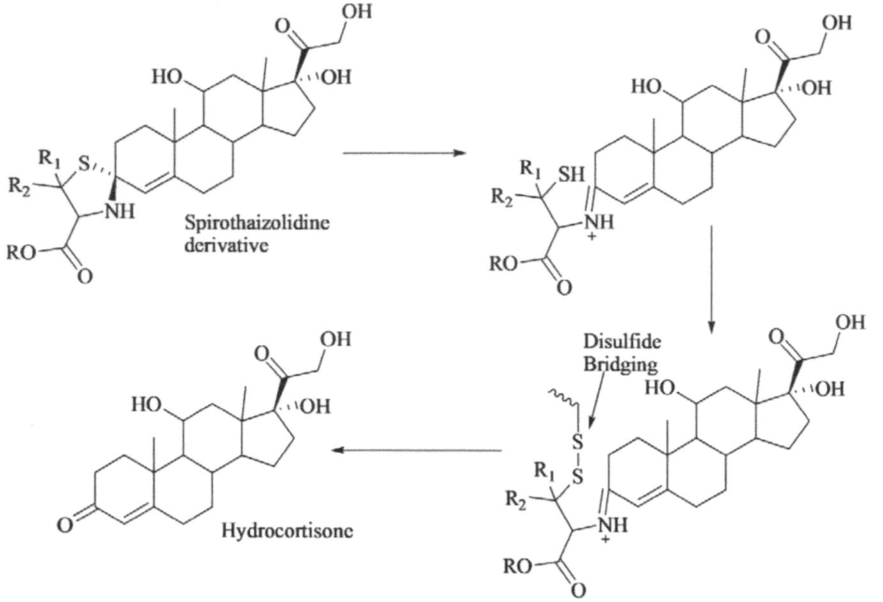

FIGURE 3.10   A spirothiazolidine derivative process for the controlled release of the hydrocortisone.

subsequently loses the activity and converted into starting inactive metabolite. N-chloramide antimicrobials are example of activated soft drugs (Kaminski et al., 1976).

Another example is soft alkylating compounds, as site specific potential antitumor agents (Figure 4.11). They are characterized by predictive in vivo metabolic destruction to nontoxic, metabolites after the achievement of the therapeutic role (Bodor and Kaminski, 1980).

3.4.5   ACTIVATED METABOLITE BASED SOFT DRUGS

In the conventional metabolism of the active drug molecule, the drug undergoes various metabolite steps and creates various intermediates and metabolites. It also uses energy and put extra burden on the body (Bodor, 1984). Sometimes these in between oxidative metabolites are as active as parent drug. The designing of active metabolites is used for the parent drug which undergoes one step deactivation and finally leads to one step detoxification. The one step deactivation can be achieved because they are already in active oxidative stage (Hwang et al., 2000).

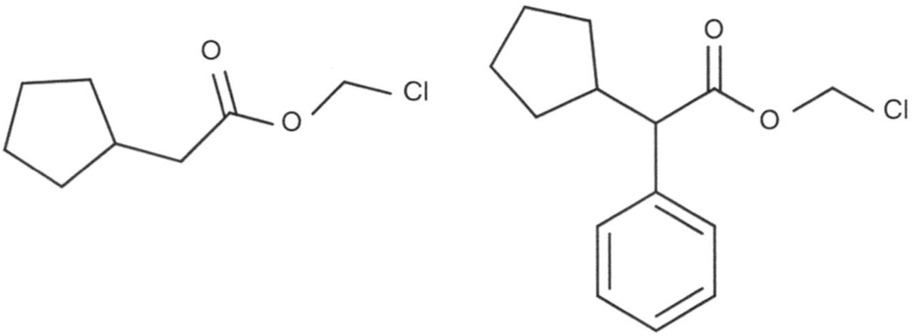

FIGURE 3.11   Soft alkylating agents.

A good example of this type of soft drug design is bufuralol (Bodor and Buchwald, 2004). The bufuralol is an active drug which undergoes various kinds of metabolic steps. The intermediates are also active, so it increases the overall activity of the drug. In the last step, it undergoes one step inactivation. The process is depicted in Figure 3.12.

The extension of the soft drug design concept is also applied to the nonpharmaceutical products. The well-known pesticide DDT (dichlorodiphenyltrichloroethane) has shown numerous *in-vivo* oxidation, oxidative dehaloformation and iterative dehydrohalogenation metabolites. The inactive acid metabolite formed by this hydrolytic transformation of DDT is less toxic and water soluble which is easily excreted in urine and feces. It is therefore the ideal candidate for the soft design approach of chemicals. The ester containing chlorobenzilate of DDT is also active as DDT, but having lower carcinogenicity. The ester group can undergo rapid metabolism leads to the nontoxic carboxylic acid (Armstrong, 2002).

3.5   COMPUTER AIDED DRUG DESIGN

The role of computational drug design approaches in the designing of potent and effective drugs are well renowned now days. Computational models exist in every level of drug discovery including drug binding affinity (Cheng et al., 2007), pharmacokinetic/pharmacodynamics calculation (Meibohm and Derendorf, 1997) and clinical trial design (Rooney et al., 2001). By incorporating some structure generating rules, that are specific to soft drug design with an analogy based ranking algorithm of metabolic properties, you can assist in selecting the most promising new candidates (Bodor et al., 1999; Kubinyi, 1999).

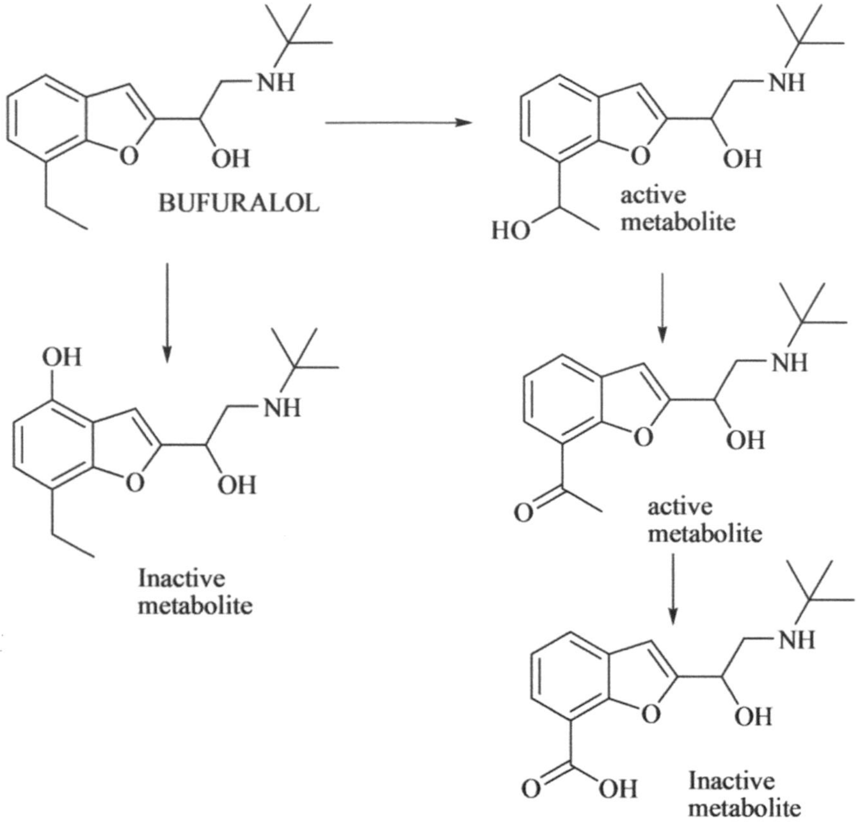

FIGURE 3.12 Structure of bufuralol and its metabolites.

Computer Aided Drug Design is a useful tool which calculates various molecular properties such as hydrolytic liability (Bodor et al., 1989;
tBodor and Huang, 1992) and introduces quantitative measures to rank. It
generates ranking order on the basis of various steric and electronic properties compared to the original lead (Buchwald and Bodor, 2000; Bodor,
1982). The whole process can be done by fully computerized expert system. The system combines various structure generating rules of soft drug
design with predictive quantitative structure–activity/structure–metabolism
relationship (QSAR/QSMR) models. The final step consists of the analogy based ranking (Bodor, 1982). After the whole process, the molecules
which are unlikely to have reasonable therapeutic activity and high toxicity
are eliminated from synthesis. The process results to avoid experimental

testing time, labor and financial savings of the eliminated molecules (Bodor et al., 1998).

The overall process for the QSMR involves following main steps:

- Lead identification
- Structure generation
- Descriptor calculation
- Candidate ranking
- Hydrolytic liability

Lead Identification: The overall approach for the QSMR is general by concept, and it can be started from any possible lead where soft drug design applies. Depending on the lead structure, structure can generate whole libraries of the compounds, which can make the quantitative study better for the analogy based scoring.

Structure generation: Available software can generate a new library of the compounds depending upon the input lead structure. The new structure can be generated by the two most successful soft drug design strategies: inactive metabolite based and soft analog approaches. It can generate common oxidative metabolites of the lead compounds and generate new analogs. For example, instead of $-CH_3$ and $-CH_2OH$ or other alkyl groups, it can replace them with $-COOH$ or $-(CH_2)n-COOH$ groups and then it can generate new soft drugs by functionalizing those oxidized metabolites to generate active soft drugs. In order to follow the rules of soft drug design, the software will track methylene/hydroxyalkyl groups and replace them with corresponding ester $-O-CO-$, reverse esters ($-CO-O-$) or other isosteric functions.[44]

Descriptor calculation: Ranking of the compounds is on the basis of the closeness of the calculated properties (Descriptors) of analogs to the lead compound. Mostly, the compounds are close steric and electronic analogs of the original lead (Dewar et al., 1985). The geometries are obtained from various advanced semi-empirical quantum chemical calculations. In order to define proper ranking, as many possible descriptors are selected (Bender and Glen, 2004; Willett, 2006). Various types of descriptors are: (i) molecular size/shape descriptors (S) (volume, $[V]$; surface area, $[S]$; ovality, $[O]$; (ii) electric/electronic descriptors (E) (dipole moment, $[D]$; average polarizability, $[\alpha]$; ionization potential, $[I]$); (iii) predicted solubility/partition properties (P) (log water solubility, $[\log rw]$; log octanol–water partition coefficient, $[\log Po/w]$); and (iv) atomic charge distribution on the unchanged portions

(Q). Amongst the computable descriptors, size/shape-, electrostatic- and H-bond related descriptors consistently seem to be the most important factors for describing physicochemical properties (Brüstle et al., 2002).

Candidate ranking: As all these properties are measured in different units and vary over different ranges, ranking factors (RF) that are all on the same dimensionless scale are used for each property and their weighted average is used for the final ranking (Peng et al., 2010).

Hydrolytic liability: Hydrolytic liability is considered as a major relevance for the ranking of the soft drug candidates as it measures the metabolic stability of the compound (softness) (Buchwald and Bodor, 2000) and for most of the metabolic study of compounds, hydrolysis is the primary focus for controlling the metabolic pathways (Testa et al., 2012).

3.6 CONCLUSION

During the drug discovery process, it is important for a lead to have better therapeutic index in addition to increased potent activity. Retrometabolic drug design approach focuses on developing compounds by predicting the metabolic pathways of the current lead compound, and to design novel analogs having predicted metabolic route with same activity. The novel compounds are designed in a way to have better therapeutic index and fewer side effects. Retrometabolic drug design focus on designing the compounds mainly by Chemical Delivery System (CDS) and Soft Drugs (SD) approaches. Both approaches are different by concept but general by nature, and can be applied to all drug classes. Various computer softwares and expert systems are developed for this application.

KEYWORDS

- **chemical delivery**
- **computational**
- **drug design**
- **metabolic**
- **retrometabolic**
- **soft drug**

REFERENCES

Armstrong, D., (2002). In: *Oxidative Stress Biomarkers and Antioxidant Protocols*. Springer, Vol. 186.

Bender, A., & Glen, R. C. (2004). Molecular similarity: a key technique in molecular informatics. *Organic & Biomolecular Chemistry 2* (22), 3204–3218.

Bhardwaj, Y. R., Pareek, A., Jain, V., & Kishore, D. (2014). Chemical delivery systems and soft drugs: Retrometabolic approaches of drug design. *Saudi Pharmaceutical Journal. SPJ: The Official Publication of the Saudi Pharmaceutical Society 22* (4), 290–302.

Bodor, N. (1984). Novel approaches to the design of safer drugs: soft drugs and site-specific chemical delivery systems. *Advances in Drug Research 13*, 255–331.

Bodor, N. (1987). Redox Drug Delivery Systems for Targeting Drugs to the Braina. *Annals of the New York Academy of Sciences 507* (1), 289–306.

Bodor, N. (1994). Designing safer ophthalmic drugs by soft drug approaches. *Journal of Ocular Pharmacology 10* (1), 3–15.

Bodor, N. (1995). Retrometabolic approaches to drug targeting. *NIDA Research Monograph1 54*, 1–27.

Bodor, N. (1997). Retrometabolic approaches for drug design and targeting. *Die Pharmazie 52* (7), 491–4.

Bodor, N. (1999). Recent advances in retrometabolic design approaches. *Journal of Controlled Release: Official Journal of the Controlled Release Society 62* (1–2), 209–222.

Bodor, N. (2000). Recent advances in retrometabolic drug design and targeting approaches. *Die Pharmazie 55* (3), 163–166.

Bodor, N., & Buchwald, P. (1997). Drug targeting via retrometabolic approaches. *Pharmacology & Therapeutics 76* (1–3), 1–27.

Bodor, N., & Buchwald, P. (2000). Soft drug design: general principles and recent applications. *Medicinal Research Reviews 20* (1), 58–101.

Bodor, N., & Buchwald, P. (2004). Designing safer (soft) drugs by avoiding the formation of toxic and oxidative metabolites. *Molecular Biotechnology 26* (2), 123–32.

Bodor, N., & Buchwald, P. (2008). Retrometabolic drug design: principles and recent developments. *Pure and Applied Chemistry 80* (8), 1669–1682.

Bodor, N., & Buchwald, P. (2012). Retrometabolic drug design. *Retrometabolic Drug Design and Targeting*, 71–76.

Bodor, N., & Kaminski, J. J. (1980). Soft drugs. 2. Soft alkylating compounds as potential antitumor agents. *Journal of Medicinal Chemistry 23* (5), 566–569.

Bodor, N., & Prokai, L. (1990). Site-and stereospecific ocular drug delivery by sequential enzymatic bioactivation. *Pharmaceutical Research 7* (7), 723–725.

Bodor, N., Buchwald, P., & Huang, M. J. (1998). Computer-assisted design of new drugs based on retrometabolic concepts. *SAR and QSAR in Environmental Research 8* (1–2), 41–92.

Bodor, N., Buchwald, P., & Huang, M.-J. (1999). The role of computational techniques in retrometabolic drug design strategies. *Theoretical and Computational Chemistry 8*, 569–618. (a) Bodor, N., Gabanyi, Z., & Wong, C. K. (1989). A new method for the estimation of partition coefficient. *Journal of the American Chemical Society 111* (11), 3783–3786; (b) Bodor, N., & Huang, M. J. (1992). A new method for the estimation of the aqueous solubility of organic compounds. *Journal of Pharmaceutical Sciences 81* (9), 954–960.

Bodor, N., Kaminski, J. J., & Selk, S., (1980). Soft drugs. 1. Labile quaternary ammonium salts as soft antimicrobials. *Journal of medicinal Chemistry 23* (5), 469–474.

Brouillette, G., Kawamura, M., Kumar, G. N., & Bodor, N. (1996). Soft drugs. 21. Design and evaluation of soft analogs of propantheline. *Journal of Pharmaceutical Sciences 85* (6), 619–623.

Brüstle, M., Beck, B., Schindler, T., King, W., Mitchell, T., & Clark, T. (2002). Descriptors, physical properties, and drug-likeness. *Journal of Medicinal Chemistry 45* (16), 3345–3355.

Buchwald, P., & Bodor, N. (1999). Quantitative structure-metabolism relationships: steric and nonsteric effects in the enzymatic hydrolysis of noncongener carboxylic esters. *Journal of Medicinal Chemistry 42* (25), 5160–5168.

Buchwald, P., & Bodor, N. (2000). Structure-based estimation of enzymatic hydrolysis rates and its application in computer-aided retrometabolic drug design. *Die Pharmazie 55* (3), 210–217.

Buchwald, P., & Bodor, N. (2006). Soft quaternary anticholinergics: comprehensive quantitative structure-activity relationship (QSAR) with a linearized biexponential (LinBiExp) model. *Journal of Medicinal Chemistry 49* (3), 883–891.

Buchwald, P., & Bodor, N. (2014). Recent advances in the design and development of soft drugs. *Die Pharmazie 69* (6), 403–13.

Buris, L. F., Bodor, N., & Buris, L. (1999). Loteprednol etabonate, a new soft steroid is effective in a rabbit acute experimental model for arthritis. *Die Pharmazie 54* (1), 58–61.

Calvani, M., Critelli, L., Gallo, G., Giorgi, F., Gramiccioli, G., Santaniello, M., Scafetta, N., Tinti, M. O., & De Angelis, F. (1998). L-Carnitine esters as "soft", broad-spectrum antimicrobial amphiphiles. *Journal of Medicinal Chemistry 41* (13), 2227–2233.

Cheng, A. C., Coleman, R. G., Smyth, K. T., Cao, Q., Soulard, P., Caffrey, D. R., Salzberg, A. C., & Huang, E. S. (2007). Structure-based maximal affinity model predicts small-molecule druggability. *Nature Biotechnology 25* (1), 71–5.

Corey, E. (1988). Robert Robinson Lecture. Retrosynthetic thinking—essentials and examples. *Chem. Soc. Rev. 17*, 111–133.

Dewar, M. J., Zoebisch, E. G., Healy, E. F., & Stewart, J. J. (1985). Development and use of quantum mechanical molecular models. 76. AM 1: a new general purpose quantum mechanical molecular model. *Journal of the American Chemical Society 107* (13), 3902–3909.

Druzgala, P., Hochhaus, G., & Bodor, N. (1991). Soft drugs—10. Blanching activity and receptor binding affinity of a new type of glucocorticoid: loteprednol etabonate. *The Journal of Steroid Biochemistry and Molecular Biology 38* (2), 149–154.

Goskonda, V. R., Khan, M. A., Bodor, N. S., & Reddy, I. K. (1999). Chemical delivery systems: evaluation of physicochemical properties and enzymatic stability of phenylephrone derivatives. *Pharmaceutical Development and Technology 4* (2), 189–198.

Gupta, R. K., & Bhattacharjee, A. K. (2007). Discovery and design of new arthropod/insect repellents by computer-aided molecular modeling. *Insect Repellents: Principles, Methods, and Uses* 195–228.

Hwang, S. K., Juhasz, A., Yoon, S. H., & Bodor, N. (2000). Soft drugs. 12. Design, synthesis, and evaluation of soft bufuralol analogs. *Journal of Medicinal Chemistry 43* (8), 1525–32.

Kaminski, J. J., Maureen Huycke, M., Selk, S. H., Bodor, N., & Higuchi, T. (1976). N-halo derivatives V: Comparative antimicrobial activity of soft N-chloramine systems. *Journal of Pharmaceutical Sciences 65* (12), 1737–1742.

Kubinyi, H. (1999). Chance favors the prepared mind-from serendipity to rational drug design. *Journal of Receptors and Signal Transduction 19* (1–4), 15–39.

Liederer, B. M., & Borchardt, R. T. (2006). Enzymes involved in the bioconversion of ester-based prodrugs. *Journal of Pharmaceutical Sciences 95* (6), 1177–1195.

Little, R. J., Bodor, N., & Loftsson, T. (1999). Soft drugs based on hydrocortisone: the inactive metabolite approach and its application to steroidal antiinflammatory agents. *Pharmaceutical Research 16* (6), 961–967.

Meibohm, B., & Derendorf, H. (1997). Basic concepts of pharmacokinetic/pharmacodynamic (PK/PD) modeling. *International Journal of Clinical Pharmacology and Therapeutics 35* (10), 401–413.

Peng, J., Macdonald, C., & Ounis, I., (2010). Learning to select a ranking function. In: *Advances in Information Retrieval*, Springer, pp. 114–126.

Prokai-Tatrai, K., Prokai, L., & Bodor, N. (1996). Brain-targeted delivery of a leucine-enkephalin analog by retrometabolic design. *Journal of Medicinal Chemistry 39* (24), 4775–4782.

Rooney, K. F., Snoeck, E., & Watson, P. H. (2001). Modeling and simulation in clinical drug development. *Drug Discovery Today 6* (15), 802–806.

Simpson, D. D., & Sells, S. B., (1990). In: *Opioid Addiction and Treatment: A 12-year Follow-Up.* Krieger Pub Co.

Smith, J. R., & Simon, E. J. (1980). Selective protection of stereospecific enkephalin and opiate binding against inactivation by N-ethylmaleimide: evidence for two classes of opiate receptors. *Proceedings of the National Academy of Sciences 77* (1), 281–284.

Somogyi, G., Buchwald, P., & Bodor, N. (2002). Targeted drug delivery to the central nervous system via phosphonate derivatives (anionic delivery system for testosterone). *Die Pharmazie 57* (2), 135–137. (a) Bodor, N. (1982). Designing safer drugs based on the soft drug approach. *Trends in Pharmacological Sciences 3*, 53–56; (b) Sloan, K., & Bodor, N. (1982). Hydroxymethyl and acyloxymethyl prodrugs of theophylline: enhanced delivery of polar drugs through skin. *International Journal of Pharmaceutics 12* (4), 299–313.

Testa, B., Pedretti, A., & Vistoli, G. (2012). Reactions and enzymes in the metabolism of drugs and other xenobiotics. *Drug Discovery Today 17* (11), 549–560.

Willett, P. (2006). Similarity-based virtual screening using 2D fingerprints. *Drug Discovery Today 11* (23), 1046–1053.

PART II

PHARMACOLOGICAL ASPECTS OF NATURAL PRODUCTS

CHAPTER 4

INSIGHT INTO THE PHARMACOLOGICAL POTENTIALS OF *CAMELLIA SINENSIS* LINN.

PRABODH CHANDER SHARMA,[1] AMIT YADAV,[1]
ARCHANA SHARMA,[1] and AAKASH DEEP[2]

[1]*Institute of Pharmaceutical Sciences, Kurukshetra University, Kurukshetra – 136119, India*

[2]*Department of Pharmaceutical Sciences, Ch. Bansi Lal University, Bhiwani – 127021, India*

CONTENTS

ABSTRACT

Tea is the second most consumed beverages in the world, next to water. Green tea has been considered as a healthful drink since ancient times. Green tea is a 'nonfermented' tea, and contains more catechins, than black tea or oolong tea. Various animal and human studies have established the beneficial effects of green tea for a wide variety of diseases like cancer, diabetes, cardiovascular disease etc. It has anti microbial activity, helps in body weight control, protects from solar radiations and also possesses neuroprotective properties. The main component of green tea which is responsible for most of these activities is (–) epigallocatechin gallate. The information obtained from various research literatures on green tea shows very promising results, but still there is need to conduct more scientific studies to establish various parameters like bioavailability, effective dose, mechanism of action etc.

4.1　INTRODUCTION

Tea is an infusion made by steeping processed leaves, buds or twigs of tea bush, *Camellia sinensis*; Family: Theaceae, in hot water for several minutes after, which it is drunk (Mughal et al., 2010). Tea is the second most consumed beverages in the world, next to water (Cheng, 2006) and well ahead of coffee, beer, wine and carbonated soft drinks (Rietveld and Wiseman, 2003). Today, 3 billion cups of tea are consumed every day by millions of people all over the world (Ng et al., 2008).

Tea was first consumed as a beverage in China sometime between 2700 BC and 220 AD. Traditional styles of green, black and oolong teas first made an appearance in the Ming Dynasty in China (1368–1644 AD). Tea began to travel as a trade item as early as the fifth century with some sources indicating Turkish traders bartering for tea on the Mongolian and Tibetan borders (Chapagain and Hoekstra, 2007). The economic and social interest of tea is clear and its consumption is a part of many people's daily routine, as an everyday drink and as a therapeutic aid in many illnesses (Cabrera et al., 2006). For a long time, green tea was preferred as a beverage because of its attractive flavor and taste. But the scientific community and the popular press have recently pondered that there might be beneficial properties of green tea. Its consumption has been associated with antiinflammatory, antioxidative, antimutagenic, and anticarcinogenic effects (Kaszkin et al., 2004).

4.2 TYPES OF TEA

Alterations in the manufacturing process result in black, green, and oolong tea, which account for approximately 75%, 23%, and 2% of the global production, respectively (Carlson et al., 2007). In the production of black tea, leaves of *C. sinensis* are picked and then allowed to wither indoors, ferment, and oxidize. For green tea, the plant leaves are steamed and parched after picking to prevent oxidation of the catechins present in the leaf. Oolong tea is produced by "semifermenting" the green leaves, resulting in a tea that is chemically a mixture of green and black teas (Frei and Higdon, 2003).

Black tea is consumed principally in Europe, North America and North Africa (except Morocco) while green tea is widely drunk in China, Japan, Korea, and Morocco, oolong tea is popular in China and Taiwan (Cabrera et al., 2006). Over the last years, numerous epidemiological and clinical studies have revealed several physiological responses to green tea which may be relevant to the promotion of health and the prevention or treatment of some chronic disease (Cabrera et al., 2006). Food stuff can be regarded as functional if it satisfactorily demonstrates to affect beneficially one or more target functions in the body, beyond adequate nutritional effects in a way which is relevant to either the state of well-being and health or the reduction of the risk of a disease. So green tea has been proved to have functional properties and at present, its consumption is widely recommended (Cabrera et al., 2006).

4.3 PRODUCTION PROCESS OF GREEN TEA

Green tea and black tea are processed differently during manufacturing. To produce green tea, freshly harvested leaves are immediately steamed to prevent fermentation, yielding a dry, stable product. This steaming process destroys the enzymes responsible for breaking down the color pigments in the leaves and allows the tea to maintain its green color during the subsequent rolling and drying processes. These processes preserve natural polyphenols with respect to the health-promoting properties (Chacko et al., 2010).

4.4   CHEMICAL COMPOSITION OF GREEN TEA

The composition of the tea leaves depends on a variety of factors, including climate, season, horticultural practices, and the type and age of the plant. The chemical composition of green tea is similar to that of the leaf (Mukhtar and Ahmad, 2000). This chemical composition is complex consisting of proteins (15–20% dry weight) whose enzymes constitute an important fraction; amino acids (1–4% dry weight) such as teanine or 5-N-ethylglutamine, glutamic acid, tryptophan, glycine, serine, aspartic acid, tyrosine, valine, leucine, threonine, arginine, lysine; carbohydrates (5–7% dry weight) such as cellulose, pectins, glucose, fructose, sucrose; lipids as linoleic and α-linolenic acids; sterols as stigmasterol; vitamins (B, C, E); xanthic bases such as caffeine and theophylline; pigments as chlorophyll and carotenoids; volatile compounds such as aldehydes, alcohols, esters, lactones, hydrocarbons, etc., minerals and trace elements (5% dry weight) such as Ca, Mg, Cr, Mn, Fe, Cu, Zn, Mo, Se, Na, P, Co, Sr, Ni, K, F and Al (Cabrera et al., 2006).

Polyphenols constitute the most interesting group of green tea leaf components, and in consequence, green tea can be considered an important dietary source of polyphenols particularly flavonoids (Cabrera et al., 2006). Green tea contains primarily monomeric flavonoids termed catechins or flavanols, predominantly (2)-epigallocatechin-3-gallate (EGCG) (2)-epigallocatechin (EGC) (2)-epicatechin-3-gallate (ECG), and (2)-epicatechin (EC). Black tea additionally contains complex oligomeric and polymeric polyphenols, theaflavins and thearubigins, which are formed during black tea manufacture as products of polyphenol oxidase-mediated catechin oxidation (Wiseman et al., 2001).

There are several polyphenolic catechins in green tea, viz. (−) epicatechin (EC) (−) epicatechin 3-gallate (ECG) (−) epigallocatechin (EGC) (−) epigallocatechin-3-gallate (EGCG) (+) catechin, and (+) gallocatechin (GC). EGCG, the most abundant catechin in green tea, accounts for 65% of the total catechin content (Zaveri, 2006).

4.5 BIOAVAILABILITY OF GREEN TEA CATECHINS

To understand the bioavailability of tea catechins in humans, Yang et al. (1998) gave 18 individuals different amount of green tea and measured the time-dependent plasma concentration and urinary excretion of tea catechins. After taking 1.5, 3.0, and 4.5 g of decaffeinated green tea solids (dissolved in 500 mL of water), the maximum plasma concentration (C_{max}) of (–)-epigallocatechin −3-gallate (EGCG) was 326 ng/mL, C_{max} of (–)-epigallocatechin (EGC) was 550 ng/mL, and the C_{max} of (–)-epicatechin (EC) was 190 ng/mL. These C_{max} values were observed at 1.4–2.4 h after ingestion from 1.5 to 3.0 g, the C_{max} values increased 2.7–3.4 folds, but increasing the dose to 4.5 g did not increase the C_{max} values significantly, which suggested a saturation phenomenon. This study provided basic pharmacokinetic parameters of green tea catechins in humans.

Oral consumption of purified polyphenols obtained through extraction process depicts low bioavailability (1–20%) depending on the derivatives. However, complexation with phospholipids showed increased bioavailability of the polyphenolic fraction after oral administration (Di Pierro et al., 2008). Polyphenols administered in the form of GTS (Green Tea Extract Supplement) showed enhanced bioavailability compared to GT (Green Tea) or BT (Black Tea), which led to small but significant increase in antioxidant capacity (Henning et al., 2004). The oral bioavailability of tea catechins has been found to be low in rodents. Because a significant fraction of the orally administered green tea catechins is not absorbed or is eliminated presystemetically, small changes in factors limiting the systemic availability of green tea catechins could have a significant impact on their oral bioavailability (Chow et al., 2005). Chow et al. (2005) showed that taking Polyphenon E on an empty stomach after an overnight fasting resulted in a dramatic increase in the blood levels of free epigallocatechin gallate, epigallocatechin, and epicatechin gallate. Catechins from green and black tea are rapidly absorbed and milk does not impair bioavailability of the tea catechins (Van het Hof et al., 1998).

It is safe for healthy individuals to take green tea polyphenol products in amounts equivalent to the EGCG content in 8–16 cups of green tea once a day or in divided doses twice a day for 4 weeks. There is >60% increase in the systemic availability of free EGCG after chronic green tea polyphenol administration at a high daily bolus dose (800 mg EGCG or Polyphenon E once daily) (Chow et al., 2003). The absorption patterns of GCG, CG and

C were similar to that of corresponding precursors EGCG, ECG and EC. In addition, the intravenous dosing experiment showed that GCG and CG had no difference in clearance pattern from their precursors EGCG and ECG in the blood (Xu et al., 2004).

4.6 PHARMACOKINETICS OF GREEN TEA CATECHINS

Green tea catechin mixture was dosed to rats by intravenous or intraportal infusion. Blood samples were collected after dosing and analyzed using high-performance liquid chromatography with the coulometric electrode array detection system. The systemic clearance of epigallocatechin gallate (EGCG), epigallocatechin (EGC), and epicatechin (EC) was 8.9, 6.3, and 9.4 mL/min, respectively. The steady state volume of distribution (Vss) of EGCG, EGC, and EC was 432, 220, and 187 mL, respectively. It was found that high percentage of green tea catechins escaped first-pass hepatic elimination, with 87.0, 108.3, and 94.9% of EGCG, EGC, and EC, respectively, available in the systemic blood following intraportal infusion. The results suggested that factors within the gastrointestinal tract such as limited membrane permeability, transporter mediated intestinal secretion, or gut wall metabolism may contribute more significantly to the low oral bioavailability of green tea catechins (Cai et al., 2002). EGC and EC seemed to be absorbed faster (larger Ka) than EGCG, and EGCG had much lower bioavailability in terms of fraction of absorption. It seems that EGCG is better absorbed when given through drinking fluid than i.g. administration. In addition, the relatively higher AUC value of EGCG in the intestine samples after i.v. injection suggests that EGCG is excreted mainly through the bile. EGC and EC are likely excreted through both the urine and bile, because similar AUC values of EGC and EC were obtained in both the kidney and intestine (Chen et al., 1997). Important role for the multispecific organic anion transporter MRP2 in the bioavailability of EC and possibly other tea flavonoids was suggested in a study by Vaidyanathan and Walle (Vaidyanathan and Walle, 2001).

4.7 PHARMACOLOGICAL SIGNIFICANCE OF GREEN TEA

Green tea has been considered a medicine and a healthful beverage since ancient times. The traditional Chinese medicine has recommended this plant for headaches, body aches and pains, digestion, depression, detoxification, as an energizer and, in general, to prolong life (Cabrera et al., 2006). Green tea has recently attracted significant attention, both in the scientific and in consumer communities for its multiple health benefits for a variety of disorders, ranging from cancer to weight loss. Green tea and its constituent catechins are best known for their antioxidant properties, which has led to their evaluation in a number of diseases associated with reactive oxygen species (ROS), such as cancer, cardiovascular and neurodegenerative diseases (Zaveri, 2006). Extensive laboratory research and the epidemiologic findings of the recent past have shown that polyphenolic compounds present in tea may reduce the risk of a variety of illnesses (Mukhtar and Ahmad, 2000).

4.8 ANTI-CANCER ACTIVITY

Cancer is a disease characterized by uncontrolled division of cells and their ability to spread. This unregulated growth is caused by damage to DNA, resulting in mutations, defects in cell cycle, and apoptotic machinery. Thus, agents that can modulate apoptosis to maintain steady-state cell population by affecting one or more signaling intermediates leading to induction of apoptosis can be useful for targeted therapy of cancer. Hence, there is a need to develop novel targets and mechanism-based agents for the management of cancer (Sarfaraz et al., 2008).

In recent years, several epidemiological studies as well as studies in rodent and in vitro models have shown that green tea and its main polyphenol, EGCG, can provide protection against various malignancies including skin, breast, prostate, lung, colon, liver, stomach, and other types of cancers (Shimizu et al., 2008). Significant interest in green tea as a cancer preventive agent in humans has been in focus due to growth inhibition of various human cancer cell lines. The encouraging results of inhibition of lung metastasis in mice, wide distribution of [3H]EGCG in various organs, cancer-preventive results with humans without any severe adverse effects with green tea tablets in humans have been observed (Suganuma et al., 1999).

The chemopreventive effects of green tea depends primarily on its antioxidant action, specific induction of detoxifying enzymes, its molecular regulatory functions on cellular growth, development and apoptosis, and finally a selective improvement in the function of the intestinal bacterial flora (Cabrera et al., 2006). To show the mechanism of action of green tea polyphenols in cancer prevention, Suganuma et al. (1999) provided evidence on the endogenous tumor promoter/enhancer, that is, TNF-α. They obtained results showing that numerous inhibitors of tumor promotion and cancer preventive agents inhibited both TNF-α gene expression in the cells and TNF-α release from the cells induced by tumor promoters, resulting in reduction of amount of TNF-α in cancer cells and probably in their surrounding tissue. The tea polyphenols (−)-epicatechin gallate (ECG), EGCG, and EGC inhibited dose dependently TNF-α release from a human stomach cancer cell line, KATO III cells, treated with okadaic acid. The potency of these green tea polyphenols is closely associated with the potency of their growth inhibition. Daisuke et al. (2008) provided details on the molecular basis for the anticancer activity of EGCG both in vitro and in vivo. Through both the cell surface receptor 67LR and eEF1A, EGCG induces reduction of the MYPT1 phosphorylation at Thr-696, thus activating myosin phosphatase and inducing dephosphorylation of MRLC.

Laurie et al. (2005) designed a trial to determine the maximum tolerated dose (MTD) of green tea extract (GTE) in patients with advanced lung cancer. They registered 17 patients with advanced lung cancer which received once-daily oral dosing of GTE at a starting dose of 0.5 g/m^2 per day, with an accelerated dose-escalation scheme. The results of this study showed that the MTD of GTE was 3 g/m^2 per day with no grade 3 or 4 toxicity seen. This study concluded that at 3 g/m^2 per day, GTE likely has limited activity as a cytotoxic agent. Suganuma et al. (1999) studied synergistic effects of (−)-epigallocatechin gallate with (−)-epicatechin, sulindac, or tamoxifen on cancer-preventive activity in the human lung cancer cell line PC-9. In this study, they presented the first evidence that cotreatment with EGCG and EC, ECG and EC, and EGC and EC synergistically induced apoptosis of PC-9 cells, mediated through enhanced incorporation of tea polyphenols into the cells. Furthermore, cotreatment with EGCG and sulindac or EGCG and tamoxifen significantly enhanced induction of apoptosis by EGCG. These results strongly indicate that green tea itself is a more effective and practical cancer preventive than EGCG alone and that drinking green tea enhances the cancer-preventive activity of sulindac and tamoxifen, resulting in smaller doses of these drugs and fewer adverse effects. Mimoto et al. (2000) investigated the effect of (−)- epigallocatechin gallate (EGCG) on cisplatin induced lung tumors in A/J mice. The result showed that the tumor multiplicity was significantly reduced by adding EGCG to cisplatin-treated mice (p < 0.01). Furthermore, EGCG significantly reduced cisplatin-induced weight loss from 24.7–26.3% (cisplatin treatment) to 10.6–11.6%

(cisplatin plus EGCG treatment) ($p < 0.01$). These findings suggested that EGCG can inhibit cisplatin-induced weight loss and lung tumorigenesis in A/J mice.

Tsubono et al. (2001) and Koizumi et al. (2003) examined the relation between the risk of gastric cancer and the consumption of green tea and conducted pooled analysis of two population based prospective cohort studies in rural Northern Japan. They found no inverse association between green tea intake and the risk of gastric cancer. Sartippour et al. (2006) conducted study on the effectiveness of combination of green tea and tamoxifen against breast cancer. The study provided evidence that green tea extract inhibits breast cancer growth by a direct antiproliferative effect on the tumor cells, as well as by indirect suppressive effects on the tumor-associated endothelial cells and observed that green tea increased the inhibitory effect of tamoxifen on the proliferation of the ER (estrogen receptor)-positive MCF-7, ZR75, T47D human breast cancer cells in vitro. This combination regimen was also more potent than either agent alone at increasing cell apoptosis. In animal experiments, mice treated with both green tea and tamoxifen had the smallest MCF-7 xenograft tumor size, and the highest levels of apoptosis in tumor tissue, as compared with either agent administered alone. Green tea decreased levels of ER-α in tumors both in vitro and in vivo. They also observed that green tea blocked ER-dependent transcription, as well as estradiol-induced phosphorylation and nuclear localization of mitogen-activated protein kinase provide mechanistic evidence that the combination of green tea and tamoxifen is more potent than either agent alone in suppressing breast cancer growth.

Roy et al. (2005) used the MDA-MB-468 human breast cancer cell line as an in vitro model of ER-negative breast cancers, and found that treatment of EGCG resulted in dose-dependent (5–80 ug/mL) and time-dependent (24–72 h) inhibition of cellular proliferation (15–100%) and cell viability (3–78%) in MDA-MB-468 cells. The results of this study provide evidence that EGCG possesses anticarcinogenic effect against ER-negative breast cancer cells. Yuan et al. (2005) conducted a nested case-control study involving 297 incident breast cancer cases and 665 control subjects within the Singapore Chinese Health Study. There was no association between intake frequencies of green tea and risk of breast cancer among all women or those with low-activity ACE genotype. Among women with high-activity ACE genotype, however, intake frequency of green tea was associated with a statistically significant decrease in risk of breast cancer. The findings of this study highlight the importance of genetically determined factors in evaluating the role of green tea intake in the development of breast cancer.

Katiyar and Mukhtar (2001) tested green tea polyphenol (–)-epigallocatechin-3-gallate treatment to mouse skin and found it to prevent UVB-induced infiltration of leukocytes, depletion of antigen-presenting cells, and oxidative stress. EGCG treatment was also found to prevent UV-B-induced depletion in the number of antigen-presenting cells when immunohisto chemically detected as class II MHC1 Ia1 cells. They found that pretreatment of EGCG decreased the number of UV-B-induced increases in H_2O_2-producing cells and inducible nitric oxide synthase expressing cells. Together, these data suggested that prevention of UV-B-induced infiltrating leukocytes, antigen-presenting cells, and oxidative stress by EGCG treatment of mouse skin may be associated with the prevention of UV-B-induced immunosuppression and photocarcinogenesis, further Katiyar et al. (2001) investigated the effects of topical application of EGCG, the major polyphenol present in green tea, to human skin before UV irradiation on UV-induced markers of oxidative stress and antioxidant enzymes. They found that application of EGCG (~1 mg/cm² skin) before a single UV exposure of 4 × minimal erythema dose (MED) markedly decreases UV-induced production of hydrogen peroxide (68–90%, $p < 0.025$–0.005) and nitric oxide (30–100%, $p < 0.025$–0.005) in both epidermis and dermis in a time-dependent manner. EGCG pretreatment also inhibits UV-induced infiltration of inflammatory leukocytes, particularly CD11b⁺ cells (a surface marker of monocytes/macrophages and neutrophils), into the skin, which are considered to be the major producers of reactive oxygen species. EGCG treatment was also found to inhibit UV-induced epidermal lipid peroxidation at each time point studied (41–84%, $p < 0.05$).

Barthelman et al. (1998) using cultured human keratinocytes, showed that UVB-induced AP-1 activity is inhibited by EGCG in a dose range of 5.45 nM to 54.5 µM. EGCG is effective at inhibiting AP-1 activity when applied before, after or both before and after UVB irradiation. EGCG also inhibits AP-1 activity in the epidermis of a transgenic mouse model. Meeran et al. (2009) found that although administration of GTPs (0.2%, w/v) in drinking water significantly reduced UVB-induced tumor development in wild-type mice, this treatment had a non-significant effect in IL-12-KO mice. GTPs resulted in reduction in the levels of markers of inflammation (COX-2, PGE2, PCNA, cyclin D1) and proinflammatory cytokines (TNF- α, IL-6, IL-1β) in chronically UVB-exposed skin and skin tumors of wild-type mice but less effective in IL-12p40-KO mice. UVB-induced DNA damage (cyclobutane pyrimidine dimers) was resolved rapidly in GTPs-treated wild-type mice than untreated wild-type mice and this resolution followed the same time course as the GTPs-induced reduction in the levels of inflammatory responses. This effect of GTPs was less pronounced in IL-12-KO mice. The above results were confirmed by treatment of IL-12-KO mice with murine rIL-12 and treatment of wild-type mice with neutralizing anti-IL-12 antibody.

El-Sherry et al. (2007) reported that green tea treatment prevented the induction of rete pegs preneoplastic hyperplasia, hair follicle hyperplasia, trichofolliculocarcinoma, squamous cell carcinoma, fibropapilloma, rhabdomyosarcoma and mixed tumors. Green tea reduced the incidence and delayed the appearance of dysplastic changes and trichofolliculoma. Arsenite has no cocarcinogenic effect. It was observed that green tea has no significant effect in the arsenite UVB group. Yang et al. (2007) conducted a prospective cohort study of green tea consumption and colorectal cancer risk in women. In this large population-based prospective cohort study, they reported that regular consumption of green tea was inversely associated with the risk of CRC, particularly among women who maintained such habit over time. Longer the duration of lifetime tea consumption, lower was the risk of CRC. CRC risk also decreased as the amount of tea consumption increased. This inverse association was independent of known risk factors for CRC and consistent with animal and in vitro experiments showing potential cancer-inhibitory effects of tea and its extracts Lee et al. (2004) studied the impact of epigallocatechin-3-gallate, a known receptor tyrosine kinase (RTK) inhibitor, on VEGF receptor status and viability of CLL B cells. It was observed that VEGF165 significantly increased apoptotic resistance of CLL B cells, and immunoblotting revealed that VEGF-R1 and VEGF-R2 are spontaneously phosphorylated on CLL B cells. Moreover EGCG significantly increased apoptosis/cell death in 8 of 10 CLL samples measured by annexin V/propidium iodide (PI) staining. The increase in annexin V/PI staining was accompanied by caspase-3 activation and poly–adenosine diphosphate ribose polymerase (PARP) cleavage at low concentrations of EGCG (3 g/mL). Moreover, EGCG suppressed the proteins B-cell leukemia/lymphoma-2 protein (Bcl-2), X-linked inhibitor of apoptosis protein (XIAP), and myeloid cell leukemia-1 (Mcl-1) in CLLB cells. Finally, EGCG (3–25 µg/mL) suppressed VEGF-R1 and VEGF-R2 phosphorylation, albeit incompletely.

Kim et al. (2004) investigated the time-course of the anticancer effects of EGCG on human ovarian cancer. It was noticed that EGCG exerts a significant role in suppressing ovarian cancer cell growth. It showed dose dependent growth inhibitory effects in each cell line and induced apoptosis and cell cycle arrest. The cell cycle was arrested at the G1 phase by EGCG in SKOV-3 and OVCAR-3 cells. In contrast, the cell cycle was arrested in the G1/S phase in PA-1 cells. EGCG differentially regulated the expression of genes and proteins (Bax, p21, Retinoblastoma, cyclin D1, CDK4 and Bcl-XL) more than 2 fold, showing a possible gene regulatory role for EGCG. The continual expression in p21WAF1 suggests that EGCG acts in the same way with p53 proteins to facilitate apoptosis after EGCG treatment. Bax, PCNA and Bcl-X are also important in EGCG-mediated apoptosis. Inoue et al. (1998) conducted a study to examine the hypothesis that tea and coffee consumption have a protective effect against development of digestive tract cancers. The results suggest the odds ratio (OR) of stomach cancer decreased to 0.69 (95% confidence interval [CI] = 0.48–1.00) with high intake of green tea (seven cups or more per day). A decreased risk was also observed for rectal cancer with three cups or more daily intake of coffee.

4.9   ANTI-DIABETIC ACTIVITY

Diabetes is a disease in which the body is unable to properly use and store glucose (a form of sugar). Glucose backs up in the blood stream – causing your blood glucose or "sugar" to rise too high (Chouhan and Vyas, 2006). The two major forms of diabetes are type 1 (formerly termed juvenile-onset diabetes) and type 2 (formerly termed adult-onset diabetes). Type 1 diabetes (T1D1) represents approximately 10% of all cases of diabetes and develops secondary to autoimmune destruction of the insulin-producing β-cells of the pancreas. Type 2 diabetes (T2D) is the most common form, which represents more than 90% of all cases. Unlike T1D, T2D can be associated with elevated, normal, or low insulin levels, depending on the stage at which the levels are measured. T2D is recognized as a progressive disorder, which is associated with diminishing pancreatic function over time. Regardless of the classification, the resulting metabolic abnormalities that characterize diabetes contribute greatly to the clinical complications, and the major clinical strategy is aimed at restoring metabolic balance (Cefalu, 2006).

Various research studies have shown that green tea consumption prevents diabetes, especially Type 2 diabetes. The water extract of *Thea sinensis* (green tea cold extract) (100 mg/kg body weight) reduced the blood glucose of KK-Ay mice 4 and 8 weeks after repeated administration and intended to decrease the plasma insulin level of KK-Ay mice under similar conditions (Miura et al., 2005). Park et al. (2009) studied ambivalent role of gallated catechins in glucose tolerance in humans and reported the effects of circulating green tea catechins on blood glucose and insulin levels. These findings may suggest that the gallated catechin when it is in the circulation elevates blood glucose level by blocking normal glucose uptake into the tissues, resulting in secondary hyperinsulinemia, whereas it decreases glucose entry into the circulation when they are inside the intestinal lumen. Karaca et al. (2010) investigated the effects of oral administration of extract of green tea and ginseng (American ginseng- *Panax quinquefolium L.*), given alone or together, on pancreatic β-cells, blood glucose, insulin, cholesterol and triglyceride levels in rats with experimental diabetes induced by a single injection of streptozotocin (STZ) (60 mg/kg, i.p.). The findings of this study demonstrated that ginseng or combined ginseng and green tea decreases blood glucose levels in diabetic rats increases preservation of β-cells, perhaps by lowering oxidative stress.

Mustata et al. (2005) studied the paradoxical effects of green tea and antioxidant vitamins in diabetic rats and concluded that green tea and antioxidant vitamins improved several diabetes-related cellular dysfunctions but worsened matrix glycoxidation in selected tissues, suggesting that antioxidant treatment tilts the balance from oxidative to carbonyl stress in the extracellular compartments. Tsuneki et al. (2004) studied the effect of green tea on blood glucose levels and serum proteomic patterns in diabetic (db/db) mice and on glucose metabolism in healthy humans. This study provides evidence that green tea has an antidiabetic effect and they also found that the 4211 (4212) Da protein level that was decreased in the diabetic state was further decreased after green tea administration. This is the first report which demonstrated that a certain serum protein may be involved in the antihyperglycemic effects of green tea. Wolfram et al. (2006) investigated the antidiabetic effects of green tea catechin, epigallocatechin gallate (EGCG, TEAVIGO), rodent models of type 2 diabetes mellitus and H4IIE rat hepatoma cells. EGCG improved oral glucose tolerance and blood glucose in food-deprived rats in a dose-dependent manner. Plasma concentrations of triacylglycerol were reduced and glucose-stimulated insulin secretion was enhanced. This study showed that EGCG beneficially modifies glucose and lipid metabolism in H4IIE cells and markedly enhances glucose tolerance in diabetic rodents.

Waltner-Law et al. (2002) showed that the regulation of hepatic glucose production is decreased by EGCG. Furthermore, like insulin, EGCG increases tyrosine phosphorylation of the insulin receptor and insulin receptor substrate-1 (IRS-1), and it reduces phosphoenolpyruvate carboxykinase gene expression in a phosphoinositide-3-kinase-dependent manner. EGCG also mimics insulin by increasing phosphoinositide-3-kinase, mitogen-activated protein kinase, and p70 [s6k] activity. EGCG differs from insulin, however, in that it affects several insulin-activated kinases with slower kinetics. Furthermore, EGCG regulates genes that encode gluconeogenic enzymes and protein-tyrosine phosphorylation by modulating the redox state of the cell. These results demonstrated that changes in the redox state may have beneficial effects for the treatment of diabetes and suggest a potential role for EGCG, or derivatives, as an anti diabetic agent. Maruyama et al. (2009) conducted a cross-sectional study of 35 male volunteers, 23–63 years old and residing in Shizuoka Prefecture in Japan. Men who consumed a 3% concentration of green tea showed lower mean values of fasting blood glucose and fructosamine than those who consumed a 1% concentration. Fasting blood glucose levels were found to be significantly associated with green tea concentration ($\beta = -0.14$, $p = 0.03$). These findings suggested that the consumption of green tea at a high concentration has the potential to reduce blood glucose levels.

Roghani and Baluchnejadmojarad (2010) evaluated the effect of chronic administration of EGCG on serum glucose and lipid profile and hepatic lipid peroxidation in streptozotocin (STZ) induced diabetic rats. Results showed that treatment of diabetic rats with EGCG produced a hypoglycemic effect and there were appropriate changes regarding serum lipids in treated diabetic group. This study concluded that chronic treatment of diabetic rats with EGCG could prevent abnormal changes in blood glucose and lipid profile and attenuate hepatic lipid peroxidation. Panagiotakos et al. (2009) evaluated the link between long-term tea intake and prevalence of type 2 diabetes mellitus, in a sample of elderly adults. The findings suggested that long-term tea intake is associated with reduced levels of fasting blood glucose and lower prevalence of diabetes.

4.10   FOR TREATING OBESITY

Today, more than 1.1 billion adults worldwide are overweight, and 312 million of them are obese. In addition, at least 155 million children world-wide are overweight or obese, according to the International Obesity Task Force. In the past 20 years, the rates of obesity have tripled in developing countries that have been adopting a Western lifestyle involving decreased physical activity and overconsumption of cheap, energy-dense food. Such lifestyle changes are also affecting children in these countries; the preva-lence of overweight among them ranges from 10 to 25%, and the prevalence of obesity ranges from 2 to 10% (Hossain et al., 2007). Obesity is a medical condition in which excess body fat is accumulated to the extent that it may have an adverse affect on health, leading to reduced life expectancy. Obesity is associated with many diseases, particularly heart disease, type 2 diabetes, breathing difficulties during sleep, certain types of cancer, and osteoarthritis (Haslam and James, 2005).

Green tea catechins have been suggested to have anti obesity effects. Various research studies have shown promising results regarding green tea as an anti obesity agent. Maki et al. (2009) evaluated the influence of a green tea catechin beverage on body composition and fat distribution in over-weight and obese adults during exercise-induced weight loss. The findings of this study suggested that green tea catechin consumption enhances exer-cise-induced changes in abdominal fat and serum triglycerides. Boschmann and Thielecke (2007) conducted a pilot study on the effects of epigallocat-echin-3-gallate on thermogenesis and fat oxidation in obese men. The find-ings of this study suggested that EGCG alone has the potential to increase fat oxidation in men and may thereby contribute to the antiobesity effects of green tea. Rudelle et al. (2007) studied the effect of a thermogenic bever-age on 24-hour energy metabolism in humans. The objective of the study was to test whether consumption of a beverage containing active ingredients will increase 24-hour energy metabolism in healthy, young, lean individu-als. The study provided evidence that consumption of a beverage containing green tea catechins, caffeine, and calcium increases 24-hour energy expen-diture by 4.6%, but the contribution of the individual ingredients cannot be distinguished.

Green tea has three major components which can promote fat reduction. These components are catechins, caffeine, and theanine which can inhibit gastric and pancreatic lipase, enzymes that digest and stores fat in a form that's healthy for the human body. Drinking green teas after a meal can slow the rise of blood sugar and thus prevents fat storage within the body (Soft, 2010). Dulloo et al. (1999) investigated whether a green tea extract, by virtue of its high content of caffeine and catechin polyphenols, could increase 24-h energy expenditure (EE) and fat oxidation in humans. The results of this study showed that green tea has thermogenic properties and promotes fat oxidation beyond that explained by its caffeine content *per se*. The green tea extract mayplay a role in the control of body composition via sympathetic activation of thermogenesis, fat oxidation, or both. Katzman et al. (2007) reported that by self-administration of conjugated linoleic acid and green tea extract caused an unexpected decrease in total body fat mass, a decrease in body fat percentage and an increase in lean body mass of each patient. Dulloo et al. reported that a green tea extract stimulates brown adipose tissue thermogenesis to an extent which is much greater that can be attributed to its caffeine content *per se*, and that its thermogenic properties could reside primarily in an interaction between its high content in catechin-polyphenols and caffeine with sympathetically released noradrenaline (NA). Since catechin-polyphenols are known to be capable of inhibiting catechol-O-methyl-transferase (the enzyme that degrades NA), and caffeine to inhibit trancellular phosphodiesterases (enzymes that breakdown NA-induced cAMP), it is proposed that the green tea extract, via its catechin-polyphenols and caffeine, is effective in stimulating thermogenesis by relieving inhibition at different control points along the NA-cAMP axis (Dulloo et al., 2004).

Bose et al. (2008) investigated the effects of the major green tea polyphenol (–) epigallocatechin-3-gallate (EGCG), on high-fat–induced obesity, symptoms of the metabolic syndrome, and fatty liver in mice. The results indicated that long-term EGCG treatment attenuated the development of obesity, symptoms associated with the metabolic syndrome, and fatty liver. Short-term EGCG treatment appeared to reverse preexisting high-fat-induced metabolic pathologies in obese mice. These effects may be mediated by decreased lipid absorption, decreased inflammation, and other mechanisms (Dulloo et al., 2000). Ahmida and Abuzogaya (2009) conducted a study to investigate the effects of water extracts of green tea, ginger, or a combination of both on serum and hepatic total cholesterol (TC), low density lipoprotein-cholesterol (LDL-c), high density lipoprotein-cholesterol (HDL-c), triglycerides (TG), and total phospholipids in induced-hyperlipidemic Wistar albino rats. They found that as compared to positive control animals, the total body weight of groups whose diets were supplemented with the extracts was reduced. Likewise, a significant reduction in serum TC, LDL-c, TG, and total phospholipids was observed, accompanied by an increase in HDL-c levels. In the liver, a slight reduction in TC and TG was observed, though total phospholipid levels remained relatively similar. Importantly, no synergism was observed between the two extracts. Together, the data suggested that consumption of green tea or ginger could aid in the treatment of obesity and other diseases related to cardiovascular disease (CVD). Nagao et al. (2007) investigated the body fat reducing effect and reduction of risks for cardiovascular disease by a green tea extract (GTE) high in catechins in humans with typical lifestyles. The results showed that the continuous ingestion of a GTE high in catechins led to a reduction in body fat, systolic blood pressure (SBP), and LDL cholesterol, suggesting that the ingestion of such an extract contributes to a decrease in obesity and cardiovascular disease risks.

Zheng et al. (2004) investigated the antiobesity effects of three major components of green tea, catechins, caffeine and theanine, in mice. The body weight increase and weight of intraperitoneal adipose tissues (IPAT) were significantly reduced by the diets containing green tea, caffeine, theanine, caffeine and catechins, caffeine and theanine and caffeine along with catechins and theanine. These results indicated that at least caffeine and theanine were responsible for the suppressive effects of green tea powder (GTP) on body weight increase and fatty accumulation. Moreover, it was shown that catechins and caffeine were synergistic in antiobesity activities.

4.11   ANTI-OXIDANT ACTIVITY

Free radicals are chemical species possessing an unpaired electron that can be considered as fragments of molecules and which are generally very reactive. Free radicals are the natural byproducts of many biological processes within and among the cells. They are also produced by exposure to various environmental factors, tobacco, smoke and radiation. Free radicals attack healthy cells, spilling cytoplasm and subjecting the cells to infection, genetic damage and mutations. Antioxidants are known to scavenge these free radicals thereby protecting the body against the deleterious effects of harmful diseases (Young and Woodside, 2001) An antioxidant can be defined as any substance which when present at low concentrations compared to those of an oxidizable substrate, significantly delays or prevents oxidation of that substrate. Anti oxidants are also known as free radicals scavengers (Zapora et al., 2009).

Green tea is considered a dietary source of antioxidant nutrients. It is rich in polyphenols (catechins and gallic acid, particularly), and it also contains carotenoids, tocopherols, ascorbic acid (vitamin C), minerals such as Cr, Mn, Se or Zn, and certain phytochemical compounds. These scientific compounds could increase significantly the GTP antioxidant potential (Cabrera et al., 2006). Several studies have proved the antioxidant potential of green tea polyphenols. Flavanol absorption was enhanced when tea polyphenols were administered as a green tea supplement in capsule form and led to a small but significant increase in plasma antioxidant activity (Henning et al., 2004). Zapora et al. (2009) studied effect of green tea on antioxidant status of erythrocytes and on hematological parameters in rats and found that green tea, due to its antioxidant properties, partially protected erythrocytes against ethanol, acetaldehyde, and tertiary-butylhydroperoxide (*t*BOOH) action (Young and Woodside, 2009). Chrostek et al. (2005) conducted a study to investigate the effect of green tea consumption during chronic ethanol intake on the activity of aldehyde dehydrogenase in the liver of rats during maturation and aging.

Khalaf et al. (2008) screened the methanolic crude extracts of some commonly used medicinal plants for their free radical scavenging properties using ascorbic acid as standard antioxidant. The results showed that the overall antioxidant activity of green tea was the strongest. Yamamoto et al. (2003) examined the oxidative status of normal epithelial, normal salivary gland, and oral carcinoma cells treated with EGCG, using ROS measurement and catalase and superoxide dismutase activity assays. The results demonstrated that high concentrations of EGCG induced oxidative stress only in tumor cells. In contrast, EGCG reduced ROS in normal cells to background levels. Mustata et al. (2005) tested the hypothesis that green tea prevents diabetes- related tissue dysfunctions attributable to oxidation and concluded that green tea and antioxidant vitamins improved several diabetes-related cellular dysfunctions but worsened matrix glycoxidation in selected tissues, suggesting that antioxidant treatment tilts the balance from oxidative to carbonyl stress in the extracellular compartment. Maeta et al. (2007) demonstrated that both EGCG and green tea extract (GTE) cause oxidative stress-related responses in the budding yeast *Saccharomyces cerevisiae* and the fission yeast *Schizosaccharomyces pombe* under weak alkaline conditions in terms of the activation of oxidative-stress-responsive transcription factors. They concluded that tea polyphenols are able to act as prooxidants to cause a response to oxidative stress in yeasts under certain conditions. Erba et al. (1999) evaluated the effect of the supplementation of the Jurkat T-cell line with green tea extract on oxidative damage and found that green tea has protective effect against oxidative damage. Bhimani et al. (1993) showed in their study that naturally occurring compounds like EGCG, along with some other chemopreventive agents, are capable of preventing the TPA-mediated oxidative events in HeLa cells.

Xu et al. (2004) examined the antioxidant activity and bioavailability of tea epicatechins compared with their corresponding epimers. The antioxidant activity of each epimer with its corresponding GTE precursor was conducted in the three in vitro systems, namely human LDL oxidation, ferric reducing–antioxidant power (FRAP), andanti-2,2-diphenyl-1-picrylhydrazyl (DPPH) free radical assays. It was concluded that the epimerization reaction occurring in manufacturing canned and bottled tea drinks would not significantly affect antioxidant activity and bioavailability of total tea polyphenols. Chyu et al. (2004) evaluated the effect of epigallocatechin gallate (EGCG), the main antioxidant derived from green tea, on evolving and established atherosclerotic lesions in hypercholesterolemic apolipoprotein E-null mice. The result suggested that antioxidant activity of EGCG differentially reduces evolving atherosclerotic lesions without influencing established atherosclerosis in the apolipoprotein E-null mice. Atoui et al. (2004) studied tea and herbal infusions for their polyphenolic content, antioxidant activity and phenolic profile. The antioxidant activity was evaluated by two methods, DPPH and chemiluminescence assays, using Trolox and quercetin as standards. The EC_{50} of Chinese green tea was found to be 0.151 ± 0.002 mg extract/mg DPPH (0.38 quercetin equivalents and 0.57 Trolox equivalents).

El-Shahat et al. (2009) investigated the protective effect of GTE against testes damage induced by Cadmium in experimental rats and concluded that the rats received GTE + Cd could enhance antioxidant/detoxification system which consequently reduced the oxidative stress in rat testes. Chen et al. (2002) elucidated, effects of the antioxidant (–)-epigallocatechin-3-gallate (EGCG), a major (and the most active) component of green tea extracts, on cultured HSC growth and activation. The results indicated that EGCG was a novel and effective inhibitor for activated HSC growth and activation in vitro. Leung et al. (2001) compared the antioxidant activities of individual theaflavins (TF) with that of each catechin using human LDL oxidation as a model. The results demonstrated that theaflavins in black tea and cate-chins in green tea are equally effective antioxidants. Mildner-Szkudlarz et al. (2009) investigated the effect of green tea extract (GTE) on biscuits lipid fraction oxidative stability. The antioxidant activity of GTE was compared with commonly used synthetic antioxidant butylated hydroxyanisole (BHA). Phenolic compounds of GTE characterized powerful antioxidant activities evaluated using free radical, 2,2-diphenyl-1-picrylhydrazyl method, com-pared with gallic acid and was found to be significantly better than BHA.

Ostrowska and Skrzydlewska (2006) examined the influence of green tea extract, epicatechin (EC), epicatechin galate (ECG) as well as epigallo-catechin galate (EGCG) on oxidative modifications of LDL of human blood serum. The results revealed that peroxidation of LDL is markedly prevented by green tea extract and in a slightly weaker way by catechins (EGCG in particular), which is manifested by a decrease in concentration of conjugated dienes, lipid hydroperoxides, MDA, dityrosine and by an increase in trypto-phan content. Both green tea as well as catechins (EGCG in particular) has also been revealed to prevent decrease in concentration of α-tocopherol in oxidating conditions. Abdel-Raheem et al. (2010) investigated the possible protective effects of GTE against gentamicin-induced nephrotoxicity in rats. The results showed that the simultaneous administration of GTE plus gen-tamicin protected kidney tissues against nephrotoxic effect of gentamicin as evidenced from amelioration of histopathological alterations and normal-ization of kidney biochemical parameters. Yamanaka et al. (1997) studied effects of (–)-epicatechin (EC) and (–)-epigallocatechin (EGC) on Cu^{2+}-induced low density lipoprotein (LDL) oxidation in initiation and propaga-tion phases. The results indicated that catechins such as EC and EGC can act as free radical terminators (reducing agents) or accelerators (oxidizing agents) under oxidation circumstances, which is a different character from NO. Hamden et al. (2009) evaluated the antioxidant effect of green tea on cadmium-induced hepatic dysfunction and stress oxidant in rats. The study showed that green tea extract significantly increased ($p < 0.05$) the enzy-matic antioxidants activities (SOD, catalase, GPX) in of rats liver compared with those given cadmium alone.

Abbas and Wink investigated the in vivo and in vitro antioxidant proper-ties of an aqueous extract of green tea (GTE). 2,2-Diphenyl-1-picrylhydrazyl (DPPH) and superoxide anion radical (O_2) assays were used to estimate the GTE antioxidant activity. To investigate the protective effects of GTE against oxidative stress, wild-type N2 and transgenic strains (TJ374, hsp-16.2/GFP) of the model organism, *Caenorhabditis elegans* (C. elegans), were chosen. In the current study, the following catechins were identified by LC/ESI-MS: catechin, epicatechin, epicatechin gallate, gallocatechin, epigallocatechin and epigallocatechin gallate. GTE exhibited a free radical scavenging activity of DPPH and O_2 with IC50 8.37 and 91.34 µg/mL, respectively. In the *C. elegans* strain (TJ374, hsp-16.2/GFP), the expression of hsp-16.2/GFP was induced by a nonlethal dose of juglone, and the fluorescence density of hsp-16.2/GFP was measured. The hsp-16.2/GFP was reduced by 68.43% in the worms pretreated with 100 µg/mL GTE. N2 worms pretreated with 100 µg/mL GTE exhibited an increased survival rate of 48.31% after a lethal dose application of juglone. The results suggest that some green tea constituents are absorbed by the worms and play a substantial role to enhance oxidative stress resistance in *C. elegans* (Abbas and Wink, 2014)

4.12 ANTIMICROBIAL ACTIVITY

Microbes are tiny organisms that cannot be seen without a microscope and include viruses, fungi, and some parasites as well as bacteria. Infectious diseases are one of the leading causes of death worldwide. During the past few decades, new infectious diseases have appeared and old ones previously thought to be controlled have reemerged. The treatment of infectious diseases still remains an important and challenging problem because of a combination of factors including emerging infectious diseases and the increasing number of multidrug resistant microbial pathogens (Chawla et al., 2010). During the past 25 years, antimicrobial agents have been introduced at a rate exceeding our ability to integrate them into clinical practice (Sharma et al., 2010). In spite of a large number of antibiotics and chemotherapeutics available for medical use, the emergence of old and new antibiotic resistance developed in the last decades, has created a substantial medical need for new classes of antibacterial agents (Kharab et al., 2010)

Drugs from natural origin are being used by about 80% of the world population primarily in the developing countries due to their safety, efficacy, cultural acceptability, lesser side effects, and most particularly their cost effectiveness and easy accessibility. There is an increased use of herbal medicines in western countries also since synthetic agents can exert more unwanted side effects when used too often indiscriminately and irrationally (Sharma et al., 2006).

The antimicrobial activity of tea, suggested for many years by unreliable evidence, was first demonstrated almost 100 years ago in the laboratory by McNaught (1906), a major in the British Army Medical Corps. He showed that brewed black tea killed *Salmonella typhi* and *Brucella melitensis* and recommended that the water bottles of troops be filled with tea in order to prevent outbreaks of infections due to these agents (Taylor et al., 2005). A series of well-conducted, systematic studies, mainly from Japan, suggests that tea extracts show several useful antimicrobial effects (Hamilton-Miller, 1995). Sakanaka et al. (2000) examined the inhibitory action of tea poly-phenols towards the development and growth of bacterial spores. The main component of tea polyphenols (–)-epigallocatechin gallate, showed strong activity against both*B. stearothermophilus* and *C. thermoaceticum*. Extracts of *C. sinensis* can reverse methicillin in methicillin-resistant *Staphylococcus aureus* (MRSA) and also, to some extent, penicillin resistance in β-lactamase producing *S. aureus* (Yam et al., 1998). Chakraborty and Chakraborty (2010) studied the antibacterial activity of methanolic extract of green tea leaves (*C. sinensis*) on four different bacteria namely *Escherichia coli, Bacillus cereus, Pseudomonas aeruginosa* and *Staphylococcus aureus* and two fungi of *Aspergillus* species and it was found that the alcoholic extract of the leaves of *C. sinensis* was most effective against *Bacillus cereus*. *Ruggiero* et al. investigated whether red wine and green tea could exert anti-*H. pylori* or anti-VacA activity in vivo in a mouse model of experimental infection. It was concluded from this investigation that Red wine and green tea are able to prevent *H. pylori* -induced gastric epithelium damage, possibly involving VacA inhibition (Ruggiero et al., 2007). *Zhang and Rock* focused on vali-dating the inhibition of the bacterial type II fatty acid synthesis system as a mechanism for the antibacterial effects of EGCG and related plant polyphe-nols. It was found that EGCG and most of the related polyphenols inhibited both reductase steps, possessed antibacterial activity, and inhibited cellular fatty acid synthesis (Zhang and Rock, 2004).

Yee and Koo (2000) conducted an in vitro study on anti-*H. pylori* activ-ity of Chinese tea (Lung Chen tea), and two tea catechins, epigallocatechin gallate and epicatechin and their minimum inhibitory concentrations (MICs) were determined. Results showed that Lung Chen, epigallocatechin gal-late and epicatechin, all inhibited the growth of *H. pylori*. It was found that EGCG, which is also an active ingredient of green tea, is probably the active ingredient responsible for most of the anti *H. pylori* activity of Chinese tea. Soekanto and Mangundjaja (2004) determined the sensitivity of polyphenols of green tea on mutans Streptococci and concluded that green tea leaves polyphenol shows antimicrobial activity against local strains of mutans of *Streptococcus mutans*, isolated from humans harboring species, in vitro. Juneja et al. (2004) investigated the inhibition of *Clostridium perjringens* spore germination and outgrowth by two green tea extracts with low and high catechin levels during abusive chilling of retail cooked ground beef, chicken, and pork. The results of this study suggested that widely consumed catechins from green tea can reduce the potential risk of *C. perjringens* spore germination and outgrowth during abusive cooling from 54.4 to 7.2°C in different time period of cooling for ground beef, chicken, and pork. Becker et al. (2009) tested green tea at a higher concentration against *Streptococcus mutans*, a common cariogenic bacterium. Results showed that green tea, at a concentration of 40 mg/mL, brewed at 90°C at 5, 20, and 40 min, was determined to be moderately effective against *S. mutans*. Hirasawa and Takada (2004) evaluated the susceptibility of *Candida albicans* to catechin under varying pH conditions and the synergism of the combination of catechin and antimycotics. The results indicated that EGCG enhances the antifungal effects of amphotericin B or fluconazole against antimycotic-susceptible and -resistant *C. albicans*. Combined treatment with catechin allows the use of lower doses of antimycotics and induces multiple antifungal effects (Hirasawa et al., 2002).

Navarro-Martínez et al. (2005) studied the antifolate activity of epigallocatechin gallate against *Stenotrophomonas maltophilia* andshowed that epigallocatechin gallate is an efficient inhibitor of *S. maltophilia* dihydrofolate reductase, a strategic enzyme that is considered an attractive target for the development of antibacterial agents. Hara-Kudo et al. (2005) studied antibacterial effects of catechins, the major green tea polyphenols, using *Clostridium* and *Bacillus* spores. The results showed that low concentrations of catechins, although requiring a long exposure time, inhibited the growth of bacterial spores. However, the effects of the purified derivatives of the catechins were not the same and GCG and EGCG were found to be the most potent. Mughal et al. (2009) performed a comparative study on antibacterial activity of green tea and lemon grass extracts against *Streptococcus mutans* and its synergism with antibiotics. Synergistic activity of tea samples with antibiotics (Chloroamphenicol, Tetracycline, Levofolxacin and Gentamycin) showed best response against the bacteria. West et al. (2001) investigated the effect of green tea on the growth and morphology of methicillin-resistant and methicillin-susceptible *Staphylococcus aureus*. The results showed that green tea inhibits the growth of both MRSA and MSA. Ando et al. (1999) studied the antibacterial activity of epigallocatechin gallate against *Staphylococci*. Results of this study showed that EGCG appeared to be effective against not only MRSA strains, but also against MSSA, MR-CNS, and MS-CNS. Mbata et al. (2008) investigated the antibacterial activity of the methanol and aqueous extract of *C. sinensis* on *Listeria Monocytogenes*. The results obtained showed that methanol and water extract exhibited anti-bacterial activities against *L.monocytogenes*. Epigallocatechin gallate is also found to reduce halotolerence in *Staphylococcus aureus* (Stapleton et al., 2006). Matsubara et al. (2003) investigated the suppression of helicobacter pylori-induced gastritis by green tea extract in Mongolian gerbils. Results showed that green tea extract (GTE) showed the strongest inhibition of *H. pylori* urease, with an IC_{50} value of 13 μg/mL. Active principles were identi-fied to be catechins, the hydroxyl group of 5'-position appearing important for urease inhibition.

Kim et al. (2008) determined the antimicrobial and antifungal effects of a GTE on the vaginal pathogens, *Proteus mirabilis*, *Streptococcus pyogenes*, and *Candida albicans*. Results of this study showed that the growth inhibi-tory effects of a GTE and the ethyl acetate fraction against *P. mirabilis* and *S. pyogenes* were stronger as compared to the antimicrobial activity of the H_2O fraction. Among the catechins, epigallocatechin gallate had the stron-gest antimicrobial activity. Imanishi et al. (2002) investigated whether GTE exerts an additional inhibitory effect on the acidification of intracellular compartments such as endosomes and lysosomes (ELS) and thereby inhib-its the growth of influenza A and B viruses in Madin-Darby canine kidney cells. The study showed that GTE inhibited acidification of ELS in a dose dependent manner and the growth of influenza A and B viruses was equally inhibited. EGCG binds to the CD4 molecule at the gp120 attachment site and inhibits gp120 binding at physiologically relevant levels, thus acts as a potential therapeutic treatment for HIV-1 infection (Williamson et al., 2006). Hauber et al. (2009) demonstrated that epigallocatechin-3-gallate (EGCG), the major active constituent of green tea, targets the fibrillar structures, termed semen-derived enhancer of virus infection (SEVI) for degradation. It was also shown that EGCG inhibits SEVI activity and abrogates semen-mediated enhancement of HIV-1 infection in the absence of cellular toxicity. Song et al. (2005) evaluated polyphenolic compound catechins from green tea for their ability to inhibit influenza virus replication in cell culture and for potentially direct virucidal effects. The EGCG and ECG were found to be potent inhibitors of influenza virus replication in MDCK cell culture and this effect was observed in all influenza virus subtypes tested, including A/H1N1, A/H3N2 and B virus. This study suggested that the antiviral effect of catechins on influenza virus is mediated not only by specific interaction with HA, but altering the physical properties of viral membrane.

Oral candidiasis is an infection caused by commensal fungi of Candida species in the oral cavity that serves as an opportunistic pathogen with *Candida albicans* (*C. albicans*) as the most frequent (80%) etiology. The prevalence of oral candidiasis tends to increase due to the increasing population of immuno-compromised patients and resistance to antifungal. This requires alternative treatment to enhance the efficacy of anti fungal medications such as green tea (*C. sinensis*) with (–)-epigallocatechin-3-gallate (EGCG) as the main polyphenol component and most potent (59–65%). The effect of EGCG green tea in obliterating oral candidiasis through neutrophil count and infected cells by *C. albicans* is not clearly determined yet up to now. *Objective*: To analyze the effect of EGCG green tea in a murine model of oral candidiasis through neutrophil count and infected cells by *C. albicans* by Prasetyo et al. (2015). True laboratory experimental study with randomized post-test only control group design, using Wistar male rats. The rats were grouped into 1 control and 3 intervention groups. EGCG was given with the dosage of 0, 1, 2 and 4 mg/kgBW/day as the intervention. Result: immunohistochemistry and Hematoxyllineos in stain showed that EGCG green tea increased neutrophil count and decreased infected cells by *C. albicans*.

4.13   ANTI-HYPERTENSIVE AND CARDIOVASCULAR ACTIVITY

Hypertension is likely the most common disease on Earth. It is associated with an increased risk of morbidity and mortality from cardiovascular disease (CVD) and represents the single greatest preventable cause of death in humans. The standard definition of hypertension as blood pressure (BP) $\geq 140/90$ mm Hg is based on the observation that the risk of CVD increases sharply above this level (Grotto et al., 2006). According to WHO, hypertension has been classified as given in Table 4.1 (World health organization, 2005).

The estimated total number of adults with hypertension in 2000 was 972 million. Of these, 333 million were estimated to be in economically

TABLE 4.1 Classification of Hypertension

Category	Systolic BP (mmHg)	Diastolic BP (mmHg)
Optimal	<120	<80
Normal	120–129	80–84
High normal	130–139	85–89
Grade 1 (mild hypertension)	140–159	90–99
Grade 2 (moderate hypertension)	160–179	100–109
Grade 3 (severe hypertension)	≥180	≥110
Isolated systolic hypertension	≥140	<90

developed countries and 639 million in economically developing countries. By 2025, the number of people with hypertension is expected to increase by about 60% to a total of 1.56 billion as the proportion of elderly people will increase significantly. Other reasons are the continuing population increase and changes in lifestyle, which includes a diet rich in sugar and high-fat processed foods and sedentary behavior, mediated by televisions, computers and cars (Kearney et al., 2005).

Epidemiologic studies suggested that tea polyphenols that can be derived from black and green tea may protect against cardiovascular diseases. Therefore, the physiologic effects of tea and its components on cardiovascular disease risk factors such as hypertension are of interest. Yang et al. (2004) investigated the protective effect of habitual tea consumption on hypertension and found that habitual moderate strength green or oolong tea consumption, 120 mL/d or more for 1 year, significantly reduces the risk of developing hypertension in the Chinese population. Negishi et al. (2004) determined the effect of black and green tea on lowering of blood pressure in stroke-prone spontaneously hypertensive rats (SHRSP). The data demonstrate that both black and green tea polyphenols attenuate blood pressure increases through their antioxidant properties in SHRSP. Henry and Stephens-Larson (1984) investigated the reduction of chronic psychosocial hypertension in mice by decaffeinated green tea and the results showed reduction in blood pressure in mice when given decaffeinated green tea. Green tea is also very beneficial in cardiovascular diseases and various studies support this fact. Potenza et al. (2007) investigated effects of EGCG treatment to simultaneously improve cardiovascular and metabolic function in spontaneously hypertensive rats. This study concluded that acute actions of EGCG to stimulate production of nitric oxide from endothelium using PI 3-kinase-dependent pathways may explain, in part, beneficial effects of EGCG therapy to simultaneously improve metabolic and cardiovascular pathophysiology in SHR. Hao et al. (2007) investigated whether EGCG supplementation could reduce in vivo pressure overload mediated cardiac hypertrophy and found that EGCG prevents the development of left ventricular concentric hypertrophy by pressure overload and may be a useful therapeutic modality to prevent cardiac remodeling in patients with pressure overload myocardial diseases. Chen et al. (2000) conducted a study to investigate the vasorelaxant and antiproliferative responses to purified green tea epicatechin mixture (–)-epicatechin and (–) epigallocatechin gallate on rat arterial smooth muscle cells and concluded that the purified epicatechin derivatives from jasmine green tea relaxed the isolated rat arteries preconstricted by phenylephrine and inhibited aortic smooth muscle cell proliferation.

Kuriyama et al. (2006) conducted a population-based prospective cohort study (the Ohsaki Study) to examine the association between green tea consumption and mortality from cardiovascular disease (CVD), cancer, and all causes with 40,530 persons in Miyagi prefecture, in northern Japan. This study found an inverse relationship between green tea consumption and mortality due to cardiovascular diseases. Papparella et al. (2008) investigated the effects of green tea on Angiotensin II-induced cardiac hypertrophy in rats by modulating reactive oxygen species production and the Src/epidermal growth factor receptor/Akt signaling pathway. The results showed that GTE blunted Angiotensin II-induced blood pressure increase and cardiac hypertrophy by regulating ROS production and the Src/EGFR/Akt signaling pathway activated by Angiotensin II. Lorenz et al. (2004) investigated the effects of epigallocatechin-3-gallate (EGCG), the major constituent of green tea, on vasorelaxation and on eNOS expression and activity in endothelial cells. Results indicated that EGCG-induced endothelium-dependent vasodilation is primarily based on rapid activation of eNOS by a phosphatidylinositol 3-kinase-, PKA-, and Akt-dependent increase in eNOS activity, independently of an altered eNOS protein content. Hofmann and Sonenshein (2003) examined the effects of GTPs on aortic smooth muscle cell (SMC) proliferation and observed that EGCG inhibits growth and induces death of SMCs in a p53- and NF-κB-dependent manner. Sato et al. (1989) investigated the possible effects of green tea on prevention of stroke and found that the proportion of persons with stroke history decreased as a function of increasing intake of green tea from the highest (over 2%) among those who had no habit of drinking green tea to the lowest (0.4%) among those who took 3–4 cups or more of the tea daily.

4.14   MISCELLANEOUS ACTIVITIES

Green tea have beneficial effects in preventing oral disease including dental caries, periodontal disease, and tooth loss which can significantly impact a person's overall health. Experimental studies have shown that green tea can help in improving oral health. Tea leaves can be used as a convenient, slow-release source of catechins and theaflavins and can be used in the prevention of oral cancer and dental caries (Lee et al., 2004). Hirasawa et al. (2002) determined the usefulness of green tea catechin for the improvement of periodontal disease. Green tea catechin showed a bactericidal effect against BPR and the combined use of mechanical treatment and the application of green tea catechin using a slow release local delivery system was effective in improving periodontal status (Hirasawa and Takada, 2004). Green tea consumption is also known to have neuroprotective effects which can be beneficial in diseases like Parkinson's disease and Alzheimer's disease. Levites et al. (2003) showed that EGCG enhances (~6-fold) the release of the nonamyloidogenic soluble form of the amyloid precursor protein (sAPPα) into the conditioned media of human SHSY5Y neuroblastoma and rat pheochromocytoma PC12 cells. EGCG markedly increased PKCα and PKCε in the membrane and the cytosolic fractions of mice hippocampus. Thus, EGCG has protective effects against Aβ-induced neurotoxicity and regulates secretory processing of nonamyloidogenic APP via PKC pathway. EGCG reduces Aβgeneration in both murine neuron-like cells (N2a) transfected with the human "Swedish" mutant amyloid precursor protein (APP) and in primary neurons derived from Swedish mutant APP-overexpressing mice (Tg APPsw line 2576) (Rezai-Zadeh et al., 2005). (−)-Epigallocatechin-3-gallate (EGCG) dramatically suppresses experimental autoimmune encephalomyelitis (EAE) induced by proteolipid protein 139–151 (Aktas et al., 2004). Sutherland et al. (2005) demonstrated that the neuroprotective effects of EGCG are, in part, due to modulation of NOS isoforms and preservation of mitochondrial complex activity and integrity and concluded that the in vivo neuroprotective effects of EGCG are not exclusively due to its antioxidant effects but involve more complex signal transduction mechanisms.

Green tea also protects eyes from various infections. Chan et al. (2010) investigated the inhibitory effects of EGCG on retinal pigment epithelial (RPE) cell migration induced by PDGF-BB, an isoform of platelet-derived growth factor (PDGF), and adhesion by fibronectin and found that EGCG can inhibit PDGF-BB-induced human RPE cell migration and, in a dose-dependent manner, RPE cell adhesion to fibronectin. This study provided the first evidence that EGCG is an effective inhibitor of RPE cell migration and adhesion to fibronectin and, therefore, may prevent epiretinal membrane formation. EGCG also protects human lens epithelial (HLE) cells from the mitochondria-mediated apoptosis induced by H_2O_2 through the modulation of caspases, the Bcl-2 family, and the mitogen activated protein kinase (MAPK) and Akt pathways (Yao et al., 2008). EGCG, due to its antioxidant activity, protects the lens by significantly reducing the inactivation of catalase due to UVA exposure (Zigman et al., 1999). Chan et al. (2008) studied the protective effects of (–)-epigallocatechin gallate on UVA-induced damage in ARPE19 cells and demonstrate that EGCG inhibits UVA-induced H_2O_2 production, mitogen activating protein kinase activation, and expression of COX-2. *Blanco* et al. investigated the effects of subMICs of EGCG on biofilm formation by 20 different ocular staphylococcal isolates and suggested that EGCG interferes with the polysaccharides that form the glycocalyx, disrupting their interactions either reciprocally or with the cell wall and thus reducing the amount of slime that accumulates (Blanco et al., 2005). Green tea (–)-epigallocatechin-gallate modulates early events in huntingtin misfolding and reduces toxicity in Huntington's disease models. EGCG potently inhibits the aggregation of mutant huntingtin (mhtt) exon 1 protein in a dose-dependent manner (Ehrnhoefer et al., 2006).

4.15   CONCLUSION

Green tea has been used for its beneficial effects since ancient times. Commendable research on green tea is going on around the world to find its applications in various areas of health care. Various systematic studies carried out on green tea have depicted promising results. Green tea polyphenols, especially EGCG, possesses various activities such as anticancer, antioxidant, antidiabetic, antihypertensive, antimicrobial etc. Experimental studies have emphasized that green tea consumption is helpful in weight loss, cardiovascular diseases, protection from harmful effects of UV radiation, and improving oral health etc. Recently, green tea has been reported to possess neuroprotective properties which are helpful in treatment of neurodegenerative diseases like Parkinson's disease and Alzheimer's disease. Although a number of preclinical studies have demonstrated the promising effects of green tea in various diseases but the human clinical data is still limited. In order to use green tea compounds in treatment of diseases, significant clinical data regarding their bioavailability, safe dose, mechanism of action in various diseases need to be explored. Hence, a lot more attention should be given towards carrying out the in-depth clinical studies of green tea components. Since beneficial health effects of green tea are being increasingly proved, it could be advisable to encourage the regular consumption of this widely available, tasty and inexpensive beverage as an interesting alternative to other drinks.

KEYWORDS

- **Green Tea**
- **EGCG**
- **polyphenols**
- **catechins**

REFERENCES

Abbas, S., & Wink, M. (2015). *Antioxidants. 3*, 129.

Abdel-Raheem, I. T., El-Sherbiny, G. A., & Taye, A. (2010). *Pak. J. Pharm. Sci. 23*, 21.

Ahmida, M. H., & Abuzogaya, M. H. (2009). *J. App. Sci. Res. 5*, 1709.

Aktas, O., Prozorovski, T., Smorodchenko, A., Savaskan, N. E., Lauster, R., Kloetzel, P-M., Infante-Duarte, C., Brocke, S., & Zipp, F. (2004). *J. Immunol. 173*, 5794.

Ando, C., Kono, K., Tarara, I., Takeda, S., Arakawa, K., & Hara, Y. (1999). *Med. Bull. Fukuoka. Univ. 26*, 195.

Atoui, A. K., Mansouri, A., Boskou, G., & Kefalas, P. (2004). *Food Chem. 89*, 27.

Barthelman, M., Bair, W. B., Stickland, K. K., Chen, W., Timmermann, B. N., Valcic, S., Dong, Z., & Bowden, G. T. (1998). *Carcinogenesis. 19*, 2201.

Becker, R., Hirsh, S., Hu, E., Jamil, A., Mathew, S., Newcomb, K., Shaikh, S., Sharon, D., Triantafillou, V., & Yeong, V. (2009). Available from: http://depts.drew.edu/govschl/NJGSS2009/Journal/TeamPapers/team3.pdf, Accessed 2010 May 20.

Bhimani, R. S., Troll, W., Grunberger, D., & Frenkel, K. (1993). *Cancer Res. 53*, 4528.

Blanco, A. R., Sudano-Roccaro, A., Spoto, G. C., Nostro, A., & Rusciano, D. (2005). *Antimicrob. Agents. Chemother. 49*, 4339.

Boschmann, M., & Thielecke, F. (2007). *J. Am. Coll. Nutr. 26*, 389S.

Bose, M., Lambert, J. D., Ju, J., Reuhl, K. R., Shapses, S. A., & Yang, C. S. (2008). *J. Nutr. 138*, 1677.

Cabrera, C., Artacho, R., & Giménez, R. (2006). *J. Am. Coll. Nutr. 25*, 79.

Cai, Y., Anavy, N. D., & Chow, H-H. S. (2002). *Drug Metab. Dispos. 30*, 1246.

Carlson, J. R., Bauer, B. A., Vincent, A., Limburg, P. J., & Wilson, T. (2007). *Mayo. Clin. Proc. 82*, 725.

Cefalu, W. T. (2006). *ILAR Journal. 47*, 186.

Chacko, S. M., Thambi, P. T., Kuttan, R., & Nishigaki, I. (2010). *Chinese Medicine. 5*, 13.

Chakraborty, D., & Chakraborty, S. (2010). *Res. J. Phytochem. 4*, 78.

Chan, C-M., Huang, J-H., Chiang, H-S., Wu, W-B., Lin, H-H., Hong, J-Y., & Hung, C-F. (2010). *Mol. Vis. 16*, 586.

Chan, C-M., Huang, J-H., Lin, H-H., Chiang, H-S., Chen, B-H., Hong, J-Y., & Hung, C-F. (2008). *Mol. Vis. 14*, 2528.

Chapagain, A. K., & Hoekstra, A. Y. (2007). 64,109.

Chawla, R., Sahoo, U., Arora, A., Sharma, P. C., & Radhakrishnan, V. (2010). *Acta. Pol. Pharm. 67*, 55.

Chen, A., Zhang, L., Xu, J., & Tang, J. (2002). *Biochem. J. 368*, 695.

Chen, L., Lee, M-J., Li, H., & Yang, C. S. (1997). *Drug Metab. Dispos. 25*, 1045.

Chen, Z-Y., Law, W-I., Yao, X-Q., Lau, C-W., Ho, W. K. K., & Huang, Y. (2000). *Acta. Pharmacol. Sin. 9*, 835.

Cheng, T. O. (2006). *Int. J. Cardio. 108*, 301.

Chouhan, V. L., & Vyas, S. (2006). *J Indian Acad App Psych. 32*, 106.

Chow, H-H. S., Cai, Y., Hakim, I. A., Crowell, J. A., Shahi, F., Brooks, C. A., Dorr, R. T., Hara, Y., & Alberts, D. S. (2003). *Clin. Cancer Res. 9*, 3312.

Chow, H-HS., Hakim, I. A., Vining, D. R., Crowell, J. A., Ranger-Moore, J., Chew, W. M., Celaya, C. A., Rodney, R. S., Hara, Y., & Alberts, D. S. (2005). *Clin. Cancer Res. 11*, 4627.

Chrostek, L., Tomaszewski, W., & Szmitkowski, M. (2005). *Roczniki Akademii Medycznej w Białymstoku. 50*, 220.

Chyu, K-Y., Babbidge, S. M., Zhao, X., Dandillaya, R., Rietveld, A. G., Yano, J., Dimayuga, P., Cercek, B., & Shah, P. K. (2004). *Circulation. 109*, 2448.

Daisuke, U., Satomi, Y., Koji, Y., & Hirofumi, T. (2008). *J. Biol. Chem. 283*, 3050.

Di Pierro, F., Menghi, A. M. B., Barreca, A., Lucarelli, M., & Calandrelli, A. (2008). *Integr. Nutr. 11*, 1.

Dulloo, A. G., Duret, C., Rohrer, D., Girardier, L., Mensi, N., Fathi, M., Chantre, P., & Vandermander, J. (1999). *Am. J. Clin. Nutr. 70*, 1040.

Dulloo, A. G., Seydoux, J., & Jacquet, J. (2004). Adaptive thermogenesis and uncoupling proteins: a reappraisal of their roles in fat metabolism and energy balance. *Physiol. Behav., 83*(4), 587-602.

Dulloo, A. G., Seydoux, J., Girardier, L., Chantre, P., & Vandermander, J. (2000). *Int. J. Obes. 24*, 252.

Ehrnhoefer, D. E., Duennwald, M., Markovic, P., Wacker, J. L., Engemann, S., Roark, M., Legleiter, J., Marsh, J. L., Thompson, L. M., Lindquist, S., Muchowski, P. J., & Wanker, E. E. (2006). *Hum. Mol. Genet. 15*, 2743.

El-Shahat, A. E., Gabr, A., Meki, A-R., & Mehana, E-S. (2009). *Int. J. Morphol. 27*, 757.

El-Sherry, M. I., Zaher, M. A., Youssef, MSE-DM., & El-Amir, Y. O. (2007). *Cancer Ther. 5*, 301.

Erba, D., Riso, P., Colombo, A., & Testolin, G. (1999). *J. Nutr. 129*, 2130.

Frei, B., & Higdon, J. V. (2003). *J. Nutr. 133*, 3275S.

Grotto, I., Grossman, E., Huerta, M., & Sharabi, Y. (2006). *Hypertension. 48*, 254.

Hamden, K., Carreau, S., Ellouz, F., Masmoudi, H., & Feki, A. E. (2009). *Afr. J. Biotechnol. 8*, 4233.

Hamilton-Miller, J. M. T. (1995). Antimicrob. *Agents Chemother. 39*, 2375.

Hao, J., Kim, C-H., Ha, T-S., & Ahn, H-Y. (2007). *J. Vet. Sci. 8*, 121.

Hara-Kudo, Y., Yamasaki, A., Sasaki, M., Okubo, T., Minai, Y., Haga, M., Kondo, K., & Sugita-Konishi, Y. (2005). *J. Sci. Food Agri. 85*, 2354.

Haslam, D. W., & James, W. P. (2005). *Lancet. 366*, 1197.

Hauber, I., Hohenberg, H., Holstermann, B., Hunstein, W., & Hauber, J. (2009). *PNAS. 106, 9033.*

Henning, S. M., Niu, Y., Lee, N. H., Thames, G. D., Minutti, R. R., Wang, H., Go, V. L. W., & Heber, D. (2004). *Am. J. Clin. Nutr. 80*, 1558.

Henry, J. P., & Stephens-Larson, P. (1984). *Hypertension. 6*, 437.

Hirasawa, M., & Takada, K. (2004). *J. Antimicrob. Chemother. 53*, 225.

Hirasawa, M., Takada, K., Makimura, M., & Otake, S. (2002). *J. Periodont. Res. 37, 433.*

Hofmann, C. S., & Sonenshein, G. E. (2003). *FASEB. J. 17*, 702.

Hossain, P., Kawar, B., & El Nahas, M. (2007). *N. Engl. J. Med. 356*, 213.

Imanishi, N., Tuji, Y., Katada, Y., Maruhashi, M., Konosu, S., Mantani, N., Terasawa, K., & Ochiai, H. (2002). *Microbiol. Immunol. 46*, 491.

Inoue, M., Tajima, K., Hirose, K., Hamajima, N., Takezaki, T., Kuroishi, T., & Tominaga, S. (1998). *Cancer Causes and Control. 9*, 209.

Juneja, V. K., Bari, M. L., Inatsu, Y., Kawamoto, S., & Friedman, M. (2007). *J. Food. Prot. 70*, 1429.

Karaca, T., Yoruk, M., Yoruk, I. H., & Uslu, S. (2010). *J. Anim. Vet. Adv. 9*, 102.

Kaszkin, M., Beck, K., Eberhardt, W., & Pfeilschifter, J. (2004). *Mol. Pharmacol. 65*, 15.

Katiyar, S. K., & Mukhtar, H. (2001). J. Leukoc. Biol. 69, 719.

Katiyar, S. K., Afaq, F., Perez, A., & Mukhtar, H. (2001). *Carcinogenesis. 22*, 287.

Katzman, M. A., Jacobs, L., Marcus, M., Vermani, M., & Logan, A. C. (2007). *Lipids in Health and Disease. 6*, 87.

Kearney, P. M., Whelton, M., Reynolds, K., Muntner, P., Whelton, P. K., & He, J. (2005). *Lancet. 365*, 217.

Khalaf, N. A., Shakya, A. K., Al-Othman, A., El-Agbar, Z., & Farah, H. (2008). *Turk. J. Biol. 32*, 51.

Kharab, R., Sharma, P. C., & Yar, M. S. (2010). *J. Enz. Inhib. Med. Chem. 26*(1), 1-21.

Kim, S. H., Lee, L. S., Bae, S. M., Han, S. J., Lee, B. R., & Ahn, W. S. (2008). Antimicrobial and antifungal effects of a green tea extract against vaginal pathogens. *J Womens Med 1*, 27–36.

Kim, Y. W., Bae, S. M., Lee, J. M., Namkoong, S. E., Han, S. J., Lee, B. R., Lee, I. P., Kim, S. H., Lee, Y. J., Kim, C. K., Kim, Y-W., & Ahn, W. S. (2004). *Cancer Res. Treat. 36*, 315.

Koizumi, Y., Tsubono, Y., Nakaya, N., Nishino, Y., Shibuya, D., Matsuoka, H., & Tsuji, I. (2003). *Cancer Epidemiol. Biomark. Prev. 12*, 472.

Kuriyama, S., Hozawa, A., Ohmori, K., Shimazu, T., Matsui, T., Ebihara, S., Awata, S., Nagatomi, R., Arai, H., & Tsuji, I. (2006). *Am. J. Clin. Nutr. 83*, 355.

Laurie, S. A., Miller, V. A., Grant, S. C., Kris, M. G., & Ng, K. K. (2005). *Cancer Chemother. Pharmacol. 55, 33.*

Lee, M-J., Lambert, J. D., Prabhu, S., Meng, X., Lu, H., Maliakal, P., Ho, C-T., & Yang, C. S. (2004). *Cancer Epidemiol. Biomark. Prev. 13*, 132.

Lee, Y. K., Bone, N. D., Strege, A. K., Shanafelt, T. D., Jelinek, D. F., & Kay, N. E. (2004). *Blood. 104*, 788.

Leung, L. K., Su, Y., Chen, R., Zhang, Z., Huang, Y., & Chen, Z-Y. (2001). *J. Nutr. 131*, 2248.

Levites, Y., Amit, T., Mandel, S., & Youdim, M. B. H. (2003). *FASEB. J. 17*, 952.

Lorenz, M., Wessler, S., Follmann, E., Michaelis, W., Dusterhoft, T., Baumann, G., Stangl, K., & Stangl, V. (2004). *J. Biol. Chem. 279*, 6190.

Maeta, K., Nomura, W., Takatsume, Y., Izawa, S., & Inoue, Y. (2007). *Appl. Environ. Microbiol. 73*, 572.

Maki, K. C., Reeves, M. S., Farmer, M., Yasunaga, K., Matsuo, N., Katsuragi, Y., Komikado, M., Tokimitsu, I., Wilder, D., Jones, F., Blumberg, J. B., & Cartwright, Y. (2009). *J. Nutr. 139*, 264.

Maruyama, K., Iso, H., Sasaki, S., & Fukino, Y. (2009). *J. Clin. Biochem. Nutr. 44*, 41.

Matsubara, S., Shibata, H., Ishikawa, F., Yokokura, T., Takahashi, M., Sugimura, T., & Wakabayashi, K. (2003). *Biochem. Biophy. Res. Commun. 310*, 715.

Mbata, T. I., Debiao, L. U., & Saikia, A. (2008). *Afr. J. Biotechnol. 7*, 1571.

Meeran, S. M., Akhtar, S., & Katiyar, S. K. (2009). *J. Invest. Dermatol. 129*, 1258.

Mildner-Szkudlarz, S., Zawirska-Wojtasiak, R., Obuchowski, W., & Góslínski, M. (2009). *J. Food Sci. 74*, S362.

Mimoto, J., Kiura, K., Matsuo, K., Yoshino, T., Takata, I., Ueoka, H., Kataoka, M., & Harada, M. (2000). *Carcinogenesis. 21*, 915.

Miura, T., Koike, T., & Ishida, T. (2005). *J health Sci. 51*, 708.

Mughal, T., Tahir, A., Aziz, M. T., & Rasheed, M. (2009). *J. App. Pharm. 2*, 16.

Mughal, T., Tahir, A., Qureshi, S., Nazir, T., & Rasheed, M. (2010). *J. App. Pharm. 2*, 60.

Mukhtar, H., & Ahmad, N. (2000). *Am. J. Clin. Nutr. 71*, 1698S.

Mustata, G. T., Rosca, M., Biemel, K. M., Reihl, O., Smith, M. A., Viswanathan, A., Strauch, C., Du, Y., Tang, J., Kern, T. S., Lederer, M. O., Brownlee, M., Weiss, M. F., & Monnier, V. M. (2005). *Diabetes. 54*, 517.

Nagao, T., Hase, T., & Tokimitsu, I. (2007). Obesity. 15, 1473.

Navarro-Martínez, M. D., Navarro-Perán, E., Cabezas-Herrera, J., Ruiz-Gómez, J., García-Cánovas, F., & Rodríguez-López, J. N. (2005). *Antimicrob. Agents Chemother. 49*, 2914.

Negishi, H., Xu, J-W., Ikeda, K., Njelekela, M., Nara, Y., & Yamori, Y. (2004). *J. Nutr. 134*, 38.

Ng, T-P., Feng, L., Niti, M., Kua, E-H., & Yap, K-B. (2008). *Am. J. Clin. Nutr. 88*, 224.

Ostrowska, J., & Skrzydlewska, E. (2006). *Adv. Med. Sci. 51*, 298.

Panagiotakos, D. B., Lionis, C., Zeimbekis, A., Gelastopoulou, K., Papairakleous, N., Das, U. N., & Polychronopoulos, E. (2009). *Yonsei. Med. J. 50*, 31.

Papparella, I., Ceolotto, G., Montemurro, D., Antonello, M., Garbisa, S., Rossi, G. P., & Semplicini, A. (2008). J. Nutr. 138, 1596.

Park, J. H., Jin, J. Y., Baek, W. K., Park, S. H., Sung, H. Y., Kim, Y. K., Lee, J., & Song, D. K. (2009). *J Physiol. Pharmacol. 60*, 101.

Potenza, M. A., Marasciulo, F. L., Tarquinio, M., Tiravanti, E., Colantuono, G., Federici, A., Kim, J., Quon, M. J., & Montagnani, M. (2007). *Am. J. Physiol. Endocrinol. Metab. 292*, E1378.

Prasetyo, P. A., Nasronudin, & Rahayu, R. P. (2015). *World J Pharm Pharm Sci. 6*, 89.

Rezai-Zadeh, K., Shytle, D., Sun, N., Mori, T., Hou, H., Jeanniton, D., Ehrhart, J., Townsend, K., Zeng, J., Morgan, D., Hardy, J.. Town, T., & Tan, J. (2005). *J. Neurosci. 25*, 8807.

Rietveld, A., & Wiseman, S. (2003). *J. Nutr. 133*, 3285S.

Roghani, M., & Baluchnejadmojarad, T. (2010). *Pathophysiol. 17*, 55.

Roy, A. M., Baliga, M. S., & Katiyar, S. K. (2005). *Mol. Cancer Ther. 4*, 81.

Rudelle, S., Ferruzzi, M. G., Cristiani, I., Moulin, J., Macé, K., Acheson, K. J., & Tappy, L. (2007). *Obesity. 15*, 349.

Ruggiero, P., Rossi, G., Tombola, F., Pancotto, L., Lauretti, L., Giudice, G. D., & Zoratti, M. (2007). *World J. Gastroenterol. 13*, 349.

Sakanaka, S., Juneja, L. R., & Taniguchi, M. (2000). *J. Biosci. Bioeng. 90*, 81.

Sarfaraz, S., Adhami, V. M., Syed, D. N., Afaq, F., & Mukhtar, H. (2008). *Cancer Res. 68*, 339.

Sartippour, M. R., Pietras, R., Marquez-Garban, D. C., Chen, H-W., Heber, D., Henning, S. M., Sartippour, G., Zhang, L., Lu, M., Weinberg, O., Rao, J. Y., & Brooks, M. N. (2006). *Carcinogenesis 27*, 2424.

Sato, Y., Nakatsuka, H., Watanabe, T., Hisamichi, S., Shimizu, H., Fujisaku, S., Hinowatariy, O., Da, S., Suda, S., Kato, K., & Eda, M. (1989). *Tohoku. J. Exp. Med. 157*, 337.

Sharma, P. C., Jain, A., Jain, S., Pahwa, R., & Yar, M. S. (2010). *J. Enz. Inhib. Med. Chem. 25*, 577.

Sharma, P. C., Sharma, O. P., Vasudeva, N., Mishra, D. N., & Singh, S. K. (2006). *Nat. Prod. Rad. 5*, 70.

Shimizu, M., Shirakami, Y., & Moriwaki, H. (2008). *Int. J. Mol. Sci. 9*, 1034.

Soekanto, S. A., Mangundjaja, S. (2004). Available from http://*staff.ui.ac.id/ internal/ 130366445 /publikasi /Greentea2004.pdf Accessed on May 2 2010.*

Soft, W. (2010). Available from: http://www.prlog.org/10647439-green-tea-and-weight-loss-synonymous.pdf Accessed May 20, 2010.

Song, J-M., Lee, K-H., & Seong, B-L. (2005). *Antivir. Res. 68*, 66.

Stapleton, P. D., Gettert, J., & Taylor, P. W. (2006). *Int. J. Food Microbiol. 111*, 276.

Suganuma, M., Okabe, S., Kai, Y., Sueoka, N., Sueoka, E., & Fujiki, H. (1999). *Cancer Res. 59*, 44.

Sutherland, B. A., Shaw, O. M., Clarkson, A. N., Jackson, D. M., Sammut, I. A., & Appleton, I. (2005). *FASEB. J. 19*, 258.

Taylor, P. W., Hamilton-Miller, J. M. T., & Stapleton, P. D. (2005). *Food Sci. Technol. Bull. 2*, 71.

Tsubono, Y., Nishino, Y., Komatsu, S., Chung-Cheng, H., Kanemura, S., Tsuji, I., Nakatsuka, H., Fukao, A., Satoh, H., & Hisamichi, S. (2001). *N. Eng. J. Med. 344*, 632.

Tsuneki, H., Ishizuka, M., Terasawa, M., Wu, J-B., Sasaoka, T., & Kimura, I. (2004). *BMC Pharmacol. 4*, 4.

Vaidyanathan, J. B., & Walle, T. (2001). *Pharm. Res. 18*, 1420.

Van Het Hof, K. H., Kivits, G. A. A., Weststrate, J. A., & Tijburg, L. B. M. (1998). *Eur. J. Clin. Nutr. 52*, 356.

Waltner-Law, M. E., Wang, X. L., Law, B. K., Hall, R. K., Nawano, M., & Granner, D. K. (2002). *J Biol. Chem. 277*, 34940.

West, P. W. J., Mathew, T. C., Miller, N. J., & Electricwala, Q. (2001). *J. Nutr. Environ. Med. 11*, 263.

Williamson, M. P., McCormick, T. G., Nance, C. L., & Shearer, W. T. (2006). *J. Allergy Clin. Immunol. 118*, 1369.

Wiseman, S., Mulder, T., & Rietveld, A. (2001). *Antioxid. Redox. Signal. 3*, 1009.

Wolfram, S., Raederstorff, D., Preller, M., Wang, Y., Teixeira, S. R., Riegger, C., & Weber, P. (2006). *J. Nutr. 136*, 2512.

World Health Organization, Regional office for the Eastern Mediterranean, Cairo (2005). Clinical guidelines for the management of hypertension. EMRO Technical Publications Series 29.

Xu, J. Z., Yeung, S. Y. V., Chang, Q., Huang, Y., & Chen, Z. Y. (2004). *Br. J. Nutr. 91*, 873.

Yam, T. S., Hamilton-Miller, J. M. T., & Shah, S. (1998). *J. Antimicrob. Chemother. 42*, 211.

Yamamoto, T., Hsu, S., Lewis, J., Wataha, J., Dickinson, D., Singh, B., Bollag, W. B., Lockwood, P., Ueta, E., Osaki, T., & Schuster, G. (2003). *J. Pharmacol. Exp. Ther. 307,* 230.

Yamanaka, N., Oda, O., & Nagao, S. (1997). *FEBS Letters. 401*, 230.

Yang, C. S., Chen, L., Lee, M-J., Balentine, D., Kuo, M. C., & Schantz, S. P. (1998). *Cancer Epidemiol Biomark Prev. 7,* 351.

Yang, G., Shu, X-O., Li, H., Chow, W-H., Ji, B-T., Zhang, X., Gao, Y-T., & Zheng, W. (2007). *Cancer Epidemiol. Biomark. Prev. 16*, 1219.

Yang, Y-C., Lu, F-H., Wu, J-S., Wu, C-H., & Chang, C-J. (2004). *Arch. Intern. Med. 164,* 1534.

Yao, K., Ye, P-P., Zhang, L., Tan, J., Tang, X-J., & Zhang, Y-D. (2008). *Mol. Vis. 14*, 217.

Yee, Y-K., & Koo, M. W-L. (2000). *Aliment. Pharmacol. Ther. 14*, 635.

Young, I. S., & Woodside, J. V. (2001). *J. Clin. Pathol. 54*, 176.

Yuan, J-M., Koh, W-P., Sun, C-L., Lee, H-P., & Yu, M. C. (2005). *Carcinogenesis. 26*, 1389.

Zapora, E., Hołub, M., Waszkiewicz, E., Dąbrowska, M., & Skrzydlewska, E. (2009). *Bull. Vet. Inst. Pulawy. 53,* 139.

Zaveri, N. T. (2006). *Life Sci. 78*, 2073.

Zhang, Y-M., & Rock, C. O. (2004). *J. Biol. Chem. 279*, 30994.

Zheng, G., Sayama, K., Ohkubo, T., Juneja, L. R., & Oguni, I. (2004). *In vivo. 18*, 3.

Zigman, S., Rafferty, N. R., Rafferty, K. A., & Lewis, N. (1999). *Biol. Bull. 197*, 285.

CHAPTER 5

PHARMACEUTICAL IMPORTANCE OF NATURAL APHRODISIAC DRUGS

RAM KUMAR SAHU and AMIT ROY

Department of Pharmaceutics, Columbia College of Pharmacy, Raipur, Chhattisgarh, India

CONTENTS

5.1   INTRODUCTION

Sexual dysfunction is an inability to achieve a normal sexual intercourse, including premature ejaculation, retrograded, retarded or inhibited ejaculation, erectile dysfunction, arousal difficulties (reduced libido), compulsive sexual behavior, orgasmic disorder and failure of detumescence. It is increasing world wide due to etiological factors and aging. Men and women have since ages been curious to improve, revive and maintain their sexual efficiency, and aphrodisiacs substances are employed for this purpose (Patel et al., 2011). An aphrodisiac is defined as any food or drug that arouses the sexual instinct, induces venereal desire and increases pleasure and performance. This word is derived from 'Aphrodite' the Greek goddess of love and these substances are derived from plants, animals or minerals and since time immemorial they have been the passion of man. Presently, the aphrodisiac substances are used to treat sexual dysfunction or having sexual activity enhancing power. Sexual desire is controlled and regulated by the central nervous system which integrates tactile, olfactory, auditory, and mental stimuli. The aphrodisiac drugs act by altering the level of specific neurotransmitters or specific sex hormone into the body. Mostly act through alteration in testosterone concentration in the body but other effective in both sexes (Malviya et al., 2011)

5.2   ERECTILE DYSFUNCTION

The physiological mechanism of erection is a complex neurovascular phenomenon that depends on neural, vascular, hormonal, and psychological factors. Integrated function of these factors is essential for production of a normal erectile response (Agarwal et al., 2006). Erectile dysfunction is defined as a difficulty in initiating or maintaining penile erection adequate for sexual relations (Lasker et al., 2010). It is one of the most common sexual dysfunctions in men. Normal erectile function depends on a precise balance between psychological, hormonal, neurological, vascular, and cavernosal factors. Therefore, an alteration in any one or combination of these factors may lead to erectile dysfunction (Agarwal et al., 2006).

Althougherectile dysfunction can be primarily psychogenic in origin, most patients have an organic disorder, commonly with some psychogenic overlay. Some men assume that erectile failure is a natural part of the aging process and tolerates it; for others it is devastating. Withdrawal from sexual intimacy because of fear of failure can damage relationships and have a profound effect on overall wellbeing for the couple. Since erectile dysfunction often accompanies chronic illnesses, such as diabetes mellitus, heart disease, hypertension, and a variety of neurological diseases. Physicians from many medical disciplines can see patients with this disorder (Wagner et al., 1998). There are two main types of aphrodisiacs, psychophysiological stimuli (visual, tactile, olfactory and aural) preparations and internal preparations (food, alcoholic drinks and love portion) (Malviya et al., 2011).

Several types of treatment are claimed in the modern medicine but due to serious side effects and higher cost, search of natural supplement from medicinal plants as an aphrodisiac substance is significantly increased (Bella et al., 2013). Plants are an important source of medicines and play a key role in the health of the world's population. The use of plant materials to treat sexual disorders has a long history in most countries, and plant materials have proven effective in improving sexual desire and sexual behavior in male animals (Jian Feng et al., 2012).

There exists no accurate classification of available aphrodisiacs. Generally, aphrodisiacs are classified on the basis of working, where they are divided into natural aphrodisiacs, made from natural sources, and non-natural aphrodisiacs, made from synthetic compounds. Furthermore, natural aphrodisiacs can be further classified into plant aphrodisiacs and nonplant aphrodisiacs (Bella et al., 2013). The structures of chemical constituents are illustrated in the Table 5.1.

5.3 NATURAL APHRODISIAC

5.3.1 ANIMAL BASED APHRODISIAC

5.3.1.1 Ambrein

Ambrein, a triterpenoid is a chief constituent of Ambergris, a secretion from the gut of sperm whales (*Phyzeter catadon*) which is used for different applications, including better sexual performance. Animal studies have indicated its stimulatory effects on pituitary and antagonizing effects on various vasoconstricting agents such as noradrenaline, acetylcholine, prostaglandin and oxytocin etc., which ultimately increases libido by altering related hormones and increasing blood flow by relaxing smooth muscles of corpus cavernosum (Taha et al., 1995).

5.3.1.2 Bufo Toad

Bufotenine or its O-methylated derivative, 5-MeO-DMT (5-Methoxy N,N-dimethyltryptamine), a tryptamine alkaloid, are widely spread in Anuran family (for example, example toad and frog) as a component of their chemical defense system and act as putative hallucinogen congener of serotonin (Lyttle et al., 1996). It is popularly used as an active ingredient in West Indian aphrodisiac "*love stone*" and Chinese medication "*chan su*" (Barry

TABLE 5.1   Medicinal Plants Exhibiting Aphrodisiac Activity Along with Active
Constituents and It Structure

Biological sources	Active constituents	Structure
Phyzeter catadon	Ambrein	
Bufo toad	Bufotenine	
Lytta vesicatoria	Cantharidin	
Allium tuberosum	Linalool	
	α-phellandrene	

TABLE 5.1   (Continued)

Biological sources	Active constituents	Structure
	Geraniol	
Alpinia calcarata	Astragalin	
	Kaempferol-3-*O*-glucuronide	
Astercantha longifolia	Lupeol (R – H) Butelin (R – CH$_2$OH)	

TABLE 5.1 (Continued)

Biological sources	Active constituents	Structure
	Lupenone	
Argyreia nervosa	1-triacontanol	
	Ergine	
Caesalpinia benthamiana	Benthaminins	
	Gallic acid	

TABLE 5.1 (Continued)

Biological sources	Active constituents	Structure
Curculigo orchioides	Yuccagenin	
Crocus sativus	Picrocrocin	
	Safranal	
Coryanthe yohimbe	Yohimbine	

TABLE 5.1   (Continued)

Biological sources	Active constituents	Structure
Camellia sinensis	Epigallocatechin gallate	
	Epigallocatechin	
	Epicatechin	
Dactylorhiza hatagirea	Dactylorhins A	
	Dactylorhins B	

TABLE 5.1 (Continued)

Biological sources	Active constituents	Structure
Eurycoma longifolia	Eurycomanone	
	Eurycomanol	
	9-hydroxycanthin-6-one	
Ferula hermonis	Ferutinin (R1-H, R2-OH, R3-H)	
	Teferdin (R1-H, R2-H, R3-H)	
	Teferin (R1-H, R2–OH, R3- OCH₃)	

TABLE 5.1 (Continued)

Biological sources	Active constituents	Structure
Panax quinquefolium	Ginsenoside (Rb1)	
	Ginsenoside (Re)	
Glycyrrhiza glabra	Glycyrrhizin	
	Licoisoflavanone	

TABLE 5.1 (Continued)

Biological sources	Active constituents	Structure
	Glabridin	
	Glycyrin	
	Glycyrrhetinic acid	
	Liquiritigenin	

TABLE 5.1   (Continued)

Biological sources	Active constituents	Structure
Lycium barbarum	Zeaxanthin monopalmitate	
	Quercetin	
	Mutatoxanthin	
Leptadenia reticulate	Stigmasterol	
	Apigenin	
	Hentriacontanol	

TABLE 5.1 (Continued)

Biological sources	Active constituents	Structure
	Reticulin	
Microdesmis keayana	Keayanine	
Mimusa pudica	Nor-epinephrine	
	Quinines	
Ocimum gratissimum	Eugenol	

TABLE 5.1 (Continued)

Biological sources	Active constituents	Structure
	Limonene	
	Methyl eugenol	
Passiflora incarnate	Benzoflavone	
Phoenix dactylifera	Catechin	
	Epicatechin	

TABLE 5.1 (Continued)

Biological sources	Active constituents	Structure
	Cinnamic acids	
Securidaca longepedunculata	2-hydroxy-1, 7-dimethoxy xanthone	
Tribulus terrestris	Protodioscin	
Turnera diffusa	Arbutine	
	Luteolin	

TABLE 5.1   (Continued)

Biological sources	Active constituents	Structure
Tinospora cordifolia	Tinosporide	
	Berberine	

et al., 1996). It shows activity similar to Lyzergic acid diethylamide but the mechanism of its action in not yet established. It is however, assumed to act through its effect on central nervous system (Sandroni, 2001).

5.3.1.3   Spanish Fly

Cantharidin (3α, 7α-dimethylhexahydro-4, 7-epoxyisobenzofuran-1, 3-dione), a compound for sexual stimulation, is derived from emerald – green beetle (*Lytta vesicatoria*) of family Meloidae. Since ancient times cantharidin finds its value for improving sexual vigor but with adverse side effects (Karras et al., 1996). Cantharides are clinically restricted in human subjects due to their unavoidable side effects like renal toxicity and associated acute tubular necrosis, gastrointestinal hemorrhages and cardiac complications. It is suggested that cantharides act through inhibition of PDE-5, protein phosphatase and activation of β-receptors which ultimately leads to sexual sensation/arousal (Sandroni, 2001). However, the exact mechanism of action and its clinical efficacy are still the subject of exploration.

5.3.2 PLANT BASED APHRODISIAC

5.3.2.1 Allium tuberosum

Allium tuberosum (Family – Alliaceae) comprising steroidal saponins, alkaloids and sulfur containing compounds, etc. *Allium tuberosum* exhibited aphrodisiac activity by improving reduction in mounting latency, ejaculatory latency, intromission latency and post ejaculatory interval and increase in mounting frequency, intromission frequency and ejaculatory frequency (Kumar et al., 2010). The linalool, citral, α-phellandrene, geraniol, propionic aldehyde and valenaldehyde are present in *Allium tuberosum* (Khalid et al., 2014).

5.3.2.2 Alpinia calcarata

Alpinia calcarata (Family – Zingiberaceae) is a perennial herb that mainly thrives in China and South-east Asia. The hot water extracts of this plant revealed the male sexual competence and fertility in rats by inhibiting the intromission and mounting latencies. It has been observed no change in libido and penile erection at lower doses whereas the increased level of testosterone and rapid penile erections found at a higher dose (Ratnasooriya et al., 2006). The active constituents present in the *Alpinia calcarata* are flavone glycosides, astragalin and kaempferol-3-*O*-glucuronide (Ahmed et al., 2015).

5.3.2.3 Astercantha longifolia

Astercantha longifolia belongs to family Acanthaceae, its ethanolic extract elevated the testosterone, fructose and sperm count (validated with histological architecture) as well as increases in weight of secondary sex organs and body weight. *Astercantha longifolia* has the capacity to improve the male sexual behavior (Chauhan et al., 2009). The chief chemical constituent available in *Astercantha longifolia* are lupeol, lupenone, stigmasterol and butelin, an isoflavone glycoside, an alkaloid, sterol I, II, III, and IV, asteracanthine and asteracanthicine (Chauhan et al., 2010).

5.3.2.4 Anacyclus pyrethrum

The roots of *Anacyclus pyrethrum* (Family – *Asteraceae*) improved sexual behavior, orientation behavior, penile erection and associated parameters.

Additionally, *Anacyclus pyrethrum* effective in improving sperm count and fructose levels concentration. Thus this plant imparts effective aphrodisiac activity (Chauhan et al., 2009).

5.3.2.5 *Argyreia nervosa*

Argyreia nervosa (Family – Convolvulaceae) has been potential to developed effective medicine for stimulating male sexual activity with an influence on sex ratio favoring males (Subramoniam et al., 2007). The 1-triacontanol, ß-sitosterol, epifriedeline, epifriedelinol, acetate, epifriedelinol, β-sitosterol, chanoclavine, ergine and ergonovine are present in *Argyreia nervosa*.

5.3.2.6 *Asparagus racemosus*

The aqueous root extracts of *Asparagus racemosus* (Family – *Liliaceae*) exhibited enhancement of body weight and reproductive organs, penile erection, mount frequency in rats, and indicates an improvement in sexual behavior. Moreover, on mixing the *Asparagus racemosus* and milk it increases the potency of aphrodisiac activity by reducing mount latency, ejaculation latency, post ejaculatory latency, intromission latency, and an increase of mount frequency. Further penile erection is also considerably enhanced (Thakur et al., 2009).

5.3.2.7 *Abelmoschus manihot*

Abelmoschus manihot (Family – Malvaceae) commonly referred to as "Junglee bhindi" showed pronounced anabolic and spermatogenic effect in Swiss albino mice. There was a remarkable increased in sperm count and penile erection index and also improved sexual behavior of male mice by increased mount and intromission frequency. Further it was noticed that the performance rate enhances without any side effect (Rewatkar et al., 2010).

5.3.2.8 *Bulbus natalensis*

Bulbus natalensis is a member of family Asphodelaceae. The secondary metabolites namely, saponins, cardiac glycosides, tannins, alkaloids and anthraquinones are present in *Bulbus natalensis*. *Bulbus natalensis* reduce mounting latency, intromission latency and post-ejaculatory interval. Indices of sexual behavior, penile erection and penile reflexes are significantly improved making this plant a potent aphrodisiac. The sex stimulating property of this plant is thought to be because of either its alkaloid content, which helps in dilation of blood vessels in the reproductive organs or saponin fraction which enhances androgen production, or it could be a combined effect of two or some more constituents (Yakubu et al., 2009).

5.3.2.9 *Butea frondosa*

Butea frondosa, belongs to family Papillionaceae exhibited significant aphrodisiac activity in male rats. The extract produces effective outcomes by significant reduction of mounting latency, intromission latency, ejaculatory latency and post ejaculatory interval and significant improvement in mounting frequency, intromission frequency and ejaculatory frequency in a dose dependent manner in male rats. The mechanism by which *Butea frondosa* enhances sexual activity is thought to be mediated by an increase in testosterone level and change in neurotransmitters level and may involve dopaminergic and adrenergic receptors (Ramachandran et al., 2004).

5.3.2.10 *Chlorophytum borivillianum*

Chlorophytum borivilianum belongs to family Papillionaceae, in India it is well known aphrodisiac agent and revitalizer with good spermatogenic property and is considered to be alternative Viagra. Roots of this plant are used, which contain The steroidal and triterpenoidal saponins, sapogenins, fructans, magnesium, potassium, calcium along with mucilage, polysaccharides and good amount of simple sugars present in the *Chlorophytum borivilianum*. This plant is found most effective in improving sexual behavior parameters in male albino rats in combination with *Asparagus racemosus* and *Curculigo orchoides*. This combination probably works through testosterone like effects or nitric oxide mediated arousal mechanism (Kenjale et al., 2008).

5.3.2.11 *Caesalpinia benthamiana*

Caesalpinia benthamiana (Fabaceae), the aqueous extract of this plant exhibits significant aphrodisiac activity. The enhanced sexual activity is considered to act through nitric oxide production which leads to vaso-relaxation (Zamble et al., 2008). The active constituents cassane diterpenoids, benthaminins 1 and 2, caesaldekarin C, gallic acid and furanoditerpenoids are present in *Caesalpinia benthamiana*.

5.3.2.12   *Curculigo orchioides*

Curculigo orchioides belongs to family Amarylladaceae and is a well-known aphrodisiac since ancient times and has a reputation in Indian traditional medicinal system especially in Ayurveda and in Chinese medicine. *Curculigo orchioides* extract of rhizome produces remarkable improvement in parameters of mating performance, sexual behavior and penile erection index in male rats. *Curculigo orchioides* probably works through the increase in the level of testosterone or some neurotransmitters; however, the mechanism of action is yet to be established. It has spermatogenic potential and increases weight of reproductive organs due to its anabolic effect. It improves sperm count in heat-exposed rats as compared to heat exposed positive control group. This inferred that *Curculigo orchioides* has potential aphrodisiac properties (Chauhan et al., 2010). The yuccagenin, corchiocide A, 25-hydroxy-33- methylpentatriconta-6-one, 21-hydroxytetracontan-20-one, 27-hydroxy-tricontan-6-one, 2-methoxy-4-acetyl-5-methytricontane, linoleic, linoleinic, arachidic, 4-methyheptadecanoic acid, curculigol, curculigenin A, cycloartenol, sitosterol, and sigmastirol are present in *Curculigo orchioides* (Asif et al., 2012).

5.3.2.13   *Crocus sativus*

Crocus sativus belongs to family Iridaceae and its dried stigma known as saffron are used for medicinal purposes since ages and is famous as strong aphrodisiac among common man. *Crocus sativus* contains crocin, picrocrocin and safranal as its major constituents. The safranal and crocin present in *Crocus sativus* and produces aphrodisiac activity. Later crocin has been developed as effective constituent exhibiting aphrodisiac activity (Hosseinzadeh et al., 2008).

5.3.2.14   *Coryanthe yohimbe*

Yohimbine is obtained from the bark of *Coryanthe yohimbe*, a member of family Rubiaceae. It has been a source of aphrodisiacs since long back.

Yohimbine history is old and in last three decades, several studies have been conducted to establish yohimbine as a star aphrodisiac. Yohimbine hydrochloride is effective in decreasing latency time and increasing mounts aged male rats. It is an alpha 2-adrenoceptor blocker and used against erectile impotence as evidenced by increased mounts and increased ejaculation in mating test with rats as well as increased copulatory behavior in sexually inactive rats. Meta – analysis of few studies has supported yohimbine as an aphrodisiac with fewer side effects in human subjects (Clark et al., 1984; Morales, 2000).

5.3.2.15 *Camellia sinensis*

Camellia sinensis (Theaceae), Sri Lankan herbal practitioners claimed to be a potent herbal aphrodisiac. Black tea brew of *Camellia sinensis* is a good sexual stimulant and delays premature ejaculation, shortens mounting and intromission latencies and elevates the serum testosterone level as tested during sexual behavior experiments. *Camellia sinensis* is nontoxic and probably works through suppression of anxiety and increase in testosterone level (Ratnasooriya et al., 2008). *Camellia sinensis* contains mainly flavanols or catechins of epigallocatechin gallate, epigallocatechin, epicatechin gallate and epicatechin.

5.3.2.16 *Chenopodium album*

Chenopodium album a member of family Chenopadiaceae, is used to enhance the desire of sex by significant increase in the mounting frequency, intromission frequency, intromission latency, erection as well as aggregate of penile reflexes and caused the reduction in the mounting latency and post ejaculatory interval. The β-sitosterol, lupeol, 3 hydroxy nonadecyl henicosanoate, ascorbic acid, b-carotene, catechin, gallocatechin, caffeic acid, p-coumaric acid, ferulic acid, campesterol, xanthotoxin, stigmasterol, imperatorin, ecdysteroid, cinnamic acid amide alkaloid, phenol, saponin, apocarotenoids, crytomeridiol, n-transferuloyl-4-O-methyl dopamine and syringaresinol are present in *Chenopodium album* (Pande et al., 2008).

5.3.2.17 *Crossandra infundibuliformis*

Crossandra infundibuliformis belong to family Acanthaceae, extract demonstrated aphrodisiac behavior by increasing mounting frequency, intromission frequency and ejaculatory latency, and reduced mounting latency and intromission latency. Moreover, enhancement in serum testosterone could be seen (Kumar et al., 2010; Agrawal et al., 2014).

5.3.2.18 *Dactylorhiza hatagirea*

Dactylorhiza hatagirea is a plant of family Orchidaceae and its bulbous roots are used as the sexual stimulant as well as nutritive and restorative tonic. *Dactylorhiza hatagirea* decreased post ejaculatory interval and considered to involve nitric oxide based mechanism for aphrodisiac activity. The reduction in mount latency, intromission latency and post ejaculatory latency may be observed during sexual behavior study (Thakur et al., 2007). Its tubers contain a glucoside, a bitter substance, starch, mucilage, albumen, a trace of volatile oil and ash. Chemically, dactylorhins A–E, dactyloses A and B and lipids, etc. are found as major constituents (Dutta, 2007).

5.3.2.19 *Eurycoma longifolia*

Eurycoma longifolia is a shrub of family Simaroubaceae, over the years this plant has been reported as a potent aphrodisiac. *Eurycoma longifolia* promotes the growth of the accessory reproductive gland, for example, prostate and seminal vesicle and is effective in improving sexual performance (Ang et al., 2000). The eurycomanone, eurycomanol, eurycomalactone, canthine-6-one alkaloid, 9-hydroxycanthin-6-one, 14,15β-dihydroxyklaineanone, phenolic components, tannins, quanissoids, and triterpenes are chief constituents of *Eurycoma longifolia* (Morita et al., 1993; Ismail et al., 1999).

5.3.2.20 *Ferula hermonis*

Ferula hermonis is a small shrub of family Apiaceae and is very famous in natives of Lebanon, Syria and Jordan. Its active constituents, ferutinin, teferdin and teferin impart chief role in producing aphrodisiac activity. The ferutinin reduced mounting and intromission latencies in sexually potent rats. Additionally, it enhances the testosterone production in a significant manner (Zanoli et al., 2005).

5.3.2.21 *Fadogia agrestis*

Fadogia agrestis is a shrub that belongs to family Rubiaceae and can be seen in Nigerian region and is claimed to be an aphrodisiac helpful in management of erectile dysfunction. *Fadogia agrestis* increases in mounting frequency and intromission frequency, and reduction in mounting latency and intromission latency along with prolonged ejaculatory latency (Yakubu et al., 2005).

5.3.2.22 *Ginseng*

Panax quinquefolium (American ginseng) and *Panax ginseng* (Asian / Korean/Chinese ginseng) belongs to family of family Araliaceae. Ginseng contains ginsenosides namely Rb1 and Re (derived from *Panax ginseng*) and trilinolein which are a triacyl glycerol with linolenic acid. The ginsenosides and trilinolein are effective in nitric oxide mediated vasorelaxation. A clinical study on 90 patients confirmed that Korean red ginseng is effective in enhancing libido, penile erection and sexual performance (Choi et al., 1995). Ginsenosides act through i-NOS mediated nitric oxide release in endothelial cells along with its direct effect on the same via increased Ca^{2+} ion concentration (Li et al., 2000).

5.3.2.23 *Glycyrrhiza glabra*

Glycyrrhiza glabra is a member of family of Leguminoaceae, the extract reduced mounting latency and intromission latency level. The extract also increased significantly mounting frequency and intromission frequency (Awate et al., 2012). *Glycyrrhiza glabra* has Glycyrrhizin as the major water-soluble constituent responsible for its sweet taste. Glycyrrhizin is a triterpenoid saponin that is present within a range of 2–14% in different species. Other phytochemicals present in *Glycyrrhiza glabra* are Liquiritigenin, Liquiritin (Flavanones), Isoliquiritigenin, Isoliquiritin (Chalcones), Genistein, Glicoricone, Glisoflavone, Isoangustone A (Isoflavones); Glycyrrhizoflavanone, Glyasperin F, Licoisoflavanone (Isoflavanones); Glyasperin C, Glyasperin D, Glabridin, Licoricidin (Isoflavans); Glycocoumarins, Lipocoumarins, Glycyrin (3-arylcoumarins) and others (Licocoumarone, Licoriphenone, Isoglycyrol) (Ammosov et al., 2003).

5.3.2.24   *Lycium barbarum*

Lycium barbarum is a Chinese medicinal plant that belongs to family Solanaceae and has earned the reputation throughout Asia and is commercially described as Red Diamonds. This plant has been a traditional remedy for male sexual dysfunction. Polysaccharide is an important phytochemical constituent of this plant, which is considered to be responsible for its medicinal properties (Luo et al., 2006). The Zeaxanthin dipalmitate, β-Cryptoxanthin palmitate, zeaxanthin monopalmitate, small amount of free zeaxanthin and β-carotene, mutatoxanthin, aglycones myricetin, quercetin, and kaempferol are the chief constituents present in different parts of *Lycium barbarum*.

5.3.2.25   *Leptadenia reticulate*

Leptadenia reticulate belongs to family Asclpiadaceae, the extract increases in mount, intromission interval, number of ejaculations and decreased latency of first mount as well as the increase in post ejaculation times. Significant weight gain in testis, seminal vesicles, prostate gland, vasdeferences, and epididymis may be seen. The major phytochemical compound is stigmasterol. It also contains ß-sitosterol α-amyrin, β-amyrin, ferulic acid, luteolin, diosmetin, rutin, stigmasterol, hentriacontanol, a triterpene alcoholsimiarenol and apigenin. The Reticulin, Deniculatin and Leptaculatin are three novel pregnane glycosides isolated from *Leptadenia reticulate* (Santosh et al., 2011).

5.3.2.26   *Landolphia dulcis*

Landolphia dulcis is a member of family of Apocynaceae, the extract significant increase in mount, intromission and ejaculation frequencies. This extract also significantly reduces the mount and intromission latencies and prolonged ejaculation latency (Ilodigwe et al., 2013).

5.3.2.27   *Microdesmis keayana*

Microdesmis keayana a plant from Pandaceae family has history of its use for erectile dysfunction. The isolated and identified major alkaloids from roots of *Microdesmis keayana* are Keayanidine B and Keayanine (Zamble et al., 2007). The aphrodisiac activity of *Microdesmis keayana* is reactive oxygen species scavenging and vasorelaxation property through e-NOS pathway, and it is considered to be its mechanism of aphrodisiac action (Zamble et al., 2009).

5.3.2.28 *Montanoa tomentosa*

Montanoa tomentosa belongs to family of Asteraceae and has been used as an aphrodisiac in ethno-medicine. *Montanoa tomentosa* extract is effective in sexual arousal and stimulation of mounting behavior in animals compromised by anesthesia in the genital area as well as in noncopulating males (Carro-Juarez et al., 2004). *Montanoa tomentosa* contains terpenoids, mainly diterpenes but also including mono-, sesqui-, and triterpenoids as well as some interesting flavonoids.

5.3.2.29 *Mimusa pudica*

Mimusa pudica a plant from Mimosae family, its extracts significantly increase mounting frequency, intromission frequency, intromission latency, erections as well as aggregate of penile reflexes and caused significant reduction in mounting latency and intromission latency. The nor-epinephrine, d-pinitol (3-mono-methyl ether of inositol), β-sitosterol, lavonoids, glycosides, alkaloids, quinines, phenols, tannins, saponins, and coumarins are present in *Mimusa pudica* (Pande et al., 2009).

5.3.2.30 *Mondia whitei*

Mondia whitei is an aromatic woody climber belonging to family Periplocaceae. *Mondia whitei* extract is effective in stimulating a significant increase in serum and testicular testosterone levels, testicular protein and sperm density in cauda epididymis in rats. *Mondia whitei* has been found to reduce alpha-adrenergically stimulated contraction in corpus cavernosum tissue in guinea pig and has the property of inducing and maintaining penile erection as it relaxes cavernous smooth muscles. The cavernosum smooth muscle relaxation does not work through the nitric oxide mechanism as a nonspecific nitric oxide synthase inhibitor NG-nitro-L-arginine methyl ester (L-NAME) did not affect the relaxation (Lampiao et al., 2008). *Mondia whitei* contains steroids, triterpenes (mixture of amyrine a- and β-acetate, lupeol, β-sitosterol, and β-sitosterol glucoside) and aromatic (2-hydroxy-4-methoxybenzaldehyde, 3-hydroxy-4-methoxy-benzaldehyde (vanillin), 4-hydroxy-3-methoxy-benzaldehyde), polyholosides [a-d-glucopyranosyl (6–1)-β-d-glucopyranose and 1-methoxy-β-d-glucopyranosyl (6–1)-β-d-glucopyranose].

5.3.2.31   *Nymphaea stellata*

Nymphaea stellata belongs to family of Nymphaeceae, the extract increases in sexual behavior as evidenced by an increase in mounting frequency, intromission frequency, ejaculatory latency and a decrease in mounting latency, intromission latency and post ejaculatory interval. The astragalin, corilagin, gallic acid, gallic acid methyl ester, isokaempferide, kaempferol, quercetin-3-methyl ether, quercetin, 2,3,4,6-tetra-o-galloyl dextroglucose, and 3-o-methylquercetin-3'-o-beta dextroxylopyranoside are present in various parts of *Nymphaea stellata* (Maruga Raja et al., 2012).

5.3.2.32   *Ocimum gratissimum*

Ocimum gratissimum is a member of Lamiaceae family; its extract increases mounting frequency, intromission frequency, ejaculatory latency and erections as well as aggregate of penile reflexes and caused significant reduction in mounting latency, intromission latency and post ejaculatory interval. The eugenol, linalool, limonene, methyl eugenol, β-caryophyllene, farnesene, α-terpineol, β-salinene, methyl isoeugeneol, geraniol, α-copaene, bisabolol, α-pinene, p-cymene, fenchone, cubenene, camphene, T-cadinol, γ-eudesmol, sabinene, myrcene, β-bisoboline, α-humelene and β-elemene are present in *Ocimum gratissimum* (Pande et al., 2009).

5.3.2.33   *Orchis latifolia*

Orchis latifolia belongs to family of Orchidaceae, the extract increases sexual behavior of diabetic rats (Thakur et al., 2008).

5.3.2.34 *Passiflora incarnate*

Passiflora incarnate is member of Passifloraceae family, is well known for its aphrodisiac activity. The Benzoflavone moiety isolated from *Passiflora incarnate* is effective in protection of fertility activities against THC (Δ 9-tetrahydrocannabinol) induced reduction in libido, sperm count and mating performance parameters. Benzoflavone is strongest aromatase enzyme inhibitor, and it prevents the metabolic breakdown of testosterone into estradiol; thus making available the free plasma testosterone which in turn stimulates gonadotropins (*Luteinizing hormone* and *Follicle-stimulating hormone*), resulting ultimately in improved spermatogenesis, libido and other fertility parameters. Benzoflavone moiety of *Passiflora incarnate* is also capable of reversing addictable drug induced suppression in sexuality. It is thought to act through neuro-steroidal mechanism (Dhawan, 2003).

5.3.2.35 *Phoenix dactylifera*

Phoenix dactylifera is belonging to Arecaceae family, is used in the traditional medicine for male infertility. *Phoenix dactylifera* improve sperm quality; enhance fertility in the male adult rat. The phenolic profile of the plant revealed the presence of mainly cinnamic acids (ferulic, sinapic and coumaric acids and their derivatives, such as 5-o-caffeoylshikimic acid also called as dactyliferic acid), flavonoid glycosides (luteolin, methyl luteolin, quercetin, and methyl quercetin), flavanols (catechin, epicatechin) (Bahmanpour et al., 2006; Biglari et al., 2008).

5.3.2.36 *Paederia fetida*

Paederia fetida (Rubiaceae) extract in albino rats showed pronounced anabolic and spermatogenic effects in animals in the treated groups. The extract significantly increased both mount and intromission frequency (Soni et al., 2012).

5.3.2.37 *Piper guineense*

Piper guineense belongs to Piperaceae family; its dry fruits significant increase the level of testosterone in the serum and testes, cholesterol in testes, α-glucosidase in the epididymis in the seminal vesicles. The essential oil namely α-pinene, sabinene, β-pinene, myrcene, α-phellandrene, δ-3-carene, limonene (Z)-β-ocimene, linalool, safrole, α-cubebene, α-copaene, β-elemene, α-gurjunene, β-caryophyllene, α-humulene (E)-β-farnesene, germacrene D, β- selinene, asaricin and α-zingiberene are present in *Piper guineense* (Mbongue et al., 2005).

5.3.2.38   *Polygonatum verticillatum*

Polygonatum verticillatum is a member of Liliaceae family, its extract reduced significantly mount latency, intromission latency, ejaculation latency and post ejaculatory interval. The extract also increased significantly mounting frequency, intromission frequency and ejaculation frequency (Kazmi et al., 2012).

5.3.2.39   *Securidaca longepedunculata*

Securidaca longepedunculata is a herb that belongs to family Polygalaceae. This plant exhibited aphrodisiac activity by the relaxation of corpus cavernosum smooth muscle (Rakuambo et al., 2006). Later, among the fractions of this plant, 2-hydroxy-1, 7-dimethoxy xanthone found potent to relax cavernosum smooth muscle cell in a specific in vitro assay (Meyer et al., 2008). The secondary metabolites such as alkaloids, flavonoids, coumarins, glycosides, gums, polysaccharides, phenols, tannins, terpenes and terpenoids are present in *Securidaca longepedunculata*. Moreover following chemical constituents 1-methyl 2-hydroxybenzoate, methyl salicylate, methyl 3-hydroxybenzoate, methyl 4-hydroxybenzoate, methyl 2-hydroxy-6-methoxybenzoate and benzyl 2-hydroxy-6-methoxybenzoate are reported in *Securidaca longepedunculata* (Jayasekara et al., 2002).

5.3.2.40   *Tribulus terrestris*

Tribulus terrestris (Family Zygophyllaceae) has been used in traditional system of medicine for treatment of various diseases particularly sexual dysfunctions in men. Efficacy of its extract, protodioscin suggested that it works probably through androgen enhancement and nitric oxide release from nerve endings of corpus cavernosum tissue (Gauthaman et al., 2002; Gauthaman et al., 2003). Another study on male rats reported increased androgen receptor and NADPH-d positive neurons, which were probably due to its androgen enhancing properties (Gauthaman et al., 2005). The furostanol and spirostanol saponins of tigogenin, neotigogenin, gitogenin, neogitogenin, hecogenin, neohecogenin, diosgenin, chlorogenin, ruscogenin, and sarsasapogenin types of chemical constituents are found in this plant. The caffeoyl derivatives, quercetin glycosides, including rutin, kaempferol, kaempferol-3-glucoside, kaempferol-3-rutinoside and tribuloside are the flavonoids present in the various parts of *Tribulus terrestris*.

5.3.2.41 *Turnera diffusa*

Turnera diffusa is an aromatic plant of family Turneraceae and thrives in Mexico, Central America, some regions of South America and West Indies. Caffeine, arbutine, luteolin, apigenin, damianine, acacetin, sodium gluconate, gonzalitozin and flavonoids are the principle constituents found in the active extract of *Turnera diffusa*. A study more than a decade back established it as an aphrodisiac and supported its traditional reputation as a sexual stimulant when its fluid extract used alone or in combination with a shrub *Pfaffia paniculata* improved the sexual performance of the sexually sluggish or impotent animals (Arletti et al., 1999). Aqueous extract of *Turnera diffusa* found effective in recovery of sexual desire/motivation in sexually exhausted rats (Estrada-Reyes et al., 2009). Probably its flavonoid content affects the libido and sexual performance via their effects on central nervous systems; however, the exact mechanism of action is yet to be established.

5.3.2.42 *Tinospora cordifolia*

Tinospora cordifolia (Menispermaceae), hydroalcoholic extract of *Tinospora cordifolia*stem showed significant aphrodisiac activity on male wistar albino rats as evidenced by an increase in number of mounts and mating performance (Wani et al., 2011). The tinosporone, tinosporic acid, cordifolisides A to E, syringen, berberine, giloin, gilenin, crude giloininand, arabinogalactan polysaccharide, picrotene, bergenin, gilosterol, tinosporol, tinosporidine, sitosterol, cordifol, heptacosanol, octacosonal, tinosporide, columbin, chasmanthin, palmarin, palmatosides C and F, amritosides, cordioside, tinosponone, ecdysterone, makisterone A, hydroxyecdysone, magnoflorine, tembetarine, syringine, glucan polysaccharide, syringine apiosylglycoside, isocolumbin, palmatine, tetrahydropalmaitine and jatrorrhizine are the chief constituent present in *Tinospora cordifolia* (Singh et al., 2003).

5.3.2.43   *Trichopus zeylanicus*

Trichopus zeylanicus belongs to family Trichopodaceae, the extract of leaves of *Trichopus zeylanicus* increased the number of mounts and mating performance (Subramoniam et al., 1997).

5.3.2.44   *Turnera aphrodisiaca*

Turnera aphrodisiaca (Turneraceae) extract exhibited significant aphrodisiac activity due to presence of alkaloidal compound. The tetraphyllin B (cyanoglycoside), gonzalitosin I (flavonoid), arbutin (phenolic glycoside), damianin, tricosan-2-one, hexacosanol (hydrocarbons); a volatile oil containing α-pinene, β-pinene, p-cymene and 1,8-cineole and β-sitosterol are present in *Turnera aphrodisiaca* (Kumar et al., 2008; Kumar et al., 2009).

5.3.2.45   *Vanda tessellate*

Vanda tessellate is belong to family Orchidaceae and the extract increases mating performance, and tend to increase the male/female ratio resulting offspring (Kumar et al., 2000). The plant contains an alkaloid, a glucoside, tannins, β-sitosterol, γ-sitosterol and a long-chain aliphatic compound, fatty oils, resins and coloring matters. Roots contain tetracozyl ferrulate and β-sitosterol-D-glucoside. Some bioactive compounds have been isolated from *V. tessellata* are 2,7,7-tri methyl bicyclo [2.2.1] heptanes, 17-β-hydroxy-14,20-epoxy-1-oxo-[22R]-3β-[O-β-d-glucopyranosyl]-5,24-withadienolide, and melianin (Chowdhury et al., 2014).

5.4 CONCLUSION AND FUTURE PERSPECTIVE

The search for natural supplement from medicinal plants is being intensified probably because of its fewer side effects, its ready availability, and less cost. The available drugs and treatments have limited efficacy, unpleasant side effects, and contraindications in certain disease conditions. A variety of botanicals are known to have a potential effect on the sexual functions, supporting older claims and offering newer hopes. While evaluating various factors that control sexual function, identifies a variety of botanicals that may be potentially useful in treating sexual dysfunction. All the plants in this chapter have exhibited significant pharmacological activity. Demands of natural aphrodisiacs require increasing studies to understand their effects on humans and safety profile. Due to unavailability of the safety data, unclear mechanisms, and lack of knowledge to support the extensive use of these substances, uses of these products may be risky to the human being. With more clinical data, exact mechanisms of action, safety profile, and drug interaction with other uses of these aphrodisiacs plant materials, treating sexual disorder can become fruitful.

KEYWORDS

- aphrodisiac
- dysfunction
- erectile
- natural
- pharmacology
- sexual

REFERENCES

Agarwal, A., Nandipat, K. C., Sharma, R. K., Zippe, C. D., & Raina, R. (2006). Role of Oxidative Stress in the Pathophysiological Mechanism of Erectile Dysfunction. *J. Andrology. 27*, 3.

Agrawal, M. Y., Agrawal, Y. P., & Shamkuwar, P. B. (2014). Phytochemical and Biological Activities of *Chenopodium album*. *Int. J. Pharm Tech Research. 6*(1), 383–391.

Ahmed, A. M. A., Sharmen, F., Mannan, A., & Rahman, M. A. (2015). Phytochemical, analgesic, antibacterial and cytotoxic effects of *Alpinia nigra* Burtt leaf extract. *J. Trad. and Compl. Med.* [Online].

Ammosov, S., & Litvinenko, V. I. (2003). Triterpenoids of Plants of Glycyrrhiza L. and Meristotropis Fisch. Et Mey Genuses (A REVIEW). *Pharm Chem J. 37*, 83–94.

Ang, H. H., Cheang, H. S., & Yusof, A. P. (2000). Effects of *Eurycoma longifolia* Jack (Tongkat Ali) on the initiation of sexual performance of inexperienced castrated male rats. *Exp. Anim. 49*(1), 35–38.

Arletti, R., Benelli, A., Cavazzuti, E., Scarpetta, G., & Bertolini, A. (1999). Stimulating property of *Turnera diffusa* and *Pfaffia paniculata* extracts on the sexual-behavior of male rats. *Psychopharmacology. 143*(1), 15–19.

Asif M. (2012). A Review on Phytochemical and Ethnopharmacological Activities of *Curculigo orchioides*. *Mahidol University Journal of Pharmaceutical Sciences. 39*(3–4), 1–10.

Awate, S. A., Patil, R. B., Ghode, P. D., Patole, V., Pachauri, D., & HajaSherief, S. (2012). Aphrodisiac activity of aqueous extract of *Glycyrrhiza glabra* in male wistar rats. *World J Pharm Res. 1*(2), 371–378.

Bahmanpour, S., Talaei, T., Vojdani, Z., Panjehshahin, M. R., Poostpasand, A., Zareei, S., & Ghaeminia, M. (2006). Effect of *Phoenix Dactylifera* Pollen on Sperm Parameters and Reproductive system of Adult Male Rats. *Iran. J. Med. Sci. 31*(4), 208–212.

Barry, T. L., Petzinger, G., & Zito, S. W. (1996). GC/MS comparison of the West Indian aphrodisiac "Love Stone" to the Chinese medication "chan su": bufotenine and related bufadienolides. *J. Forensic Sci. 41*(6), 1068–1073.

Bella, A. J., & Shamloul, R. (2013). Traditional Plant Aphrodisiacs and Male Sexual Dysfunction. *Phytother. Res. 28*, 831–835.

Biglari, F., Abbas, F. M., Al Karkhi, & Azhar, M. E. (2008). Antioxidant activity and phenolic content of various date palm (*Phoenix dactylifera*) fruits from Iran. *Food Chemistry. 107*, 1636–1641.

Carro-Juarez, M., Cervantes, E., Cervantes-Méndez, M., & Rodríguez-Manzo G. (2004). Aphrodisiac properties of *Montanoa tomentosa* aqueous crude extract in male rats. *Pharmacol. Biochem. Behav. 78*(1), 129–134.

Chauhan, A., Singh, N., & Dixit, V. K. (2010). *Asteracantha longifolia* (L.) Nees, Acanthaceae:chemistry, traditional, medicinal uses and its pharmacological activities – areview. *Rev. bras. farmacogn.* [online], *20*(5), 812–817.

Chauhan, N. S., Sharma, V., Thakur, M., & Dixit, V. K. (2010). *Curculigo orchioides:* the black gold with numerous health benefits. *J. Chin. Integr. Med. 8*(7), 613–623.

Chauhan, N. S., Sharma, V., & Dixit, V. K. (2009). Effect of *Astercantha longifolia* seeds on the sexual behavior of male rats. *Nat. Prod. Res. 14*, 1–9.

Choi, H. K., Seong, D. H., & Rha, K. H. (1995). Clinical efficacy of Korean red ginseng for erectile dysfunction. *Int. J. Impot. Res. 7*(3), 181–186.

Chowdhury, M. A., Rahman, M. M., Chowdhury, R. H., Uddin, M. J., Sayeed, M. A., & Hossain, M. A. (2014). Antinociceptive and cytotoxic activities of an epiphytic medicinal orchid: *Vanda tessellata* Roxb. *BMC Complementary and Alternative Medicine,14*, 464.

Clark, J. T., Smith, E. R., & Davidson, J. M. (1984). Enhancement of sexual motivation in male rats by yohimbine. *Science. 225*(4664), 847–849.

Dhawan, K. (2003). Drug/substance reversal effects of a novel tri-substituted benzoflavone moiety (BZF) isolated from *Passiflora incarnata* Linn. – a brief perspective. *Addict. Biol. 8*(4), 379–386.

Dutta, I. C. (2007). Non Timber Forest Products of Nepal: identification, classification, ethnic uses & cultivation. Hill Side Press, Kathmandu, Nepal.

Estrada-Reyes, R., Ortiz-López, P., Gutiérrez-Ortíz, J., & Martínez-Mota L. (2009). *Turnera diffusa* Wild (Turneraceae) recovers sexual behavior in sexually exhausted males. *J. Ethnopharmacol. 123*(3), 423–429.

Gauthaman, K., & Adaikan, P. G. (2005). Effect of *Tribulus terrestris* on nicotinamide adenine dinucleotide phosphate-diaphorase activity and androgen receptors in rat brain. *J. Ethnopharmacol. 96*(1–2), 127–132.

Gauthaman, K., Adaikan, P. G., & Prasad, R. N. (2002). Aphrodisiac properties of *Tribulus Terrestris* extract (Protodioscin) in normal and castrated rats. *Life Sci. 71*(12), 1385–1396.

Gauthaman, K., Ganesan, A. P., & Prasad, R. N. (2003). Sexual effects of puncturevine (*Tribulus terrestris*) extract (protodioscin): an evaluation using a rat model. *J. Altern. Complement. Med. 9*(2), 257–265.

Hosseinzadeh, H., Ziaee, T., & Sadeghi, A. (2008). The effect of saffron, *Crocus sativus* stigma, extract and its constituents, safranal and crocin on sexual behaviors in normal male rats. *Phytomedicine. 15*(6–7), 491–495.

Ilodigwe, E. E., Igbokwe, E. N., Ajaghaku, D. L., & Ihekwereme, C. P. (2013). Aphrodisiac activity of ethanol root extract and fractions of *Landolphia dulcis* (Sabine) Pichon. *IJPSR. 4*(2), 809–814.

Ismail, Z., Ismail, N., & Lassa, J. (1999). Malaysian Herbal Monograph, vol. 1, Malaysian *Monograph Committee*, Kuala Lumpur, Malaysia.

Jayasekara, T. K., Stevenson, P. C., Belmain, S. R., Farman, D. I., & Hall, D. R. (2002). Identification of methyl salicylate as the principal volatile component in the methanol extracts of root bark of *Securidaca longepedunculata* Fers. *J. Mass Spectrom. 37*(6), 577–580.

JianFeng, C., PengYing, Z., ChengWei, X., TaoTao, H., YunGui, B., & KaoShan, C. (2012). Effect of aqueous extract of Arctium lappa (burdock) roots on the sexual behavior of male rats. *Complementary and Alternative Medicine. 12*, 8.

Karras, D. J., Farrell, S. E., Harrigan, R. A., Henretig, F. M., & Gealt, L. (1996). Poisoning from "Spanish fly" (cantharidin). *Am J Emerg Med. 14*(5), 478–483.

kazmi, I., Afzal, M., Rahman, M., Gupta, G., & Anwar, F. (2012). Aphrodisiac properties of *Polygonatumverticillatum* leaf extract. *Asian Pac. J. Trop. Dis.* 841–845.

Kenjale, R., Shah, R., & Sathaye, S. (2008). Effects of *Chlorophytum borivilianum* on sexual behavior and sperm count in male rats. *Phytother. Res. 22*(6), 796–801.

Khalid, N., Ahmed, I., Latif, M. S. J., Rafique, T., & Fawa, S. A. (2014). Comperision of Antimicrobial activity, Phytochemical profile and Minerals composition of Garlic *Allium sativum* and *Allium tuberosum. J. Korean Soc. Appl. Biol. Chem. 57*(3), 311–317.

Kumar, P. K. S., Subramoniam, A., & Pushpangadan, P. (2000). Aphrodisiac activity of *Vanda tessellate* (Roxb.) extract in male mice. *Indian J. Pharmacol. 32*, 300–304.

Kumar, S., Madaan, R., & Sharma, A. (2009). Evaluation of aphrodisiac activity of *Turneraaphrodisiaca. Int. J. Pharmacogn. Phytochem. Res. 1*(1), 1–4.

Kumar, S., Madaan, R., & Sharma, A. (2008). Pharmacological evaluation of Bioactive Principle of*Turnera aphrodisiaca. Indian Journal of Pharmaceutical Sciences. 70*(6), 740–744.

Kumar, S., Sumalatha, K., & Mohana Lakshmi, S. (2010). Aphrodisiac activity of Crossandrainfundibuliformis Linn. onetanol induced testicular toxicity in male rats. *Pharmacologyonline. 2*, 812–817.

Lampiao, F., Krom, D., & Du Plessis, S. S. (2008). The in vitro effects of *Mondia whitei* on human sperm motility parameters. *Phytother. Res. 22*, 1272–1273.

Lasker, G. F., Maley, J. H., & Kadowitz, P. J. (2010). A Review of the Pathophysiology and Novel Treatments for Erectile Dysfunction. *Advances in Pharmacological Sciences.* 2010, 1–10.

Li, Z., Niwa, Y., Sakamoto, S., Shono, M., Chen, X., & Nakaya, Y. (2000). Induction of inducible nitric oxide synthase by ginsenosides in cultured porcine endothelial cells. *Life Sci. 67*(24), 2983–2989.

Luo, Q., Li, Z., Huang, X., Yan, J., Zhang, S., & Cai, Y. Z. (2006). *Lycium barbarum* polysaccharides: Protective effects against heat-induced damage of rat testes and H_2O_2-induced DNA damage in mouse testicular cells and beneficial effect on sexual behavior and reproductive function of hemicastrated rats. *Life Sci. 79*(7), 613–621.

Lyttle, T., Goldstein, D., & Gartz, J. (1996). Bufo toads and bufotenine: fact and fiction surrounding an alleged psychedelic. *J. Psychoactive Drugs. 28*(3), 267–290.

Malviya, N., Jain, S., Gupta, V. B., & Vyas, S. (2011). Recent studies on aphrodisiac herbs for the management of male sexual dysfunction – A Review. *Acta Poloniae Pharmaceutica – Drug Research. 68*, 3–8.

Mbongue, F. G. Y., Kamtchouing, P., Essame, O. J. L., Yewah, P. M., Dimo, T., & Lontsi, D. (2005). Effect of the aqueous extract of dry fruits of *Piper guineenseon* the reproductive function of adult male rats. *Indian J. Pharmacol. 37*(1), 30–32.

Meyer, J. J., Rakuambo, N. C., & Hussein, A. A. (2008). Novel xanthones from *Securidaca longepedunculata* with activity against erectile dysfunction. *J. Ethnopharmacol. 119*(3), 599–603.

Mohan Maruga Raja, M. K., Agilandeswari, D., Madhu, B. H., Mallikarjuna Math, M., & SaiSowjanya, P. J. (2012). Aphrodisiac activity of ethanolic extract of *Nymphaeastellata* leaves in male rats. *Contemp. Invest. Observations Pharm. 1*(1), 24–30.

Morales, A. (2000). Yohimbine in erectile dysfunction: the facts. *Int. J. Impot. Res. 12*, 70–74.

Morita, H., Kishi, E., Takeya, K., Itokawa, H., & Iitaka, Y. (1993). Squalene derivatives from *Eurycoma longifolia. Phytochemistry. 34*(3), 765–771.

Pande, M., & Pathak, A. (2009). Aphrodisiac activity of roots of *Mimosa pudica* Linn. ethanolic extract in mice. *Int. J. Pharm. Sci. Nanotechnol. 2*(1), 477–486.

Pande, M., & Pathak, A. (2009). Effect of ethanolic extract of *Ocimumgratissimum* (Ram tulsi) on sexual behavior in male mice. *Int. J. Pharmaceutical. Tech. Res. 1*, 468–473.

Pande, M., & Pathak, A. (2008). Sexual function improving effect of *Chenopodium album* (bathu sag) in normal male mice. *Biomed. Pharmacol. J. 1*, 325–332.

Patel, D. K., Kumar, R., Prasad, S. K., & Hemalatha, S. (2011). Pharmacologically screen edaphrodisiac plant: A review of current scientific literature. *Asian Pacific Journal of Tropical Biomedicine.* 131–138.

Rakuambo, N. C., Meyer, J. J., Hussein, A., Huyzer, C., Mdlalose, S. P., & Raidani, T. G. (2006). *In vitro* effect of medicinal plants used to treat erectile dysfunction on smooth muscle relaxation and human sperm. *J. Ethnopharmacol. 105*(1–2), 84–88.

Ramachandran, S., Sridhar, Y., Sam, S. K., Saravanan, M., Leonard, J. T., Anbalagan, N., & Sridhar, S. K. (2004). Aphrodisiac activity of *Butea frondosa* Koen. ex Roxb. extract in male rats. *Phytomedicine. 11*(2–3), 165–168.

Ratnasooriya, W. D., & Fernando, T. S. (2008). Effect of black tea brew of *Camellia sinensis* on sexual competence of male rats. *J. Ethnopharmacol. 118*(3), 373–377.

Ratnasooriya, W. D., & Jayakody, J. R. (2006). Effects of aqueous extract of *Alpinia calcarata* rhizomes on reproductive competence of male rats. *Acta. Biol. Hung. 57*(1), 23–35.

Rewatkar, K. K., Ahmed, A., Khan, M. I., & Ganesh, N. (2010). A landmark approach to aphrodisiac property of *Abelmoschus manihot* (L.). *Int. J. Phytomed.* 2, 312–319.

Sandroni, P. (2001). Aphrodisiacs past and present: a historical review. *Clin. Auton. Res.* 11(5), 303–307.

Sharma, V., Thakur, M., Chauhan, N. S., & Dixit, V. K. (2010). Effects of petroleum ether extract of *Anacyclus pyrethrum* DC on sexual behavior in male rats. *Zhong Xi Yi Jie He Xue Bao.* 8(8), 767–773.

Singh, S. S., Pandey, S. C., Srivastava, S., Gupta, V. S., & Patro, B. (2003). Chemistry and Medicinal Properties of *Tinospora cordifolia* (Guduchi). *AC Ghosh Indian Journal of Pharmacology, 35*, 83–91.

Soni, D. K., Sharma, V., Chauhan, N. S., & Dixit, V. K. (2012). Effect of ethanolic extract of *Paederiafetida* Linn. leaves on sexual behavior and spermatogenesis in male rats. *J. Men's Health.* 9(4), 268–276.

Subramoniam, A., Madhavachandran, V., Rajasekharan, S., & Pushpangadan, P. (1997). Aphrodisiac property of *Trichopuszeylanicus* extract in male mice. *J. Ethnopharmacol. 57*, 21–27.

Subramoniam, A., Madhavachandran, V., Ravi, K., & Anuja, V. S. (2007). Aphrodisiac property of the elephant creeper *Argyreia nervosa. J. Endocrinol Reprod. 2*(11), 82–85.

Taha, S. A., Islam, M. W., & Ageel, A. M. (1995). Effect of ambrein, a major constituent of ambergris, on masculine sexual behavior in rats. *Arch. Int. Pharmacodyn. Ther. 329*(2), 283–294.

Thakur, M., Bhargava, S., & Dixi, V. K. (2009). Effect of *Asparagus racemosus* on sexual dysfunction in hyperglycemic male rats. *Pharm. Biol. 47*, 390–395.

Thakur, M., & Dixit, V. K. (2008). Ameliorative effect of Fructo-Oligosaccharide rich extract of *Orchislatifolia* Linn. on sexual dysfunction in hyperglycemic male rats. *Sex Disabil. 26*, 37–46.

Thakur, M., & Dixit, V. K. (2007). Aphrodisiac Activity of *Dactylorhiza hatagirea* (D. Don) Soo in Male Albino Rats. *Evid Based Complement Alternat. Med. 4*(1), 29–31.

Wagner, G., & Saenz de Tejada, I. (1998). Update on male erectile dysfunction. *BMJ. 316*, 678–685.

Wani J. A., Achur, R. N., & Nema, R. K. (2011). Phytochemical screening and aphrodisiac property of *Tinosporacordifolia. Int. J. Pharm. Clin. Res. 3*(2), 21–26.

Yakubu, M. T., & Afolayan, A. J. (2009). Effect of aqueous extract of *Bulbine natalensis* (Baker) stem on the sexual behavior of male rats. *Int. J. Androl. 32*(6), 629–636.

Yakubu, M. T., Akanji, M. A., & Oladiji, A. T. (2005). Aphrodisiac potentials of the aqueous extract of *Fadogia agrestis* (Schweinf. Ex Hiern) stem in male albino rats. *Asian J. Androl. 7*(4), 399–404.

Zamble, A., Hennebelle, T., Sahpaz, S., & Bailleul F. (2007). Two new quinoline and tris (4-hydroxycinnamoyl) spermine derivatives from *Microdesmis keayana* roots. *Chem. Pharm. Bull. 55*(4), 643–645.

Zamble, A., Martin-Nizard, F., Sahpaz, S., Hennebelle, T., Staels, B., Bordet, R., Duriez, P., Brunet, C., & Bailleul, F. (2008). Vasoactivity, antioxidant and aphrodisiac properties of *Caesalpinia benthamiana* roots. *J. Ethnopharmacol. 116*(1), 112–119.

Zamble, A., Martin-Nizard, F., Sahpaz, S., Reynaert, M. L., Staels, B., Bordet, R., Duriez, P., Gressier, B., & Bailleul F. (2009). Effects of *Microdesmis keayana* alkaloids on vascular parameters of erectile dysfunction. *Phytother. Res. 23*(6), 892–895.

Zanoli, P., Rivasi, M., Zavatti, M., Brusiani, F., Vezzalini, F., & Baraldi, M. (2005). Activity of single components of *Ferula hermonis* on male rat sexual behavior. *Int. J. Impot. Res. 17*(6), 513–518.

PART III

CHEMICAL MEDIATORS: DRUGGABLE TARGETS

CHAPTER 6

SEROTONIN: CHEMICAL, BIOLOGICAL, AND THERAPEUTIC ASPECTS

BIBHUDATTA MISHRA[1] and GUNJAN JOSHI[2]

[1]Wilmer Eye Institute, John Hopkins University School of Medicine, 400 N Broadway, Baltimore, MD 20287, USA

[2]Center for Biomedical Engineering and Technology, School of Medicine, University of Maryland, Baltimore, USA

CONTENTS

6.1   INTRODUCTION

Serotonin (5-hydroxytryptamine; 5-HT) is a monoamine neurotransmitter primarily present in the enterochromaffin cells (ECs) of gastrointestinal tract (GT), Central Nervous System (CNS), blood platelets and of gastrointestinal mucosa (Saxena PR, 1995). It was first isolated and characterized by Rapport et al. (1948a–c). Serotonin exerts its effects through a variety of membrane-bound receptors, present in the central and peripheral nervous system, and in nonneuronal tissues in the gut, cardiovascular system and blood (Hoyer et al., 2002). It belongs to a complex family of neurotransmitters consisting of seven families of 5-HT (5-HT$_1$ −5-HT$_7$) receptors (Hoyer et al., 2002), and at least 14 different serotonin receptor subtypes (Nichols DE and Nichols CD, 2008). Except for 5-HT$_3$ receptor, which is a ligand-gated ion channel mediating fast depolarization (Sugita et al., 1992), all other 5-HT recep-tors belong to the G-protein-coupled receptor (GPCR) superfamily (Hoyer et al., 2002). The 5-HT$_2$-receptor family is positively linked to phospholi-pase C, whereas 5-HT$_1$ and 5-HT$_5$ receptors inhibit adenylate cyclase and 5-HT$_4$, 5-HT$_6$ and 5-HT$_7$ receptors are known to stimulate adenylate cyclase (Hannon and Hoyer, 2008; Millan et al., 2008; Pytliak et al., 2011). These receptors are classified according to their structure, their pharmacology and the transduction signal initiated by them. Serotonin present in CNS is respon-sible for a number of behavioral functions, including the regulation of mood, sleep or wakefulness, appetite, nociception, and sexual behavior (Jacobs and Azmitia, 1992). Other than CNS, it is involved in the regulation of GI, endo-crine, respiratory, cardiovascular system (Berger et al., 2009). Because of its multiple regulatory roles in the human body, serotonin is an important target for a variety of therapeutic diseases including depression, migraines, schizo-phrenia, anxiety, and learning and memory disorders (Durham and Russo, 2002; Lesch, 2001; Mann et al., 2001).

6.2 BIOSYNTHESIS AND METABOLISM OF SEROTONIN

Serotonin is synthesized in ECs and neurons from its precursor amino acid tryptophan by two enzymatic steps. The first rate-limiting hydroxylation step involves the enzyme tryptophan hydroxylase (TPH), which transfers a hydroxyl group to the benzyl ring of tryptophan producing 5-hydroxy-tryptophan (5-HTP) (Lesurtel M et al., 2008). Two forms of TPH, a periph-eral form (TPH$_1$) and a central form (TPH$_2$) are known, providing different sources of peripheral and central 5-HT (Walther and Bader, 2003; Walther et al., 2003). In the second step, the enzyme aromatic amino acid decarboxyl-ase decarboxylases 5-HTP to produce serotonin (Gershon and Tamir, 1981).

Metabolism of serotonin in tissues is very rapid. Monoamine oxidase A (MAOA) and aldehyde dehydrogenase enzymes metabolize serotoninto 5-hydroxyindole acetic acid (5-HIAA) in the kidney and the liver, which ultimately gets excreted in the urine (Lesurtel et al., 2008). MAOA is an intracellular enzyme, hence serotonin reuptake transporter (SERT) and nor-adrenaline transporter (NET) facilitate the intake of 5-HT inside the cell (Kawasaki H et al., 1987). Any cell that can intake 5-HT and inhabit MAOA is able to metabolize 5-HT, mainly the lung, intestine and endothelial cells (Ni and Watts, 2006).

6.3 SEROTONIN RECEPTORS

6.3.1 THE 5-HT$_1$ RECEPTOR CLASS

The 5-HT1-receptor class forms the largest subclass of 5-HT receptor sub-types, comprising of five receptor subtypes (5-HT$_{1A}$, 5-HT$_{1B}$, 5-HT$_{1D}$, 5-HT$_{1E}$ and 5-HT$_{1F}$) (Hoyer et al., 2002). They are mostly (but not exclusively) cou-pled to Gi/G0 proteins and inhibit cAMP production (Pytliak et al., 2011). 5-HT1 receptors are distinguished by the absence of introns in their coding sequence (Saudou et al., 1992), their binding to G-protein and inhibiting cAMP (Gerhardt and van Heerikhuizen, 1997). 5-HT$_{1A}$, 5-HT$_{1B}$ and 5-HT$_{1D}$ receptors are present in many tissues of various species (Hoyer and Martin, 1997), whereas the physiological role of 5-HT$_{1E}$ and 5-HT$_{1F}$ receptors is still unknown (Hoyer et al., 2002).

6.3.1.1 5-HT$_{1A}$ Receptors

The gene encoding the human 5-HT$_{1A}$ receptor was first named G21 and was categorized as aG-protein-coupled receptor-encoding gene (Kobilka et al., 1987). The 5-HT$_{1A}$ receptors are widely distributed of all the 5-HT recep-tors, being present in the cerebral cortex, hippocampus, amygdala, raphe nucleus, basal ganglia and thalamus (El Mestikawy et al., 1993). They are known to activate phospholipase C pathway to some extent (Gerhardt and van Heerikhuizen, 1997). 5HT$_{1A}$ receptors play an important role in the regulation of body temperature and feeding behavior (Wang et al., 2009). They are also involved in psychiatric disorders (Gerhardt and van Heerikhuizen, 1997) and regulation of adrenocorticotrophic hormone (ACTH) (Jorgensen et al., 2001).

6.3.1.2   5-HT$_{1B}$ Receptors

These receptors are present in CNS and mainly concentrated in the basal ganglia and the striatum, where they act as autoreceptors inhibiting serotonin release and in the frontal cortex where they act as a terminal receptor inhibiting dopamine release (Pytliak et al., 2011). They are also found in cerebral and other arteries (Jin et al., 1992), where they are involved in arterial contraction (Hamel et al., 1993). They are also involved in behavior and anxiety responses (Wilkinson and Dourish, 1991). They also control release of other neurotransmitters like dopamine, acetylcholine, glutamate etc. (Pytliak et al., 2011). The activation of 5-HT$_{1B}$ receptor inhibits adenylyl cyclase (Gerhardt and van Heerikhuizen, 1997).

6.3.1.3   5-HT$_{1D}$ Receptors

The first 5-HT$_{1D}$ encoding gene was isolated as an 'orphan' G-protein-coupled receptor, called RDC4 in dog (Libert et al., 1989). The 5-HT1D receptor is coupled to second messenger systems and negatively regulates adenylate cyclase (Hamblin and Metcalf, 1991; Zgombick et al., 1991). Both 5-HT$_{1D}$ and 5-HT$_{1B}$ receptors show 63% structural homology, but the expression of 5-HT1D receptor is low as compared to 5-HT$_{1B}$ (Pytliak et al., 2011). 5-HT1D receptors are found in the CNS, where they cause vascular vasoconstriction and play a role in anxiety, locomotion and in some movement disorders as well as in migraine (Pytliak et al., 2011). They also stimulate serotonin release in the heart (Pullar et al., 2004).

6.3.1.4   5-HT$_{1E}$ Receptors

The 5-HT$_{1E}$ receptor was first identified as 5HT$_1$ receptor gene S31 (Levy et al., 1992a; Gudermann et al., 1993; McAllister et al., 1992; Zgombick et al., 1992). 5-ht1E receptor is extensively present in the frontal cortex, hippocampus and olfactory bulb, and functions in memory regulation (Shimron-Abarbanell et al., 1995). 5-HT$_{1E}$ activation inhibits adenylase cyclase by coupling to Gi (Levy et al., 1992; McAllister et al., 1992; Zgombick et al., 1992; Gudermann et al., 1993), but to a lesser extent than other 5-HT$_1$ receptors (Adham et al., 1994). It also stimulates cAMP production by coupling to Gs (Dukat et al., 2004).

6.3.1.5   5-HT$_{1F}$ Receptors

5-HT$_{1F}$ receptors are found in the brain mainly concentrated in the dorsal raphe, hippocampus, cortex, striatum, thalamus and hypothalamus (Hoyer et al., 2002). They are also present in the uterus and coronary arteries where they function in vascular contraction (e.g., Nilsson et al., 1999). 5-HT$_{1F}$ receptors have sequence homology to other 5-HT$_1$ receptors like 5-HT$_{1A}$, 5-HT$_{1D}$ and 5-HT$_{1E}$ (Pytliak et al., 2011). Human 5-HT$_{1F}$ receptors which were expressed in NIH-3T3 cells were shown to inhibit adenylase cyclase (Adham et al., 1993b), whereas those expressed in Ltk-fibroblasts caused phospholipase C activation and Ca^{2+} mobilization (Adham et al., 1993a).

6.3.2   THE 5-HT$_2$-RECEPTOR CLASS

This receptor class includes the 5-HT$_{2A}$, 5-HT$_{2B}$ and 5-HT$_{2C}$ receptors, with 46–50% sequence homology (Hoyer et al., 2002). They are coupled to Gq11 protein, and increase the inositol trisphosphate hydrolysis and intracellular Ca^{2+} concentration (Pytliak et al., 2011). 5-HT$_2$ receptor is the main excitatory G-protein coupled receptor subtype for serotonin (5-HT), but it has inhibitory role in the visual cortex and the orbitofrontal cortex (Hannon and Hoyer, 2002).

6.3.2.1   5-HT$_{2A}$ Receptors

5-HT$_{2A}$ receptor is expressed in many central and peripheral tissues (Pytliak et al., 2011). In the CNS, these receptors are mainly localized to the cortex, claustrum and basal ganglia (Hoyer et al., 2002). Its function in the central and peripheral tissues involves mediating contractile responses in many vascular smooth muscles, platelet aggregation (Cook et al., 1994), regulation of sleep and motor behavior (Gerhardt and van Heerikhuizen, 1997). 5-HT2A receptor activation also stimulates neurotransmitter and hormone secretion (Van de Kar et al., 2001, Bortolozzi et al., 2005, Feng et al., 2001). It is also involved in some psychiatric disorders like epilepsy, schizophrenia, migraine, anxiety and depression (Gerhardt and van Heerikhuizen, 1997).

6.3.2.2   5-HT$_{2B}$ Receptors

5-HT$_{2B}$ receptors are localized to the cerebellum, hypothalamus, lateral septum and amygdala (Cox et al., 1995; Schmuck et al., 1994). Its activation causes smooth muscle contraction of stomach fundus (Hoyer et al., 2002), as well as longitudinal muscle contraction in the human intestine (Ellis et al., 1995; Borman et al., 2002). In mouse fibroblasts, it is known to activate MAP kinase (Nebigil et al., 2000).

6.3.2.3   5-HT2C receptors

A minimum of 14 different functional isoforms of the 5-HT2C receptor have been identified so far (Burns et al., 1997; Fitzgerald et al., 1999). Little is known regarding the function of this receptor. 5-HT2C receptors may be involved in the initiation of migraine, hypoactivity, hypophagia and oral dyskinesia (Buhot, 1997).

6.3.3   THE 5-HT$_3$ RECEPTOR CLASS: AN INTRINSIC LIGAND-GATED CHANNEL

5-HT$_3$ receptors are present in various areas of the brain triggering depolarization (Laporte et al., 1992) and neurotransmitter release (Pytliak et al., 2011). The receptor consists of 5 subunits arranged around a central ion-conducting pore, which is permeable to potassium, calcium and sodium ions. Serotonin binding opens the channel, causing excitatory response in neurons (Pytliak et al., 2011). After activation, these receptors are known to release dopamine in brain, and also play a role in schizophrenia and anxiety (Hoyer et al., 2002). They also have an effect on the cardiovascular system and in regulation of motility and intestinal secretion (De Ponti and Tonini, 2001).

6.3.4   5-HT$_4$ RECEPTORS

Seven variants of 5-HT$_4$ receptor are known which couple preferentially to Gs and promote cAMP formation (Hoyer et al., 2002). 5-HT$_4$ receptors are also linked to potassium and calcium channels (Pauwels, 2003). These receptors are distributed specifically in certain areas like human intestine

(Pytliak et al., 2011). In the CNS, they stimulate the release of other neurotransmitters and increase synaptic transmission (Ciranna, 2006). 5-HT_4 receptors also play a role in cardiac, urinary and endocrine system (Bockaert et al., 1992).

6.3.5   5-HT$_5$ RECEPTORS

Two subtypes of 5-ht5 receptors exist, the 5-HT_{5A} and 5-HT_{5B} receptor subtypes with 5-HT_{5A} subtypes exclusively found in human brain (Grailhe et al., 2001), and are involved in motor control, behavior, anxiety, depression, learning and memory (Thomas, 2006).

6.3.6   5-HT$_6$ RECEPTORS

Like 5-HT_4 receptors, 5-HT_6 receptors also couple preferentially to Gs and promote cAMP formation (Hoyer et al., 2002). The receptor is located in amygdala, hippocampus, cortex and olfactory tubercle (Hoyer et al., 2002). Two subtypes of 5-HT_6 receptor are known so far, but their exact role is unknown (Pytliak et al., 2011). 5-HT_6 receptors may be involved in behavior and psychiatric disorders and in control of cholinergic neurotransmission (Bourson et al., 1995).

6.3.7   5-HT$_7$ RECEPTORS

5-HT_7 receptors have seven transmembrane domains similar to GPCRs, and are known to stimulate adenylase cyclase and promote cAMP formation via Gs protein pathway (Pytliak et al., 2011; Bard et al., 1993; Shen et al., 1993). The cAMP then activates protein kinase A (PKA) leading to phosphorylation of different proteins (Leopoldo et al., 2011). It is also known to stimulate the mitogen-activated protein kinase extracellular signal-regulated kinases (ERK) (Erricoetal, 2001). Four 5-HT_7-receptor isoforms (5-HT_7A-D) are known, each differing in their C-termini (Heidmann et al., 1997), but not in pharmacological and tissue distribution (Jasper et al., 1997; Heidmann et al., 1998). 5-HT_7 receptors are expressed in the vessels and in extravascular smooth muscles (e.g., in the gastrointestinal tract) and CNS, with the highest expression in thalamus, hypothalamus, amygdala and hippocampus (Hagan et al., 2000; Terrón and Martínez-García, 2007). 5-HT$_7$ receptor activation stimulates the activity of GABAergic interneurons in the hippocampus (Tokarski et al., 2011).

6.4 PATHOPHYSIOLOGICAL FUNCTIONS OF SEROTONIN

6.4.1   SEROTONIN IN BRAIN

Serotonin (5-HT) in the CNS acts as neurotransmitter and neurohormone and is known to play an important role in neural transmission (Turlejski, 1996), neurogenesis (Lauder and Krebs, 1978), cell migration, hormone secretion, synaptic plasticity, sleep, mood, aggressive behavior and sexual behavior (Chubakov et al., 1993; Lauder, 1990; Hannon and Hoyer, 2008; Nichols and Nichols, 2008; Daubert and Condron, 2010; Lesch and Waider, 2012). Serotonin helps in neuronal plasticity by maintaining the synaptic connections in the cortex and hippocampus (Azmitia et al., 1995; Mazer et al., 1997) and altering the length of dendrites (Faber and Haring, 1999; Mazer et al., 1997; Wilson et al., 1998). It also helps in maintaining synapses in the adult brain (Azmitia, 1999; Okado et al., 2001). The activity of 5-HT$_2$ receptors is seen higher in the developing brain (Claustre et al., 1988). Serotinergic insults like degeneration of serotonergic fibers in neurodegenerative disease are known to cause cognitive and psychological defects (Aucoin et al., 2005; Liu et al., 2008; Azmitia and Nixon, 2008). 5-HT also regulates neurite outgrowth and maintains of neuronal connectivity required for brain development (Daubert and Condron, 2010; Lesch and Waider, 2012). Any dysfunction in the serotonergic system leads to a number of neurodegenerative diseases like autism (Betancur et al., 2002; Marazziti et al., 2000), Down's syndrome (Geldmacher et al., 1997) schizophrenia (Dayer, 2014), depression and anxiety (McKeith et al., 1987). 5-HT receptors also play an important role in learning and memory (Buhot et al., 2003; Schmitt et al., 2006).

6.4.2   SEROTONIN IN CARDIOVASCULAR SYSTEM

Serotonin is known to regulate vascular tone by contraction and relaxation of blood vessels (Nilsson et al., 1999; Yildiz et al., 1998). The 5-HT$_{2A}$ receptor is found in arterial smooth muscle (Ullmer et al., 1995) and play a role in vasoconstriction (Kaumann et al., 1993), whereas it is absent on endothelial cells (Ullmer et al., 1995). In platelets it plays a role in platelet aggregation (De Clerck et al., 1984; De Chaffoy de Courcelles et al., 1985). Activation of 5-HT_{1A} receptors leads to a reduction in blood pressure and heart rate (Ramage, 2001). Unlike 5-HT_{1A}, 5-HT_{1B} receptor is found on both smooth and endothelial muscle cells (Ullmer et al., 1995) and facilitates vasoconstriction (Kaumann et al., 1993). The 5-HT_{2B} receptor exclusively plays a role in cardiac proliferating cells and in neural crest-derived progenitors (Choi and Maroteaux 1996; Choi et al., 1997). Little is known about the role of the $5\text{-HT}_{1D}/5\text{-HT}_E$ receptor in cardiovascular system. $5\text{-HT}_{1B}/5\text{-HT}_{1D}$ receptors have opposing action in rats, with the activation of 5-HT_{1B} receptors is known to cause a rise in blood pressure as against the activation of 5-HT_{1D} receptor, which leads to a decrease in blood pressure (Ramage, 2001). 5-HT_4 receptors are present in cardiac atria (Kaumann and Sanders, 1998) and ventricles (Brattelid et al., 2004) and facilitate contraction. No role of 5-HT_6 receptor in cardiovascular system is known. 5-HT_7 receptors are present in coronary arteries (Bard et al., 1993) and on vascular smooth muscle to facilitate relaxation (Ullmer et al., 1995). 5-HT receptors have been implicated in several vascular diseases, including hypertension, pre-eclampsia, migraine (Kaumann and Levy, 2006), depression (Hoefgen et al., 2005) and coronary artery disease (Vikenes et al., 1999). Increased 5-HT levels lead to sinus tachycardia, atrial fibrillation (Yusuf et al., 2003) and in patients with coronary lesions cause severe angina (Rubanyi et al., 1987; Van den Berg et al., 1989). Serotonin levels are also elevated in patients with congenital thrombocytopathy, a defect in the platelet-serotonin storage (Herve et al., 1990).

6.4.3 SEROTONIN IN GATROINTESTINAL TRACT

About 95% of serotonin in the body is found in the gastrointestinal tract (GI) formed by intestinal EC cells in the GI tract, and remaining 5% is found in the brain (Lesurtel et al., 2008). Hence, it functions both as a neurotransmitter and as a local hormone in the peripheral vascular system and in the gut (Gershon, 2004; Gershon and Tack, 2007). 5-HT is released from EC via a number of receptor-mediated mechanisms (McLean et al., 2007). The stimuli for serotonin release include food intake, adrenaline stimulation, lower intralumen pH, circulating amino acids or short-chain fatty acids and obstruction of gut motility (McLean et al., 2007). GI tract is the source of almost all of the serotonin in the blood (Bertaccini, 1960). Serotonin is known to act in the gut performing various functions like epithelial secretion, smooth muscle activation or relaxation and stimulation of sensory neurons (Gershon, 1991; Wallis et al., 1982). 5-HT$_{2B}$ receptors cause smooth muscle contraction in the human ileum (Borman and Burleigh, 1995), and activation of 5-HT$_7$ receptors and obstruction of 5-HT$_4$ receptors cause relaxation of human colon smooth muscle (Prins et al., 1999). 5-HT has been implicated in various GI disorders, such as inflammatory bowel disease (IBD) (Coates et al., 2004) irritable bowel syndrome (IBS) (Coates et al., 2004; Faure et al., 2010), nausea, vomiting, and peristalsis (Lesurtel et al., 2008). Altered 5-HT signaling is associated with celiac disease (Coleman et al., 2006), and colorectal cancer (Ataee et al., 2010). All these effects of serotonin in the gut are generated by multiple receptor subtypes present on smooth muscle, enteric neurons, and immune cells (Manocha and Khan, 2012).

6.4.4 SEROTONIN IN RESPIRATION

Serotonin plays an important role in controlling a large variety of sensory and motor functions. The role of serotonin in the control of breathing was noticed for the first time by Reid and Rand (1951). Later, many studies (Olsen Jr. et al., 1979; McCrimmon and Lalley, 1982; Morin et al., 1990; Monteau and Hilaire, 1991; Lindsay and Feldman, 1993) established its importance as a neuromodulator of breathing. Respiratory neurons contain multiple serotonin receptors and projections to multiple sites within the respiratory network (Steinbusch, 1981; Voss et al., 1990). Medullary raphe nuclei control respiration by sending 5-HT projections to the brain and spinal cord (Li et al., 1993; Manaker and Tischler, 1993). During hypoxia, 5-HT gets released by caudal raphe neurons near respiratory premotor and motor neurons (Teppema et al., 1997; Kinkead et al., 2001). 5-HT is known to restore respiration during hypoxia or ischemia (Bonham, 1995).

6.4.5 SEROTONIN IN CANCER

Serotonin (5HT) is found to be a probable regulator of cell proliferation (Vicaut et al., 2000) and is used as tumor marker of gastrointestinal, neuroendocrine, bronchial, hepatic, pancreatic and ovarian carcinoids (Jungwirth et al., 2008; De Harder, 2007). It plays an important role in the cancers of skin, gut, brain, lung, prostate, liver, breast and pancreas (Lesurtel et al., 2008; Mallon et al., 2000; Alpini et al., 2008; Ogawa et al., 2005; Siddiqui et al., 2005). It also plays an important role in tumor growth in other cancer types like cholangiocarcinoma (Alpini et al., 2008, Coufal et al., 2010), colon cancer (Nocito et al., 2008), and Hepatocellular carcinoma (HCC) (Soll et al., 2010). Various serotonin receptors are involved in tumorous growth. 5-HT_{1A}, 5-HT_{1B}, 5-HT_{2B} and 5-HT_4 receptors are found at various tumor stages of prostatic cancer (Abdul et al., 1995; Dizeyi et al., 2005), whereas 5-HT_{2B} receptor is the main target in HCC (Soll et al., 2010). 5-HT_1, 5-HT_3 and 5-HT_4 receptors play a potential role in colorectal cancer (Sonier et al., 2005). The 5-HT_{2A} receptor is expressed and localized in the plasma membrane of the human breast adenocarcinoma cell line (Pai et al., 2009). The blood vessels supplying the tumors contain many serotonin receptors, which helps promote angiogenesis and hence tumor growth (Zamani and Qu, 2012). 5-HT_{1D} and 5-HT_{2B} receptors are predominantly expressed in vascular endothelial cells in both benign and malignant prostate tissues (Dizeyi et al., 2005). Breast cancer cells shows deregulation of serotonin receptors, where some isoforms are expressed and others get suppressed (Pai et al., 2009; Kopparapu et al., 2013). Expression of 5-HT_{1A}, 5-HT_{1B}, 5-HT_{2B} and 5-HT_4 receptors are higher in benign and noninvasive cancer cells, and 5-HT_{2B} expression is reduced with disease progression (Henriksen et al., 2012).

6.4.6 SEROTONIN IN MIGRAINE

Migraine is a chronic debilitating headache disorder characterized mainly by recurrent throbbing pain on one side of the head. Serotonin is implicated in the pathogenesis of migraine with an increase in the urinary excretion of its metabolite, 5-hydroxy-indole-acetic acid (5-HIAA) during migraine attacks (Berman et al., 2006; Ferrari et al., 1989). Migraine is initiated by dilatation of blood vessels (Comings, 1994). The pain gets relieved on vomiting as it stimulates intestinal motility and raises blood serotonin (Aggarwal et al., 2012). The neuronal 5-HT_3 receptors, which are found in areas in brainstem and peripheral nervous system, are responsible for the mechanism of vomiting and nociception (Cubeddu et al., 1990; Pratt et al.,

1990). Blood serotonin is mostly present in platelets, and increased plate-let activity is present in migraine patients during attacks especially in with migraine with aura (D'Andrea et al., 1984). Serotonin gets released from the platelets in migraine, causing reduction in platelet serotonin content and increases in serotonin metabolites in the blood (Ferrari et al., 1989). Other studies (Jernej et al., 2002; Srikiatkhachorn and Anthony, 1996) did not show any role of platelet in migraine attacks suggesting indefi-nite relationship between platelet serotonin and migraine. These platelets release serotonin in a number of migraine cases (Lance, 1991; Panconesi and Sicuteri, 1997). Serotonin is an important regulator of pain sensation (Sommer, 2004). Serotonin is a potent vasoconstrictor of nerve endings and blood vessels and cause nociceptive pain (Taylor and Basbaum, 1995), inflammation with plasma protein extravasation and edema (Koo and Balaban, 2006). Out of all serotonin receptors, $5\text{-}HT_1$, $5\text{-}HT_2$, and $5\text{-}HT_3$ receptors (especially $5\text{-}HT_1$ are responsible for most of the migraine activ-ity and are present in the trigeminal nerve endings (Berman et al., 2006; Hu et al., 2004). It is seen that migraine is more prevalent in females after puberty, and estrogen plays a role in that, because a drop in estrogen level is seen during menstruation, which causes a decrease in serotonin lead-ing to migraine (Somerville, 1975). Moreover, more serotonin in plate-lets is found during the follicular phase in female migraineours, than in women without migraine in any phase of the menstrual cycle (Fioroni et al., 1996). Serotonin and other migraine drugs like triptans activate sero-tonin receptors $5\text{-}HT_{1B}$ and $5\text{-}HT_{1D}$, and cause vasoconstriction (Parsons and Whalley, 1989). Intravenous injections of serotonin reduce migraine headache (Kimball et al., 1960). Serotonin agonists like sumatriptan have been used to reduce migrainous headache (Anthony et al., 1967) by block-ing pain transmission via $5\text{-}HT_{1B}/5\text{-}HT_{1D}$ receptors on trigeminal nerve endings (Moskowitz, 1992).

6.5   SEROTONIN AGONISTS AND ANTAGONISTS

6.5.1   5-HT$_{1A}$ RECEPTOR AGONIST AND ANTAGONISTS

$5\text{-}HT_{1A}$ receptors have therapeutic potential in several neuropsychiatric dis-orders (Barnes and Neumaier, 2015), and for that many ligands, which are highly specific to partial agonists/antagonists, have been developed and tried successfully. 5-HT$_{1A}$ receptor partial agonists are known to act on the autoreceptors to reduce serotonergic activity, and function as antipsychotic drugs (Barnes and Neumaier, 2015).

Buspirone is a partial agonist that is used as a target for anxiety and depression (Tunnicliff, 1991; Den Boer et al., 2000). Gepirone and MDL 72832 are partial agonists (Hoyer et al., 2002). Xaliproden and S-14506 are selective agonists. 8-hydroxy-di-n-propylamino tetralin (8-OH-DPAT) is another important selective 5-HT$_{1A}$ receptor agonist (Arvidsson et al., 1981; Sleight and Peroutka, 1991; Lanfumey and Hamon, 2004) which share receptor affinity for 5-HT$_7$ receptor also in the brain areas controlling memory (Meneses and Hong, 1999; Perez-Garcia et al., 2006). 8-OH-DPAT is known to suppress dorsal raphe cell firing (Meneses and Perez-Garcia, 2007). Since both 5-HT$_{1A}$ and 5-HT$_7$ receptors have opposing actions in memory functions, sharing a common ligand with both the receptor has an important functional implication (Meneses and Hong, 1999, 2007; Romano et al., 2006). Amongst the antagonists, WAY-100635 is a highly selective neutral antagonist of 5-HT$_{1A}$ receptor (Forster et al., 1995; Fletcher et al., 1996).

6.5.2   5-HT$_{1B}$ RECEPTOR AGONIST AND ANTAGONISTS

5-HT$_{1B}$ receptor agonists have several behavioral effects, whereas selective antagonists are known to have procognitive potential (Clark and Neumaier, 2002; Sari, 2004). The nonselective 5-HT$_{1B}$/5-HT$_{1D}$ receptor agonist sumatriptan prolongs the migrating motor complex (MMC) cycle (Calvert et al., 2004), which helps to clear undigested material. With its antimigraine properties, sumatripan has and increase in interest in the clinical trials (Pytliak et al., 2011). In patients with functional dyspepsia, it relaxes the gastric fundus, reduces gastric emptying and early satiety (Sanger, 1998). Some other selective agonists of 5-HT$_{1B}$ receptor includes CP-93129 and the more brain-penetrant CP-94253 (Barnes and Neumaier, 2015), MK-462 (rizatriptan), BW-311C90 (zolmitriptan), GR-46611 and L-694247 (Hoyer et al., 2002). Some agonists have affinity for multiple receptors, for eg. sumatriptan has significant affinity to 5-HT$_{1F}$ receptors (Hoyer et al., 2002), and L-694247 recognizes both 5-HT$_{1B}$ and 5-HT$_{1D}$ receptors equally (Hoyer et al., 2002). SB-224289 is a known 5-HT$_{1B}$ receptor antagonist (Barnes and Neumaier, 2015).

6.5.3   5-HT$_{1D}$ RECEPTOR AGONIST AND ANTAGONISTS

5-HT$_{1D}$ receptor ligands share high affinity for more than one receptor, preferably the 5-HT$_{1B}$ or 5-HT$_{1A}$ receptors (Barnes and Neumaier, 2015). Interestingly, selective antagonist ketanserin, which has highest affinity for 5-HT$_{2A}$ receptors, is known to have ~100-fold higher affinity for human 5-HT$_{1D}$ than 5-HT$_{1B}$ receptors (Barnes and Neumaier, 2015). Another antagonist, GR-127935, blocks the effect of antidepressants in the mouse tail suspension test (O'Neill et al., 1996). Selective 5-HT$_{1D}$ receptor agonist like PNU-109291 play an important role in suppression of trigeminal nociception as well as neurogenic inflammation in trigeminic ganglia, making 5-HT$_{1D}$ receptor subtype an important therapeutic target for migraine and related headaches (Cutrer et al., 1999).

6.5.4   5-HT$_{1E}$ RECEPTOR AGONIST AND ANTAGONISTS

BRL-54443 is a selective agonist for the 5-*ht1E* and 5-*ht1E* serotonin receptor subtypes (McKune and Watts, 2001; Janssen et al., 2004). One of the weak competitive antagonists of 5-ht1E known is methiothepin (Zgombick et al., 1992). Other than these, not much information is available about 5-ht1E receptor agonist and antagonists.

6.5.5   5-HT$_{1F}$ RECEPTOR AGONIST AND ANTAGONISTS

Both 5-HT$_{1B}$ and 5-HT$_{1F}$ receptors are expressed in trigeminal ganglion and vestibular nuclei neurons, and have high affinity for triptan drugs which are useful for migraine treatment. Hence, more selective drugs need to be tested to get selective outcomes (Barnes and Neumaier, 2015). One such selective 5-HT$_{1F}$-receptoragonist LY-334370, has ~100-fold higher affinity for 5-HT$_{1F}$ over 5-HT$_{1B}$ receptors, shows prominent binding in the cortical areas, striatum, hippocampus and olfactory bulb and is used for antimigraine activity (Shepheard et al., 1999). Some other selective 5-HT$_{1F}$ agonists like LY-344864 and BRL-54443, also have affinity for 5-ht1E receptors (Phebus et al., 1997; McKune and Watts, 2001). Interestingly, the antimigraine 5-HT1B/1D agonists sumatriptan and naratriptan also shows high affinity to 5-HT$_{1F}$ receptors (Hoyer et al., 2002). To date, no selective 5-HT$_{1F}$ antagonists have been recognized.

6.5.6   5-HT2A RECEPTOR AGONIST AND ANTAGONISTS

8-hydroxy-2-(di-n-propylamino) tetralin (DOI) and lyzergic acid diethylamide (LSD) are the two 5-HT$_{2A}$ receptor agonists with high affinity for 5-HT$_{2A}$ receptors, whereas TCB-2 is a selective agonist of 5-HT$_{2A}$ receptor (Barnes and Neumaier, 2015). Highly selective antagonist of 5-HT$_{2A}$ receptors includes ketanserin and MDL-100907 (Barnes and Neumaier, 2015). One nonspecific antagonist of 5-HT$_{1A}$ and 5-HT$_{2A}$ receptors is cyproheptadine (Nisijima et al., 2001). Some of the 5-HT$_{2A}$ antagonists with divergent selectivity include risperidone, ritanserin, seroquel, olanzapine, have been developed for the treatment of schizophrenia (Kim et al., 2009). Ritanserin has comparable affinity for both 5-HT$_{2A}$ and 5-HT$_{2C}$ receptors, whereas ketanserin has seven-fold greater affinity for 5-HT$_{2A}$ receptor (Graeff, 1997). Ritanserin is an atypical receptor antagonist because it downregulates, instead of upregulating 5-HT$_{2A}$/5-HT$_{2C}$ receptors (Graeff, 1997).

6.5.7   5-HT$_{2B}$ RECEPTOR AGONIST AND ANTAGONISTS

SB-204741 is the first reported selective 5-HT$_{2B}$ receptor antagonist (Hoyer et al., 2002). Amongst the other few highly selective 5-HT$_{2B}$ receptor antagonists is RS-127445 (Barnes and Neumaier, 2015). With other functions, it is also known to affect colonic motility, reducing peristalsis in mouse isolated colon and fecal output in rats (Bassil et al., 2009). SB-200646 and SB-206553 are selective 5-HT$_{2C}$/5-HT$_{2B}$ receptor antagonists, with low affinity for 5-HT2A (Kennett et al., 1994,1996). 5-methoxytryptamine acts as a full agonist with high affinity for the 5-HT$_{2B}$ sites, and α-Methyl-5-hydroxytryptamine (α-Me-5-HT) is its selective agonist (Jerman et al., 2001).

6.5.8   5-HT$_{2C}$ RECEPTOR AGONIST AND ANTAGONISTS

To date, no highly selective 5-HT$_{2C}$ agonists have been developed (Barnes and Neumaier, 2015). Partial agonist of 5-HT$_{2C}$ includes Lorcaserin (Fletcher et al., 2009). MK-212 and Ro60–0175 are some of the moderately selective agonists of 5-HT$_{2C}$ receptors. SB-242084 is a selective 5-HT$_{2C}$ receptor antagonist with anxiolytic activity in various animal models (Kennett et al., 1997b). LY-53857, ritanserin, mianserin and mesulergine are some of the nonselective antagonists of 5-HT$_{2C}$ receptors (Hoyer et al., 1994).

6.5.9 5-HT$_3$ RECEPTOR AGONIST AND ANTAGONISTS

A number of 5-HT$_3$ antagonists including ondansetron, granisetron, dolasetron, and ramosetron are effective in the treatment of chemotherapy and nausea associated with chemotherapy (Callahan, 2002; Gyermek, 1995). These antagonists may enhance memory and could be beneficial in migraine, anxiety, depression, pain and dementia (Pytliak et al., 2011). Partial agonists of 5HT$_3$ receptor include pumosetrag (DDP733) and mCPBG (m-Chlorophenyl biguanidine). SR-57227A is a full agonist and although pumosetrag is a partial agonist of 5HT$_3$ receptor, it shows relatively high levels of agonistactivity causing emesis in some patients (Evangelista, 2007). Pumosetrag is effective in patients with constipation and low intestinal motility and constipation-induced irritable bowel syndrome (cIBS) (Sanger, 2008). The 5-HT$_3$-receptor antagonist alosetron is used for relief in irritable bowel syndrome (IBS). It was temporarily withdrawn from the market for its side effects including an increase in ischemic colitis and mesenteric ischemia (Chang et al., 2006). Later, aldosterone was reintroduced in the USA on restricted basis (Sanger, 2008). Alosetron and cilansetron are also effective in reducing symptoms in diarrhea-predominant intestinal bowel syndrome (D-IBS) (Andresen et al., 2008), although cilansetron did not receive FDA-approval.

6.5.10 5-HT$_4$ RECEPTOR AGONIST AND ANTAGONISTS

Selective 5-HT$_4$ receptor ligands possess therapeutic potential in cardiac arrhythmia (Kaumann and Sanders, 1994; Rahme et al., 1999), neurodegenerative diseases (Reynolds et al., 1995; Wong et al., 1996) and urinary incontinence (Boyd and Rohan, 1994; Hegde and Eglen, 1996). Selective high affinity antagonists of 5-HT$_4$ receptor include GR-113808, SB-204070 and RS-100235 (Bonhaus et al., 1994; Clark, 1998), whereas nontryptamine selective agonists include RS-67506, ML-10302 and BIMU8 (Eglen, 1997). 5-methoxytryptamine is a nonselective agonist of the 5-HT$_4$ receptor. Cisapride, a benzamide derivatives and gastrokinetic agent has agonist actions, but was withdrawn due to arrhythmogenic activity (Barnes and Neumaier, 2015) (Kii Y et al., 2001). It acts on 5-HT$_{2B}$, 5-HT$_3$, D$_2$, and α$_1$ adrenergic receptors (Beattie DT, Smith JA, 2008). In addition to relief in gastroesophageal reflux, cisapride has an affinity for the cardiac hERG (human ether-a-go-go) potassium channel that predisposed patients to sudden cardiac death from cardiac dysrhythmias (De Maeyer JH et al., 2008).Unlike cisapride, another 5-HT$_4$ receptor agonist, tegaserod (HTF-919) has little effect on the hERG channel (Pasricha, 2007), and was also withdrawn due to its cardiovascular side effects similar to cisapride (Barnes and Neumaier, 2015). It is a partial 5-HT$_4$ agonist and a 5-HT$_{2B}$ antagonist, inhibited serotonin transporter (SERT), and bind to 5-HT$_{1A}$, 5-HT$_{1B}$, 5-HT$_{1D}$, 5-HT$_{2A}$, and 5-HT$_7$ receptors (De Maeyer et al., 2008; Beattie, Smith, 2008). Tegaserod is known to decrease constipation and provide comfort in heartburn (De Maeyer et al., 2008). Metoclopramide is a 5-HT$_4$ agonist that also acts as a 5-HT$_3$ and dopamine D$_2$ antagonist (Hasler, 2009). It is known to treat upper gut disorders (Mohammad-Zadeh et al., 2008). Because of its associated toxicity after prolonged use, metoclopramide was advised to be used only for short-term basis (Hasler, 2009). Cisapride is thought to be a stronger prokinetic than metoclopramide for colonic movement (Mohammad-Zadeh et al., 2008). TD-5108 (Velusetrag) is another 5-HT$_4$ agonist, which is selective for 5-HT$_4$ receptors over 5-HT$_{2B}$ and 5-HT$_3$ subtypes, and does not affect hERG channels (Hasler, 2009). Other 5-HT$_4$ agonists include renzapride and mosapride (Kim et al., 2008; Spiller et al., 2008). Pumosetrag is a partial 5-HT3 agonist that enhances intestinal motor function and relieves chronic constipation and c-IBS (Evangelista, 2007). Prucalopride is a benzofuran-derived 5-HT4-receptor agonist that stimulates stomach (De Schryver et al., 2002), and increases spontaneous bowel movements (Coremans et al., 2003).

6.5.11   5-HT$_5$ RECEPTOR AGONIST AND ANTAGONISTS

SB-699551 is a potent selective antagonist of 5-HT$_5$ receptor, with higher selectivity towards 5-HT$_5$ receptor than other 5-HT receptor types (Thomas, 2006). Its relatively low specificity towards rodent 5-HT$_5$ receptors limits its use in further clinical trials.

6.5.12   5-HT$_6$ RECEPTOR AGONIST AND ANTAGONISTS

2-Ethyl-5-methoxy-N,N-dimethyltryptamine   (EMDT),   EMD-386088, and WAY-181,187 are relatively selective 5-HT$_6$ receptor agonists, whereas SB-399885, SB-258585 and SB-271046 are some potent selective 5-HT$_6$ antagonists (Glennon et al., 2000; Bromidge et al., 1999; Barnes and Neumaier, 2015). The selective 5-HT$_6$ receptor antagonist, Ro 04-6790 plays a role in behavioral syndrome by involving an increase in acetylcholine neurotransmission (Bourson et al., 1995; Sleight et al., 1998).

6.5.13 5-HT$_7$ RECEPTOR AGONIST AND ANTAGONISTS

AS-19 and LP-12 are moderately selective agonists, and SB-258719 and SB-269970 are very selective antagonists of 5HT$_7$ receptors (Thomas et al., 1998a; Hagan et al., 2000). SB-269970 is known to block all reflex-evoked vagal bradycardias in rats (37,38). 5-carboxamidotryptamine (5-CT) is another 5-HT$_7$ agonist with reasonable affinity for 5-HT$_{1A}$, 5-HT$_{1B}$, 5-HT$_{1D}$ and 5-HT$_{5A}$ receptors (Ramage and Villalón, 2008). 5-CT-induced hypothermia in guinea pigs was blocked by both SB 269970 and the nonselective 5-HT$_7$ receptor antagonist, metergoline (Hoyer et al., 2002).

The structures of agonists and antagonists are described in Figures 6.1 and 6.2. The IUPAC nomenclature of all these serotonin receptors agonists and antagonists are shown in Table 6.1.

6.6 CONCLUSION

Serotonin functions through distinct G-protein-coupled receptors and a ligand-gated ion channel. It has diverse biological functions in CNS, GI, respiratory, endocrine and cardiovascular system. Any defect in the serotonergic system may lead to memory, cognition, mood regulation, psychiatric and other disorders. Over the years, diverse physiological roles of serotonin have been identified making it an important target for a variety of diseases. The functions of many of its subtypes in various diseases still need to be explored. This diversity in its functions and insufficient availability of selective ligands makes drug targeting difficult for serotonin receptors. With the development of new tools for specific targeting of serotonin receptors, the therapeutic potential of serotonin receptors can be improved.

TABLE 6.1   Receptor Agonist and Antagonists and Their IUPAC Nomenclature

Receptor type	Agonists	IUPAC
5-HT1A	Xaliproden	1-[2-(2-naphthyl)ethyl]-4-[3-(trifluoromethyl) phenyl]-1,2,3,6-tetrahydropyridine
	Busiprone	8-[4-(4-pyrimidin-2-ylpiperazin-1-yl)butyl]-8-azaspiro[4.5]decane-7,9-dione
	Gepirone	4,4-dimethyl-1-[4-(4-pyrimidin-2-ylpiperazin-1-yl) butyl]piperidine-2,6-dione
	MDL-72832	8-[4-(l,4- benzodioxan-2-ylmethylamino)butyl]-8-azaspiro[4.5]decane-7,9-dione
	Xaliproden	1-[2-(2-naphthyl)ethyl]-4-[3-(trifluoromethyl) phenyl]-1,2,3,6-tetrahydropyridine
	8-OH-DPAT	7-(Dipropylamino)-5,6,7,8-tetrahydronaphthalen-1-ol
5-HT1B	Sumatriptan	1-[3-(2-Dimethylaminoethyl)-1H-indol-5-yl]-N-methyl-methanesulfonamide
	CP-93129	3-(1,2,3,6-tetrahydropyridin-4-yl)-1,4-dihydropyrrolo[3,2-b]pyridin-5-one
	CP-94253	3-(1,2,5,6-tetrahydro-4-pyridyl)-5-propoxypyrrolo[3,2-b]pyridine
	MK-462 (rizatriptan)	N,N-dimethyl-2-[5-(1H-1,2,4-triazol-1-ylmethyl)-1H-indol-3-yl]ethanamine
	BW-311C90 (zolmitriptan)	(S)-4-({3-[2-(dimethylamino)ethyl]-1H-indol-5-yl} methyl)-1,3-oxazolidin-2-one
	GR-46611	3-[3-(2-Dimethylaminoethyl)-1H-indol-5-yl]-N-(4-methoxybenzyl)acrylamide
	L-694247	(2-[5-[3-(4-methylsulphonylamino)benzyl-1,2,4-oxadiazol-5-yl]-1H-indole-3-yl]ethylamine)
5-HT1D	PNU-109291	(S)-3,4-Dihydro-1-[2-[4-(4-methoxyphenyl)-1-piperazinyl]ethyl]-N-methyl-1H-2-benzopyran-6-carboxamide
5-HT1E	BRL-54443	3-(1-methylpiperidin-4-yl)-1H-indol-5-ol
5-HT1F	LY-344864	N-[(3R)-3-(Dimethylamino)-2,3,4,9-tetrahydro-1H-carbazol-6-yl]-4-fluorobenzamide hydrochloride
	BRL-54443	3-(1-methylpiperidin-4-yl)-1H-indol-5-ol
	LY-334370	4-fluoro-N-[3-(1-methyl-4-piperidinyl)-1H-indol-5-yl]benzamide
5-HT2A	Lyzergic acid diethylamide (LSD)	6aR,9R)-N,N-diethyl-7-methyl-4,6,6a,7,8,9-hexahydroindolo-[4,3-fg]quinoline-9-carboxamide
	TCB-2	1-[(7R)-3-bromo-2,5-dimethoxybicyclo[4.2.0]octa-1,3,5-trien-7-yl]methanamine

TABLE 6.1 (Continued)

Receptor type	Agonists	IUPAC
5-HT2B	5-methoxytryptamine	2-(5-Methoxy-1H-indol-3-yl)ethanamine
	α-Me-5-HT	3-(2-aminopropyl)-1H-indol-5-ol
5-HT2C	Lorcaserin	(1R)-8-chloro-1-methyl-2,3,4,5-tetrahydro-1H-3-benzazepine
	MK-212	2-Chloro-6-(1-piperazinyl)pyrazine
	Ro60-0175	(S)-6-Chloro-5-fluoro-1H-indole-2-propanamine
5-HT3	Pumosetrag	N-[(3R)-1-azabicyclo[2.2.2]octan-3-yl]-7-oxo-4H-thieno[3,2-b]pyridine-6-carboxamide
	mCPBG	2-(3-chlorophenyl)-1-(diaminomethylidene)guanidine
	SR-57227A	1-(6-chloropyridin-2-yl)piperidin-4-amine
5-HT4	RS-67506	N-[2-[4-[3-(4-amino-5-chloro-2-methoxyphenyl)-3-oxopropyl]piperidin-1-yl]ethyl]methanesulfonamide
	ML-10302	4-Amino-5-chloro-2-methoxybenzoic acid 2-(1-piperidinyl)ethyl ester hydrochloride
	BIMU8	N-[(1R,5S)-8-methyl-8-azabicyclo[3.2.1]oct-3-yl]-2-oxo-3-(propan-2-yl)-2,3-dihydro-1H-benzimidazole-1-carboxamide hydrochloride
	Cisapride	(±)-cis-4-amino-5-chloro-N-(1-[3-(4-fluorophenoxy)propyl]-3-methoxypiperidin-4-yl)-2-methoxybenzamide
	Tegaserod	(2E)-2-[(5-Methoxy-1H-indol-3-yl)methylene]-N-pentylhydrazinecarboximidamide
	Metoclopramide	4-amino-5-chloro-N-(2-(diethylamino)ethyl)-2-methoxybenzamide
	Velusetrag	N-[8-[(2R)-2-hydroxy-3-[methyl(methylsulfonyl)amino]propyl]-8-azabicyclo[3.2.1]octan-3-yl]-2-oxo-1-propan-2-ylquinoline-3-carboxamide
	Renzapride	4-amino-N-[(4S,5S)-1-azabicyclo[3.3.1]non4-yl]-5-chloro-2-methoxybenzamide
	Mosapride	(RS)-4-amino-5-chloro-2-ethoxy-N-{[4-(4-fluorobenzyl)morpholin-2-yl]methyl}benzamide
	Pumosetrag	N-[(8R)-1-azabicyclo[2.2.2]oct-8-yl]-2-oxo-9-thia-5-azabicyclo[4.3.0]n ona-3,7,10-triene-3-carboxamide hydrochloride; N-[(3R)-1-azabicyclo[2.2.2]oct-3-yl]-7-oxo-4,7-dihydrothieno[3,2-b]pyridine-6-carboxamide hydrochloride
	Prucalopride	4-Amino-5-chloro-N-[1-(3-methoxypropyl)piperidin-4-yl]-2,3-dihydro-1-benzofuran-7-carboxamide

TABLE 6.1 (Continued)

Receptor type	Agonists	IUPAC
5-HT6	EMDT	2-(2-ethyl-5-methoxy-1H-indol-3-yl)-*N*,*N*-dimethylethanamine
	EMD 386088	5-chloro-2-methyl-3-(1,2,3,6-tetrahydro-4-pyridinyl)-1*H*-indole
	WAY-181,187	2-(1-{6-chloroimidazo[2,1-b][1,3]thiazole-5-sulfonyl}-1H-indol-3-yl)ethan-1-amine
5-HT67	AS-19	(2S)-N,N-dimethyl-5-(1,3,5-trimethylpyrazol-4-yl)-1,2,3,4-tetrahydronaphthalen-2-amine
	LP-12	4-(2-diphenyl)-N-(1,2,3,4-tetrahydronaphthalen-1-yl)-1-piperazinehexanamide
	5-carboxamido-tryptamine (5-CT)	3-(2-Aminoethyl)-1*H*-indole-5-carboxamide

	Antagonists	**IUPAC**
5-HTA1	WAY-100635	*N*-[2-[4-(2-methoxyphenyl)-1-piperazinyl]ethyl]-*N*-(2-pyridyl)cyclohexanecarboxamide
5-HT1B	SB-224289	1'-Methyl-6,7-dihydro-5H-spiro[furo[2,3-f]indole-3,4'-piperidin]-5-yl)[2'-methyl-4'-(5-methyl-1,2,4-oxadiazol-3-yl)-4-biphenylyl]methanone
5-HT1D	Ketanserin	3-{2-[4-(4-fluorobenzoyl)piperidin-1-yl]ethyl}quinazoline-2,4(1*H*,3*H*)-dione
	GR127935	*N*-[4-methoxy-3-(4-methyl-1-piperazinyl)phenyl]-2'-methyl-4'-(5-methyl-1,2,4-oxadiazol-3-yl)-1–1'-biphenyl-4-carboxamide
5-HT1E	Methiothepin	1-methyl-4-(8-methylsulfanyl-5,6-dihydrobenzo[b][1]benzothiepin-6-yl)piperazine
5-HT1F	LY-344864	*N*-[(3*R*)-3-(Dimethylamino)-2,3,4,9-tetrahydro-1*H*-carbazol-6-yl]-4-fluorobenzamide hydrochloride
	BRL-54443	3-(1-methylpiperidin-4-yl)-1H-indol-5-ol
	LY-334370	4-fluoro-*N*-[3-(1-methyl-4-piperidinyl)-1*H*-indol-5-yl]benzamide
5-HT2A	Ketanserin	3-{2-[4-(4-fluorobenzoyl)piperidin-1-yl]ethyl}quinazoline-2,4(1*H*,3*H*)-dione
	Risperidone	4-[2-[4-(6-fluorobenzo[*d*]isoxazol-3-yl)-1-piperidyl]ethyl]-3-methyl-2,6-diazabicyclo[4.4.0]deca-1,3-dien-5-one
	Ritanserin	6-[2-[4-[bis(4-fluorophenyl)methylidene]piperidin-1-yl]ethyl]-7-methyl-[1,3]thiazolo[2,3-b]pyrimidin-5-one
	Seroquel	2-(2-(4-dibenzo[*b*,*f*][1,4]thiazepine-11-yl-1-piperazinyl)ethoxy)ethanol

TABLE 6.1   (Continued)

Receptor type	Agonists	IUPAC
	Olanzapine	2-Methyl-4-(4-methyl-1-piperazinyl)-10*H*-thieno[2,3-*b*][1,5]benzodiazepine
5-HT2B	SB-204741	*N*-(1-Methyl-1*H*-indol-5-yl)-*N*'-(3-methylisothiazol-5-yl)urea
	RS-127445	4-(4-fluoro-1-naphthyl)-6-isopropylpyrimidin-2-amine
	SB-200646	*N*-(1-Methyl-1*H*-indol-5-yl)-*N*'-3-pyridinylurea
	SB-206553	5-methyl-1-(3-pyridylcarbamoyl)-1,2,3,5-tetrahydropyrrolo[2,3-f]indole
5-HT2C	SB-242084	6-chloro-5-methyl-*N*-{6-[(2-methylpyridin-3-yl)oxy]pyridin-3-yl}indoline-1-carboxamide
	LY-53857	6-Methyl-1-(1-methylethyl)ergoline-8β-carboxylic acid 2-hydroxy-1-methylpropyl ester maleate salt
	Mianserin	(±)-2-methyl-1,2,3,4,10,14b-hexahydrodibenzo[*c,f*]pyrazino[1,2-*a*]azepine
	Mesulergine	*N*'-[(8α)-1,6-dimethylergolin-8-yl)-*N,N*-dimethylsulfamide
5-HT3	Ondansetron	(*RS*)-9-methyl-3-[(2-methyl-1*H*-imidazol-1-yl)methyl]-2,3-dihydro-1*H*-carbazol-4(9*H*)-one
	Granisetron	1-methyl-*N*-((1*R*,3*r*,5*S*)-9-methyl-9-azabicyclo[3.3.1]nonan-3-yl)-1*H*-indazole-3-carboxamide
	Dolasetron	(3*R*)-10-oxo-8-azatricyclo[5.3.1.0^{3,8}]undec-5-yl 1*H*-indole-3-carboxylate
	Ramosetron	(1-methyl-1*H*-indol-3-yl)[(5*R*)-4,5,6,7-tetrahydro-1*H*-benzimidazol-5-yl]methanone
	Alosetron	5-methyl-2-[(4-methyl-1*H*-imidazol-5-yl)methyl]-2,3,4,5-tetrahydro-1*H*-pyrido[4,3-*b*]indol-1-one
	Cilansetron	(10R)-10-[(2-methyl-1*H*-imidazol-1-yl)methyl]-5,6,9,10-tetrahydro-4*H*-pyrido(3,2,1-jk)carbazol-11-one
5-HT4	GR-113808	1-(2-methylsulfonylaminoethyl-4-piperidinyl)methyl-1-methyl-1H-indole-3-carboxylate
	SB-204070	(1-butyl-4-piperidinyl)methyl-8-amino-7-chloro-1,4-benzodioxane-5-carboxylate
	RS-100235	1-(5-amino-6-chloro-2,3-dihydro-1,4-benzodioxin-8-yl)-3-[1-[3-(3,4-dimethoxyphenyl)propyl]piperidin-4-yl]propan-1-one

TABLE 6.1 (Continued)

Receptor type	Agonists	IUPAC
5-HT5	SB-699551	3-cyclopentyl-N-[2-(dimethylamino)ethyl]-N-[(4'-([[(2-phenylethyl)amino]methyl)-4-biphenylyl)methyl]propanamide
5-HT6	SB-399885	N-(3,5-Dichloro-2-methoxyphenyl)-4-methoxy-3-(1-piperazinyl)benzenesulfonamide
	SB-271046	5-chloro-*N*-(4-methoxy-3-piperazin-1-ylphenyl)-3-methyl-1-benzothiophene-2-sulfonamide
5-HT7	SB-258719	(1R)-3,N-dimethyl-N-[1-methyl-3-(4-methylpiperidin-1-yl)propyl]benzenesulfonamide
	SB-269970	(2R)-1-[(3-Hydroxyphenyl)sulfonyl]-2-(2-(4-methyl-1-piperidinyl)ethyl)pyrrolidine
	Metergoline	{[(8β)-1,6-dimethylergolin-8-yl]methyl}carbamate

KEYWORDS

- **biosynthesis**
- **function**
- **inhibitor**
- **mechanism**
- **metabolism**
- **serotonin**

REFERENCES

Abdul, M., Logothetis, C. J., & Hoosein, N. M. (1995). Growth-inhibitory effects of serotonin uptake inhibitors on human prostate carcinoma cell lines. *J Urol 154*, 247–250.

Adham, N., Borden, L. A., Schechter, L. E., Gustafson, E. L., Cochran, T. L., Vaysse, P. J., Weinshank, R. L., & Branchek, T. A. (1993a). Cell-specific coupling of the cloned human 5-HT1F receptor to multiplesignal transduction pathways. *Naunyn Schmiedebergs Arch Pharmacol. 348*(6), 566–575.

Adham, N., Kao, H. T., Schecter, L. E., Bard, J., Olsen, M., Urquhart, D., Durkin, M., Hartig, P. R., Weinshank, R. L., & Branchek, T. A. (1993). Cloning of another human serotonin receptor (5-HT1F): a fifth 5-HT1 receptor subtype coupled to the inhibition of adenylate cyclase. *Proc Natl Acad Sci USA. 90*(2), 408–412.

Adham, N., Vaysse, P. J., Weinshank, R. L., & Branchek, T. A. (1994). The cloned human 5-HT1E receptor couples to inhibition and activation of adenylylcyclase via two distinct pathways in transfected BS-C-1 cells. *Neuropharmacology. 33(3–4)*, 403–410.

Aggarwal, M., Puri, V., & Puri, S. (2012). Serotonin and CGRP in migraine. *Ann Neurosci. 19*(2), 88–94.

Alpini, G., Invernizzi, P., Gaudio, E., Venter, J., Kopriva, S., & Bernuzzi, F., et al. (2008). Serotonin metabolism is dysregulated in cholangiocarcinoma, which has implications for tumor growth. *Cancer Res 68*, 9184–9193.

Andresen, V., Montori, V. M., Keller, J., West, C. P., Layer, P., & Camilleri, M. (2008). Effects of 5-hydroxytryptamine (serotonin) type 3 antagonists on symptom relief and constipation in nonconstipated irritable bowel syndrome: a systematic review and meta-analysis of randomized controlled trials. *Clin Gastroenterol Hepatol. 6*(5), 545–555.

Anthony, M., Hinterberger, H., & Lance, J. W. (1967). Plasma serotonin in migraine and stress. *Arch Neurol. 16*(5), 544–552.

Arvidsson, L. E., Hacksell, U., Nilsson, J. L., Hjorth, S., Carlsson, A., Lindberg, P., Sanchez, D., & Wikstrom, H. (1981). 8-Hydroxy-2-(di-n-propylamino)tetralin, a new centrally acting 5-hydroxytryptamine receptor agonist. *J Med Chem. 24*(8), 921–923.

Ataee, R., Ajdary, S., Zarrindast, M., Rezayat, M., & Hayatbakhsh, M. R. (2010). Anti-mitogenic and apoptotic effects of 5-HT1B receptor agonist on HT29 color ectalcancercelline. *J Cancer Res Clin Oncol. 136*(10), 1461–1469.

Aucoin, J. S., Jiang, P., Aznavour, N., Tong, X. K., Buttini, M., Descarries, L., & Hamel, E. (2005). Selective cholinergic denervation, independent from oxidative stress, in a mouse model of Alzheimer's disease. *Neuroscience. 132*(1), 73–86.

Azmitia, E. C., & Nixon, R. (2008). Dystrophic serotonergic axons in neurodegenerative diseases. *Brain Res. 1217*, 185–194.

Azmitia, E. C., Rubinstein, V. J., Strafaci, J. A., Rios, J. C., & Whitaker-Azmitia, P. M. (1995). 5-HT1 agonist and dexamethasone reversal of para-chloroamphetamine induced loss of MAP-2 and synaptophys in immunoreactivity in adult rat brain. *Brain Res. 677*(2), 181–192.

Azmitia, E. C. (1999). Serotonin neurons, neuroplasticity, and homeostasis of neural tissue. *Neuropsychopharmacology. 21*(2), 33S-45S.

Bard, J. A., Zgombick, J., Adham, N., Vaysse, P., Branchek, T. A., & Weinshank, R. L. (1993). Cloning of a novel human serotonin receptor (5-HT7) positively linked to adenylatecyclase. *J Biol Chem. 268*(31), 23422–23426.

Bard, J. A., Zgombick, J., Adham, N., Vaysse, P., Branchek, T. A., & Weinshank, R. L. (1993). Cloning of a novel human serotonin receptor (5-HT7) positively linked to adenylatecyclase. *J Biol Chem. 268*(31), 23422–23426.

Barnes, N. M., & Neumaier, J. F. (2015). Neuronal 5-HT Receptors and SERT. Tocris Bioscience Scientific Review Series. www.tocris.com. 1–16.

Bassil, A. K., Taylor, C. M., Bolton, V. J., Gray, K. M., Brown, J. D., Cutler, L., Summerfield, S. G., Bruton, G., Winchester, W. J., Lee, K., & Sanger, G. J. (2009). Inhibition of colonic motility and defecation by RS-127445 suggests an involvement of the 5-HT2B receptor in rodent large bowel physiology. *Br J Pharmacol. 158*(1), 252–258.

Beattie, D. T., & Smith, J. A. (2008). Serotonin pharmacology in the gastrointestinal tract: a review. Naunyn Schmiedebergs *Arch Pharmacol. 377*(3), 181–203.

Berger, M., Gray, J. A., & Roth, B. L. (2009). The expanded biology of serotonin. *Annu Rev Med. 60*, 355–366.

Berman, N. E., Puri, V., Chandrala, S., Puri, S., Macgregor, R., Liverman, C. S., & Klein, R. M. (2006). Serotonin in trigeminal ganglia of female rodents: relevance to menstrual migraine. Headache. *46*(8), 1230–1245.

Bertaccini, G. (1960). Tissue 5-hydroxytryptamine and urinary 5-hydroxyindoleacetic acid after partial or total removal of the gastro-intestinal tract in the rat. *J Physiol. 153*(2), 239–249.

Betancur, C., Corbex, M., Spielewoy, C., Philippe, A., Laplanche, J. L., Launay, J. M., Gillberg, C., Mouren-Siméoni, M. C., Hamon, M., Giros, B., Nosten-Bertrand, M., & Leboyer, M. (2002). Serotonin transporter gene polymorphisms and hyperserotonemia in autistic disorder. *Mol Psychiatry. 7*(1), 67–71.

Bockaert, J., Fozard, J. R., Dumuis, A., & Clarke, D. E. (1992). The 5-HT4 receptor: a place in the sun. *Trends Pharmacol Sci. 13*(4), 141–145.

Bonham, A. C. (1995). Neurotransmitters in the CNS control of breathing. *Respir Physiol. 101*(3), 219–230.

Borman, R. A., & Burleigh, D. E. (1995). Functional evidence for a 5-HT2B receptor mediating contraction of longitudinal muscle in human smallintestine. *Br J Pharmacol. 114*(8), 1525–1527.

Borman, R. A., Tilford, N. S., Harmer, D. W., Day, N., Ellis, E. S., Sheldrick, R. L., Carey, J., Coleman, R. A., & Baxter, G. S. (2002). 5-HT(2B) receptors play a key role in mediating the excitatory effects of 5-HT in human colon in vitro. *Br J Pharmacol. 135*(5), 1144–1151.

Bortolozzi, A., Díaz-Mataix, L., Scorza, M. C., Celada, P., & Artigas, F. (2005). The activation of 5-HT receptors in prefrontal cortex enhances dopaminergic activity. *J Neurochem. 95*(6), 1597–1607.

Bourson, A., Borroni, E., Austin, R. H., Monsma, F. J., & Sleight, A. J. (1995). Determination of the role of the 5-ht 6 receptor in the rat brain: a study using antisense oligonucleotides. *J Pharmacol Exp Ther. 274*, 173–180.

Boyd, I. W., & Rohan, A. P. (1994). Urinary disorders associated with cisapride. Adverse Drug Reactions Advisory Committee. *Med J Aust. 160*, 579–580.

Brattelid, T., Qvigstad, E., Lynham, J. A., Molenaar, P., Aass, H., Geiran, O., Skomedal, T., Osnes, J. B., Levy, F. O., & Kaumann, A. J. (2004). Functional serotonin 5-HT4 receptors in porcine and human ventricular myocardium with increased 5-HT4 mRNA in heart failure. *Naunyn Schmiedebergs Arch Pharmacol. 370*(3), 157–166.

Bromidge, S. M., Brown, A. M., Clarke, S. E., Dodgson, K., Gager, T., Grassam, H. L., Jeffrey, P. M., Joiner, G. F., King, F. D., Middlemiss, D. N., Moss, S. F., Newman, H., Riley, G., Routledge, C., & Wyman, P. J. (1999). 5-chloro-N-(4-methoxy-3-piperazin-1-yl-phenyl)-3-methyl-2-benzothiophene sulfonamide (SB-271046): A Potent, Selective, and Orally Bioavailable 5-HT6 Receptor Antagonist. *Med Chem. 42*, 202–205.

Buhot, M. C., Wolff, M., & Segu, L. (2003). Serotonin. In: Gernot, R., & Bettina, P. (Eds.). Memories are Made of These: From Messengers to Molecules. www.Eurekah.com and Kluwer Academic/Plenum Publishers, pp. 1–19.

Buhot, M. C. (1997). Serotonin receptors in cognitive behaviors. *Curr Opin Neurobiol. 7*(2), 243–254.

Burns, C. M., Chu, H., Rueter, S. M., Hutchinson, L. K., Canton, H., Sanders-Bush, E., & Emeson, R. B. (1997). Regulation of serotonin-2C receptor G-protein coupling by RNA editing. *Nature. 387*(6630), 303–308.

Callahan, M. J. (2002). Irritable bowel syndrome neuropharmacology. A review of approved and investigational compounds. *J Clin Gastroenterol.. 35*, S58–67.

Calvert, E. L., Whorwell, P. J., & Houghton, L. A. (2004). Inter-digestive and postprandial antro-pyloro-duodenal motor activity in humans: effect of 5-hydroxytryptamine 1 receptor agonism. *Aliment Pharmacol Ther.. 19*(7), 805–815.

Chang, L., Chey, W. D., Harris, L., Olden, K., Surawicz, C., & Schoenfeld, P. (2006). Incidence of is chemiccolitis and serious complications of constipation among patients using aldosterone: systematic review of clinical trials and postmarketing surveillance data. *Am J Gastroenterol.. 101*(5), 1069–1079.

Chang, L., Chey, W. D., Harris, L., Olden, K., Surawicz, C., & Schoenfeld, P. (2006). Incidence of ischemic colitis and serious complications of constipation among patients using alosetron: systematic review of clinical trials and post marketing surveillance data. *Am J Gastroenterol.. 101*(5), 1069–1079.

Choi, D. S., & Maroteaux, L. (1996). Immunohistochemical localization of the serotonin 5-HT2B receptor in mouse gut, cardiovascular system, and brain. *FEBS Lett.. 391(1–2)*, 45–51.

Choi, D. S., Ward, S. J., Messaddeq, N., Launay, J. M., & Maroteaux, (1997). L. 5-HT2B receptor-mediated serotonin morphogenetic functions in mouse cranial neural crest and myocardiac cells. *Development.. 124*(9), 1745–1755.

Chubakov, A. R., Tsyganova, V. G., & Sarkisova, E. F. (1993). The stimulating influence of the raphé nuclei on the morphofunctional development of the hippocampus during their combined cultivation. *Neurosci Behav Physiol.. 23*(3), 271–276.

Ciranna, L. (2006). Serotonin as a Modulator of Glutamate- and GABA-Mediated Neurotransmission: Implications in Physiological Functions and in Pathology. *Curr Neuropharmacol.. 4*(2), 101–114.

Clark, M. S., & Neumaier, J. F. (2001). The 5-HT1B receptor: behavioral implications. Psychopharmacol Bull. *Autumn; 35*(4), 170–185.

Claustre, Y., Rouquier, L., Scatton, & B. (1988). Pharmacological characterization of serotonin-stimulated phosphoinositide turnover in brain regions of the immaturerat. *J Pharmacol Exp Ther.. 244*(3), 1051–1056.

Coates, M. D., Mahoney, C. R., Linden, D. R., Sampson, J. E., Chen, J., Blaszyk, H., Crowell, M. D., Sharkey, K. A., Gershon, M. D., Mawe, G. M., & Moses, P. L. (2004). Molecular defects in mucosal serotonin content and decreased serotonin up take transporter in ulcerative colitis and irritable bowel syndrome. *Gastroenterology.. 126*(7), 1657–1664.

Coleman, N. S., Foley, S., Dunlop, S. P., Wheatcroft, J., Blackshaw, E., Perkins, A. C., Singh, G., Marsden, C. A., Holmes, G. K., & Spiller, R. C. (2006). Abnormalities of serotonin metabolism and their relation to symptoms in untreated celiac disease. *Clin Gastroenterol Hepatol. 4*(7), 874–881.

Comings, D. E. (1994). Serotonin: a key to migraine disorders? Nutrition Health Review, *Health and Fitness Magazine.*

Cook, E. H. Jr, Fletcher, K. E., Wainwright, M., Marks, N., Yan, S. Y., & Leventhal, B. L. (1994). Primary structure of the human platelet serotonin 5-HT2A receptor: identify with frontal cortex serotonin 5-HT2A receptor. *J Neurochem.. 63*(2), 465–469.

Coremans, G., Kerstens, R., De Pauw, M., & Stevens, M. (2003). Prucalopride is effective in patients with severe chronic constipation in whom laxatives fail to provide adequate relief. Results of a double-blind, placebo-controlled clinical trial. *Digestion.. 67*, 82–89.

Coufal, M., Invernizzi, P., Gaudio, E., Bernuzzi, F., Frampton, G. A., & Onori, P., et al. (2010). Increased local dopamine secretion has growth-promoting effects in cholangiocarcinoma. Int J Cancer. *126*, 2112–2122.

Cox, D. A., & Cohen, M. (1995). L.5-Hydroxytryptamine 2B receptor signaling in rat stomach fundus: role of voltage-dependent calcium channels, intracellular calcium release and protein kinase, *C. J Pharmacol Exp Ther.. 272*(1), 143–150.

Cubeddu, L. X., Hoffmann, I. S., Fuenmayor, N. T., & Finn, A. L. Efficacy of ondansetron (GR 38032F) and the role of serotonin in cisplatin-induced nausea and vomiting.

Cutrer, F. M., Yu, X-J, Ayata, G., Moskowitz, M. A., & Waeber, C. (1999). Effects of PNU 109291, a selective 5-HT1D receptor agonist, on electrically induced dural plasma extravasation and capsaicin-evoked c-fos immunoreactivity within trigeminal nucleus caudalis. *Neuropharmacology.. 38,* 1043–1053.

D'Andrea, G., Toldo, M., Cananzi, A., & Ferro-Milone, F. (1984). Study of platelet activation in migraine: control by low doses of aspirin. *Stroke.. 15*(2), 271–275.

Damaso, E. L., Bonagamba, L. G., Kellett, D. O., Jordan, D., Ramage, A. G., & Machado, B. H. (2007). Involvement of central 5-HT7 receptors in modulation of cardiovascular reflexes in awake rats. *Brain Res. 1144,* 82–90.

Daubert, E. A., (2010). Condron BG Serotonin: a regulator of neuronal morphology and circuitry. *Trends Neurosci.. 33*(9), 424–434

de Chaffoy de Courcelles, D., Leyzen, J. E., De Clerck, F., Van Belle, H., & Janssen, P. A. (1985). Evidence that phospholipid turnover is the signal transducing system coupled to serotonin-S2 receptor sites. *J Biol Chem.. 260*(12), 7603–7608.

De Clerck, F., Xhonneux, B., Leyzen, J., & Janssen, P. A. (1984). Evidence for functional 5-HT2 receptor sites on human blood platelets. *Biochem Pharmacol.. 33*(17), 2807–2811.

De Herder, (2007). W. W. Biochemistry of neuroendocrine tumors. *Best Pract Res Clin Endocrinol Metab 21,* 33–41.

De Maeyer, J. H., Lefebvre, R. A., & Schuurkes, J. A. (2008). 5-HT4 receptor agonists: similar but not the same. *Neurogastroenterol Motil. 20*(2), 99–112.

De Ponti, F., & Tonini, (2001). M. Irritable bowel syndrome: new agents targeting serotonin receptor subtypes. *Drugs. 61,* 317–332.

De Schryver, A. M., Andriesse, G. I., Samsom, M., Smout, A. J., Gooszen, H. G., Akkermans, L. M. (2002). The effects of the specific 5HT(4) receptor agonist, prucalopride, on colonic motility in healthy volunteers. *Aliment Pharmacol Ther. 16,* 603–612.

Den Boer, J. A., Bosker, F. J., & Slaap, B. R. (2000). Serotonergic drugs in the treatment of depressive and anxiety disorders. *Hum Psychopharmacol. 15,* 315–336.

Dizeyi, N., Bjartell, A., Hedlund, P., Tasken, K. A., Gadaleanu, V., & Abrahamsson, P. A. (2005). Expression of serotonin receptors 2B and 4 in human prostate cancer tissue and effects of their antagonists on prostate cancer cell line. *Eur Urol (2005). 47,* 895–900.

Dukat, M., Smith, C., Herrick-Davis, K., Teitler, M., & Glennon, R. A. (2004). Binding of tryptamineanalogs at H5-HT1E receptors: a structure-affinity investigation. *Bioorg Med Chem. 12*(10), 2545–2552.

Durham, P. L., & Russo, A. F. (2002). New insights into the molecular actions of serotonergicant imigraine drugs. *Pharmacol Ther.. 94(1–2),* 77–92.

Eglen, R. M. (1997). 5-Hydroxytryptamine (5-HT)4 receptors and central nervous system function: an update. *Prog Drug Res.. 49,* 9–24.

El Mestikawy, S., Fargin, A., Raymond, J. R., Gozlan, H., & Hnatowich, M. (1991). The 5-HT1A receptor: an overview of recent advances. *Neurochem Res. 16*(1), 1–10.

Ellis, E. S., Byrne, C., Murphy, O. E., Tilford, N. S., & Baxter, G. S. (1995). Mediation by 5-hydroxytryptamine 2B receptors of endothelium-dependent relaxation in rat jugular vein. *Br J Pharmacol. 114*(2), 400–404.

Errico, M., Crozier, R. A., Plummer, M. R., & Cowen, D. S. (2001). 5-HT(7) receptors activate the mitogen activated protein kinase extracellular signal related kinase in culture drat hippocampal neurons. *Neuroscience. 102*(2), 361–367.

Evangelista, S. (2007). Drug evaluation: Pumosetrag for the treatment of irritable bowel syndrome and gastro esophageal reflux disease. *Curr Opin Investig Drugs. 8*(5), 416–422.

Faber, K. M., & Haring, J. H. . (1999). Synaptogenesis in the postnatal rat fasciadentata is influenced by 5-HT1 a receptor activation. *Brain Res Dev Brain Res 114*(2), 245–252.

Faure, C., Patey, N., Gauthier, C., Brooks, E. M., & Mawe, G. M. (2010). Serotonin signaling is altered in irritable bowel syndrome with diarrhea but not in functional dyspepsia in pediatric age patients. *Gastroenterology. 139*(1), 249–258.

Feng, J., Cai, X., Zhao, J., & Yan, Z. (2001). Serotonin receptors modulate GABA(A) receptor channels through activation of anchored protein kinase C in prefrontal cortical neurons. *J Neurosci. 21*(17), 6502–6511.

Ferrari, M. D., Odink, J., Tapparelli, C., Van Kempen, G. M., Pennings, E. J., & Bruyn, G. W. (1989). Serotonin metabolism in migraine. *Neurology. 39*(9), 1239–1242.

Fioroni, L., Andrea, G. D., Alecci, M., Cananzi, A., & Facchinetti, F. (1996). Platelet serotonin pathway in menstrual migraine. *Cephalalgia. 16*(6), 427–430.

Fitzgerald, L. W., Iyer, G., Conklin, D. S., Krause, C. M., Marshall, A., Patterson, J. P., Tran, D. P., Jonak, G. J., & Hartig, P. R. (1999). Messenger RNA editing of the human serotonin 5-HT2C receptor. *Neuropsychopharmacology. 21(2 Suppl)*, 82S-90S.

Fletcher, A., Forster, E. A., Bill, D. J., Brown, G., Cliffe, I. A., Hartley, J. E., Jones, D. E., McLenachan, A., Stanhope, K. J., Critchley, D. J., Childs, K. J., Middlefell, V. C., Lanfumey, L., Corradetti, R., Laporte, A. M., Gozlan, H., & Hamon, M., Dourish, (1996). C. T. Electrophysiological, biochemical, neurohormonal and behavioral studies with WAY-100635, a potent, selective and silent 5-HT1A receptor antagonist. *Behav Brain Res. 73*, 337–353.

Fletcher, P. J., Tampakeras, M., Sinyard, J., Slassi, A., Isaac, M., & Higgins, G. A. (2009). Characterizing the effects of 5-HT(2C) receptor ligands on motor activity and feeding behavior in 5-HT(2C) receptor knockout mice. *Neuropharmacology. 57*(3), 259–267.

Forster, E. A., Cliffe, I. A., Bill, D. J., Dover, G. M., Jones, D., Reilly, Y., & Fletcher, (1995). A. A pharmacological profile of the selective silent 5-HT1A receptor antagonist, WAY 100635. *Eur J Pharmacol. 281*, 81–8.

Geldmacher, D. S., Lerner, A. J., Voci, J. M., Noelker, E. A., Somple, L. C., & Whitehouse, P. J. (1997). Treatment of functional decline in adults with Down syndrome using selective serotonin-reuptake inhibitor drugs. *J Geriatr Psychiatry Neurol.. 10*(3), 99–104.

Gerhardt, C. C., & van Heerikhuizen, H. (1997). Functional characteristics of heterologously expressed 5-HT receptors. *Eur J Pharmacol. 334*(1), 1–23.

Gershon, M. D., & Tack, J. (2007). The serotonin signaling system: from basic understanding to drug development for functional GI disorders. *Gastroenterology. 132*(1), 397–414.

Gershon, M. D., & Tamir, H. (1981). Release of endogenous 5-hydroxytryptamine from resting and stimulated entericneurons. *Neuroscience.. 6*(11), 2277–2286.

Gershon, M. D. (2004). Serotonin receptors and transporters -- roles in normal and abnormal gastrointestinal motility. *Aliment Pharmacol Ther. 20*(7), 3–14

Gershon, M. D. (1991). Serotonin: its role and receptors in enteric neurotransmission. *Adv Exp Med Biol.. 294*, 221–230.

Glennon, R. A., Lee, M., Rangisetty, J. B., Dukat, M., Roth, B. L., Savage, J. E., McBride, A., Rauser, L., Hufeisen, S., & Lee DKH. (2000). 2-Substituted tryptamines: agents with selectivity for 5-HT6 serotonin receptors. *J Med Chem. 43*, 1011–1018.

Graeff, F. G. (1997). Serotonergic systems. *Psychiatr Clin North Am. 20*(4), 723–739.

Grailhe, R., Grabtree, G. W., & Hen, R. (2001). Human 5-HT(5) receptors: the 5-HT(5A) receptor is functional but the 5-HT(5B) receptor was lost during mammalian evolution. *Eur J Pharmacol. 418*(3), 157–167.

Gudermann, T., Levy, F. O., Birnbaumer, M., Birnbaumer, L., & Kaumann, A. J. (1993). Human S31 serotonin receptor clone codes a 5-hydroxytryptamine 1E-like serotonin receptor. *Mol Pharmacol. 43*(3), 412–418.

Gyermek, (1995). L. 5-HT3 receptors: pharmacologic and therapeutic aspects. *J Clin Pharmacol. 35*(9), 845–855.

Hagan, J. J., Price, G. W., Jeffrey, P., Deeks, N. J., Stean, T., Piper, D., Smith, M. I., Upton, N., Medhurst, A. D., Middlemiss, D. N., Riley, G. J., Lovell, P. J., Bromidge, S. M., & Thomas, D. R. (2000). Characterization of SB-269970-A, a selective 5-HT(7) receptor antagonist. *Br J Pharmacol. 130*(3), 539–548.

Hagan, J. J., Price, G. W., Jeffrey, P., Deeks, N. J., Stean, T., Piper, D., Smith, M. I., Upton, N., Medhurst, A. D., Middlemiss, D. N., Riley, G. J., Lovell, P. J., Bromidge, S. M., & Thomas, D. R. (2000). Characterization of SB-269970-A, a selective 5-HT7 receptor antagonist. *Br J Pharmacol. 130*, 539–548.

Hamblin, M. W., & Metcalf, M. A. (1991). Primary structure and functional characterization of a human 5-HT1D-type serotonin receptor. *Mol Pharmacol. 40*(2), 143–148.

Hannon, J., & Hoyer D, (2002). Serotonin receptors and systems: endless diversity? *Acta Biolog Szeged. 46*, 1–12.

Hannon, J., & Hoyer, D. (2008). Molecular biology of 5-HT receptors. *Behav Brain Res. 195*(1), 198–213.

Hartig, P. R., Hoyer, D., Humphrey, P. P., & Martin, G. R. (1996). Alignment of receptor nomenclature with the human genome: classification of 5-HT1B and 5-HT1D receptor subtypes. *Trends Pharmacol Sci. 17*(3), 103–105.

Hasler, W. L. (2009). Serotonin and the GI tract. *Curr Gastroenterol Rep. 11*(5), 383–391.

Hegde, S. S., & Eglen, R. M. (1996). Peripheral 5-HT4 receptors. *FASEB J. 10*, 1398–1408.

Heidmann, D. E. A., Metcalf, M. A., Kohen, R., & Hamblin, M. (1997). W. Four 5-hydroxytryptamine 7 (5-HT7) receptor isoforms in human and rat produced by alternative splicing: species differences due to altered intron-exon organization. *J Neurochem. 68*, 1372–1381.

Heidmann, D. E. A., Szot, P., Kohen, R., & Hamblin, M. W. (1998). Function and distribution of three rat 5-hydroxytryptamine 7 (5-HT7) receptor isoforms produced by alternative splicing. *Neuropharmacology.. 37*, 1621–1632.

Henriksen, R., Dizeyi, N., & Abrahamsson, P. A. (2012). Expression of serotonin receptors 5-HT1A, 5-HT1B, 5-HT2B and 5-HT4 in ovary and in ovarian tumors. *Anticancer Res. 32*, 1361–1366.

Herve, P., Drouet, L., Dosquet, C., Launay, J. M., Rain, B., Simonneau, G., Caen, J., & Duroux, P. (1990). Primary pulmonary hypertension in a patient with a familial platelet storage pool disease: role of serotonin. *Am J Med. 89*(1), 117–120.

Hoefgen, B., Schulze, T. G., Ohlraun, S., von Widdern, O., Höfels, S., Gross, M., Heidmann, V., Kovalenko, S., Eckermann, A., Kölsch, H., Metten, M., Zobel, A., Becker, T., Nöthen, M. M., Propping, P., Heun, R., Maier, W., & Rietschel, M. . (2005). The power of sample size and homogenous sampling: association between the 5-HTTLPR serotonin transporter polymorphism and major depressive disorder. *Biol Psychiatry 57*(3), 247–251.

Hoyer, D., Clarke, D. E., Fozard, J. R., Hartig, P. R., Martin, G. R., Mylecharane, E. J., Saxena, P. R., & Humphrey PPA. (1994). International Union of Pharmacology classification of receptors for 5-hydroxytryptamine (serotonin). *Pharmacol Rev.. 46*, 157–204.

Hoyer, D., Hannon, J. P., & Martin, G. R. (2002). Molecular, pharmacological and functional diversity of 5-HT receptors. *Pharmacol Biochem Behav.* 71(4), 533–554.

Hoyer, D., & Martin, (1997). G. 5-HT receptor classification and nomenclature: towards a harmonization with the human genome. *Neuropharmacology.* 36(4–5), 419–428.

Hu, W. P., Guan, B. C., Ru, L. Q., Chen, J. G., & Li, Z. W. (2004). Potentiating of 5-HT3 receptor function by the activation of coexistent 5-HT2 receptors in trigeminal ganglion neurons of rats. *Neuropharmacology.* 47(6), 833–840.

Jacobs, B. L., & Azmitia, E. C. (1992). Structure and function of the brain serotonin system. *Physiol Rev.. 72*(1), 165–229.

Janssen, P., Tack, J., Sifrim, D., Meulemans, A. L., & Lefebvre, R. A. (2004). Influence of 5-HT1 receptor agonists on feline stomach relaxation. *Eur J Pharmacol. 492(2–3),* 259–267.

Jasper, J. R., Kosaka, A., To, Z. P., Chang, D. J., & Eglen, R. M. (1997). Cloning, expression and pharmacology of a truncated splice variant of the human 5-HT7 receptor (h5-HT7b). *Br J Pharmacol. 122,* 126–132.

Jerman, J. C., Brough, S. J., Gager, T., Wood, M., Coldwell, M. C., Smart, D., & Middlemiss, D. N. (2001). Pharmacological characterization of human 5-HT2 receptor subtypes. *Eur J Pharmacol.. 414,* 23–30.

Jernej, B., Banović, M., Cicin-Sain, L., Hranilović, D., Balija, M., Oresković, D., & Folnegović-Smalc, V. (2000). Physiological characteristics of platelet/circulatory zerotonin: study on a large human population. *Psychiatry Res.. 94*(2), 153–162.

Jin, H., Oksenberg, D., Ashkenazi, A., Peroutka, S. J., Duncan, A. M., Rozmahel, R., Yang, Y., Mengod, G., Palacios, J. M., & O'Dowd, B. F. (1992). Characterization of the human 5-hydroxytryptamine 1B receptor. *J Biol Chem. 25; 267*(9), 5735–5738.

Jørgensen, H., Kjaer, A., Warberg, J., & Knigge, U. (2001). Differential effect of serotonin 5-HT(1A) receptor antagonists on the secretion of corticotrophin and prolactin. *Neuroendocrinology.. 73*(5), 322–333.

Jungwirth, N., Haeberle, L., Schrott, K. M., & Wullich, B. (2008). Krause FS Serotonin used as prognostic marker of urological tumors. *World J Urol 26,* 499–504.

Kaumann, A. J., & Levy, F. O. (2006). 5-hydroxytryptamine receptors in the human cardiovascular system. *Pharmacol Ther. 111*(3), 674–706.

Kaumann, A. J., Parsons, A. A., & Brown, A. M. (1993). Human arterial constrictor serotonin receptors. *Cardiovasc Res.. 27*(12), 2094–2103.

Kaumann, A. J., & Sanders, (1998). L. 5-Hydroxytryptamine and human heart function: the role of 5-HT4 receptors. In: Eglen, R. M. (Ed.), 5-HT4 Receptors in the Brain and Periphery; Heidelberg: Springer, 127–148.

Kaumann, A. J., & Sanders, L. (1994). 5-Hydroxytryptamine causes rate-dependent arrhythmias through 5-HT4 receptors in human atrium: facilitation by chronic b-adrenoceptor blockade. *Naunyn-Schmiedeberg's Arch Pharmacol 349,* 331–337.

Kawasaki, H., Urabe, M., & Takasaki, K. (1987). Enhanced 5-hydroxytryptamine release from vascular adrenergic nerves in spontaneously hypertensive rats. *Hypertension. 10*(3), 321–327.

Kellett, D. O., Ramage, A. G., & Jordan, D. (2005). Central 5-HT7 receptors are critical for reflex activation of cardiac vagal drive in anesthetized rats. *J Physiol. 563(Pt 1),* 319–331.

Kennett, G. A., Wood, M. D., Bright, F., Cilia, J., Piper, D. C., Gager, T., Thomas, D., Baxter, G. S., & Forbes, I. T. (1996). In vitro and in vivo profile of SB 206553, a potent 5-HT2C/5-HT2B receptor antagonist with anxiolytic-like properties. *Br J Pharmacol. 117,* 427–434.

Kennett, G. A., Wood, M. D., Bright, F., Trail, B., Riley, G., Holland, V., Avenell, K. Y., Stean, T., Upton, N., Bromidge, S., Forbes, I. T., Brown, A. M., Middlemiss, D. N., & Blackburn, T. P. (1997). SB 242084, a selective and brain penetrant 5-HT2C receptor antagonist. *Neuropharmacology. 36,* 609–620.

Kennett, G. A., Wood, M. D., Glen, A., Grewal, S., Forbes, I., Gadre, A., & Blackburn, T. P. (1994). In vivo properties of SB 200646A, a 5-HT2C/2B receptor antagonist. *Br J Pharmacol.. 111,* 797–802.

Kii, Y., Nakatsuji, K., Nose, I., Yabuuchi, M., Mizuki, Y., & Ito, T. (2001). Effects of 5-HT(4) receptor agonists, cisapride and mosapridecitrate on electrocardiogram in anesthetized rats and guinea-pigs and conscious cats. *Pharmacol Toxicol. 89*(2), 96–103.

Kim, H. S., Choi, E. J., & Park, H. (2008). The effect of mosapridecitrate on proximal and distalcolonicmotor function in the guinea-pig in vitro. *Neurogastroenterol Motil.. 20*(2), 169–176.

Kim, S. W., Shin, I. S., Kim, J. M., Youn, T., Yang, S. J., Hwang, M. Y., & Yoon, J. S. (2009). The 5-HT2 receptor profiles of antipsychotics in the pathogenesis of obsessive-compulsive symptoms in schizophrenia. *Clin Neuropharmacol. 32*(4), 224–226.

Kimball, R. W., Friedman, A. P., & Vallejo, E. (1960). Effect of serotonin in migraine patients. *Neurology. 10,* 107–111.

Kinkead, R., Bach, K. B., Johnson, S. M., Hodgeman, B. A., & Mitchell, G. S. (2001). Plasticity in respiratory motor control: intermittent hypoxia and hypercapnia activate opposing serotonergic and noradrenergic modulatory systems. *Comp Biochem Physiol A Mol Integr Physiol.. 130*(2), 207–218.

Kobilka, B. K., Frielle, T., Collins, S., Yang-Feng, T., Kobilka, T. S., Francke, U., Lefkowitz, R. J., & Caron, M. G. (1987). An intronlessgeneen coding a potential member of the family of receptors coupled to guanine nucleotide regulatory proteins. *Nature.. 329(6134),* 75–79.

Koo, J. W., & Balaban, C. D. (2006). Serotonin-induced plasma extravasation in the murineinnerear: possible mechanism of migraine-associated innereardys function. *Cephalalgia. 26*(11), 1310–1319.

Kopparapu, P. K., Tinzl, M., Anagnostaki, L., Persson, J. L., & Dizeyi, N. (2013). Expression and localization of serotonin receptors in human breast cancer. *Anticancer Res. 33,* 363–370.

Lance, J. W (1991). 5-Hydroxytryptamine and its role in migraine. Eur *Neurol.. 31*(5), 279–281.

Lanfumey, L., & Hamon, M. (2004). 5-HT1 receptors. *Curr Drug Targets CNS Neurol Disord.. 3*(1), 1–10.

Laporte, A. M., Kidd, E. J., Verge, D., Gozlan, H., & Hamon, M. (1992). Autoradiographic mapping of central 5-HT3 receptors. In: Hamon, M., Ed. Central and Peripheral 5-HT3 Receptors. London, UK: Academic Press, pp. 157–187.

Lauder, J. M., Krebs, H. (1978). Serotonin as a differentiation signal in early neurogenesis. *Dev Neurosci. 1*(1), 15–30.

Lauder, J. M. (1990). Ontogeny of the serotonergic system in the rat: serotonin as a developmental signal. *Ann N Y Acad Sci. 600,* 297–313; discussion 314.

Leopoldo, M., Lacivita, E., Berardi, F., Perrone, R., & Hedlund, P. B. (2011). Serotonin 5-HT7 receptor agents: Structure-activity relationships and potential therapeutic applications in central nervous system disorders. *Pharmacol Ther. 129*(2), 120–148.

Lesch, K. P., Waider, J. (2012). Serotonin in the modulation of neural plasticity and networks: implications for neurodevelopmental disorders. *Neuron. 76*(1), 175–191.

Lesch, K. P. (2001). Serotonergic gene expression and depression: implications for developing novel antidepressants. *J Affect Disord. 62(1–2)*, 57–76.

Lesurtel, M., Soll, C., Graf, R., Clavien, P. A. (2008). Role of serotonin in the hepato-gastro Intestinaltract: an old molecule for new perspectives. *Cell Mol Life Sci. 65(6)*, 940–952.

Levy, F. O., Gudermann, T., Birnbaumer, M., Kaumann, A. J., & Birnbaumer, L. (1992). Molecular cloning of a human gene (S31) encoding a novel serotonin receptor mediating inhibition of adenylyl cyclase. *FEBS Lett.. 296(2)*, 201–206.

Li, Y. Q., Takada, M., Mizuno, N. (1993). The sites of origin of serotoninergic afferent fibers in the trigeminal motor, facial, and hypoglossal nuclei in the rat. *Neurosci Res. 17(4)*, 307–313.

Libert, F., Parmentier, M., Lefort, A., Dinsart, C., Van Sande, J., Maenhaut, C., Simons, M. J., Dumont, J. E., & Vassart, G. (1989). Selectiveamplification and cloning of four new members of the G protein-coupled receptor family. *Science. 244(4904)*, 569–572.

Lindsay, A. D., & Feldman, J. L. (1993). Modulation of respiratory activity of neonatal rat phrenicmotoneurones by serotonin. *J Physiol. 461*, 213–233.

Liu, Y., Yoo, M. J., Savonenko, A., Stirling, W., Price, D. L., Borchelt, D. R., Mamounas, L., Lyons, W. E., Blue, M. E., & Lee, M. K. (2008). Amyloid pathology is associated with progressive monoaminergic neurode generation in a transgenic mouse model of Alzheimer'sdisease. *J Neurosci. 28(51)*, 13805–13814.

Mallon, E., Osin, P., Nasiri, N., Blain, I., Howard, B., & Gusterson, B. (2000). The basic pathology of human breast cancer. *J Mammary Gland Biol Neoplasia. 5(2)*, 139–63.

Manaker, S., & Tischler, L. J. (1993). Origin of serotoninergic afferents to the hypoglossal nucleus in the rat. *J Comp Neurol. 334(3)*, 466–476.

Mann, J. J., Brent, D. A., & Arango, V. (2001). The neurobiology and genetics of suicide and attempted suicide: a focus on the serotonergic system. *Neuropsychopharmacology. 24(5)*, 467–477.

Manocha, M., & Khan, W. I. (2012). Serotonin and GI Disorders: An Update on Clinical and Experimental Studies. *Clin Transl Gastroenterol 3*, e13.

Marazziti, D., Muratori, F., Cesari, A., Masala, I., Baroni, S., Giannaccini, G., Dell'Osso, L., Cosenza, A., Pfanner, P., & Cassano, G. B. (2000). Increased density of the platelet serotonin transporter in autism. *Pharmacopsychiatry. 33(5)*, 165–168.

Mazer, C., Muneyyirci, J., Taheny, K., Raio, N., Borella, A., & Whitaker-Azmitia, P. (1997). Serotonin depletion during synaptogenesis leads to decreased synaptic density and learning deficits in the adultrat: a possible model of neurodevelopmental disorders with cognitive deficits. *Brain Res. 760(1–2)*, 68–73.

McAllister, G., Charlesworth, A., Snodin, C., Beer, M. S., Noble, A. J., Middlemiss, D. N., Iversen, L. L., & Whiting, P. (1992). Molecular cloning of a serotonin receptor from human brain (5HT1E): a fifth 5HT1-like subtype. *Proc Natl Acad Sci USA. 89(12)*, 5517–5521.

McCrimmon, D. R., & Lalley, P. M. (1982). Inhibition of respiratory neural discharges by clonidine and 5-hydroxytryptophan. *J Pharmacol Exp Ther.. 222(3)*, 771–777.

McKeith, I. G., Marshall, E. F., Ferrier, I. N., Armstrong, M. M., Kennedy, W. N., Perry, R. H., Perry, E. K., & Eccleston, D. (1987). 5-HT receptor binding in postmortem brain from patients with affective disorder. *J Affect Disord. 13(1)*, 67–74.

McKune, C. M., & Watts, S. W. (2001). Characterization of the serotonin receptor mediating contraction in the mousethoracicaorta and signal pathway coupling. *J Pharmacol Exp Ther. 297(1)*, 88–95.

McKune, C. M., & Watts, S. W. (2001). Characterization of the serotonin receptor mediating contraction in the mouse thoracic aorta and signal pathway coupling. *J Pharmacol Exp Ther. 297*, 88–95.

McLean, P. G., Borman, R. A., & Lee, K. (2007). 5-HT in the enteric nervous system: gut-function and neuropharmacology. *Trends Neurosci. 30*(1), 9–13.

Meneses, A., & Hong, E. (1999). 5-HT1A receptors modulate the consolidation of learning in normal and cognitively impaired rats. *Neurobiol Learn Mem. 71*(2), 207–218.

Meneses, A., & Perez-Garcia, G. (2007). 5-HT(1A) receptors and memory. *Neurosci Biobehav Rev. 31*(5), 705–727.

Millan, M. J., Marin, P., Bockaert, J., & Mannoury la Cour, C. (2008). Signaling at G-protein-coupled serotonin receptors: recent advances and future research directions. *Trends Pharmacol Sci. 29*(9), 454–464.

Mohammad-Zadeh, L. F., Moses, L., & Gwaltney-Brant, S. M. (2008). Serotonin: a review. *J Vet Pharmacol Ther. 31*(3), 187–199.

Monteau, R., Hilaire, G. (1991). Spinal respiratory motoneurons. *Prog Neurobiol. 37*(2), 83–144.

Morin, D., Hennequin, S., Monteau, R., & Hilaire, G. (1990). Serotonergic influences on central respiratory activity: an in vitro study in the newborn rat. *Brain Res. 535*(2), 281–287.

Moskowitz, M. A. (1992). Neurogenic versus vascular mechanisms of sumatriptan and ergot alkaloids in migraine. *Trends Pharmacol Sci. 13*(8), 307–311.

Nebigil, C. G., Launay, J. M., Hickel, P., Tournois, C., & Maroteaux, L. (2000). 5-hydroxytryptamine 2B receptor regulates cell-cycle progression: cross-talk with tyrosine kinase pathways. *Proc Natl Acad Sci USA. 97*(6), 2591–2596.

Ni, W., & Watts, S. W. (2006). 5-hydroxytryptamine in the cardiovascular system: focus on the serotonin transporter (SERT). *Clin Exp Pharmacol Physiol. 33*(7), 575–583.

Nichols, D. E., & Nichols, C. D. (2008). Serotonin receptors. *Chem Rev. 108*(5), 1614–1641.

Nilsson, T., Longmore, J., Shaw, D., Pantev, E., Bard, J. A., Branchek, T., & Edvinsson, L. (1999). Characterization of 5-HT receptors in human coronary arteries by molecular and pharmacological techniques. *Eur J Pharmacol. 372*(1), 49–56.

Nisijima, K., Yoshino, T., Yui, K., & Katoh, S. (2001). Potent serotonin (5-HT)(2A) receptor antagonists completely prevent the development of hyperthermia in an animal model of the 5-HT syndrome. *Brain Res. 890*(1), 23–31.

Nocito, A., Dahm, F., Jochum, W., Jang, J. H., Georgiev, P., & Bader, M., et al. (2008). Serotonin regulates macrophage-mediated angiogenesis in a mouse model of colon cancer allografts. *Cancer Res 68*, 5152–5158.

O'Neill, M. F., Fernández, A. G., & Palacios, J. M. (1996). GR127935 blocks the locomotor and antidepressant-like effects of RU24969 and the action of antidepressants in the mouse tail suspension test. *Pharmacol Biochem Behav. 53*(3), 535–539.

Ogawa, T., Sugidachi, A., Tanaka, N., Fujimoto, K., Fukushige, J., Tani, Y., & Asai, F. (2005). Effects of R-102444 and its active metabolite R-96544, selective 5-HT2A receptor antagonists, on experimental acute and chronic pancreatitis: Additional evidence for possible involvement of 5-HT2A receptors in the development of experimental pancreatitis. *Eur J Pharmacol. 521(1–3)*, 156–163.

Okado, N., Narita, M., & Narita, N. (2001). A biogenic amine-synapse mechanism for mental retardation and developmental disabilities. *Brain Dev. 23 Suppl 1:S11–5.*

Olson EB Jr, Dempsey, J. A., & McCrimmon, D. R. (1979). Serotonin and the control of ventilation in awake rats. *J Clin Invest 64*(2), 689–693.

Pai, V. P., Marshall, A. M., Hernandez, L. L., Buckley, A. R., & Horseman, N. D. (2009). Altered serotonin physiology in human breast cancers favors paradoxical growth and cell survival. *Breast Cancer Res. 11*(6), *R81.*

Panconesi, A., & Sicuteri, R. (1997). Headache induced by serotonergic agonists--a key to the interpretation of migraine pathogenesis.? *Cephalalgia. 17*(1), 3–14.

Parsons, A. A., & Whalley, E. T. (1989). Characterization of the 5-hydroxytryptamine receptor which mediates contraction of the human isolated basilar artery. *Cephalalgia. 9*(9), 47–51.

Pasricha, P. J. (2007). Desperatelyzeekingserotonin. A commentary on the withdrawal of tegaserod and the state of drug development for functional and motility disorders. *Gastroenterology 132*(7), 2287–2290.

Pauwels, & P. J. (2003). 5-HT receptors and their ligands. *Tocris Rev. 25,* 1–12.

Pérez-García, G., Gonzalez-Espinosa, C., & Meneses, A. (2006). An mRNA expression analysis of stimulation and blockade of 5-HT7 receptors during memory consolidation. *Behav Brain Res. 169*(1), 83–92.

Phebus, L. A., Johnson, K. W., Zgombick, J. M., Gilbert, P. J., Van Belle, K., Mancuso, V., Nelson, D. L. G., Calligaro, D. O., Kiefer, A. D., Branchek, T. A., & Flaugh, M. E. (1997). Characterization of LY 344864 as a pharmacological tool to study 5-HT1F receptors: binding affinities, brain penetration and activity in the neurogenic dural inflammation model of migraine. *Life Sci. 61,* 2117–2126.

Pratt, G. D., Bowery, N. G., Kilpatrick, G. J., Leslie, R. A., Barnes, N. M., Naylor, R. J., Jones, B. J., Nelson, D. R., Palacids, J. M., & Slater, P., et al. (1990). Consensus meeting agrees distribution of 5-HT3 receptors in mammalian hindbrain. *Trends Pharmacol Sci. 11*(4), 135–137.

Prins, N. H., Briejer, M. R., Van Bergen, P. J., Akkermans, L. M., & Schuurkes, J. A. (1999). Evidence for 5-HT7 receptors mediating relaxation of human colonic circular smooth muscle. *Br J Pharmacol. 128*(4), 849–852.

Pullar, I. A., Boot, J. R., Broadmore, R. J., Eyre, T. A., Cooper, J., Sanger, G. J., Wedley, S., & Mitchell, S. N. (2004). The role of the 5-HT1D receptor as a presynaptic auto receptor in the guinea pig. *Eur J Pharmacol. 493(1–3),* 85–93.

Pytliak, M., Vargová, V., Mechírová, V., & Felšöci, M. (2011). Serotonin receptors –from molecular biology to clinical applications. *Physiol Res 60*(1), 15–25.

Rahme, M. M., Cotter, B., Leistad, E., Wadhwa, M. K., Mohabir, R., Ford APDW, Eglen, R. M., & Feld, G. K. (1999). Electrophysiological and antiarrhythmic effects of the atrial selective 5-HT4 receptor antagonist RS-100302 in experimental atrial flutter and fibrillation. *Circulation. 100,* 2010–2017.

Ramage, A. G., & Villalón, C. M. (2008). 5-hydroxytryptamine and cardiovascular regulation. *Trends Pharmacol Sci. 29*(9), 472–481.

Ramage, A. G. (2001). Central cardiovascular regulation and 5-hydroxytryptamine receptors. *Brain Res Bull. 56*(5), 425–439.

Rapport, M. M., & Green, A. A., (1948a). Page, I. H. *Crystalline Serotonin. Science. 108(2804),* 329–330.

Rapport, M. M., & Green, A. A., (1948b). Page, I. H. Partial purification of the vasoconstrictor in beef serum. *J Biol Chem. 174*(2), 735–741.

Rapport, M. M., Green, A. A., & Page, I. H. (1948c). Serum vasoconstrictor, serotonin; isolation and characterization. *J Biol Chem. 176*(3), 1243–1251.

Reid, G., & Rand, M. (1951). Physiological actions of the partially purified serum vasoconstrictor (serotonin). *Aust J Exp Biol Med Sci. 29*(6), 401–415.

Reynolds, G. P., Mason, S. L., Meldrum, A., De-Keczer, S., Parnes, H., Eglen, R. M., & Wong EHF. (1995). 5-Hydroxytryptamine (5-HT) 4 receptors in post mortem human brain tissue: distribution, pharmacology and effects of neurodegenerative diseases. *Br J Pharmacol. 114*, 993–998.

Romano, A. G., Quinn, J. L., Liu, R., Dave, K. D., Schwab, D., Alexander, G., Aloyo, V. J., & Harvey, J. A. (2006). Effect of serotonin depletion on 5-HT2A-mediated learning in the rabbit: evidence for constitutive activity of the 5-HT2A receptor in vivo. *Psychopharmacology (Berl). 184*(2), 173–181.

Rubanyl, G. M., Frye, R. L., Holmes DR Jr, & Vanhoutte, P. M. (1987). Vasoconstrictor activity of coronary sinus plasma from patients with coronary artery disease. *J Am Coll Cardiol. 9*(6), 1243–1249.

Sanger, G. J. (2008). 5-hydroxytryptamine and the gastrointestinal tract: where next? *Trends Pharmacol Sci. 29*(9), 465–471.

Sanger, G. J. (1998). Different pathophysiological functions of 5-HT4 and 5-HT3 receptors in small and large intestine. *Curr. Res. Serotonin. 3*, 99–104.

Sari, Y. (2004). Serotonin 1B receptors: from protein to physiological function and behavior. *Neurosci Biobehav Rev. 28*(6), 565–582.

Saudou, F., Boschert, U., Amlaiky, N., Plassat, J. L., & HenR. (1992). A family of Drosophila serotonin receptors with distinct intracellular signaling properties and expression patterns. *EMBO J. 11*(1), 7–17.

Saxena, P. R. (1995). Serotonin receptors: subtypes, functional responses and therapeutic relevance. *Pharmacol Ther 66*(2), 339–368.

Schmitt, J. A., Wingen, M., Ramaekers, J. G., Evers, E. A., & Riedel, W. J. (2006). Serotonin and human cognitive performance. *Curr Pharm Des. 12*(20), 2473–2486.

Schmuck, K., Ullmer, C., Engels, P., & Lübbert, H. (1994). Cloning and functional characterization of the human 5-HT2B serotonin receptor. *FEBS Lett. 342*(1), 85–90.

Shen, Y., Monsma, F. J. Jr., Metcalf, M. A., Jose, P. A., Hamblin, M. W., & Sibley, D. R. (1993). Molecular cloning and expression of a 5-hydroxytryptamine 7 serotonin receptor subtype. *J Biol Chem. 268*(24), 18200–18204.

Shepheard, S., Edvinsson, L., Cumberbatch, M., Williamson, D., Mason, G., Webb, J., Boyce, S., Hill, R., & Hargreaves, R. (1999). Possible antimigraine mechanisms of action of the 5HT1F receptor agonist LY334370. *Cephalalgia. 19*(10), 851–858.

Shimron-Abarbanell, D., Nöthen, M. M., Erdmann, J., & Propping, P. (1995). Lack of genetically determined structural variants of the human serotonin-1E (5-HT1E) receptor protein points to its evolutionary conservation. *Brain Res Mol Brain Res. 29*(2), 387–90.

Siddiqui, E. J., Thompson, C. S., Mikhailidis, D. P., & Mumtaz, F. H. (2005). The role of serotonin in tumor growth (review). *Oncol Rep. 14*(6), 1593–1597.

Sleight, A. J., Boess, F. G., Bos, M., Levet-Trafit, B., Riemer, C., & Bourson, A. (1998). Characterization of Ro 04–6790 and Ro 63–0563: potent and selective antagonists at human and rat 5-ht6 receptors. *Br J Pharmacol. 124*, 556–562.

Sleight, A. J., & Peroutka, S. J. (1991). Identification of 5-hydroxytryptamine 1A receptor agents using a composite pharmacophore analysis and chemical database screening. *Naunyn Schmiedebergs Arch Pharmacol.. 343*(2), 109–116.

Soll, C., Jang, J. H., Riener, M. O., Moritz, W., Wild, P. J., & Graf, R., et al. (2010). Serotonin promotes tumor growth in human hepatocellular cancer. *Hepatology 51*, 1244–1254.

Somerville, B. W. (1975). Estrogen-withdrawal migraine. Duration of exposure required and attempted prophylaxis by premenstrual estrogen administration. *Neurology. 25*(3), 239–244.

Sommer, C. (2004). Serotonin in pain and analgesia: actions in the periphery. *Mol Neurobiol.* *30*(2), 117–125.

Sonier, B., Lavigne, C., Arseneault, M., Ouellette, R., & Vaillancourt, C. (2005). Expression of the 5-HT2A serotoninergic receptor in human placenta and choriocarcinoma cells : mitogenic implications of serotonin. *Placenta 26,* 484–490.

Spiller, R. C., Meyers, N. L., & Hickling, R. I. (2008). Identification of patients with nond, nonCirritable bowel syndrome and treatment with renzapride: an exploratory, multi-center, randomized, double-blind, placebo-controlled clinical trial. Dig Dis Sci. *53*(12), 3191–3200.

Srikiatkhachorn, A., & Anthony, M. (1996). Platelet serotonin in patients with analgesic-induced headache. *Cephalalgia. 16*(6), 423–436.

Steinbusch, H. W. (1981). Distribution of serotonin-immunore activity in the central nervous system of the rat-cell bodies and terminals. *Neuroscience. 6*(4), 557–618.

Sugita, S., Shen, K. Z., North, R. A. (1992). 5-hydroxytryptamine is a fast excitatory transmitter at 5-HT3 receptors in ratamygdala. *Neuron. 8*(1), 199–203.

Taylor, B. K., Basbaum, A. I. (1995). Neurochemical characterization of extracellular serotonin in the rostral ventromedial medulla and its modulation by noxious stimuli. *J Neurochem. 65*(2), 578–589.

Teppema, L. J., Veening, J. G., Kranenburg, A., Dahan, A., Berkenbosch, A., & Olievier, C. (1997). Expression of c-fos in the rat brainstem after exposure to hypoxia and to normoxic and hyperoxichypercapnia. *J Comp Neurol. 388*(2), 169–190.

Terrón, J. A., & Martínez-García, E. (2007). 5-HT7 receptor-mediated dilatation in the middle meningeal artery of anesthetized rats. *Eur J Pharmacol. 560*(1), 56–60.

Thomas, D. R., Gittins, S. A., Collin, L. L., Middlemiss, D. N., Riley, G., Hagan, J., Gloger, I., Ellis, C. E., Forbes, I. T., & Brown, A. M. (1998). Functional characterization of the human cloned 5-HT7 receptor (long form); antagonist profile of SB 258719. *Br J Pharmacol. 124,* 1300–1306.

Thomas, D. R. (2006). 5-ht5A receptors as a therapeutic target. *Pharmacol Ther. 111*(3), 707–714.

Tokarski, K., Kusek, M., Hess, G. (2011). 5-HT7 receptors modulate GABAergic transmission in rat hippocampal CA1 area. *J Physiol Pharmacol.. 62*(5), 535–540.

Tunnicliff, G. (1991). Molecular basis of buspirone's anxiolytic action. *Pharmacol Toxicol. 69,* 149–156.

Turlejski, K. (1996). Evolutionary ancient roles of serotonin: long-lasting regulation of activity and development. *Acta Neurobiol Exp (Wars). 56*(2), 619–636.

Ullmer, C., Schmuck, K., Kalkman, H. O., & Lübbert, H. (1995). Expression of serotonin receptor mRNAs in blood vessels. *FEBS Lett.. 370*(3), 215–221.

Van de Kar, L. D., Javed, A., Zhang, Y., Serres, F., Raap, D. K., & Gray, T. S. (2001). 5-HT2A receptors stimulate ACTH, corticosterone, oxytocin, renin, and prolactin release and activate hypothalamic CRF and oxytocin-expressing cells. *J Neurosci. 21*(10), 3572–3579.

van den Berg, E. K., Schmitz, J. M., Benedict, C. R., Malloy, C. R., Willerson, J. T., Dehmer, G. J. (1989). Transcardiac serotonin concentration is increased in selected patients with limiting angina and complex coronary lesion morphology. *Circulation. 79*(1), 116–124.

Vicaut, E., Laemmel, E., & Stücker, O. (2000). Impact of serotonin on tumor growth. *Ann Med. 32*(3), 187–194.

Vikenes, K., Farstad, M., & Nordrehaug, J. E. (1999). Serotonin is associated with coronary artery disease and cardiac events. *Circulation. 100*(5), 483–489.

Voss, M. D., De Castro, D., Lipski, J., Pilowsky, P. M., & Jiang, C. (1990). Serotonin immunoreactive boutons formcloseap positions with respiratory neurons of the dorsal respiratory group in the cat. *J Comp Neurol. 295*(2), 208–218.

Wallis, D. I., Stansfeld, C. E., & Nash, H. L. (1982). Depolarizing responses recorded from no dose ganglion cells of the rabbit evoked by 5-hydroxytryptamine and other substances. *Neuropharmacology. 21*(1), 31–40.

Walther, D. J., & Bader, M. A. (2003). Unique central tryptophan hydroxylase isoform. *Biochem Pharmacol. 66*(9), 1673–1680.

Walther, D. J., Peter, J. U., Bashammakh, S., Hörtnagl, H., Voits, M., Fink, H., & Bader, M. (2003). Synthesis of serotonin by a second tryptophan hydroxylase isoform. *Science.. 299(5603),* 76.

Wang, H. T., Han, F., & Shi, Y. X. (2009). Activity of the 5-HT1A receptor is involved in the alteration of glucocorticoid receptor in hippocampus and corticotrophin-releasing factor in hypothalamus in SPSrats. *Int J Mol Med. 24*(2), 227–231.

Wilkinson, L. O., Dourish, C. T. In: Peroutka, S. (Ed). (1991). Serotonin Receptor Subtypes: Basic and Clinical Aspects. Wiley, New York, NY. pp. 147–210.

Wilson, C. C., Faber, K. M., & Haring, J. H. (1998). Serotonin regulates synaptic connections in the dentate molecular layer of adultratsvia 5-HT1a receptors: evidence for a glial mechanism. *Brain Res. 782(1–2),* 235–239.

Wong, E. H., Reynolds, G. P., Bonhaus, D. W., Hsu, S., Eglen, R. M. (1996). Characterization of [3H]GR 113808 binding to 5-HT4 receptors in brain tissues from patients with neurodegenerative disorders. *Behav Brain Res. 73,* 249–252.

Yildiz, O., Smith, J. R., Purdy, R. E. (1998). Serotonin and vasoconstrictorsynergism. *Life Sci. 62*(19), 1723–1732.

Yusuf, S., Al-Saady, N., & Camm, A. J. (2003). 5-hydroxytryptamine and atrial fibrillation: how significant is this piece in the puzzle?. *J Cardiovasc Electrophysiol. 14*(2), 209–214.

Zamani, A., & Qu, Z. (2012). Serotonin activates angiogenic phosphorylation signaling in human endothelial cells. *FEBS Lett 586,* 2360–2365.

Zgombick, J. M., Borden, L. A., Cochran, T. L., Kucharewicz, S. A., Weinshank, R. L., & Branchek, T. A. (1993). Dual coupling of cloned human 5-hydroxytryptamine-1-D-alpha and 5-hydroxytryptamine-1-D-beta receptors stably expressed in murine fibroblasts: inhibition of adenylatecyclase and elevation of intracellularcalciumconcentrationsviapertussistoxin-sensitiveGprotein(s). *Mol Pharmacol. 44*(3), 575–582.

Zgombick, J. M., Schechter, L. E., Macchi, M., Hartig, P. R., Branchek, T. A., & Weinshank, R. L. (1992). Human gene S31 encodes the pharmacologically defined serotonin 5-hydroxytryptamine 1E receptor. *Mol Pharmacol. 42*(2), 180–185.

Zgombick, J. M., Weinshank, R. L., Macchi, M., Schechter, L. E., Branchek, T. A., & Hartig, P. R. (1991). Expression and pharmacological characterization of a canine 5-hydroxytryptamine 1D receptor subtype. *Mol Pharmacol. 40*(6), 1036–1042.

CHAPTER 7

EICOSANOIDS: CONSTITUTION AND PATHOPHYSIOLOGICAL FUNCTIONS

GUNJAN JOSHI

Center for Biomedical Engineering and Technology, School of Medicine, University of Maryland, Baltimore, USA

CONTENTS

7.1 INTRODUCTION

Eicosanoids belongs to a large family of metabolites formed by oxygenation of 20 carbon (eicosa) fatty acids such as dihomolinolenic acid, arachidonic acid, and eicosanopentaenoic acid, the main precursor being the essential fatty acidarachidonic acid (cis 5,8,11,14-eicosatetraenoic acid) in mammalian cells (Smith et al., 2000). There are three major pathways within the arachidonic cascade, which metabolizes arachidonic acid to form the eicosanoids. These pathways include the cyclooxygenase, lipoxygenase, and epoxygenase pathways (Smith and Murphy, 2002). There are various subfamilies of eicosanoids. These include prostanoids (products of a cyclooxygenase pathway), the leukotrienes (eoxins) and hydroxyeicosatetraenoic acid (HETE) (products of lipoxygenase pathway), and epoxyeicosatrienoic acid (EETs) (product of epoxygenase pathway). Prostanoids subfamily includes the prostaglandins (PG) and thromboxanes (TX) (Noverr et al., 2003). Eicosanoids are quiet specific in stereochemical precision and recognition and have unique biological properties (Funk et al., 2001). They have various roles in nervous system (Tassoni et al., 2008), inflammation (Khanapure et al., 2007), cardiovascular system (Wennmalm et al., 1988), immunity (Harizi et al., 2005), reproduction (Zahradnik et al., 1992), diabetes (Luo and Wang, 2011), endochrine and neuroendocrine system (Bhathena, 2006), and in tissue regeneration, repair, and wound healing (Kalish et al., 2013).

7.2 CHEMISTRY, BIOSYNTHESIS, DEGRADATION OF EICOSANOIDS

7.2.1 CHEMISTRY OF PROSTANOIDS: THE PROSTAGLANDINS AND THROMBOXANES

The chemical structures of various prostaglandins and thromboxanes are shown in Figure 7.1. Each prostaglandins (PG) has a cyclopentane ring is named by the functional groups on the ring (A to K) as PGA to PGK, and number (1, 2, or 3) of double bonds in the side chain. Thromboxanes (TX) are similar, with the cyclopentane ring replaced by a tetrahydropyran system. Functional group A to and J has a keto group in various positions on the ring. Functional group F has two hydroxyl groups while K has two keto substituents on the ring. Functional groups and H are bicyclic endoperoxides. Prostacyclin (PGI) is denoted by an oxygen bridge between carbons 6 and 9. Greek subscripts are used to denote the orientation of ring hydroxyl groups at carbon 9 (e.g., PGFα). All prostaglandins have a hydroxyl group on carbon 15 and a transdouble bond at carbon 13 (Nelson, 1973; Lands WEM 1979; Smith and Murphy, 2002). Prostanoids are highly lipid-soluble with pH below pH 3.0 and prostanoids like PGH_2 often have very poor water solubility. PGE, PGF, and PGD derivatives are relatively stable in aqueous solution at pH 4–9. PGI_2 is unstable below pH 8.0 (Smith and Murphy, 2002; The Eicosanoids).

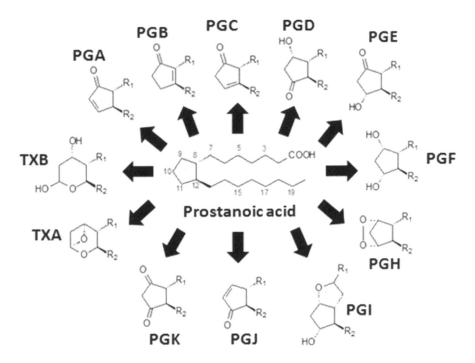

FIGURE 7.1 The chemical structures of prostaglandins and thromboxanes.

7.2.2 BIOSYNTHESIS OF PROSTANOIDS (THE CYCLOOXYGENASE PATHWAY)

Eicosanoids are synthesized and released in response to extracellular hormonal stimuli. Prostanoid formation occurs in three steps: (a) mobilization of free arachidonic acid from membrane phospholipids by lipases; (b) conversion of arachidonic acid to the prostaglandin endoperoxide PGH_2 by PGH synthase; and (c) conversion of PGH2 to other major prostanoids by isomerases or reductases (PGD_2, PGE_2, $PGF_{2\alpha}$, PGI_2), or thromboxanes (TXA_2) (Smith WLet al., 1991; Smith, 1992) (Figure 7.2).

7.2.2.1 Mobilization of Arachidonic Acid

Prostanoid formation occurs by stimulated mobilization of arachidonic acid from membrane phospholipids. The arachidonic acid that is released is then

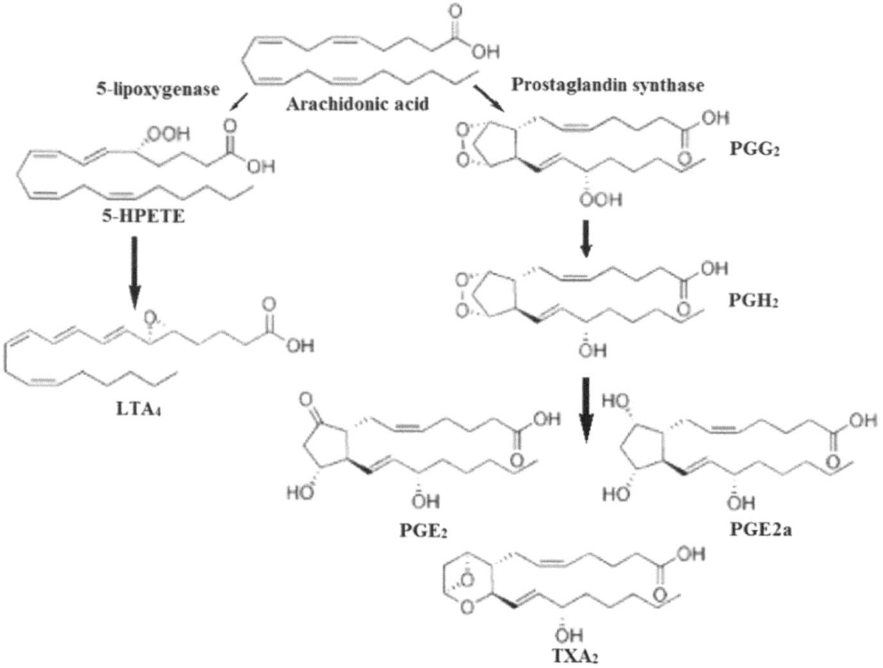

FIGURE 7.2   Biosynthesis of prostanoids.

converted to PGH$_2$ by the action of PGH synthase (Smith et al., 1991). A large number of hormones initiate prostanoids synthesis, which is triggered by mobilization of AA. AA cellular levels are kept under check by esterification of AA until phospholipases (PLA$_2$) acts on it (Smith, 1992). PLA$_2$ are enzymes that mobilize free fatty acids from the sn-2 position of certain phospholipids (Crofford, 2001). Current consensus suggests that cytosolic PLA$_2$ (cPLA$_2$) is involved directly in mobilizing arachidonic acid for constitutive prostaglandin endoperoxide H synthase-1 (PGHS-1), while cPLA$_2$ is indirectly, and secretory PLA$_2$ (sPLA$_2$) is directly, involved in mobilizing arachidonate for the inducible PGHS-2 (Gijon, et al., 2000; Murakami, et al., 1999; Smith and Langenbach, 2001). It is also stated that among all the PLA$_2$ members, type IV cPLA$_2$ is mainly involved in eicosanoid production, as cells devoid of cPLA$_2$ do not synthesize eicosanoid (Crofford, 2001). cPLA$_2$ regulate lipid mediator formation as soon as the cells get activated (Murakami et al., 1997). AA metabolism is initiated by Ca^{2+} mediated translocation of cPLA$_2$, and binding of cPLA$_2$ to phospholipid membranes (Leslie, 1997). AA metabolism by cPLA$_2$ occurs in the nuclear envelope and

ER, which are also sites for localization of COX enzymes and some of the terminal synthases (Murakami et al., 1997). The activity of cytosolic PLA_2 is triggered by phosphorylation by kinases (Gijon et al., 2000).

7.2.2.2 Conversion of Arachidonic Acid to the Prostaglandin Endoperoxide PGH_2 by PGH Synthase

The AA mobilized by $cPLA_2$ is acted on by prostaglandin H synthase (PGHS or COX) at the ER and nuclear membrane (Morita et al., 1995; Spencer et al., 1998), and is metabolized to prostaglandin PGH_2 (Funk, 2001; Thuresson et al., 2001). PGHS exists as two isoforms referred to as PGHS-1 (COX-1) and PGHS-2 (COX-2) (Smith et al., 2000). AA is the sole substrate of PGHS-1, whereas PGHS-2 can use both fatty acids or 2-AG (2-arachidonyl-glycerol) as substrate (Kozak et al., 2000). The PGHSs possess both cycloxygenase and peroxidase enzyme activities. Cyclooxygenase activity is required for the formation of PGG_2 (or 2-PGG_2-glycerol) and a peroxidase activity for the formation of PGH_2 (or 2-PGH_2-glycerol) (Smith et al., 1991; Smith and Murphy, 2002). Both enzymes catalyze the same two reactions at different sites. In a cyclooxygenase reaction, two oxygen molecules are added at C-11 to AA after removal of 13-proS hydrogen from AA to form a bicyclic endoperoxide. Serial cyclization of the 11-hydroperoxyl radical yields a bicyclic peroxide. Second O_2 molecule reacts at C-15 to form PGG_2. The peroxidase reaction then reduces the hydroperoxide by two-electron reduction to form prostaglandin PGH_2 (Smith et al., 1991; Smith and Murphy, 2002).

7.2.2.3 Conversion of PGH_2 to Other Major Prostanoids

Prostanoid synthesis is cell-specific (Smith et al., 2000), with endothelial cells producing mainly PGI_2, dense tubular membranes of platelets form TXA_2 and PGE_2 is the major prostanoid formed by renal collecting tubule cells. After synthesis, PGH_2 is converted to one of several prostanoids by a terminal synthase (Smith and DeWitt, 1995). PGE_2, PGD_2, $PGF_{2\alpha}$, PGI_2, and T_XA_α from PGH_2 are catalyzed by their respective synthases (Smith et al., 2000). PGD, and PGE, are formed by simple nonoxidative rearrangement of PGH_2 (Smith and DeWitt, 1995). $PGF_{2\alpha}$ production requires a net two-electron reduction of PGH_2, catalyzed by a PGF_α synthase using NADPH. Rest of the prostanoids is formed by isomerization reactions. PGF_2 derivatives are formed by three mechanisms: (a) reduction of PGH_2 to $PGF_{2\alpha}$ by an endo-peroxide reductase (b) reduction of PGD_2 to $9\alpha,11\beta$-PGF_2 by 11-ketoreduc-tase, and (c) reduction of PGE_2 to $PGF_{2\alpha}$ by a 9-keto-reductase (Smith et al., 1991). PGH_2 conversion to PGI_2 requires an acid catalyzed heterolytic cleav-age of PGH_2, and a reaction of a transiently positive oxygenat C-6 (Smith et al., 1991). TxA_2 is formed by the transient formation of an electropositive oxygen at C-11 and subsequent cleavage of the 9,11-peroxido group (Pace-Asciak and Smith, 1983; Smith et al., 1991).

7.2.3 DEGRADATION OF PROSTANOIDS

After synthesis in the ER, the prostanoids are diffused tothe cell membrane and exits the cell via carrier-mediated transport (Smith, 1992). Unlike other circulating hormones that exits from one major endocrine site, prostanoids are synthesized and released by almost all organs (Smith, 1992). The first step in catabolism of prostaglandins is the oxidation of the allylic-OH group at the carbon 15 position to yield the corresponding 15-keto prostaglandin metabolite by 15-hydroxyprostaglandin dehydrogenase (15-OH-PGDH). The enzyme requires NAD^+ or $NADP^+$ as cofactor and is specific for the $C_{15}(S)$ alcohol group (Hansen, 1976; Smith, 1992) The 15-keto prostaglan-din metabolites are further degraded to the 13,14 dihydro-15-keto PG by prostaglandin Δ^{13}-reductase (Anggard and Samuclsson, 1964). These metab-olites are further processed by β/and/or ω-oxidation and the metabolites are finally excreted in the urine (Smith, 1992). PGE and PGF may be intercon-verted enzymatically. Prostaglandin 9-ketoreductase catalyzes the oxidation of the ketone group at C_9 of PGE_2 to yield $PGF_{2\alpha}$. The reverse reaction is catalyzed by prostaglandin 9-hydroxydehydrogenase (9-OH-PGDH) (Leslie and Levine, 1973). TXA_2 with a very short half-life is nonenzymatically hydrolyzed to TxB_2, which isthen excreted out after getting metabolized by 15-PGDH and prostaglandin Δ^{13}-reductase. Prostacyclin (PGI_2) is degraded by 15-PGDH to form 6,15-diketo $PGF_{1\alpha}$ (Machleidt et al., 1981).

7.2.4 CHEMISTRY OF LEUKOTRIENES

The structure of leukotrienes (LTs) and lipoxins are conjugated chemicals that have double bonds that are saturated. Leukotrienes are derivatives of arachidonic acid and other polyunsaturated fatty acids. The name leukotriene is derived since they were discovered in white blood cells (polymorphonuclear Leucocytes), and has three conjugated double bonds (triene) (Mathews et al., 1996). In most LTs, the conjugated triene is in a trans, trans, cis arrangement, and are labeled by letters A-E, with subscripts indicating number of double bonds present (Bhagavan, 2002). Amongst the leukotrienes, leukotriene B4 (LTB$_4$) and leukotriene C$_4$ (LTC$_4$) play an important role. LTB$_4$ is a very potent chemotactic and chemokinetic agent for the human polymorphonuclear leukocyte, and LTC$_4$ is a specific smooth muscle constrictor, thus assist in edema by leakage of vascular fluid (Ford Hutchinson, 1994).

7.2.5 BIOSYNTHESIS OF LEUKOTRIENES (THE LIPOXYGENASE PATHWAY)

Leukotrienes are synthesized predominantly by inflammatory cells like polymorphonuclear leukocytes, macrophages, and mast cells (Funk, 2001). Being primarily formed in various types of leukocytes, they are named Leukotrienes (Haeggstrom et al., 2002). Like other eicosanoids, leukotrienes are paracrine mediators exerting their actions locally in a cell. The leukotrienes are classified as the dihydroxy acid leukotriene B$_4$ (LTB$_4$), and the cysteinyl-leukotrienes (cys-LTs) such as LTC$_4$, LTD$_4$, LTE$_4$ and LTF$_4$ (Samuelsson, 1983). Leukotrienes are produced by the action of calcium and ATP-dependent enzyme, 5–lipoxygenase (5-LO) on AA (Dighe et al., 2010; Rouzer et al., 1986; Shimizu et al., 1986; Hogaboom et al., 1986). The enzyme carries out the stereospecific abstraction of H at C7 and insertion of an oxygen molecule at C5 of AA yielding 5(S)-hydroperoxyeicosatetraenoic acid (5-HpETE) (Panossian et al., 1982). 5-LO then catalyzes a dehydration reaction forming biological precursor of all leukotrienes, the leukotriene A$_4$ (LTA$_4$). LTA$_4$ undergoes hydrolysis, conjugation with glutathione, or transcellular metabolism to generate other leukotrienes (Gronert et al., 1999). LTA$_4$ is then hydrolyzed by leukotriene A$_4$ hydrolase (LTA$_4$) in cytoplasm and nucleus to yield LTB$_4$, a potent neutrophil chemoattractant and stimulator of leukocyte adhesion to endothelial cells (Samuelsson, 1983; Peters-Golden, 2001). LTA$_4$ also conjugates with glutathione to yield LTC$_4$ by LTC$_4$ synthase (Penrose and Austen, 1999) at the nuclear envelope (Samuelsson et al., 1991). LTC_4 undergoes extracellular metabolism yielding LTD_4 and LTE_4 (Funk, 2001).

7.2.6 DEGRADATION OF LEUKOTRIENES

Leukotrienes are metabolized rapidly from the body with a half-life less than five minutes (Levi et al., 1988). In contrast to the prostaglandins, the leukotrienes are not degraded from the carbon-l-carboxyl group but are degraded from the ω-end of the ω-carboxy metabolites (Jedlitschky et al., 1993). LTB_4 is rapidly metabolized through both oxidative and reductive pathways (Murphy and Wheelan, 1998). Both LTB_4 and LTC_4 are metabolized by cytochrome P-450 s of the CYP4F family followed by β-oxidation. Ten different enzymes carry the ω-oxidation of leukotrienes (Kikuta et al., 1999). Amongst them, LTB_4 ω-hydroxylase (CYP4F3A) converts LTB_4 into 20-hydroxy-LTB_4 in the human neutrophils, and then 20-hydroxy-LTB_4 is converted by CYP4F3A to 20-carboxy-LTB_4 in human neutrophils (Kikuta et al., 1998; Shak and Goldstein, 1985), or by alcohol dehydrogenase (ADH) or aldehyde dehydrogenase (AldDH) in the hepatocytes (Sutyak et al., 1989). 20-carboxy-LTB_4 is further degraded to 18-and 16-carboxy-LTB_4 by β-oxidation.

7.3   RECEPTORS AND MECHANISM OF ACTIONS OF EICOSANOIDS

7.3.1   PROSTANOIDS RECEPTORS

Prostanoids act like hormones, both in an autocrine way targeting the same cell, and in paracrine way by acting on the neighboring cells (Tilley et al., 2001; Narumiya, 2001). Prostanoids work through G-protein-coupled prostanoid receptors (GPCRs) (Narumiya and FitzHerald, 2001). Most of the prostaglandin GPCRs are localized at the plasma membrane, although some are present at the nuclear envelope as well (Bhattacharya et al., 1998). Eight types and subtypes of membrane prostanoid receptors are known, which includes the PGD receptor (DP), four subtypes of the PGE receptor (EP1, EP2, EP3, and EP4), the PGF receptor (FP), the PGI receptor (IP), and the TxA receptor (TP) (Narumiya and Sugimoto, 1999; Coleman et al., 1994). All these receptors are G protein–coupled rhodopsin-type receptors with seven transmembrane domains. The IP, DP, EP_2, and EP_4 receptors are named "relaxant" receptors as they facilitate cAMP rise, whereas the TP, FP, and EP_1 receptors are known as "contractile" receptors and are known to induce calcium mobilization and the EP_3 receptor has been named "inhibitory" receptors as it induces a decline in cAMP levels (Narumiya and Fitz Herald, 2001).

7.3.1.1 DP/DP1 Receptor

DP1 receptor is the most studied PGD_2 receptor, and is coupled to adenylate cyclase via a G_s protein its activation leads to Gs mediated increase in cyclic AMP (Alvarez et al., 1981; Ito et al., 1990; Trist et al., 1989). Activation of DP1 receptors causes inhibition of platelet (Matsuoka et al., 2000) activation (Whittle et al., 1985), vasodilatation (Coleman, 1990; Giles, 1989) and bronchodilatation (Matsuoka et al., 2000). DP_1 receptors are expressed in the brain, where they may be involved in the regulation of sleep (Urade and Hayaishi, 2011). DP1 is expressed by certain leukocytes and it controls various functions including cytokine production (Faveeuw et al., 2003; Gosset et al., 2005; Tanaka et al., 2004).

7.3.1.2 DP2 (CRTH2) Receptor

It was known as an orphan receptor expressed by T helper 2 (TH_2) lymphocytes and was named the chemoattractant receptor-homologous molecule ($CRTH_2$). It is receptor for PGD_2 and helps in activation of eosinophil, basophil, monocytes and Th2 cells for chemotaxis, proinflammatory responses (Hirai et al., 2001). Although DP2 binds PGD_2 with a similar affinity as DP1, it is not structurally related to DP1 and signals through a different mechanism. In contrast to DP1 receptor, DP_2 receptor is coupled to adenylate cyclase by Gi and it inhibits cAMP formation (Gallant et al., 2007; Sawyer et al., 2002).

7.3.1.3 EP Receptors

PGE_2 has four different PGE (EP) receptors know as EP_1, EP_2, EP_3, and EP_4 receptors (Smith and Murphy, 2002). The EP receptor undergoes alternative splicing at the carboxyl terminus to yield the isoforms. Each EP receptor is coupled to different second messenger systems (Crofford, 2001). Each EP receptor has selective agonist. The EP_1 receptor is coupled to Ca^{2+} channels, which on getting activated results in an increased intracellular Ca^{2+} concentration (Smith and Murphy, 2002; Crofford, 2001). EP_1 is coupled through G_q to activate phospholipase C (Smith and Murphy, 2002. The EP_2 and EP_4 receptors are coupled to G_s subunits to stimulate adenylate cyclase and increase cyclic adenosine monophosphate (cAMP) production (Crofford, 2001). The EP_3 receptor is coupled to a G_i subunit to inhibit and of adenylate cyclase and decreases cAMP (Narumiya and FitzGerald, 2001).

7.3.1.4   TP Receptors

TP receptors can be found in platelets, lung, vascular smooth muscle, endothelial cells, kidney, brain, spleen, thymus, monocytes, uterus, and placenta (Blackman et al., 2002; D'Angelo et al., 1996; Kitanaka et al., 1995; Hirata et al., 1991; Swanson et al., 1992; Masuda et al., 1991; Brown and Venuto, 1999; Ushikubi et al., 1993; Allan and Halushka, 1994). The TP/TXA_2 receptor belongs to the G protein-coupled rhodopsin (G_q-type receptor), and on stimulation leads to an increase in intracellular Ca^{2+} and activation of phospholipase C (Stichtenoth et al., 2001). This Gq-coupled prostanoid receptor binds thromboxane with high affinity, promoting platelet aggregation and constriction of both vascular and airway smooth muscle (Pettipher, 2007). A total of nine G proteins are known to couple to TP receptors, Gαq, Gαi2, Gαs, Gα11, Gα12, Gα13, Gα15, Gα16, and Gh that couple to TP receptors (Shenker et al., 1991; Kinsella et al., 1997; Offermans et al., 1994; Djellas et al., 1999; Manganello et al., 1999). Signal transduction occurs through coupling of G proteins to stimulate PI hydrolysis (Itoh et al., 1985).

7.3.1.5   FP Receptor

FP receptors are members of G-protein-coupled receptors (GPCR) and are coupled to Gq to activate phospholipase C and stimulate protein kinase C (PKC) and Ca^{2+} signaling (Abramovitz et al., 1994). FP receptors also activate Rho and focal adhesion kinase to stimulate the formation of actin stress fibers (Pierce et al., 1999). By activation of FP receptor, $PGF_{2\alpha}$ mediates luteolysis and smooth muscle contraction in uterus and gastrointestinal tract sphincters (Biancani et al., 1997). FP receptor-stimulation also known to reduce intraocular pressure (Stjernschantz and Resul, 1992).

7.3.1.6 IP/Prostacyclin (PGI_2) Receptor

The prostacyclin (IP/PGI_2) receptor stimulation of adenylate cyclase by coupling to a G_s protein leading to increased intracellular cyclic AMP (cAMP) (Alvarez et al., 1981). IP is coupled to a guanosine nucleotide-binding α-stimulatory protein (Gαs). The IP receptor is capable of coupling to multiple G proteins by unknown mechanism (Miggin and Kinsella, 2002; Vassaux et al., 1992). Phosphorylation of key proteins by activation of protein kinase A (PKA) by cAMP rise cause relaxation of vascular smooth muscle, vasodilation and inhibition of platelet aggregation and reduced cell proliferation (Gleim et al., 2009; Smyth et al., 2009). Human IP receptors are present on variety of cell types including platelets, neutrophils, eosinophils, dendritic cells, T regulatory cells, activated T cells, and many cell types in lung (El-Haroun et al., 2008; Mohite et al., 2011; Soberman and Christmas, 2006). There is a growing evidence of some nuclear receptors for eicosanoids, which are named as the peroxisome proliferator-activated receptors (PPARs) (Serhan et al., 1996) which functions as a transcription factor after activation by PGI_2 (Mohite et al., 2011; Gurgul-Convey and Lenzen, 2010). The various PPAR isoforms are α, δ, β, and γ (Mohite et al., 2011). PPARs are activated by a broad range of ligands other than PGI_2 (Mohite et al., 2011). The agonist and antagonists of prostanoid receptors are mentioned in Table 7.1.

7.3.2. LEUKOTRIENES RECEPTORS

7.3.2.1 BLT Receptors

LTB_4 is a potent chemotactic agent and it exerts its action through two specific G-protein-coupled receptors termed BLT1 and BLT2 (Serhan and Prescott, 2000, Yokomizo et al., 2000; Kamohara et al., 2000; Tryzelius et al., 2000). BLTI is exclusively expressed inhuman inflammatory cells, including lymphocytes and mast cells and has high specificity for LTB_4 (Kato et al., 2000; Lundeen et al., 2006). BLT2 receptor is homologous to the BLT1 receptor but has a higher specificity for LTB_4 and also broad ligand specificity for

TABLE 7.1 Agonist and Antagonists of Prostanoids and Leukotrienes

Receptors	Agonists	Reference	Antagonists	Reference
Prostanoid receptors				
DP1	BW 245C, RS-93520; PGD2	Town et al. (1983); Whittle et al. (1983); Alvarez et al. (1991); Crider et al. (1999); Keery and Lumley (1988)	Laropiprant; S-5751; AH-6809; BW-A868C; MK-0524	Cheng et al. (2006); Paolini et al. (2009); Shichijo et al. (2009); Wheeldon and Vardey (1993); Lydford et al. (1996); Sturino et al. (2007)
DP2/CRTH$_2$	Indomethacin; PGD2	Sawyer (2002); Hirai (2002); Mathiesen et al. (2006)	Ramatroban; TM-30089; AM 156; K-117	Uller et al. (2007); Stebbins et al. (2010); Mimura et al. (2005)
EP1	Iloprost; Sulprostone; PGE2; ONO-DI-004	Durocher et al. (2000); Walch et al. (2001); Suzawa et al. (2000); Giblin et al. (2007)	ONO-8711; ONO-8713; SC-51089; MK-0524	Ikeda et al. (2006); Ahmad et al. (2006); Matsuo et al. (2004); Sturino et al. (2007)
EP2	Misoprostol, Butaprost; 13205; PGE2	Ushikubi (1995); Duckworth et al. (2002); Nials et al. (1993); Woodward et al. (1995)	AH-6809	Woodward et al. (1995)
EP3	ONO-AE-248; 28767; PGE2; Sulprostone	Kunikata et al. (2005); Nakagawa et al. (1995); Singh et al. (2009); Clark et al. (2008)	DG-041; L-826266; Compound 18; Compound 19	Singh et al. (2009); Gallant et al. (2002); Juteau et al. (2001); Belley et al. (2005)
EP4	ONO-AE1–329; AGN 205203 and its methyl ester; 4819-CD; ONO-AE3–208; PGE2; ONO-AE1–329	Nitta et al. (2002); Jiang et al. (2007); Xiao et al. (2004); Yang et al. (2006); Davis et al. (2004); Foudi et al. (2008)	AH23848; BGC-20–1531; CJ-023423	Ushikubi (1995); Maubach et al. (2009); Nakao et al. (2007)
TP	U-46619; U-44069; STA2	Itoh et al. (1993); Imura et al. (1990); Nakahata et al. (1990)	KP-496; ONO-3708; S18886 (terutroban); STA2, U-46619	Ishimura et al. (2009); Andoh et al. (2007); Sebeková et al. (2007); Nakahata et al. (1990); Coleman and Sheldrick (1989)

Receptor	Agonist(s)	Reference	Antagonist(s)	Reference
FP	PGF2a	Cirillo et al. (2007)	AS-604872; THG-113	Cirillo et al. (2007); Peri et al. (2006)
IP	Carbacyclin; Cicaprost	Bley et al. (2006); Jones et al. (2006)	RO-1138452; RO-3244794; Compound 24	Bley et al. (2006); Brescia et al. (2007)
Leukotriene receptor				
BLT	20-hydroxy-LTB4. Order of potency of agonists LTB4>12(R)-HETE (LTC4 and LTD4 are mainly inactive)	Smith and Murphy (2002); Dighe (2010)	Amelubant (BIIL-284), Etalocib sodium (LY-293111), Moxilubant maleate (CGS-25019C), and the highly BLT1 selective biphenylyl-substituted chroman carboxylic acid CP-105696	Yokomizo (2001); Birke (2001); Marder (1995); Raychaudhuri (1995); Koch (1994)
$CysLT_1$	LTD4=LTC4>LTE4 (LTE4 being a partial agonist in some tissues)	Dighe (2010)	Montelukast, Zafirlukast, Pranlukast; BAY u9773 (nonselective), MK-571	Jones (1995); Krell (1990); Obata (1987); Heise (2000)
$CysLT_2$	LTC4>LTD4>LTE 4 (LTE4 being a partial agonist in some tissues)	Dighe (2010)	HAMI 3379; BAY u9773 (nonselective),	Zhang (2013); Heise (2000)

12-HETE and 15-HETE (Yokomizo et al., 2000). In contrast to the BLT1 receptor, BLT2 is ubiquitously expressed in many tissues. BLT2 is known to signal antiinflammatory functions (Iizuka et al., 2010).

7.3.2.2 CysLT Receptor

The biological functions of cysteinyl leukotriene (Cys-LTs) are mediated through two receptors, CysLTI and CysLT2. The CysLTI receptor is a G-protein-coupled receptor with seven trans membrane regions (Lynch et al., 1999) and is found in the spleen, leukocytes, lung tissue, airway smooth muscle cells, macrophages, and mast cells (Lynch et al., 1999; Sjostrom et al., 2002; Mellor et al., 2001). The preferred ligands for the CysLT1 receptor are LTD_4 followed by LTC_4 and LTE_4. The CysLT2 receptor has sequence identity to the CysLT1 receptor (Heise et al., 2000; Nothacker et al., 2000). CysLT2 receptor binds both LTC_4 and LTD_4, with lower affinity towards. It is found in Purkinje fibers of heart, brain, leukocytes, spleen, placenta, and lymph nodes. CysLTs assists in contractile responses in the lung, coronary artery and distal and mesenteric pulmonary artery, with no role in most systemic large arteries (Smith and Murphy, 2002). The agonist and antagonists of prostanoid receptors are mentioned in Table 7.1.

7.4   ACTION AND PATHOPHYSIOLOGICAL FUNCTIONS OF EICOSANOIDS

7.4.1   EICOSANOIDS IN ALLERGY AND INFLAMMATORY RESPONSES

Prostaglandin D_2 is responsible for a number of allergic responses like blood flow changes, recruitment of eosinophils and T-helper (Th) 2 lymphocytes and Th2 cytokine production. PGD_2 is produced by dendritic cells, macrophages, eosinophils, type 2 T helper (Th2) cells and endothelial cells (Tajima et al., 2008; Söderström et al., 2003; Vinall et al., 2007; Hyo et al., 2007; Soler et al., 2000). It is known to enhance LT_4 synthesis by eosinophils during allergic inflammation (Mesquita-Santos et al., 2006), and in controlling Th1-mediated allergic mechanisms (Ajuebor et al., 2000; Trivedi et al., 2006). CRTH2 is expressed in basophils and are known to migrate towards PGD_2 (Hirai et al., 2001; Royer et al., 2008). CRTH2 present in the macrophages and mast cells are found in large numbers in the nasal mucosa of allergic patients (Shirasaki et al., 2009). PGD_2 and its metabolites interact with DP_1, CRTH2 and other intracellular pathways to produce opposing roles in inflammatory responses by suppression of leukocyte activation (Kostenis and Ulven, 2006; Sandig et al., 2007). This DP1-mediated suppression of leukocyte activation is responsible for increasing the allergic responses. PGD_2 is also responsible for the pathological blood flow changes observed in allergic diseases, leading to nasal congestion, enhance leakage of plasma protein and therefore increased nasal secretions. PGD_2 at sites of mast cell activation cause increase Th2 and IgE, which ultimately enhance mast cell production (Caron et al., 2001; Mazzoni et al., 2001). PGD_2 promotes chemotaxis of eosinophils through DP_2 receptor (Monneret et al., 2001). It also induces CRTH2-dependent production of interleukin 4, 5 and 13 in the absence of allergen (Xue et al., 2005). Activation of antigen-specific T cells results in accumulation of Th2 cells and CRTH2 helps in the recruitment and accumulation of these Th2 cells (Vinall et al., 2007).

PGI_2 is an immunoinhibitory prostanoid found in human lungs during allergies (Dahlen et al., 1983; Schulman et al., 1981). Inhibition of PGI_2 by COX-2 inhibitors causes increase in Th2 cytokine expression in lungs and decreased IL-10 formation leading to enhanced airway inflammation (Jaffar et al., 2002).

Phospholipase A_2 enzymes (PLA_2) are important enzyme catalyzing AA synthesis and thereby catalyzing eicosanoids production (Kudo and Murakami, 2002). A subgroup of PLA_2, $sPLA_2$-IIA has been found to be involved in a number of inflammatory responses, like asthma, theumatoid arthritis, psoriasis, Crohn's disease, etc. (Seilhamer et al., 1989; Murakami et al., 1997; Pruzanski and Vadas, 1991; Minami, 1994).

TXA_2 is involved in the inflammatory responses in several tissues (Martin et al., 2001; Miller et al., 2007) like heart where it is known to cause inflammatory tachycardia (Takayama et al., 2005). TXA_2 is also responsible for allergic responses in asthma, rhinitis and atopic dermatitis (Tanaka et al., 2002; Nagai et al., 2006; Shirasaki et al., 2007).

Similar to prostaglandins, leukotrienes are also responsible for a number of inflammatory and immunomodulatory responses and are implicated in the pathogenesis of many human inflammatory diseases such as asthma, rheumatoid arthritis, psoriasis, dermatitis, and cancer (Lewis et al., 1990; Peters-Golden et al., 2007). The levels of leukotriene are enhanced in the lungs of patients with idiopathic pulmonary fibrosis (Ozaki et al., 1992; Wilborn et al., 1996). cysLTs are produced in mast cells, eosinophils, and alveolar macrophages (Drazen et al., 1999). cysLT1 and cysLT2 receptors are expressed by tissue mononuclear cells (MCs) in nasal cavities, and in human cord blood-derived MCs (Lewis et al., 2007; Mellor et al., 2001; Mellor et al., 2003; Figueroa et al., 2003; Sousa et al., 2002). TGF-β is known to increase cysLT synthesis by macrophages (Steinhilber et al., 1993; Riddick et al., 1999). CysLTs are important modulators inflammation and for inflammatory responses in asthma (Figueroa et al., 2001; Kanaoka et al., 2004). Allergen exposure increases presence of cysLTs (LTC_4, LTD_4, and LTE_4) and LTB_4 in airways leading to airway hyper responsiveness (Mondino et al., 2004; Peters-Golden et al., 2000). The cysLTs increase the number of eosinophils in the airways of patients with asthma (Charvat et al., 1997; Mulder et al., 1997), and stimulate survival of eosinophils, mast cells, and basophils for inflammatory responses (Lee et al., 2000; Holgate et al., 2003). cysLTs also induce the formation of Th2 cytokines for allergic responses (Chibana et al., 2003; Mellor et al., 2001). Cys-LTs are also known to be involved in the maturation of dendritic cells (Leier, 1994; Wijnholds, 1997), which influence the T-cell responses to regulate adaptive immunity (Kanaoka et al., 2004).

7.4.2 EICOSANOIDS IN BRAIN

Primary prostaglandins are distributed uniformly throughout brain, with more activity in cerebrocortical gray matter than the white matter (Holmes and Horton, 1968). Cerebrospinal fluid contains prostaglandins (Ramwell, 1964), whereas thromboxane A_2 and prostacyclin are produced by cerebral microvessels and by the choroid plexus (Hagen et al., 1979. According to their distribution, primary prostaglandins are involved in brain neural activity (Maurer et al., 1980; Birkle et al., 1981; Goehlert et al., 1981), whereas prostacyclin plays a major role in the regulation of cerebral blood flow (Kontos et al., 1981). PGs have excitatory effect on the neurons (Avanzino et al., 1966; Siggins et al., 1971). An increase in the levels of PGs has been seen in a variety of neurological diseases, as PGs are known to relax the blood vessels. $PGF_{2\alpha}$ levels in the CSF are enhanced in patients with epilepsy, meningoencephalitis, hydrocephalus, and after surgical trauma (Wolfe and Coceani, 1979). Increase in PGE_2 and $PGF_{2\alpha}$ is found in patients with subarachnoid hemorrhage and stroke (Latorre, et al., 1974; Wolfe and Marner, 1974). All primary prostaglandins and PGI_2 are formed in cerebral vessels

(White et al., 1977; Hagen et al., 1978). PGE_2, an intracranial vasoconstrictor (Pickard, 1977) is implicated in the pathogenesis of migraine (Harper et al., 1977; Welch et al., 1974). PGs are potent pyretic agents (Milton and Wendland, 1970), especially PGE_2 and PGE_1 are mainly responsible for pyrogenic fever induction (Feldberg et al., 1975; Hori and Harada, 1974; Milton and Wendland, 1971; Nistico and Marley, 1976; Pittman et al., 1976). Studies have shown that PGs might also be responsible for the formation of noninfective fever (DeyPK et al., 1974). Many behavioral studies have identified a role of PGs in sedation (Horton, 1964). The PGs responsible for this are PGEs (PGE_1, PGE_2 and PGE_3), excluding $PGF_{2\alpha}$ (Horton, 1964; Gilmore and Shaikh, 1972). PGE_1 and PGE_2 (but not PGFs) raise the cAMP levels in nervous tissue in vitro (Gilman and Nirenberg, 1971, Hamprecht and Schultz, 1973) by activation of adenylate cyclase.

Astrocytes and astrocytoma cells express Thromboxane A2 receptor (TP), and stimulation of TPs leads to interleukin 6 (IL-6) and astrogliosis secretion (Obara et al., 2005; Honma et al., 2006). TPs contribute to the dipsogenic response to angiotensin II (Kitiyakara et al., 2002), the stimulation of fetal adrenocorticotropic hormone (ACTH) secretion (Wood et al., 1993) and adrenaline release (Murakami et al., 1998). TPs in hippocampus are important in neuronal excitability and synaptic transmission (Hsu et al., 1996). Presynaptic TP activation triggers glutamate release, whereas postsynaptic TPs when gets activated inhibit synaptic transmission (Hsu and Kan, 1996).

During the secondary injury phase from a brain damage from traumatic brain injury (TBI), phospholipase A_2 (PLA_2) gets activated, resulting in AA release and generation of prostaglandins, leukotrienes (LTs), and thromboxanes (Phillis et al., 2003; Leslie et al., 2004). CysLTs levels increases in CNS after brain injuries such as cerebral ischemia, brain trauma and tumors (Schuhmann et al., 2003; Ciceri et al., 2001). The increase in CysLTs is associated with edema and cellular inflammatory responses (Schuhmann et al., 2003). In stroke, the Cys-LTs triggers blood brain barrier disruption and brain edema (Rao et al., 1999; Wang et al., 2006). Neurons and astrocytes require LTA_4 for the formation of cys-LTs, which are produced by activated neutrophils and macrophages (Farias et al., 2007).

7.4.3 EICOSANOIDS IN CARDIOVASCULAR DISEASES

Eicosanoids play a role in controlling coronary artery tone and produce locally active metabolites to modify cardiac metabolism (Willerson et al., 1984; Hirsch et al., 1981). The prostaglandins are involved in various clinical conditions like myocardial ischemia and infarction, thrombosis, hypertension, and inflammation, cardiac fibrosis and atherosclerosis (Yuhki et al., 2011; Willerson et al., 1984; Vanhoutte et al., 1985). Prostanoid receptors play important role in homeostasis in thrombosis and atherosclerosis (Jabbour et al., 2006). Among the prostanoids, PGE_2 plays important role in cardiovascular inflammation (Tilley et al., 2001; Furuyashiki and Narumiya, 2011) by exerting the effects via the G-coupled receptors (Coleman et al., 1994). Renal cortex arteries and arterioles synthesize prostaglandins I_2 (PGI_2) and E_2 (PGE_2) along with prostaglandin $F_{2\alpha}$ ($PGF_{2\alpha}$) (Schlondorff et al., 2001; Terragno et al., 1978). Since both prostaglandin E, and prostacyclin are coronary vasodilators and both can inhibit platelet aggregation, they are used in the treatment of myocardial ischemia (Bugiardini et al., 1985; Henriksson et al., 1985; Chierchia et al., 1982), hence are also effective and used in patients undergoing cardiopulmonary bypass (Fish et al., 1986). The coronary endothelium synthesizes vasodilatory prostaglandins like prostacyclin (PGI_2) (Bolton et al., 1980; Needleman et al., 1977). PGI_2 is a vasodilator and a potent inhibitor of platelet aggregation (Moncada et al., 1976; Gryglewski et al., 1976). Its antiaggregatory activity is due to its stimulation of platelet adenylate cyclase, thereby elevating platelet cAMP levels (Gorman et al., 1977). On the contrary TXA_2 is an endogenous constrictor of arteries and promotes platelet aggregation by reducing cAMP levels (Bloor et al., 1973). There is an imbalance between the thromboxane/prostacyclin ratio, favoring thromboxane production during acute myocardial ischemia (Berger et al., 1977). Moreover, patients with stable angina have elevated thromboxane levels at rest compared to control patients (Tada et al., 1981). Patients with a deficiency of platelet thromboxane synthetase suffer from a bleeding deficiency and deficiency of PGI_2 causes arterial thrombosis (Weiss and Lages, 1977; Machin et al., 1980). Prostaglandins are known in the production of atherosclerosis (Smith, 1980; Stoffersen et al., 1978) with a reduction in PGI_2 formation by the coronary vascular endothelium (Dembinska-Kiec et al., 1977; D'Angelo et al., 1978; Sinzinger et al., 1980). Prostaglandins are also released during myocardial ischemia which influences myocardial adenylate cyclase (Klein and Levey, 1971) and platelet aggregation (Smith, 1980). Myocardial ischemia in patients with variant angina has been is known to be due to an imbalance between PGI_2 and TXA_2 production (Tada et al., 1981). An increase in coronary sinus TXB_2 and platelet aggregation is also seen in myocardial ischemia patients (Kumpuris et al., 1980; Green et al., 1980). TXB_2 levels are also elevated in patients with coronary artery spasm (Robertson et al., 1981) and in patients with unstable angina (Robertson et al., 1981). Similarly, in patients with acute myocardial infarction, there is a marked increase in coronary sinus and peripheral TXB_2 levels (Szczeklik et al., 1978), enhanced platelet activation, and increased incidence of platelet aggregates (Mehta and Mehta, 1979). Early depletion of PGI_2 from endothelial tissue could leads to the atherosclerosis by adipocyte lipid deposition in smooth muscle cells (Vane and Corin, 2003).

A number of studies provide evidence indicates that leukotrienes are involved in various stages and types of cardiovascular disease. Lipoxygenase 5-LO and leukotrienes, in particular LTB_4 play a role in in human atherosclerosis, myocardial infarction, and stroke (Funk, 2005). Lipoxygenases, especially human 15-LO-1, are known to participate in oxidation of LDL, conversion of macrophages into foam cells, and promoting atherosclerosis (Steinberg et al., 1989; Ylä-Herttuala et al., 1991). LTC_4 and LTD_4 induced transient constriction of arterioles (Dahlen et al., 1981), whereas LTB_4 induce adhesion of leukocytes to the endothelial cells of postcapillary venules (Dahlen et al., 1981; Bjork et al., 1982). Urinary levels of LTE_4 are elevated in patients of myocardial ischemia (Steinberg, 1991; De Caterina, 1988), and cys-LTs are known to cause coronary artery contraction (Sala et al., 1996).

7.4.4 EICOSANOIDS IN REPRODUCTION

PGs in uterus and ovaries are regulators of some important reproductive events such as ovulation, luteolysis, embryo implantation, and maintaining pregnancy (Weems et al., 2006). Prostaglandins were first discovered in seminal fluid and were found to stimulate smooth muscle contraction. This effect on uterine muscle may facilitate sperm movement within the female reproductive (Eliasson, 1959). Prostaglandins also induce LH-induced progesterone synthesis (Kuehl et al., 1970) and play a crucial role in luteolysis (Pharriss and Wyngarden, 1969). Uterine vasoconstriction in the endometrium is caused mainly by prostaglandins like $PGF_{2\alpha}$ (Clark and Brody,

1982; Clark et al., 1981). Prostaglandins may stimulate LH and ovulatory enzymes such as protease or collagenase (Beers et al., 1975). Studies indicate that prostaglandins function as central neurotransmitters (Horton, 1973) and stimulate the hypothalamic pituitary axis (Prostaglandins, 1972). They also may induce follicle wall contraction. The levels of PGE and PGF keep changing in the uterus (Singh et al., 1975; Maathuis and Kelly, 1978). During postovulatory phase, the PGF levels increase in the endometrium mainly due to progesterone withdrawal (Maathuis and Kelly, 1978), whereas during the postovulatory period, there is progesterone withdrawal with an increase in PGF-to-PGE ratios in the endometrium (Singh et al., 1975; Downie et al., 1974). PGE_2 causes relaxation in oviduct, whereas PGF_2 induces contractions (Sundberg et al., 1963; Lindblom et al., 1978). Ovarian $PGF_{2\alpha}$ production also gets elevated shortly after ovulation (Swanston et al., 1977). In some studies it was found that inhibition of prostaglandin synthesis blocks ovulation (Zanagnolo et al., 1996). PGE_2 is a major prostaglandin produced in the first trimester by human placental tissue (North et al., 1991). It acts in suppression of maternal rejection by inhibiting interleukin-2 production, hence inhibiting activation of natural killer cells and cytokine activated cells (Chouaib et al., 1985; Parhar et al., 1989). There have been conflicted reports regarding the levels of levels of vasodilator, PGE_2, and of the vasoconstrictor, PGF in preeclampsia. One study found PGE_2 levels were decreased whereas PGF levels were elevated in preeclampsia as compared to the normal pregnancy (Demers and Gabbe, 1976), another study found the opposite (Robinson et al., 1979). One more study found no difference in the levels of these prostaglandins between normal and preeclamptic patients (Hillier and Smith, 1981). Prostacyclin is known to inhibit uterine contraction (Wilhelmsson et al., 1981; Lye and Challis, 1982) and its production is increased in pregnancy (Goodman et al., 1982; Barrow et al., 1983). It is known to be produced by placenta (Makila et al., 1984; Jeremy et al., 1985), umbilical and uterine vessels (Kawano and Mori, 1983; Bjoro et al., 1986), and myometrium (Omini et al., 1979; Seed et al., 1983). Prostacyclin levels are reduced in preeclampsia (Bussolino et al., 1980; Downing et al., 1980). Thromboxane is a potent vasoconstrictor unlike prostacyclin, and in preeclampsia, there is more production of thromboxane than prostacyclin (Ylikorkala et al., 1981; Makarainen and Ylikorkala, 1984). Prostaglandin levels are known to get elevated at the site of implantation and cause changes in vascular permeability (Evans and Kennedy, 1978). One of the isoforms of cyclooxygenase, COX-1 gene cause changes in endometrium for vascular permeability, whereas the other isoform, COX-2 gene functions in angiogenesis for placental establishment (Chakraborty et al., 1996). PGE_2, on the other hand, prepares the uterus for implantation (Yang et al., 1997). Studies have been done to measure circulating levels of prostaglandins during pregnancy in humans (Whalen et al., 1978). Thromboxane levels and the ratio of vasodilator PGI_2 to vasoconstrictor TXA_2 increase during pregnancy. Prostaglandins are produced in the uterus and placenta and may be important in regulating blood flow in uterus and umbilical cord (Mitchell et al., 1980; Mitchell, 1981). PGE_1 and PGE_2 are potent vasodilators in the uterus (Clark et al., 1976). $PGF_{2\alpha}$ also induced some changes in uterine blood flow through vasoconstriction (Clark et al., 1982). PGI_2 administration cause vasodilation and leading to an increase in fetal heart rate, and a decrease in umbilical blood flow (Parisi and Walsh, 1989). Prostaglandins are used to induce early labor and abortion by promoting uterine contractions (Goldberg and Ramwell, 1975; Karim SMM 1975). The peripheral levels of most prostaglandins do not change throughout pregnancy (Mitchell, 1981; Dubin et al., 1981), but the levels of PGE_2, $PGF_{2\alpha}$, and 6-keto-$PGF_{1\alpha}$ gets elevated in amniotic fluid at the time of labor (Satoh et al., 1979).

LTs play important roles in reproduction and may enhance the action of PGs (Samuelsson, 2000). They play important role in a number of intraovarian processes such as ovulation, angiogenesis, development and regression of the corpus luteum (Lafrance and Hansel, 1992; Korzekwa et al., 2008). 5-LO and LT receptors are expressed in ovarian cell type and modify the secretory functions of ovarian cells (Korzekwa et al., 2010). LTB_4 stimulates secretions of some PGE_2 and progesterone, whereas LTC_4 stimulates the secretion of luteolytic $PGF_{2\alpha}$ (Korzekwa et al., 2010b). LTs also cause smooth muscle contractions in the uterus (Ritchie et al., 1984).

7.4.5 EICOSANOIDS IN ENDOCRINE SYSTEM

Eicosanoids works in autrochrine manner and act locally to produce biological effects, unlike the hormones, which are transported to target sites from the tissues where they are produced (Bhathena, 2000). PGs, especially PGE_2, affect secretion of many hormones of the endochrine system like ACTH, thyroid-stimulating hormone (TSH), follicle-stimulating hormone (FSH), LH, growth hormone (GH) and prolactin (Ojeda et al., 1981). PGE is known to stimulate adrenocorticotrophic hormone (ACTH) release in vivo, and could affect pituitary hormone secretion at the hypothalamic and pituitary level (De Wied et al., 1969). In the adrenal cortex of felines, AA is shown to form $PGF_{2\alpha}$, PGE_2, and PGA (Laychock and Rubin, 1975). In the adrenal cortex of rat, AA is primarily esterified to cholesterol and in phospholipids (Goodman, 1965; Walker, 1970; Vahouny et al., 1978). PGE_1 and PGE_2 have been shown to stimulate ACTH release, whereas $PGF_{1\alpha}$ and $PGF_{2\alpha}$ do not (de Wied et al., 1969; Peng et al., 1970). In one study, adreno-corticotrophic hormone (ACTH) produced a significant increase in $PGF_{2\alpha}$ release as compared to PGE_2 (Laychock and Rubin, 1977), whereas another study indicated increased formation and release of PGE_2 by ACTH in rat adrenocortical cells (Chanderbhan et al., 1979). PGE has been suggested to modulate adenyl cyclase activity (Kuehl Jr., 1974; Dazord, 1974). PGE synthesis in a study in rat adrenocorticotrophic cells is shown to be stimu-lated by ACTH or dibutyryl cyclic AMP, but not by dibutyryl cyclic GMP (Chanderbhan et al., 1979). PGE and $PGF_{2\alpha}$ incorporation to adrenocortical cells have opposing actions, with PGE mimicking the action of ACTH with an increase in the levels of cyclic AMP (Hodges et al., 1978), whereas $PGF_{2\alpha}$ addition did not increase free cyclic GMP levels (Vahouny et al., 1978).

PGs may play an important role in regulating thyroid function, the more potent being PGE and PGF (Burke and Sato, 1971; Sato et al., 1972). Thyroid stimulating hormone (TSH) increases PGs levels in thyroid cells indicating an important role of PGs in mediating TSH effects on thyroid (Burke and Sato, 1971; Sato et al., 1972).

In the neuroendochrine system, PGE may also regulate autonomic norad-renergic neurotransmission (Hedqvist, 1969; Hedqvist, 1977). PGE levels are shown to be elevated by noradrenaline (NA), and PGE restricts calcium avail-ability to reduces NA release from presynaptic nerve endings and reduce stim-ulated NA neurotransmission (Giiliner, 1983). Other PGs have less significant effects on NA neurotransmission, with $PGF_{2\alpha}$ showing completely opposite effects (Giiliner, 1983). PGs like PGE_2 have also been shown to affect pineal gland function (Cardinali, 1979). PGE increased proteolysis and the cyclic AMP levels in thyroid (Ahn and Rosenberg, 1970). PGE_1, PGE_2, and PGA_1 also increase adenyl cyclase activity and cAMP levels in the anterior pituitary gland (Zor et al., 1969). Since cAMP has been shown to release TSH23 and ACTH24, these prostaglandins in turn increase the levels of TSH and ACTH.

Gonadotropin-releasing hormone (GnRH) is is a peptide hormone that stimulates the synthesis and secretion of gonadotropins, luteinizing hormone (LH), and follicle-stimulating hormone from the anterior pituitary (Catt et al., 1984; Conn et al., 1987). LTs and some other lipoxygenase products of AA metabolism mediate LH release by GnRH (Hulting et al., 1985; Snyder et al., 1983). GnRH stimulates the formation of LTB_4 and LTC_4 in pituitary cells and GnRH receptor expressing cells acts as a major source of LT synthesis in the anterior pituitary (Kiesel et al., 1991).

PGs administration has been shown to increase levels of GH (Hertelendy et al., 1972; Ojeda et al., 1977), ACTH (Hedge and Hanson, 1972; Peng et al., 1970), prolactin (Ojeda et al., 1974a; Sato et al., 1974), FSH (Sato et al., 1974), and LH (Ratner et al., 1974; Sato et al., 1974; Tsafriri et al., 1973), PGEs being more potent than PGFs (Labrie et al., 1976). PGE_2 on acting in the hypothalamus increases prolactin secretion by inhibiting release of a prolactin-secretion-inhibiting factor (Ojeda et al., 1974).

PGs and phospholipids may influence insulin release from the pancreas (Robertson, 1983) and abnormal collagen metabolism in diabetes (Yue et al., 1985). PGE_2 has been reproducibly shown to modulate β-cell function (Tran et al., 2002; McDaniel et al., 1996). Insulin response to glucose was inhibited by PGE_2 administration in normal human subjects (Robertson and Chen, 1997). Fish oil is considered good for diabetic patients because it decreases TXA_2 and PGI_2 and increases PGI_3 and inactive TXA_3 causing reduction in the platelet aggregation, to avoid any coronary artery disease and hypertension in these diabetic patients (Knapp and FitzGerald, 1989; Nishikawa et al., 1997).

7.4.6   EICOSANOIDS IN RESPIRATION

PGE_2 plays a diverse role in a number of processed in the human system (Serhan and Levy, 2003). PGE_2 is found in airway smooth muscle, alveolar macrophages, dendritic cells, and pulmonary endothelial cells (Ozaki et al., 1987; Widdicombe et al., 1989; Sheller et al., 2000). The plasma and sputum supernatants of asthmatic patients have increased amounts of PGE_2 (Nemoto et al., 1976; Profita et al., 2003). This PGE_2 increase is due to increased activity and expression of COX-2 in airways of asthmatic patients (Kuitert et al., 1996; Sousa et al., 1997). PGE_2 causes relaxation of airway smooth muscles (Smith et al., 1975; Kawakami et al., 1973), and decreases proliferation of airway smooth muscle cells (Burgess et al., 2004). PGE_2 has been associated with diseases like rheumatoid arthritis (McCoy et al., 2002) asthma and chronic bronchitis (Melillo et al., 1994). PGE_2 relaxes the airway smooth muscles, inhibits bronchoconstriction and many inflammatory processes

(Tanaka et al., 2005; Sheller et al., 2000). It also decreases proliferation of lymphocytes, and Th2 cytokines (Jarvinen et al., 2008). Its antiinflammatory effect is due to mast cell activation (Pavord et al., 1993), and prevents allergen responses due to decreased production of cysteinyl-leukotrienes (Martin et al., 2002). Similar to PGE_2, another prostanoid, PGD_2 is also involved in a number of respiratory processes. PGD_2 is a potent bronchoconstrictors (Hardy et al., 1984), and is synthesized by activated mast cells during an allergic response and is found in pathogenesis of allergic asthma and atopic dermatitis (Hardy et al., 1984; Iwasaki et al., 2002; Murray et al., 1986). PGD_2 exerts its allergic inflammatory action by binding to two G-protein-coupled receptors (GPCRs), the D-prostanoid receptor (DP) and the chemoattractant receptor homologous molecule expressed on Th2 cells (CRTH2) (current). DP activation increases cAMP levels and calcium influx (4), leading to vasodilation and bronchodilation (Norel et al., 1999; Walch et al., 1999). On the contrary, CRTH2 inhibits cAMP and increased calcium influx (Sawyer et al., 2002). Production of interleukins by Th2 cells by PGD_2 is mediated through CRTH2 and not DP (Xue et al., 2005). LTB_4, TXA_2 and LTC_4 are produced by alveolar macrophages and blood monocyted (Martin et al., 1987; Goldyne et al., 1984). TXA_2 and LTB_4 are potent bronchoconstrictors and platelet aggregators (Henderson Jr., 1987; Henderson Jr., 1987; Sirois et al., 1980). LTB_4 stimulates leukocyte chemotaxis, degranulation, aggregation and other relevant leukocyte functions to promote inflammation in the airways (Henderson Jr., 1987; Henderson Jr., 1987; Ford-Hutchinson, 1980).

Leukotrienes are also involved in a number of allergic airway diseases including asthma and allergic rhinitis (Dahlen, 2006; Busse and Kraft, 2005; Peters-Golden et al., 2006). Human eosinophils and lung mast cells produce mainly LTC_4 (Peters et al., 1984; Henderson et al., 1984) whereas the alveolar macrophages and neutrophils produce LTB_4 (Martin et al., 1984; Henderson et al., 1983). Increased LTB_4 and LTC_4 production is also present in leukocytes of asthmatic patients (Sampson et al., 192). Urinary excretion of LTE_4 also increases in asthma (Manning et al., 1990). The cysLTs are the most potent constrictors of airway smooth muscles in humans (Dahlen et al., 1980; Weiss et al., 1982). They stimulate airway secretions (Coles et al., 1983), proliferation of airway smooth muscle (Cohen et al., 1995) and can cause eosinophil migration (Laitinen et al., 1993). cysLTs are involved in the pathogenesis of allergic rhinitis (Peters-Golden et al., 2006) and allergic rhinitis (de Graaf-in 't Veld et al., 1996). LTB_4, the main LT formed in the nasal mucosa, is also likely to have a fundamental role in the pathophysiology of allergic rhinitis (Dahlen et al., 1998).

7.5 CONCLUSION AND FUTURE DIRECTIONS

Eicosanoids are involved in a number of physiological and pathological responses in a human body. They are involved in a number of pathways including allergy and inflammation, reproduction, endocrine system, respiratory system, central and peripheral nervous system etc. A number of studies have been conducted in deciphering the importance of eicosanoid biology in human body. The biochemical and molecular biosynthetic pathways of eicosanoids provide a thorough understanding of the role of eicosanoids in therapeutic functions. Still, there is a need to identify the relationship between their structure and function. Understanding the synthesis and release of leukotrienes is still a challenge and so is the understanding of the various enzymatic pathways in eicosanoid biology. Much specific agonists and antagonists of eicosanoid receptors need to be developed. This chapter will provide a basic understanding of the current status of eicosanoid research and an insight into the various biosynthetic pathways and functions of eicosanoids.

KEYWORDS

- biosynthesis
- eicosanoids
- function
- mechanism
- metabolism
- pathophysiology

REFERENCES

Abramovitz, M., Boie, Y., Nguyen, T., Rushmore, T. H., Bayne, M. A., Metters, K. M., Slipetz, D. M., & Grygorczyk, R., (1994). Cloning and expression of a cDNA for the human prostanoid FP receptor. *J Biol Chem 269*, 2632–2636.

Ahmad, A. S., Saleem, S., Ahmad, M., & Dore, S., (2006). Prostaglandin EP1 receptor contributes to excitotoxicity and focal ischemic brain damage. *Toxicol Sci 89*, 265–270.

Ahn, C. S., & Rosenberg, I. N., (1970). Proteolysis in thyroid slices: Effects of TSH, dibutyryl cyclic 3', 5'-AMP and prostaglandin E. *Endocrinology 86*(4), 870–873.

Ajuebor, M. N., Singh, A., & Wallace, J. L., (2000). Cyclooxygenase-2-derived prostaglandin D 2 is an early antiinflammatory signal in experimental colitis. *Am J Physiol Gastrointest Liver Physiol 279*, G238–G244.

Allan, C. J., & Halushka, P. V., (1994). Characterization of human peripheral blood monocyte thromboxane A2. *J Pharmacol Exp Therap 270*, 446–452.

Alvarez, R., Eglen, R. M., Chang, L. F., Bruno, J. J., Artis, D. R., Kluge, A. F., & Whiting, R. L., (1994). Stimulation of prostaglandin D2 receptors on human platelets by analogs of prostacyclin. *Prostaglandins 42*, 105–119.

Alvarez, R., Taylor, A., Fazzari, J. J., & Jacobs, J. R. (1981). Regulation of cyclic AMP metabolism in human platelets. Sequential activation of adenylate cyclase and cyclic AMP phosphodiesterase by prostaglandins. *Mol. Pharmacol 20*, 302–309.

Andoh, T., Nishikawa, Y., Yamaguchi-Miyamoto, T., Nojima, H., Narumiya, S., & Kuraishi, Y., (2007). Thromboxane A2 induces itch-associated responses through TP receptors in the skin of mice. *J Invest Derm 127*, 2042–2047.

Anggard E. & Samuclsson, B., (1964). Prostaglandins and related factors 28. Metabolism of prostaglandin E1 in guinea-pig lung: The structures of two metabolited. *J Biol Chem 239*, 4097–4102.

Avanzino, G. L., Bradley, P. B., & Wolstencroft, J. H., (1966). Excitatory action of prostaglandin E-1 on brain-stem neurones. *Nature. 209*(5018), 87–88.

Barrow, S. E., Blair, I. A., Waddell, K. A., Shepherd, G. L., Lewis, P. J., & Dollery, C. T., (1983). Prostacyclin in late pregnancy: analysis of 6-oxo-prostaglandin F1a in maternal plasma, in: P. J. Lewis, S. Moncada, J. O'Grady (Eds.), Prostacyclin in Pregnancy, Raven Press, New York, 79–85.

Beers, W. H., Strickland, S., & Reich, E., (1975). Ovarian plasminogen activator: Relationships to ovulation and hormone *regulation. Cell 6*, 387.

Beijnen, J. H., van der Valk, M., Krimpenfort, P., & Borst, P., (1997). Increased sensitivity to anticancer drugs and decreased inflammatory response in mice lacking the multidrug resistance-associated protein. *Nat. Med. 3*, 1275–1279.

Belley, M., Gallant, M., Roy, B., Houde, K., Lachance, N., Labelle, M., Trimble, L. A., Chauret, N., Li, C., Sawyer, N., Tremblay, N., Lamontagne, S., Carrière, M. C., Denis, D., Greig, G. M., Slipetz, D., Metters, K. M., Gordon, R., Chan, C. C., & Zamboni, R. J., (2005). Structure-activity relationship studies on ortho-substituted cinnamic acids, a new class of selective EP3 antagonists. *Bioorg Med Chem Lett 15*, 527–530.

Berger, H. J., Zaret, B. L., Sperloff, L., Cohen, L. S., & Wolfson, S. (1977). Cardiac prostaglandin release during myocardial ischemia induced by atrial pacing in patients with coronary artery disease. *Am J Cardiol 39*, 481.

Bhagavan, N. V., (2002). Medical Biochemistry. Fourth edition. Academic Press, Medical, 1016 pp.

Bhathena, S. J., (2000). Relationship between fatty acids and the endocrine system. *Biofactors. 13(1–4)*, 35–39.

Bhathena, S. J., (2006). Relationship between fatty acids and the endocrine and neuroendocrine system. *Nutr Neurosci. 9(1–2)*, 1–10.

Bhattacharya, M., Peri, K. G., Almazan, G., Ribeiro-da-Silva, A., Shichi, H., Durocher, Y., Abramovitz, M., Hou, X., Varma, D. R., & Chemtob, S., (1998). *Proc Natl Acad Sci USA 95*, 15792–15797.

Biancani, P., Sohn, U. D., Rich, H. G., & Harnett, K. M., & Behar, J., (1997). Signal transduction pathways in esophageal and lower esophageal sphincter circular muscle. *Am J Med. 103(5A)*, 23S–28S.

Birke, F. W., Meade, C. J., Anderskewitz, R., Speck, G. A., & Jennewein, H. M. J., (2001). In vitro and in vivo pharmacological characterization of BIIL 284, a novel and potent leukotriene B(4) receptor antagonist. *Pharmacol Exp Ther 297*, 458–466.

Birkle, D. L., Wright, K. F., Ellis, C. K., & Ellis, E. F., (1981). Prostaglandin levels in isolated brain micro vessels and in normal and norepinephrine-stimulated cat brain homogenates. *Prostaglandins 21*, 865–877.

Bjork, J., Hedqvist, P., & Arfors, K. E., (1982). Increase in vascular permeability induced by leukotriene B4 and the role of polymorphonuclear leukocytes. *Inflammation. 6(2)*, 189–200.

Bjoro, K., Hovig, T., Stokke, K. T., & S. Stray-Pedersen, S., (1986). Formation of prostanoids in human umbilical vessels perfused in vitro. *Prostaglandins 31*, 683–698.

Blackman, S. C., Dawson, G., Antonakis, K., & Le Breton, G. C., (1998). The identification and characterization of oligodendrocyte thromboxane A2 receptors. *J Biol Chem 273*, 475–483.

Bley, K. R., Bhattacharya, A., Daniels, D. V., Gever, J., Jahangir, A., O'Yang, C., Smith, S., Srinivasan, D., Ford, A. P., & Jett, M. F., (2006). RO1138452 and RO3244794: characterization of structurally distinct, potent and selective IP (prostacyclin) receptor antagonists. *Br J Pharmacol 147*, 335–345.

Bloor, C. M., White, F. C., & Sobel, B. E., (1973). Coronary and systemic hemodynamic effects of prostaglandins in the unanesthetized dog. *Cardiovasc Res 7*, 156–166.

Bolton, H. S., Chanderbhan, R., Bryant, R. W., Bailey, J. M., Weglicki, W. B., & Vahouny, G. V. (1980). Prostaglandin synthesis by adult heart myocytes. *J Mol Cell Cardiol 12*, 1287.

Brescia, M. R., Rokosz, L. L., Cole, A. G., Stauffer, T. M., Lehrach, J. M., & Auld, D. S., (2007). Discovery and preliminary evaluation of 5-(4-phenylbenzyl)oxazole-4-carboxamides as prostacyclin receptor antagonists. *Bioorg Med Chem Lett 17*, 1211–1215.

Brown, G. P., & Venuto, R. C., (1999). Thromboxane receptors in human kidney tissues. *Prostaglandins Lipid Mediators 57*, 179–188.

Bugiardini, R., Galvani, M., Ferrini, D., Gridelli, C., Tollemeto, D., Mari, L., Puddu, P., & Lenzi, S., (1985). Myocardial ischemia induced by prostacyclin and iloprost. *Clin Pharmacol Ther 38*, 101.

Burgess, J. K., Ge, Q., Boustany, S., Black, J. L., & Johnson, P. R., (2004). Increased sensitivity of asthmatic airway smooth muscle cells to prostaglandin E2 might be mediated by increased numbers of E-prostanoid receptors. *J. Allergy Clin. Immunol. 113*, 876–881.

Burke, G., & Sato, S., (1971). Effects of long-acting thyroid stimulator and prostaglandin antagonists on adenyl cyclase activity in isolated bovine thyroid cells. *Life Sci 10*, 969.

Busse, W., & Kraft, M., (2005). Cysteinyl leukotrienes in allergic inflammation: *strategic target for therapy. Chest 127*, 1312–1326.

Bussolino, F., Benedetto, C., Massobrio, M., & Camussi, G., (1980). Maternal vascular prostacyclin activity in preeclampsia, *Lancet 2(8196)*, 702.

Cardinali, D. P., (1979). Models in neuroendocrinology: neurohormonal pathways to the pineal gland. *Trends Neurosci 2*, 250–253.

Caron, G., Delneste, Y., Roelandts, E., Duez, C., Bonnefoy, J. Y., Pestel, J., & Jeannin, P., Histamine polarizes human dendritic cells into Th2 cell promoting effector dendritic cells. *J Immunol 167*, 3682–3686.

Catt, K. J., Loumaye, E., Wynn, P., Suarez-Quian, C., Kiesel, L., Iwashita, M., Hirota, K., Morgan, R., & Chang, J., (1984). In Endocrinology, eds. Labrie, F. & Proulx, L. (Elsevier, Amsterdam), 57–65.

Chakraborty, I., Das, S. K., Wang, J., & Dey, S. K., (1996). Developmental expression of the cyclo-oxygenase-1 and cyclo-oxygenase-2 genes in the peri-implantation mouse uterus and their differential regulation by the blastocyst and ovarian steroids. *J Mol Endocrinol 16*, 107.

Chanderbhan, R., Hodges, V. A., Treadwell, C. R., & Vahouny, G. V., (1979). Prostaglandin synthesis in ratadreno cortical cells. *J Lipid Res. 20*(1), 116–124.

Cheng, K., Wu, T. J., Wu, K. K., Sturino, C., Metters, K., Gottesdiener, K., Wright, S. D., Wang, Z., O'Neill, G., Lai, E., & Waters, M. G., (2006). Antagonism of the prostaglandin D2 receptor 1 suppresses nicotinic acid-induced vasodilation in mice and humans. *Proc Natl Acad Sci USA 103*, 6682–6687.

Chibana, K., Ishii, Y., Asakura, T., & Fukuda, T., (2003). Up-regulation of cysteinyl leukotriene 1 receptor by IL-13 enables human lung fibroblasts to respond to leukotriene C4 and produce eotaxin. *J Immunol 170*, 4290–4295.

Chierchia, S., Patrono, C., Crea, F., Ciabattoni, G., De Caterina, R., Cinotti, G. A., Distante, A., & Maseri, A., (1982). Effects of intravenous prostacyclin in variant angina. *Circulation 65*, 470.

Chouaib, S., Welte, K., Mertlesmann, R., & DuPont, B., (1985). Prostaglandin & acts at two distinct pathways on T cell activation: inhibition of interleukin 2 production and downregulation of transferrin receptor expression. *J Immunol 135*, 1172–1179.

Ciceri, P., Rabuffetti, M., Monopoli, A., & Nicosia, S., (2001). Production of leukotrienes in a model of focal cerebral ischaemia in the rat. *Br J Pharmacol 133*, 1323–1329.

Cirillo, R., Tos, E. G., Page, P., Missotten, M., Quattropani, A., Scheer, A., Schwarz, M. K., & Chollet, A., (2007). Arrest of preterm labor in rat and mouse by an oral and selective nonprostanoid antagonist of the prostaglandin F2a receptor (FP). *Am J Obstet Gynecol 197*(54), e1–e9.

Clark, K. E., & Brody, M. J., Prostaglandins and uterine blood flow. In Greenburg, S., Kadowitz, P. J., (1982). Burks, T. (eds): Prostaglandins: Organ and Tissue Specific Actions, p 107. New York, *Marcel Dekker*,

Clark, K. E., Austin, J. E., & Seeds, A. E., (1982). Effect of bisenoic prostaglandins and arachidonic acid on the uterine vasculature of pregnant sheep. *Am J Obstet Gynecol 142*, 261.

Clark, K. E., Austin, J. E., & Stys, S. J., (1981). Effect of bisenoic prostaglandins on the uterine vasculature of the nonpregnant sheep. *Prostaglandins 22*, 333.

Clark, K. E., Ryan, M. J., & Brody, M. J., (1976). Effect of prostaglandins on vascular resistance and adrenergic vasoconstriction responses in the canine uterus. *Prostaglandins 12*, 71.

Clark, P., Rowland, S. E., Denis, D., Mathieu, M. C., Stocco, R., Poirier, H., Burch, J., Han, Y., Audoly, L., & Therien, A. G., (2008). Xu, D., MF498 [N-{[4-(5,9-Diethoxy-6-oxo-6,8-dihydro-7H-pyrrolo[3,4-g]quinolin-7-yl)-3-methylbenzyl]sulfonyl}-2-(2-methoxyphenyl)acetamide], a selective E prostanoid receptor 4 antagonist, relieves joint inflammation and pain in rodent models of rheumatoid and osteoarthritis. *J Pharmacol Exp Ther 325*, 425–434.

Cohen, P., Noveral, J. P., Bhala, A., Nunn, S. E., Herrick, D. J., & Grunstein, M. M., (1995). Leukotriene D4 facilitates airway smooth muscle cell proliferation via modulation of the IGF axis. *Am J Physiol Lung Cell Mol Physiol 269*, L151–157.

Coleman, R. A., Kennedy, I., Humphrey, P. P. A., Bunce, K., & Lumley, P., (1990). Prostanoids and their receptors in comprehensive Medicinal Chemistry Edited by Hansch, C., Sammes, P. G., *Taylor JB Pergamon Press* 643–714.

Coleman, R. A., Sheldrick, R. L., (1989). Prostanoid-induced contraction of human bronchial smooth muscle is mediated by TP-receptors. *Br J Pharmacol 96*, 688–692.

Coleman, R. A., Smith, W. L., & Narumiya, S., (1994). International Union of Pharmacology classification of prostanoid receptors: properties, distribution, and structure of the receptors and their subtypes. *Pharmacol Rev 46(2)*, 205–229.

Coles, S. J., Neill, K. H., Reid, L. M., Austen, K. F., Nii, Y., Corey, E. J., et al. (1983). Effects of leukotrienes C4 and D4 on glycoprotein and lysozyme secretion by human bronchial mucosa. *Prostaglandins 25*, 155–170.

Conn, P. M., Huckle, W. R., Andrews, W. V., & McArdle, C. A., (1987). The molecular mechanism of action of gonadotropin releasing hormone (GnRH) in the pituitary. *Recent Prog Horm Res 43*, 29–68.

Crider, J. Y., Griffin, B. W., & Sharif, N. A., (1999). Prostaglandin DP receptors positively coupled to adenylyl cyclase in embryonic bovine tracheal (EBTr) cells: pharmacological characterization using agonists and antagonists. *Br J Pharmacol 127*, 204–210.

Crofford, L. J., (2001). Prostaglandin biology. *Gastroenterol Clin North Am. 30(4)*, 863–76.

D'Angelo, D., Terasawa, T., Carlisle, S. J., Dorn, G. W. I., & Lynch, K. R., (1996). Characterization of a rat kidney thromboxane A2 receptor: high affinity for the agonist ligand I-BOP. *Prostaglandins 52*, 303–316.

D'Angelo, V., Villa, S., Mysliwiec, M., Donati, M. B., & de Gaetano, G., (1978). Defective fibrinolytic and prostacyclin-like activity in human atheromatous plaques. *Thromb Haemostas. 39*, 535.

Dahlen, B, Nizankowska, E., Szczeklik, A., Zetterström, O., Bochenek, G., Kumlin, M., Mastalerz, L., Pinis, G., Swanson, L. J., Boodhoo, T. I., Wright, S., Dubé, L. M., & Dahlén, S. E., (1998). Benefits from adding the 5-lipoxygenase inhibitor zileuton to conventional therapy in aspirin-intolerant asthmatics. *Am J Resp Crit Care Med 157*, 1187–1194.

Dahlen, S. E., (2006). Treatment of asthma with antileukotrienes: first line or last resort therapy? Eur *J Pharmacol 533*, 40–56

Dahlen, S. E., Bjork, J., Hedqvist, P., Arfors, K. E., Hammarström, S., Lindgren, J. A., & Samuelsson, B., (1981). Leukotrienes promote plasma leakage and leukocyte adhesion in postcapillary venules: in vivo effects with relevance to the acute inflammatory response.*Proc Natl Acad Sci USA 78*(6), 3887–3891.

Dahlen, S. E., Hedqvist, P., Hammarstrom, S., & Samuelsson, B., (1980). Leukotrienes are potent constrictors of human bronchi. *Nature 288*, 484–486.

Dahlen, S. K., Hansson, G., Hedqvist, P., Bjorck, T., Granstrom, E., & Dahlen, B., (1983). Allergen challenge of lung tissue from asthmatic elicits bronchial contraction that correlates with the release of leukotrienes *C4, D4, and E4. PNAS USA 80(6)*, 1712–1716.

Davis, R. J., Murdoch, C. E., Ali, M., Purbrick, S., Ravid, R., Baxter, G. S., Tilford, N., Sheldrick, R. L., Clark, K. L., & Coleman, R. A., (2004). EP4 prostanoid receptor-mediated vasodilatation of human middle cerebral arteries. *Br J Pharmacol 141*, 580–585.

Dazord, A., Morera, A. M., Bertrand, J., & Saez, J. M., (1974). Prostaglandin receptors in human and ovine adrenal glands. Bending and stimulation of adenyl cyclase in subcellular preparations. *Endocrinology 95*, 352–359.

De Caterina, R., Mazzone, A., Giannessi, D., Sicari, R., Pelosi, W., & Lazzerini, G., (1988). Azzara' A, Forder, R., Carey, F., Caruso, D., et al. Leukotriene B4 production in human atherosclerotic plaques. *Biomed Biochim Acta 47(10–11)*, S182–185.

de Graaf-in t Veld, C., Garrelds, I. M., Koenders, S., & Gerth van Wijk, R., (1996). Relationship between nasal hyperreactivity, mediators and eosinophils in patients with perennial allergic rhinitis and controls. *Clin Exp Allergy 26*, 903–908.

De Wied, D., Wilter, A., Versteeg, D. H. G., & Mulder, A. H., (1969). Release of ACTH by substances of central nervous origin. *Endocrinology 85*, 561–569.

Dembinska-Kiec, A., Gryglewska, T., Zmuda, A., & Gryglewski, R. J., (1977). The generation of prostacyclin by arteries and by the coronary vascular bed is reduced in experimental atherosclerosis in rabbits. *Prostaglandins 14*, 1025–1034.

Demers, L. M., & Gabbe, S. G., (1976). Placental prostaglandin levels in preeclampsia. *Am J Obstet Gynecol 126*, 137–139.

Dey, P. K., Feldberg, W., Gupta, K. P., Milton, A. S., & Wendland, S., (1974). Further studies on the role of prostaglandin in fever. *J Physiol 241(3)*, 629–646.

Diamant, Z., Hiltermann, J. T., van Rensen, E. L., Callenbach, P. M., Veselic- Charvat, M., van der Veen, H., et al. (1997). The effect of inhaled leukotriene D4 and methacholine on sputum cell differentials in asthma. *Am J Respir Crit Care Med 155*, 1247–1253.

Dighe, N. S., Pattan, S. R., Merekar, A. N., Dighe, S. B., Chavan, P. A., Musmade, D. S., Gaware, V. M., (2010). Leucotrienes and Its Biological Activities: *A Review J Chem Pharm Res 2*(1), 338–348.

Djellas, Y., Manganello, J. M., Antonakis, K., & LeBreton, G. C., (1999). Indentification of Ga13 as one of the G-proteins that couple to human platelet thromboxane A2 receptors. *J Biol Chem 274*, 14325–14330.

Downie, J., Poyzer, N. L., & Wunderlich, M., (1974). Levels of prostaglandins in human endometrium during the normal menstrual cycle. *J Physiol (Lond) 236*, 465.

Downing, I., Shepherd, G. L., & Lewis, P. J., (1980). Reduced prostacyclin production in preeclampsia. *Lancet 2 (8208–9)*, 1374.

Drazen, J. M., Israel, E., & O'Byrne, P. M., (1999). Treatment of asthma with drugs modifying the leukotriene pathway. *N Engl J Med 340*, 197–199.

Dubin, N. H., Johnson, J. W. C., Calhoun, S., Ghodgaonkar, R. B., & Beck, J. C., (1981). Plasma prostaglandin in pregnant women with term and preterm deliveries. *Obstet Gynecol 57*, 203.

Duckworth, N., Marshall, K., & Clayton, J. K., (2002). An investigation of the effect of the prostaglandin EP2 receptor agonist, butaprost, on the human isolated myometrium from pregnant and nonpregnant women. *J Endocrinol 172(2)*, 263–269.

Durocher, Y., Perret, S., Thibaudeau, E., Gaumond, M. H., Kamen, A., & Stocco, R., Abramovitz, M., (2000). A reporter gene assay for high-throughput screening of G-protein-coupled receptors stably or transiently expressed in HEK293 EBNA cells grown in suspension culture. *Anal Biochem 284*, 316–326.

El-Haroun, H., Clarke, D. L., Deacon, K., Bradbury, D., Clayton, A., Sutcliffe, A., & Knox, A. J., (2008). IL-1β, BK, and TGF-β1 attenuate PGI 2-mediated cAMP formation in human pulmonary artery smooth muscle cells by multiple mechanisms involving p38 MAP kinase and P.K.A., *American Journal of Physiology 294 (3)*, L553–L562.

Eliasson, R., (1959). Studies on prostaglandin occurrence, formation and biological actions. *Acta physiologica scandinavica 46(158)*, 1–73.

Evans, C. A., Kennedy, T. G., (1978). The importance of prostaglandin synthesis for the initiation of blastocyst implantation in the hamster. *J Reprod Fertill 54*, 255.

Farias, S. E., Zarini, S., Precht, T., Murphy, R. C., & Heidenreich, K. A., (2007). Transcellular biosynthesis of cysteinyl leukotrienes in rat neuronal and glial cells. *J Neurochem 103*, 1310–1318.

Faveeuw, C., Gosset, P., Bureau, F., Angeli, V., Hirai, H., Maruyama, T., Narumiya, S., Capron, M., & Trottein, F., (2003). Prostaglandin D2 inhibits the production of interleukin-12 in murine dendritic cells through multiple signaling pathways. *Eur J Immunol 33*, 889–898.

Feldberg, W., & Saxena, P. N., (1975). Prostaglandins, endotoxin and lipid A on body temperature in rats. *J Physiol London 249*, 601–615

Figueroa, D. J., Borish, L., Baramki, D., Philip, G., Austin, C. P., & Evans, J. F., (2003). Expression of cysteinyl leukotriene synthetic and signaling proteins in inflammatory cells in active seasonal allergic rhinitis. *Clin Exp Allergy 33*, 1380.

Figueroa, D. J., Breyer, R. M., Defoe, S. K., Kargman, S., Daugherty, B. L., Waldburger, K., et al. (2001). Expression of the cysteinyl leukotriene 1 receptor in normal human lung and peripheral blood leukocytes. *Am J Respir Crit Care Med 163*, 226–233.

Fish, K. J., Sarnquist, F. H., van Stiennis, C., Mitchell, R. S., Hilberman, M., Jamieson, S. W., Linet, O. I., & Miller, D. C., (1986). A prospective, randomized study of the effects of prostacyclin on platelets and blood loss during coronary bypass operations. *J Thorac Cardiovasc Surg 91*, 436.

Ford Hutchinson, A. W., Gresser, M., & Young, R. N., (1994) 5-Lipoxygenase. *Annu Rev Biochem. 63*, 383–417.

Ford-Hutchinson, A. W., (1990). Leukotriene B4 in inflammation. *Crit Rev Immunol 10*, 1–12.

Foudi, N., Kotelevets, L., Louedec, L., Leséche, G., Henin, D., Chastre, E., & Norel, X., (2008). Vasorelaxation induced by prostaglandin E2 in human pulmonary vein: role of the EP4 receptor subtype. *Br J Pharmacol 154*, 1631–1639.

Funk, C. D., (2001). Prostaglandins and leukotrienes: advances in eicosanoid biology. *Science, 30, 294(5548)*, 1871–1875.

Funk, C. D., (2005). Leukotriene modifiers as potential therapeutics for cardiovascular disease. iRev *Drug Discov 4(8)*, 664–672.

Furuyashiki, T., & Narumiya, S., (2011). Stress responses: the contribution of prostaglandin E(2) and its receptors. *Nat Rev Endocrinol 7(3)*, 163–75.

Gallant, M. A., Slipetz, D., Hamelin, E., Rochdi, M. D., Talbot, S., de Brum-Fernandes, A. J., & Parent, J. L., (2007). Differential regulation of the signaling and trafficking of the two prostaglandin D2 receptors, prostanoid DP receptor and CRTH2. *Eur J Pharmacol 557 (2–3)*, 115–123.

Gallant, M., Carrière, M. C., Chateauneuf, A., Denis, D., Gareau, Y., Godbout, C., Greig, G., Juteau, H., Lachance, N., Lacombe, P., Lamontagne, S., Metters, K. M., Rochette, C., Ruel, R., Slipetz, D., Sawyer, N., Tremblay, N., & Labelle, M., (2002). Structure-activity relationship of biaryl acylsulfonamide analogs on the human EP3 prostanoid receptor. *Bioorg Med Chem Lett 12*, 583–2586.

Giblin, G. M., Bit, R. A., Brown, S. H., Chaignot, H. M., Chowdhury, A., Chessell, I. P., Clayton, N. M., Coleman, T., Hall, A., Hammond, B., Hurst, D. N., Michel, A. D., Naylor, A., Novelli, R., Scoccitti, T., Spalding, D., Tang, S. P., Wilson, A. W., & Wilson, R., (2007). The discovery of 6-[2-(5-chloro-2-{[(2,4-difluorophenyl)methyl] oxy} phenyl)-1-cyclopenten-1-yl]-2-pyridinecarboxylic acid, GW848687X, a potent and selective prostaglandin EP1 receptor antagonist for the treatment of inflammatory pain. *Bioorg Med Chem Lett 17*, 385–389.

Giiliner, H. G., (1983). Prostaglandin actions on tie adrenergic nervous system. *Klin Wochemchr 61*, 533–540.

Gijon, M. A., Spencer, D. M., Siddiqi, A. R., Bonventre, J. V., & Leslie, C. C., (2000). Cytosolic phospholipase A2 is required for macrophage arachidonic acid release by agonists that Do and Do not mobilize calcium. Novel role of mitogen-activated protein kinase pathways in cytosolic phospholipase A2 regulation. *J. Biol. Chem. 275*, 20146–20156.

Giles, H., Leff, P., Bolofo, M. L., Kelly, M. G., & Robertson, A. D., (1989). The classification of prostaglandin DP-receptors in platelets and vasculature using BW A868C, a novel, selective and potent competitive antagonist. *Br J Pharmacol 96*, 291–300.

Gilman, A. G., Nirenberg, M., (1971). Regulation of adenosine 3',5'-cyclic monophosphate metabolism in cultured neuroblastoma cells. *Nature 10, 234(5328)*, 356–358.

Gilmore, D. P., & Shaikh, A. A., (1972). The effect of prostaglandin E2 in inducing sedation in the rat. *Prostaglandins 2(2)*, 143–151.

Gleim, S., Kasza, Z., Martin, K., & Hwa, J., (2009). Prostacyclin receptor/thromboxane receptor interactions and cellular responses in human atherothrombotic disease. *Curr Atheroscler Rep 11*, 227–235.

Goehlert, U. G., Ng Ying Kin, N. M. K., & Wolfe, L. S., (1981). Biosynthesis of prostacyclin in rat cerebral microvessels and the choroid plexus. *J Neurochem 36*, 1192–1201.

Goldberg, V. J., & Ramwell, P. W., (1975). Role of prostaglandins in reproduction. *Physiol Rev. 55*, 325.

Goldyne, M. E., Burrish, G. F., Poubelle, P., & Borgeat, P., (1984). Arachidonic acid metabolism among human mononuclear leukocytes. Lipoxygenase-related pathways. *J Bio Chern 259*, 8815–8819.

Goodman, D. S., (1965). Cholesterol ester metabolism. *Physiol Rev 45*, 747–839.

Goodman, R. P., Killam, A. P., Brash, A. R., & Branch, R. A., (1982). Prostacyclin production during pregnancy: comparison of production during normal pregnancy and pregnancy complicated by hypertension. *Am J Obstet Gynecol 142*, 817–822.

Gorman, R. R., Bunting, S., & Miller, O. V., (1977). Modulation of human platelet adenylate cyclase by prostacyclin (PGX). *Prostaglandins 13*, 377–388.

Gosset, P., Pichavant, M., Faveeuw, C., Bureau, F., Tonnel, A. B., & Trottein, F., (2005). Prostaglandin D2 affects the differentiation and functions of human dendritic cells: impact on the T cell response. *Eur J Immunol 35*, 1491–1500.

Green, L. H., Seroppian, E., & Handin, R. I., (1980). Platelet activation during exercise-induced myocardial ischemia. *N Engl J Med 302*, 193–197.

Gronert, K., Clish, C. B., Romano, M., & Serhan, C. N., (1999). Transcellular regulation of eicosanoid biosynthesis. *Methods Mol Biol 120*, 119.

Gryglewski, R. J., Bunting, S., Moncada, S., Flower, R. J., & Vane, J. R., (1976). Arterial walls are protected against deposition of platelet thrombi by a substance (prostaglandin X) which they make from prostaglandin endoperoxides. *Prostaglandins 12*, 685–713.

Gurgul-Convey E., & Lenzen, S., (2010). Protection against cytokine toxicity through endoplasmic reticulum and mitochondrial stress prevention by prostacyclin synthase overexpression in insulin-producing cells. *The Journal of Biological Chemistry 285(15)*, 11121–11128.

Haeggstrom, J. Z., & Wetterholm, A., (2002). Enzymes and receptors in the leukotriene cascade. *Cell. Mol. Life Sci. 59*, 742–753.

Hagen, A. A., White, R. P., & Robertson, J. T., (1979). Synthesis of prostaglandins and thromboxane B 2 by cerebral arteries. *Stroke 10*, 306–309.

Hagen, A. A., White, R. P., Terragno, N. A., & Robertson, J. T., (1978). Synthesis of prostaglandins (PGs) by bovine cerebral arteries. *Fed Proc 37*, 384.

Hamprecht, B., & Schultz, J., (1973). Stimulation by prostaglandin E1 of adenosine 3',5'-cyclic monophosphate formation in neuroblastoma cells in the presence of phosphodiesterase inhibitors. *FEBS Lett. 34(1)*, 85–89.

Hansen, H., (1976). 15-hydroxyprostaglandin dehydrogenase. A review. *Prostaglandins 12*, 647–679.

Hardy, C. C., Robinson, C., Tattersfield, A. E., & Holgate, S. T., (1984). The bronchoconstrictor effect of inhaled prostaglandin D2 in normal and asthmatic men. *N Eng Med 311*, 209–213.

Harizi, H., & Gualde, N., (2005). The impact of eicosanoids on the crosstalk between innate and adaptive immunity: the key roles of dendritic cells. *Tissue Antigens. 65(6)*, 507–514.

Harper, A. M., McCulloch, J., MacKenzie, E. T., & Pickard, J. D., (1977). Migraine and the blood-brain barrier. *Lancet 1(8020)*, 1034–1036.

Hedge, G. A., Hanson, S. D., (1972). The effects of prostaglandins on ACTH secretion. *Endocrinology 91*, 925–933.

Hedqvist, P., (1969). Modulating effect of prostaglandin E on noradrenaline release from the isolated cat spleen. *Acta Physiol Scand 75*, 511–512.

Hedqvist, P., (1977). Basic mechanisms of prostaglandin action on autonomic neurotransmission. *Annu Rev Pharmacol Toricol 17*, 259–279.

Heise, C. E., O'Dowd, B. F., Figueroa, D. J., Sawyer, N., Nguyen, T., Im, D. S., Stocco, R., Bellefeuille, J. N., Abramovitz, M., Cheng, R., Williams, D. L. Jr., Zeng, Z., Liu, Q., Ma, L., Clements, M. K., Coulombe, N., Liu, Y., Austin, C. P., George, S. R., O'Neill, G. P., Metters, K. M., Lynch, K. R., & Evans, J. F., (2000). Characterization of the human cysteinyl leukotriene 2 receptor. *J Biol Chem. 275(39)*, 30531–30536.

Henderson, W. R. Jr. (1987). Eicosanoids and lung inflammation. *Am Rev Respir Dis 135*, 1176–1185.

Henderson, W. R. Jr. (1987). Lipid-derived and other chemical mediators of inflammation in the lung. *Allergy Clin Immunol 79*, 543–553.

Henderson, W. R., & Klebanoff, S. J., (1983). Leukotriene production and inactivation by normal, chronic granulomatous disease and myeloperoxidase-deficient neutrophils. *J Riol Chem 258*, 13522–13527.

Henderson, W. R., Harley, J. R., & Fauci, A. S., (1984). Arachidonic acid metabolism in normal and hypereosinophilic syndrome human eosinophils: generation of leukotrienes R4, C4, D4 and 15-lipoxygenase products. *Immunology 51*, 679–686.

Henriksson, P., Edhag, O., & Wennmalm, A., (1985). Prostacyclin infusion in patients with acute myocardial infarction. *Br Heart J 53*, 173.

Hertelendy, F., Todd, H., Ehrhart, K., & Blute, R., (1972). Studies on growth hormone secretion: IV. In vivo effects of prostaglandin, E., *Prostaglandins 2*, 79–91.

Hillier, K., & Smith, M. D., (1981). Prostaglandin E and F concentrations in placentae of normal, hypertensive and preeclamptic patients. *Br J Obstet Gynaecol 88*, 274–277.

Hirai, H., Tanaka, K., Takano, S., Ichimasa, M., Nakamura, M., & Nagata, K., (2002). Agonistic effect of indomethacin on a prostaglandin D2 receptor, CRTH2. *J Immunol 168*, 981–985.

Hirai, H., Tanaka, K., Yoshie, O., Ogawa, K., Kenmotsu, K., Takamori, Y., Ichimasa, M., Sugamura, K., Nakamura, M., Takano, S., & Nagata, K., (2001). Prostaglandin D 2 selectively induces chemotaxis in T helper type 2 cells, eosinophils, and basophils via seven-transmembrane receptor CRTH2. *J Exp Med 193*, 255–261.

Hirata, M., Hayashi, Y., Ushikubi, F., Yokota, Y., Kageyama, R., Nakanishi, S., & Narumiya, S., (1991). Cloning and expression of cDNA for a human thromboxane A2 receptor. *Nature 349*, 617–620.

Hirsch, P. D., Campbell, W. B., Willerson, J. T., & Hillis, L. D., (1981). Prostaglandins and ischemic heart disease. *Am J Med 71*, 1009.

Hodges, V. A., Treadwell, C. T., & Vahouny, G. V., (1978). Prostaglandin E2-induced hydrolysis of cholesterol esters in rat adrenocortical cells. *J Steroid Biochem. 9(11)*, 1111–1118.

Hogaboom, G. K., Cook, M., Newton, J. F., Varrichio, A., Shorr, R. G., & Sarau, H. M., (1986). Crooke, S. T., Purification, characterization, and structural properties of a single protein from rat basophilic leukemia (RBL-1) cells possessing 5-lipoxygenase and leukotriene A4 synthetase activities. *Mol Pharmacol. 30(6)*, 510–519.

Holgate, S. T., Peters-Golden, M., Panettieri, R. A., & Henderson, W. R. Jr. (2003). Roles of cysteinyl leukotrienes airway inflammation, smooth muscle function, and remodeling. *J Allergy Clin Immunol 111*, S18–36.

Holmes, S. W., & Horton, E. W., (1968). The distribution of tritium-labeled prostaglandin E1 injected in amounts sufficient to produce central nervous effects in cats and chicks. *Br J Pharmacol 34*, 32–37.

Honma, S., Saika, M., Ohkubo, S., Kurose, H., & Nakahata, N., (2006). Thromboxane A2 receptor-mediated G12/13-dependent glial morphological change. *Eur J Pharmacol 545*, 100–108.

Hori, T., & Harada, Y., (1974). The effects of ambient and hypothalamic temperatures on the hyperthermic responses to prostaglandins EI and E2. *Pjteugers Arch 350*, 123–134.

Horton, E. W., (1964). Actions of prostaglandins E1, E2 and E3 on the central nervous system. *Br J Pharmacol 22*, 189–192.

Horton, E. W., (1973). Prostaglandins at adrenergic nerve endings. *British Medical Bull*etin *29*, 148–151.

Hsu KS & Kan, W. M., (1996). Thromboxane A2 agonist modulation of excitatory synaptic transmission in the rat hippocampal slice. *Br J Pharmacol 118(8)*, 2220–7.

Hsu, K. S., Huang, C. C., Kan, W. M., & Gean, P. W., (1996). TXA2 agonists inhibit high-voltage-activated calcium channels in rat hippocampal CA1 neurons. *Am J Physiol 271*, C1269–1277.

Hulting, A. L., Lindgren, J. A., Hokfelt, T., Eneroth, P., Werner, S., Patrono, C., & Samuelsson, B., (1985). *Proc Nati Acad Sci USA 82*, 3834–3838.

Hyo, S., Kawata, R., Kadoyama, K., Eguchi, N., Kubota, T., Takenaka, H., & Urade Y., (2007). Expression of prostaglandin D 2 synthase in activated eosinophils in nasal polyps. *Arch Otolaryngol Head Neck Surg 133*, 693–700.

Iizuka, Y., Okuno, T., Saeki, K., Uozaki, H., Okada, S., Misaka, T., Sato, T., Toh, H., Fukayama, M., Takeda, N., Kita, Y., Shimizu, T., Nakamura, M., & Yokomizo, T., (2010). Protective role of the leukotriene B4 receptor BLT2 in murine inflammatory colitis. *FASEB J 24*, 4678–4690.

Ikeda, M., Kawatani, M., Maruyama, T., & Ishihama, H., (2006). Prostaglandin facilitates afferent nerve activity via EP1 receptors during urinary bladder inflammation in rats. *Biomed Res 27*, 49–54.

Imura, Y., Terashita, Z., Shibouta, Y., Inada, Y., & Nishikawa, K., (1990). Antagonistic action of AA-2414 on thromboxane A2/prostaglandin endoperoxide receptor in platelets and blood vessels. *Jpn J Pharmacol 52*, 35–43.

Ishimura, M., Maeda, T., Kataoka, S., Suda, M., Kurokawa, S., & Hiyama, Y., (2009). Effects of KP-496, a novel dual antagonist for cysteinyl leukotriene receptor 1 and thromboxane A2 receptor, on Sephadex-induced airway inflammation in rats. *Biol Pharm Bull 32*, 1057–1061.

Ito, S., Negishi, M., Sugama, K., Okuda-Ashitaka, K., & Hayaishi, O., (1990). Signal transduction coupled to prostaglandin D_2. *Adv. Prostaglandin Thromboxane Leukotriene Res 21*, 371–374.

Itoh, T., Ueno, H., & Kuriyama, H., (1985). Calcium-induced calcium release mechanism in vascular smooth muscles-assessments based on contractions evoked in intact and saponin-treated skinned muscles. *Experientia 41*, 989–996.

Itoh, Y., Shindoh, J., Horiba, M., Kohno, S., Ohata, K., Ashida, Y., & Tagaki, K., (1993). Inhibitory effects of AA-2414, a thromboxane (Tx) A2 receptor antagonist, on U-46619-, prostaglandin (PG) D2- and 9a,11b PGF2-induced contractions of guinea-pig tracheas and isolated human bronchi (Japanese). *Arerugi 42*, 1670–1676.

Iwasaki, M., Nagata, K., Takano, S., Takahashi, K., Ishii, N., & Ikezawa, Z., (2002). Association of a new type prostaglandin D 2 receptor CRTH2 with circulating T helper 2 cells in patients with atopic dermatitis. *J Invest Dermatol 119*, 609–616.

Jabbour, H. N., Sales, K. J., Smith, O. P., Battersby, S., & Boddy, S. C., (2006). Prostaglandin receptors are mediators of vascular function in endometrial pathologies. *Mol Cell Endocrinol. 27, 252(1–2)*, 191–200.

Jaffar, Z., Wan, K. S., & Roberts, K., (2002). A key role for prostaglandin I2 in limiting lungmucosal Th2, but not Th1, responses to inhaled allergen. *Journal of Immunology 169(10)*, 5997–6004.

Jarvinen, L., Badri, L., S. Wettlaufer, S., Ohtsuka, T., Standiford, T. J., Toews, G. B., Pinsky, D. J., Peters-Golden, M., & Lama, V. N., (2008). Lung resident mesenchymal stem cells isolated from human lung allografts inhibit T cell proliferation via a soluble mediator. *Journal of Immunology 181(6)*, 4389–4396.

Jedlitschky, G., Mayatepek, E., & Keppler, D., (1993). Peroxisomal leukotriene degradation: biochemical and clinical implications *Adv Enzyme Regul 33*, 181–194.

Jeremy, J. Y., Barradas, M. A., Craft, I. L., Mikhailidis, D. P., & Dandona, P., (1985). Does human placenta produce prostacyclin? *Placenta 6*, 45–52.

Jiang, G. L., Nieves, A., Im, W. B., Old, D. W., Dinh, D. T., & Wheeler, L., (2007). The prevention of colitis by E prostanoid receptor 4 agonist through enhancement of epithelium survival and regeneration. *J Pharmacol Exp Ther 320*, 22–28.

Jones, R. L., Wise, H., Clark, R., Whiting, R. L., & Bley, K. R., (2006). Investigation of the prostacyclin (IP) receptor antagonist RO1138452 on isolated blood vessel and platelet preparations. *Br J Pharmacol. 149(1)*, 110–120.

Jones, T. R., Labelle, M., Belley, M., Champion, E., Charette, L., Evans, J., Ford-Hutchinson, A. W., & Gauthier, J. Y., (1995). Lord, A., Masson, P., et al. Pharmacology of montelukast sodium (Singulair), a potent and selective leukotriene D4 receptor antagonist. *Physiol Pharmacol 73*, 191–201.

Juteau, H., Gareau, Y., Labelle, M., Sturino, C. F., Sawyer, N., Tremblay, N., Lamontagne, S., Carrière, M. C., Denis, D., & Metters, K. M., (2001). Structure-activity relationship of cinnamic acylsulfonamide analogs on the human EP3 prostanoid receptor. *9*, 1977–1984.

Kalish, B. T., Kieran, M. W., Puder, M., & Panigrahy, D., (2013). The growing role of eicosanoids in tissue regeneration, repair, and wound healing. *Prostaglandins Other Lipid Mediat 104–105*, 130–138.

Kamohara, M., Takasaki, J., Matsumoto, M., Saito, T., Ohishi, T., Ishii, H., & Furuichi, K. J., (2000). *Biol Chem 275(35)*, 27000–27004.

Kanaoka, Y., & Boyce, J. A., (2004). Cysteinyl leukotrienes and their receptors: cellular distribution and function in immune and inflammatory responses. *J Immunol 173*, 1503–1510.

Karim, S. M. M., (1975). Prostaglandins and Reproduction. Baltimore, University Park Press.

Kato, K., Yokomizo, T., Izumi, T., & Shimizu, T. J., (2000). Cell-specific transcriptional regulation of human leukotriene B(4) receptor gene. *Exp Med 192*, 413–420.

Kawakami, Y., & Uchiyama, K., Irie, T., & Murao, M., (1973). Evaluation of aerosols of prostaglandins E1 and E2 as bronchodilators. *Eur J Clin Pharmacol 6*, 127–132.

Kawano, M., & Mori, N., (1983). Prostacyclin producing activity of human umbilical, placental and uterine vessels, *Prostaglandins 26*, 645–662.

Keery, R. J., & Lumley, P., (1988). AH6809, a prostaglandin DP-receptor blocking drug on human platelets. *Br J Pharmacol 94*, 745–754.

Khanapure, S. P., Garvey, D. S., Janero, D. R., & Letts, L. G., (2007). Eicosanoids in inflammation: biosynthesis, pharmacology, and therapeutic frontiers. *Curr Top Med Chem. 7(3)*, 311–340.

Kiesel, L., Przylipiak, A. F., Habenicht, A. J., Przylipiak, M. S., & Runnebaum, B., (1991). Production of leukotrienes in gonadotropin-releasing hormone-stimulated pituitary cells: potential role in luteinizing hormone release. *Proc Natl Acad Sci USA 88(19)*, 8801–8805.

Kikuta, Y., Kusunose, E., Ito, M., & Kusunose, M., (1999). Purification and characterization of recombinant rat hepatic CYP4F1. *Arch Biochem Biophys. 369(2)*, 193–196.

Kikuta, Y., Kusunose, E., Sumimoto, H., Mizukami, Y., Takeshige, K., Sakaki, T., & Kusunose, M. (1998). Purification and characterization of recombinant human neutrophil leukotriene B4 ω-hydroxylase (cytochrome P450 4F3). *Arch Biochem Biophys 355(2)*, 201-205.

Kinsella, B. T., O'Mahony, D. J., & FitzGerald, G. A., (1997). The human thromboxane A2 receptor a isoform (TPa) functionally couples to the G proteins Gq and G11 in vivo and is activated by the isoprostane 8-epiprogstaglandin F2a. *J Pharmacol Exp Therap 281*, 957–964.

Kitanaka, J., Hashimoto, H., Sugimoto, Y., Sawada, M., Negishi, M., Suzumura, A., & Marunouchi, T., (1995). Ichikawa, A., Baba, A., cDNA cloning of a thromboxane A2 receptor from rat astrocytes. *Biochim Biophys Acta 1265*, 220–223.

Kitiyakara, C., Welch, W. J., Verbalis, J. G., & Wilcox, C. S., (2002). Role of thromboxane receptors in the dipsogenic response to central angiotensin II, *Am J Physiol Regul Integr Comp Physiol 282*, R865–R869.

Kitiyakara, C., Welch, W. J., Verbalis, J. G., & Wilcox, C. S., (2002). Role of thromboxane receptors in the dipsogenic response to central angiotensin II, *Am J Physiol Regul Integr Comp Physiol 282*, R865–R869.

Klein, I., & Levey, G. S., (1971). Effect of prostaglandins on guinea pig myocardial adenyl cyclase. *Metabolism 20*, 890–896.

Knapp, H. R., & FitzGerald, G., (1989). The antihypertensive effects of fish oil. A controlled study of polyunsaturated fatty acid supplements in essential hypertension. *N Engl J Med 320*, 1037–1045.

Koch, K., Melvin, L. S. Jr., Reiter, L. A., Biggers, M. S., Showell, H. J., Griffiths, R. J., Pettipher, E. R., Cheng, J. B., Milici, A. J., Breslow, R. et al. (1994). (+)-1-(3S,4R)-[3-(4-phenylbenzyl)-4-hydroxychroman-7-yl]cyclopentane carboxylic acid, a highly potent, selective leukotriene B4 antagonist with oral activity in the murine collagen-induced arthritis model. *J Med Chem 37,* 3197–3199.

Kontos, H. A., Weie, P., Ellis, E. F., Dietrich, W. D., & Povlishock, J. T., (1981). Prostaglandins in physiological and in certain pathological responses of the cerebral circulation. *Fedn Pro Fedn Am Socs Exp Biol 40,* 2326–2330.

Korzekwa, A. J., Jaroszewski, J. J., Woclawek-Potocka, I., Bah, M. M., & Skarzynski, D. J., (2008). Luteolytic effect of prostaglandin F2a on bovine corpus luteum depends on cell composition and contact. *Reproduction in Domestic Animals 43,* 464–472.

Korzekwa, A., Murakami, S., Wocławek-Potocka, I., Bah, M. M., Okuda, K., & Skarzynski, D. J., (2008). The influence of tumor necrosis factor a (TNF) on the secretory function of bovine corpus luteum: TNF and its receptors expression during the estrous cycle. *Reproductive Biology 8,* 245–262.

Kostenis, E., & Ulven, T., (2006). Emerging roles of DP and CRTH2 in allergic inflammation. *Trends Mol Med 12,* 148–158.

Kozak, K. R., Rowlinson, S. W., & Marnett, L. J., (2000). Oxygenation of the endocannabinoid, 2-arachidonylglycerol, to glyceryl prostaglandins by cyclooxygeuase-2. *J Biol Chem 275,* 33744–33749.

Krell, R. D., Aharony, D., Buckner, C. K., Keith, R. A., Kusner, E. J., Snyder, D. W., Bernstein, P. R., Matassa, V. G., Yee, Y. K., Brown, F. J., et al. (1990), The preclinical pharmacology of ICI 204, 219. A peptide leukotriene antagonist. *Am Rev Respir Dis 141,* 978–987.

Kudo, I., & Murakami, M., (2002). Phospholipase A2 enzymes. *Prostaglandins Other Lipid Mediat. 68–69,* 3–58.

Kuehl, F. A. Jr. (1974). Prostaglandins, cyclic nucleotides and cell function. *Prostaglandins. 5,* 325–340.

Kuehl, F. A. Jr., Humes, J. L., Tarnoff, J., Cirillo, V. J., & Ham, E. A., (1970). Prostaglandin receptor site: Evidence for an essential role in the action of luteinizing hormone. *Science 169,* 883–886.

Kuitert, L. M., Newton, R., Barnes, N. C., Adcock, M., & Barnes, P. J., (1996). Eicosanoid mediator expression in mononuclear and polymorphonuclear cells in normal subjects and patients with atopic asthma and cystic fibrosis. *Thorax 51,* 1223–1228.

Kumpuris, A. G., Luchi, R. J., Waddell, C. C., & Miller, R. R., (1980). Production of circulating platelet aggregates by exercise in coronary patients. *Circulation. 61,* 62–65.

Kunikata, T., Yamane, H., Segi, E., Matsuoka, T., Sugimoto, Y., Tanaka, S., Tanaka, H., Nagai, H., Ichikawa, A., & Narumiya, S., (2005). Suppression of allergic inflammation by the prostaglandin E receptor subtype EP3. *Nat Immunol 6,* 524–531.

Kusunose, M., (1998). Purification and characterization of recombinant human neutrophil leukotriene B4 omega-hydroxylase (cytochrome P450 4F3). *Arch Biochem Biophys 355(2),* 201–205.

Labrie, F., Pelletier, G., Borgeat, P., Drouin, J., Ferland, L., & Belanger, A., (1976). Mode of action of hypothalamic regulatory hormones in the adenohypophsis. In Frontiers in Neuroendocrinology. ed. Martini, L., Ganong, W. F. *4,* 63–93. NY: Raven Press.

Lafrance, M., & Hansel, W., (1992). Role of arachidonic acid and its metabolites in the regulation of progesterone and oxytocin release from the bovine corpus luteum. *Proc Soc Exp Biol Med 201,* 106–113.

Laitinen, L. A., Laitinen, A., Haahtela, T., Vilkka, V., Spur, B. W., & Lee, T. H., (1993). Leukotriene E4 and granulocytic infiltration into asthmatic airways. *Lancet 341,* 989–990.

Lands, W. E. M., (1979). The biosynthesis and metabolism of prostaglandins. *Ann Rev Physiol 41*, 633.

Latorre, E., Patrono, C., Fortuna, A., & Grossi-Belloni, D., (1974). Role of prostaglandin F2a in human cerebral vasospasm. *J. Neurosurg 41*, 293–299.

Laychock, S. G., & Rubin, R. P., (1975). ACTH-induced prostaglandin biosynthesis from 3H-arachidonic acid by adrenocortical cells. *Prostaglandins 10, 529–540*.

Laychock, S. G., & Rubin, R. P., (1977). Regulation of steroid genesis and prostaglandin formation in isolated adrenocortical cells: The effects of pregnenolone and cycloheximide. *J. Steroid Biochem 8, 663–667*.

Lee, E., Robertson, T., Smith, J., & Kilfeather, S., (2000). Leukotriene receptor antagonists and synthesis inhibitors reverse survival in eosinophils of asthmatic individuals. *Am J Respir Crit Care Med 161*, 1881–1886.

Leier, I., Jedlitschky, G., Buchholz, U., Cole, P. C., Deeley, R. G., & Keppler, D., (1994). The MRP1 gene encodes an ATP-dependent export pump for leukotriene C4 and structurally related conjugates. *J Biol Chem 269*, 27807–27810.

Leslie, C. A., & Levine, L., (1973). Evidence for the presence of a prostaglandin E 2−9-keto reductase in rat organs. *Biochem Biophys Res Comm 52(3), 717–724*.

Leslie, C. C., (1997). Properties and regulation of cytosolic phospholipase A2. *J Biol Chem 272*, 16709–16712.

Leslie, C. C., (2004). Regulation of the specific release of arachidonic acid by phospholipase A2. Prostaglandins Leukot Essent *Fatty Acids 70*, 373–376.

Levi, Roberto, & Robert D. Krell (1988). Biology of the Leukotrienes. New York: The New York Academy of Sciences. pp. 122–123, 201–205, 218–219, 252–255.

Lewis, R. A., Austen, K. F., & Soberman, R. J., (1990). Leukotrienes and other products of the 5-lipoxygenase pathway. Biochemistry and relation to pathobiology in human diseases. *N Engl J Med 323*, 645–655.

Lindblom, B., Hamberger, L., & Wiquist, N., (1978). Differentiated contractile effects of prostaglandins E and F in the isolated circular and longitudinal smooth muscle of the human oviduct. *Fertil Steril 30*, 553–559.

Lundeen, K. A., Sun, B., Karlsson, L., & Fourie, A. M., (2006). Leukotriene B4 receptors BLT1 and BLT2: expression and function in human and murine mast cells. *J Immunol 177*, 3439–3447.

Luo, P., & Wang, M. H., (2011). Eicosanoids, β-cellfunction, and diabetes. *Prostaglandins Other Lipid Mediat 95*(1–4), 1–10.

Lydford, S. J., Li, S. W., & McKechnie, K. C., (1996). Comparison of prostanoid DP receptors in the rabbit saphenous vein and human neutrophil. *Br J Pharmacol 117*, 190p.

Lye, S. J., & Challis, J. R. G., (1982). Inhibition by PGI2 of myometrial activity in vivo in nonpregnant ovariectomized sheep. *J Reprod Fertil 66*, 311–315.

Lynch, K. R., O'Neill, G. P., Liu, Q., Im, D. S., Sawyer, N., Metters, K. M., Coulombe, N., Abramovitz, M., Figueroa, D. J., Zeng, Z., Connolly, B. M., Bai, C., Austin, C. P., Chateauneuf, A., Stocco, R., Greig, G. M., Kargman, S., Hooks, S. B., Hosfield, E., Williams, D. L. Jr, Ford-Hutchinson, A. W., Caskey, C. T., & Evans, J. F., (1999). Characterization of the human cysteinyl leukotriene CysLT1 receptor. *Nature 399*, 789–793.

Maathuis, J. B., & Kelly, R. W., (1978). Concentrations of prostaglandins $F_{2\alpha}$ and E_2 in the endometrium throughout the human menstrual cycle, after administration of clomiphene or an estrogen-progesterone pill and in early pregnancy. *J Endocrinol Meta*b 77, 361–371.

Machin, S. J., Carreras, L. O., Chamone, D. A. F., Defreyn, G., Dauden, M., & Vermylen, J., (1980). Familial deficiency of thromboxane synthetase (abstr). *Acta Therapeutica 6*, 34.

Machleidt, C., Fbrstermann, U., Anhut, H., & Hertting, G., (1981). Formation and elimination of prostacyclin metabolites in the cat in vivo as determined by radioimmunoassay of unextracted plasma. *European Journal of Pharmacology 74*, 19–26.

Makarainen, L., & Ylikorkala, O., (1984). Amniotic fluid 6-keto-prostaglandin F1 alpha and thromboxane B2 during labor. *Am J Obstet Gynecol 150*, 765–768.

Makila, U. M., Viinikka, L., & Ylikorkala, O., (1984). Increased thromboxane A2 production but normal prostacyclin by the placenta in hypertensive pregnancies, *Prostaglandins 27*, 87–95.

Manganello, J. M., Djellas, Y., Borg, C., Antonakis, K., & Le Breton, G. C., (1999). Cyclic AMP-dependent phosphorylation of thromboxane A2 receptor-associated Ga13. *J Biol Chem 274*, 28003–28010.

Manning, P. J., Rokach, J., Malo, J. L., Ethier, D., Cartier, A., Girard, Y., Charleson, S., & O'Byrne, P. M., (1990). Urinary leukotriene E4 levels during early and late asthmatic responses. *J Allergy Clin Immunol 86*, 211–220.

Marder, P., Sawyer, JS, Froelich, L. L., Mann, L. L., & Spaethe, S. M., (1995). Blockade of human neutrophil activation by 2-[2-propyl-3-[3-[2-ethyl-4-(4-fluorophenyl)-5-Hydroxyphenoxy]propoxy]phenoxy] benzoic acid (LY293111), a novel leukotriene B4 receptor antagonist.Biochem. *Pharmacol 49*, 1683–1690.

Martin, C., Uhlig, S., & Ullrich, V., (2001). Cytokine-induced bronchoconstriction in precision-cut lung slices is dependent upon cyclooxygenase-2 and thromboxane receptor activation. *Am J Respir Cell Mol Biol 24*, 139–145.

Martin, J. G., Suzuki, M., Maghni, K., Pantano, R., Ramos-Barbon, D., & Ihaku, D., (2002). The immunomodulatory actions of prostaglandin E2 on allergic airway responses in the rat. *J. Immunol 169*, 3963–3969.

Martin, T. R., Altman, L. C., Albert, R. K., et al. (1984). Leukotriene R4 production by the human alveolar macrophage: a potential mechanism for amplifying inflammation in the lung. *Am Rev Respir Dis 129*, 106–111.

Martin, T. R., Raugi, G., & Merritt, T. L., (1987). Henderson WR Jr. Relative contribution of leukotriene B4 to the neutrophil chemotactic activity produced by the resident human alveolar macrophage. *J Clin Invest 80*, 1114–1124.

Masuda, A., Mais, D. E., Oatis, Jr, J. E., & Halushka, P. V., (1991). Platelet and vascular thromboxane A2/prostaglandin H2 receptors: evidence for different subclasses in the rat. *Biochem Pharmacol 42*, 537–544.

Mathews, K., Christopher, & Van Holde, K. E., (1996). Biochemistry Second Edition.

Mathiesen, J. M., Christopoulos, A., Ulven, T., Royer, J. F., Campillo, M., Heinemann, A., Pardo, L., & Kostenis, E., (2006). On the mechanism of interaction of potent surmountable and insurmountable antagonists with the prostaglandin D2 receptor CRTH2. *Mol Pharmacol 69*, 1441–1453.

Matsuo, M., Yoshida, N., Zaitsu, M., Ishii, K., & Hamasaki, Y., (2004). Inhibition of human glioma cell growth by a PHS-2 inhibitor, NS398, and a prostaglandin E receptor subtype EP1-selective antagonist, SC51089. *J Neurooncol 66*, 285–292.

Matsuoka, T., Hirata, M., Tanaka, H., Takahashi, Y., Murata, T., Kabashima, K., Sugimoto, Y., Kobayashi, T., Ushikubi, F., Aze, Y., Eguchi, N., Urade, Y., Yoshida, N., Kimura, K., Mizoguchi, A., Honda, Y., Nagai, H., & Narumiya, S., (2000). Prostaglandin D2 as a mediator of allergic asthma. *Science 287*, 2013–2317.

Maubach, K. A., Clark, D. E., Fenton, G., Lockey, P. M., Clark, K. L., Oxford, A. W., Hagan, R. M., Routledge, C., & Coleman, R. A., (2009). BCG20–1531, a novel, potent and selective prostanoid EP4 receptor antagonist; a putative new treatment for migraine headache. *Br J Pharmacol 156*, 316–327.

Maurer, P., Moskowitz, M. A., Levine, L., & Melamed, E., (1980). The synthesis of prostaglandins by bovine cerebral microvessels. *Prostaglandins Med 4*, 1531–61.

Mazzoni, A., Young, H. A., Spitzer, J. H., Visintin, A., & Segal, D. M., (2001). Histamine regulates cytokine production in maturing dendritic cells, resulting in altered T cell polarization. *J Clin Invest 108*, 1865–1873.

McCoy, J. M., Wicks, J. R., & Audoly, L. P., (2002). The role of prostaglandin E2 receptors in the pathogenesis of rheumatoid arthritis. *Journal of Clinical Investigation 110(4)*, 651–658.

McDaniel, M. L., Kwon, G., Hill, J. R., Marshall, C. A., & Corbett, J. A., (1996). Cytokines and nitric oxide in isletinflammation and diabetes. *Proc Soc Exp Biol Med. 211*(1), 24–32.

Mehta, P., Mehta, J., (1979). Platelet function studies in coronary artery disease: Evidence for enhanced platelet microthrombus formation activity in acute myocardial infarction. *Am J Cardiol 43*, 757–760.

Melillo, E., Woolley, K. L., Manning, P. J., Watson, R. M., & O'Byrne, P. M., (1994). Effect of inhaled PGE2 on exercise-induced bronchoconstriction in asthmatic subjects. *American Journal of Respiratory and Critical Care Medicine 149*(5), 1138–1141.

Mellor, E. A., Frank, N., Soler, D., Lora, J. M., Hodge, M. R., Austen, K. F., & Boyce, J. A., (2003). Expression of the type 2 cysteinyl leukotriene receptor by human mast cells; demonstration of function distinct from that of CysLTR1. *Proc Natl Acad Sci USA 100*, 11589.

Mellor, E. A., Maekawa, A., Austen, K. F., & Boyce, J. A., (2001). Cysteinyl leukotriene receptor 1 is also a pyrimidinergic receptor and is expressed by human mast cells. *Proc Natl Acad Sci USA 98*, 7964–7969.

Mellor, E., Maekawa, A., Austen, K. F., & Boyce, J. A., (2001). Cysteinyl leukotriene receptor 1 is also a pyrimidinergic receptor and is expressed by human mast cells. *Proc Natl Acad Sci USA 98*, 7964–7969.

Mesquita-Santos, F. P., Vieira-de-Abreu, A., Calheiros, A. S., Figueiredo, I. H., Castro-Faria-Neto, H. C., Weller, P. F., Bozza, P. T., Diaz, B. L., & Bandeira-Melo, C., (2006). Cutting edge: prostaglandin D2 enhances leukotriene C 4 synthesis by eosinophils during allergic inflammation: synergistic in vivo role of endogenous exotoxin. *J Immunol 176*, 1326–1330.

Miggin, S. M., & Kinsella, B. T., Investigation of the mechanisms of G protein: effector coupling by the human and mouse prostacyclin receptors. Identification of critical species-dependent differences. *J Biol Chem 277*, 27053–27064.

Miller, A. M., & Masrorpour, M., Klaus, C., & Zhang, J. X., (2007). LPS exacerbates endothelin-1 induced activation of cytosolic phospholipase A2 and thromboxane A2 production from Kupffer cells of the prefibrotic rat liver. *J Hepatol 46*, 276–285.

Milton, A. S., & Wendland, S., (1970). A possible role for prostaglandin E1 as a modulator for temperature regulation in the central nervous system of the cat. *J Physiol 207*(2), 76P–77P.

Milton, A. S., & Wendland, S., (1971). Effects on body temperature of prostaglandins of the, A., E, and F series on injection into the third ventricle of unanesthetized cats and rabbits. *J Physiol London 218,* 325–336.

Mimura, H., Ikemura, T., Kotera, O., Sawada, M., Tashiro, S., Fuse, E., Ueno, K., Manabe, H., Ohshima, E., Karasawa, A., & Miyaji, H., (2005). Inhibitory effect of the 4-aminotetrahydroquinoline derivatives, selective chemoattractant receptor-homologous molecule expressed on T helper 2 cell antagonists, on eosinophil migration induced by prostaglandin D2. *J Pharmacol Exp Ther 314*, 2442–2451.

Minami, T., Tojo, H., ShinomuraY, MatsuzawaY, & Okamoto, M., (1994). Increased group II phospholipase A2 in colonic mucosa of patients with Crohn's disease and ulcerative colitis. *Gut 35,* 1593–1598.

Mitchell, M. D., (1981). Prostaglandins during pregnancy and the perinatal period. *J Reprod Fertil 62*, 305–315.

Mitchell, M. D., Brunt, J., Clover, L., & Walker, D. W., (1980). Prostaglandins in the umbilical and uterine circulation during late pregnancy in the ewe. *J Reprod Fertil 58*, 283–287.

Mohite, A., Chillar, A., So, S. P., Cervantes, V., & Ruan, K. H., (2011). Novel mechanism of the vascular protector prostacyclin: regulating microRNA expression. *Biochemistry 50(10),* 1691–1699.

Moncada, S., Gryglewski, R. J., Bunting, S., & Vane, J. R., (1976). An enzyme isolated from arteries transforms prostaglandin endoperoxides to an unstable substance that inhibits plateletaggregation. *Nature 263*, 663–665.

Mondino, C., Ciabattoni, G., Koch, P., Pistelli, R., Trove, A., Barnes, P. J., & Montuschi, P., (2004). Effects of inhaled corticosteroids on exhaled leukotrienes and prostanoids in asthmatic children. *J Allergy Clin Immunol 114*, 761–767.

Monneret, G., Gravel, S., Diamond, M., Rokach, J., & Powell, W. S., (2001). Prostaglandin D2 is a potent chemoattractant for human eosinophils that acts via a novel DP receptor. *Blood 98,* 1942–1948.

Morita, I., Schindler, M., Regier, M. K., Otto, J. C., Hori, T., DeWitt, D. L., & Smith, W. L., (1995). Different intracellular locations for prostaglandin endoperoxide H synthase-1 and -2. *J Biol Chem 270*, 10902–10908.

Mulder, A., Gauvreau, G. M., Watson, R. M., & O'Byrne, P. M., (1997). The effect of inhaled leukotriene D4 on airway eosinophilia and airway hyperresponsiveness in asthmatic subjects. *Am J Respir Crit Care Med 159*, 1562–1567.

Murakami, M., Kambe, T., & Shimbara, S., & Kudo, I., (1999). Functional coupling between various phospholipase A2 s and cyclooxygenases in immediate and delayed prostanoid biosynthetic pathways. *J Biol Chem 274*, 3103–3015.

Murakami, M., Nakatani, Y., Atsumi, G., Inoue, K., & Kudo, I., (1997). Regulatory functions of phospholipase A2. *Crit Rev Immunol 17*, 225–283.

Murakami, Y., Yokotani, K., Okuma, Y., Osumi, Y. (1998). Thromboxane A2 is involved in the nitric oxide-induced central activation of adrenomedullary outflow in rats. *Neuroscience. 87(1)*, 197–205.

Murphy, R. C., & Wheelan, E., (1998). Pathways of leukotriene metabolism in isolated cell models and human subjects. In: J. M. Drazen, S.-E. Dahlan, & T. H. Lee (Eds.), Lung Biology in Health and Disease: Five-Lipoxygenase Products in Asthma. Marcel Dekker, New York, NY, pp. 87–123.

Murray, J. J., Tonnel, A. B., Brash, A. R., Roberts, L. J. 2nd, Gosset, P., Workman, R., Capron, A., & Oates, J. A. (1986). Release of prostaglandin D 2 into human airways during acute antigen challenge. *N Engl J Med 315*, 800–804.

Nagai, H., Teramachi, H., & Tuchiya, T., (2006). Recent advances in the development of antiallergic drugs. *Allergol Intern 55*, 35–42.

Nakagawa, T., Minami, M., Katsumata, S., Ienaga, Y., & Satoh, M., (1995). Suppression of naloxone-precipitated withdrawal jumps in morphine-dependent mice by stimulation of prostaglandin EP3 receptor. *Br J Pharmacol 116*, 2661–2666.

Nakahata, N., Sato, K., Abe, M. T., & Nakanishi, H., (1990). ONO NT-126 is a potent and selective thromboxane A2 antagonist in human astrocytoma cells. *Eur J Pharmacol 184*, 233–238.

Nakao, K., Murase, A., Ohshiro, H., Okumura, T., Taniguchi, K., Murata, Y., Masuda, M., Kato, T., Okumura, Y., & Takada, J., (2007). CJ-023, 423, a novel, potent and selective prostaglandin EP4 receptor antagonist with antihyperalgesic properties. *J Pharm Exp Ther 322*, 686–694.

Narumiya, S., & FitzGerald, G. (2001). A. Genetic and pharmacological analysis of prostanoid receptor function. *J Clin Invest 108*, 25–30.

Narumiya, S., & FitzGerald, G. (2001). A., Genetic and pharmacological analysis of prostanoid receptor function. *J Clin Invest 108*(1), 25–30.

Narumiya, S., & Sugimoto, Y., & Ushikubi, F., (1999). Prostanoid receptors: structures, properties, and functions. *Physiol Rev 79*, 1193–1226.

Narumiya, S., Carbone, D. P., (2006). Host and direct antitumor effects and profound reduction in tumor metastasis with selective EP4 receptor antagonism. *Cancer Res 66*, 9665–9672.

Needleman, P., Kulsarni, P. S., & Raz, A., (1977). Coronary tone modulation: Formation and functions of prostaglandins, endoperoxides and thromboxanes. *Science 195*, 409–412.

Nelson, N. A., (1973). Nomenclature and structure of prostaglandins. *J Med Chem 17*, 911.

Nemoto, T., Aoki, H., Ike, A., Yamada, K., (1976). Kondo, T., Serum prostaglandin levels in asthmatic patients. *J Allergy Clin Immunol 57*, 89–94.

Nials, A. T., Vardey, C. J., Denyer, L. H., Thomas, M., Sparrow, S. J., Shepherd, G. D., & Coleman, R. A., (1993). AH 13205, a selective prostanoid EP2-receptor agonist. *Cardiovasc Drug Rev 11*, 165–179.

Nishikawa, M., Hishinuma, T., Nagata, K., Koseki, Y., Suzuki, K., & Mizugaki, M., (1997). Effects of eicosapentaenoic acid (EPA) on prostacyclin production in diabetics: GC/MS analysis of PGI2 and PGI3 levels. *Methods Find Exp Clin Pharmacol 19*, 429–433.

Nistico, G., Marley, E., (1976). Central effects of prostaglandins E1 in adult fowls. *Neuropharmacology 15*, 737–741

Nitta, M., Hirata, I., Toshina, K., Murano, M., Maemura, K., Hamamoto, N., & Katsu, K., (2002). Expression of the EP4 Prostaglandin E2 Receptor Subtype with Rat Dextran Sodium Sulphate Colitis: Colitis Suppression by a Selective Agonist, ONO-AE1-329. *Scand J Immunol, 56(1)*, 66-75.

Norel, X., Walch, L., Labat, C., Gascard, J. P., Dulmet, E., & Brink, C., (1999). Prostanoid receptors involved in the relaxation of human bronchial preparations. *Br J Pharmacol 126*, 867–872.

North, R. A., Whitehead, R., & Larkms, R. G. (1991). Stimulation by human chorionic gonadotropin of prostaglandin synthesis by early human placental tissue. *J Clin Endocrinol Metab 73*, 60–70.

Nothacker, H. P., Wang, Z. W., Zhu, Y. H., Reinscheid, R. K., Lin, S. H. S., & Civelli, O., (2000). Molecular cloning and characterization of a second human cysteinyl leukotriene receptor: discovery of a subtype selective agonist. *Mol. Pharmacol 58*, 1601–1608.

Noverr, M. C., Erb-Downward, J. R., & Huffnagle, G. B., (2003). Production of eicosanoids and other oxylipins by pathogenic eukaryotic microbes. *Clin Microbiol Rev 16*(3), 517–533.

Obara, Y., Kurose, H., & Nakahata, N., (2005). Thromboxane A2 promotes interleukin-6 bio-synthesis mediated by an activation of cyclic AMP response element-binding protein in 1321N1 human astrocytoma cells. *Mol Pharmacol 68*, 670–679.

Obata, T., Nambu, F., Kitagawa, T., Terashima, H., Toda, M., Okegawa, T., & Kawasaki, A., (1987). Adv. *Prostaglandin Thromboxane Leukotriene Res. 17A*, 540.

Offermans, S., Laugwitz, K. L., Spicher, K., & Schultz, G., (1994). G proteins of the G12 family are activated via thromboxane A2 and thrombin receptors in human platelets. *Proc Natl Acad Sci 91*, 504–508.

Ojeda, J. R., Negro-Vilar, A., & McCann, S. M., (1981). Role of prostaglandins in the control of pituitary hormone secretion. *Prog Clin Bioi Res 74*, 229–247.

Ojeda, S. R., Harms, P. G., & McCann, S. M., (1974a). Central effect of prostaglandin E, on prolactin release. *Endocrinology 95*, 613–618.

Ojeda, S. R., Harms, P. G., & McCann, S. M., (1974b). Possible role of cyclic AMP and prostaglandin, E., in the dopaminergic control of prolactin release. *Endocrinology 95*, 1694–1703.

Ojeda, S. R., Jameson, H. E., & McCann, S. M., (1977). Prostaglandin E2 induced growth hormone release: effect of intrahypothalamic and intrapituitary implants. *Prostaglandins 13*, 943–955.

Omini, C., Folco, G. C., Pasargiklian, R., Fano, M., & Berti, F., (1979). Prostacyclin (PGI2) in pregnant human uterus, *Prostaglandins 17*, 113–120.

Ozaki, T., Hayashi, H., Tani, K., Ogushi, F., Yasuoka, S., & Ogura, T., (1992). Neutrophil chemotactic factors in the respiratory tract of patients with chronic airway diseases or idiopathic pulmonary fibrosis. *Am Rev Respir Dis 145*, 85–91.

Ozaki, T., Rennard, S. I., & Crystal, R. G., (1987). Cyclooxygenase metabolites are compart-mentalized in the human lower respiratory tract. *Journal of Applied Physiology 62(1)*, 219–222.

Pace-Asciak, C. R., & Smith, W. L., (1983). Enzymes in the biosynthesis and catabolism of the eicosanoids: prostaglandins, thromboxanes, leukotrienes, and hydroxy fatty acids. In: The Enzymes. Boyer, P. D. (ed.) Academic Press, New York, 16, 543–603.

Panossian, A., Hamberg, M., & Samuelsson, G. B., (1982). *FEBS Lett 150*, 511–513.

Paolini, J. F., Mitchel, Y. B., Reyes, R., Kher, U., Lai, E., Watson, D. J., Norquist, J. M., Meehan, A. G., Bays, H. E., Davidson, M., & Ballantyne, C. M., (2009). Effects of laropiprant on nicotinic acid induced flushing in patients with dyslipidemia. *Am J Cardiol 101*, 625–630.

Parhar, R. S., Yagel, S., & Lala, P. K., (1989). PGE-mediated immunosuppression by first trimester human decidual cells blocks activation of maternal leukocytes in the decidua with potential antitrophoblast activity. *Cell Immunol 120*, 61–74.

Parisi, V. M., & Walsh, S. W., (1989). Fetal vascular response to prostacyclin. *Am J Obstet Gynecol 160*, 871.

Pavord, I. D., Wong, C. S., Williams, J., Tattersfield, A. E., (1993). Effect of inhaled prosta-glandin E2 on allergen-induced asthma. *Am Rev Respir Dis 148*, 87–90.

Peng, T. C., Six, K. M., & Munson, P. L., (1970). Endocrinology effects of prostaglandin E-1 on the hypothalamo-hyophyzeal-adrenocortical *axis in rats. 86(2)*, 202–206.

Penrose, J. F., & Austen, K. F., (1999). The biochemical, molecular, and genomic aspects of leukotriene C4 synthase. *Proc Assoc Am Physicians 111*, 537–46.

Peri, K., Polyak, F., Lubell, W., Thouin, E., & Chemtob, S., (2005). Peptides and peptido-mimetics useful for inhibiting the activity of prostaglandin F2a receptor. *2006 US 2006/0211626, Quebec.*

Peters-Golden, M., & Brock, T. G., (2000). Intracellular compartmentalization of leukotriene biosynthesis. *Am J Respir Crit Care Med 161*, S36–40.

Peters-Golden, M., & Brock, T. G., (2001). Intracellular compartmentalization of leukotriene synthesis: unexpected nuclear secrets. *FEBS Lett 487*, 323–326.

Peters-Golden, M., Gleason, M. M., & Togias, A., (2006). Cysteinyl leukotrienes: multifunctional mediators in allergic rhinitis. *Clin Exp Allergy 36*, 689–703.

Peters-Golden, M., Henderson, W. R., & Leukotrienes, Jr. (2007) *N Engl J Med, 357*, 1841–1854.

Peters, S. P., MacGlashan, D. W. Jr., Schulman, E. S., Hayes, E. C., Rokach, J., Adkinson, N. F. Jr., & Lichtenstein, L. M., (1984). Arachidonic acid metabolism in purified human lung mast cells. *J Immunol 132*, 1972–1979.

Pettipher, R., Hansel, T. T., & Armer, R., (2007). Antagonism of the prostagland in D2 receptors DP1 and CRTH2 as an approach to treatallergic diseases. *Nat Rev Drug Discov 6(4)*, 313–325.

Phasrriss, B. B., & Wyngarden, L. J., (1969). The effect of prostaglandin F2 M on the progestogen content of ovaries from pseudopregnant rats. *Proceedings of the Society for Experimental Biology and Medicine 130*, 92–94.

Phillis, J. W., & O'Regan, M. H., (2003). The role of phospholipases, cyclooxygenases, and lipoxygenases in cerebral ischemic/traumatic injuries. *Crit Rev Neurobiol 15*,61–90.

Pickard, J. D., MacDonell, L. A., MacKenzie, E. T., & Harper, A. M., (1977). Prostaglandin-induced effects in the primate cerebral circulation. *Eur J Pharmacol 43*, 343–351.

Pierce, K. L., Fujino, H., Srinivasan, D., & Regan, J. W., (1999). Activation of FP prostanoid receptor isoforms leads to Rho-mediated changes in cell morphology and in the cell cytoskeleton. *J Biol Chem 274*, 35944–35949.

Pittman, Q. I., Veale, W. L., Cockeram, A. W., & Cooper, K. E., (1976). Changes in body temperature produced by prostaglandins and pyrogens in the chicken. *Am J Physiol 230*, 1284–1287.

Profita, M., Sala, A., Bonanno, A., Riccobono, L., Siena, L., & Melis, M. R., (2003). Increased prostaglandin E2 concentrations and cyclooxygenase-2 expression in asthmatic subjects with sputum eosinophilia. *J Allergy Clin Immunol 112*, 709–716.

Pruzanski, W., & Vadas, P., (1991). Phospholipase A2: a mediator between proximal and distal effectors of inflammation. *Immunol Today 12*, 143–146.

Ramwell, P. W., (1964). The action of cerebrospinal fluid on the frog rectus abdominis muscle and other isolated tissue preparations. *J Physiol (Lond) 170*, 21–23.

Rao, A. M., Hatcher, J. F., Kindy, M. S., & Dempsey, R. J., (1999). Arachidonic acid and leukotriene C4: role in transient cerebral ischemia of gerbils. *Neurochem Res 24*, 1225–1232.

Ratner, A., Wilson, M. C., Srivastava, L., & Peake, G. T., (1974). Stimulatory effects of prostaglandin, E., on rat anterior pituitary cyclic AMP and luteinizing hormone release. *Prostaglandins 5*, 165–174.

Raychaudhuri, A., Kotyuk, B., Pellas, T. C., Pastor, G., Fryer, L. R., Morrissey, M., & Main, A. J., (1995). Effect of CGS 25019C and other LTB4 antagonists in the mouse ear edema and rat neutropenia models. *Inflammation Res 44 (2)*, S141–142.

Riddick, C. A., Serio, K. J., Hodulik, C. R., et al. (1999). TGF-beta increases leukotriene C4 synthase expression in the monocyte-like cell line, THP-1. *J Immunol 162*, 1101–1107.

Ritchie, D. M., Hahn, D. W., & McGuire, J. L., (1984). Smooth muscle contraction as a model to study the mediator role of endogenous lipoxygenase products of arachidonic acid. *Life Sci. 34(6)*, 509–513.

Robertson, P. R., (1983). Prostaglandins, glucose homeostasis, and diabetes mellitus. *Annu Rev Med 34*, 1–12.

Robertson, R. M., Robertson, D., Roberts, L. J., Maas, R. L., FitzGerald, G. A., Friesinger, G. C., & Oates, J. A., (1981). Thromboxane A2 in vasotonic angina pectoris: evidence from direct measurements and inhibitor trials. *N Engl J Med 304*, 998–1003.

Robertson, R. P., & Chen, M., (1997). A role for prostaglandin E in defective insulin secretion and carbohydrate intolerance in diabetes mellitus. *J Clin Invest 60*, 747–753.

Robinson, J. S., Redman, C. W., Clover, L., Mitchell, & M. D., (1979). The concentrations of the prostaglandins E and F 13 14-dihydro-15- oxo-prostaglandin F and thromboxane B2 In tissues obtained from women with and without preeclampsia, *Prostaglandins Med 3*, 223–234.

Rouzer, C. A., Matsumoto, T., & Samuelsson, B., (1986). Single protein from human leukocytes possesses 5-lipoxygenase and leukotriene A4 synthase activities. Proc Natl Acad Sci USA *83(4)*, 857–861.

Royer, J. F., Schratl, P., Carrillo, J. J., Jupp, R., Barker, J., Weyman-Jones, C., Beri, R., Sargent, C., Schmidt, J. A., Lang-Loidolt, D., & Heinemann, A. (2008). A novel antagonist of prostaglandin D2 blocks the locomotion of eosinophils and basophils. *Eur J Clin Invest 38*, 663–671.

Sala, A., Aliev, G. M., Rossoni, G., Berti, F., Buccellati, C., Burnstock, G., & Folco, G., (1996). Maclouf, J., Morphological and functional changes of coronary vasculature caused by transcellular biosynthesis of sulfidopeptide leukotrienes in isolated heart of rabbit. *Blood. 87(5)*, 1824–1832.

Sampson, A. P., Thomas, R. U., Costello, J. F., & Piper, P. J., (1992). Enhanced leukotriene synthesis in leukocytes of atopic and asthmatic subjects. *Rr J Clin Pharmacol 33*, 423–430.

Samuelsson, B., (1983). Leukotrienes: mediators of immediate hypersensitivity reactions and inflammation. *Science 220*, 568–575.

Samuelsson, B., (2000). The discovery of the leukotrienes. *American Journal of Respiratory and Critical Care Medicine 161*, 2–6.

Samuelsson, B., Haeggström, J. Z., & Wetterholm, A., (1991). Leukotriene biosynthesis. *Ann N Y Acad Sci 629*, 89–99.

Sandig, H., Pease, J. E., & Sabroe, I., (2007). Contrary prostaglandins: the opposing roles of PGD2 and its metabolites in leukocyte function. *J Leukoc Biol 81*, 372–382.

Sato, S., Szabo, M., Kowalski, K., & Burke, G., (1972). R ole of prostaglandin in thyrotropin action on thyroid. *Endocrinology 90*, 343–356.

Sato, T., Jyujyo, T., Iesaka, T., Ishikawa, J., & Igarashi, M., (1974). Follicle stimulating hormone and prolactin release induced by prostaglandins in rat. *Prostaglandins 5*, 483–490.

Sato, T., Taya, K., Jyujyo, T., Hirono, M., & Igarashi, M., (1974). The stimulatory effect of prostaglandins on luteinizing hormone release. *Am J Obstet Gynecol 118*, 875–876.

Satoh, K., Yasumizu, T., Fukuoka, H., Kinoshita, K., & Kaneko, Y., (1979). Tsuchiya, M., Sakamoto, S., Prostaglandin $F_{2\alpha}$ metabolite levels in plasma, amniotic fluid and urine during pregnancy and labor. *Am J Obstet Gynecol 133*, 886–890.

Sawyer, N., Cauchon, E., Chateauneuf, A., Cruz, R. P., Nicholson, D. W., Metters, K. M., & O'Neill, G. P., (2002). Gervais, F. G., Molecular pharmacology of the human prostaglandin D2 receptor, CRTH2. *Br J Pharmacol 137*, 1163–1172.

Schlondorff, D., Ardailloui, R., (1986). Prostaglandins and other arachidonic acid metabolites in the kidney. *Kidney Int 29*, 108–119.

Schuhmann, M. U., Mokhtarzadeh, M., Stichtenoth, D. O., Skardelly, M., Klinge, P. M., Gutzki, F. M., et al. (2003). Temporal profiles of cerebrospinal fluid leukotrienes, brain edema and inflammatory response following experimental brain injury. *Neurol Res 25*, 481–491.

Schulman, E. S., Newball, H. H., Demers, L. M., Fitzpatrick, F. A., & Adkinson Jr. N. F. (1981). Anaphylactic release of thromboxane A2, Prostaglandin D2, and prostacyclin from human lung parenchyma. *American Review of Respiratory Disease 124 (4)*, 402–406.

Sebekova´ K, Eifert, T., Klassen, A., Heidland, A., & Amann, K., (2007). Renal effects of S18886 (Terutroban), a TP receptor antagonist, in an experimental model of type 2 diabetes. *Diabetes 56*, 968–974.

Seed, M. P., Williams, K. I., & Bamford, D. S., (1983). Influence of gestation on prostacyclin synthesis by the human pregnant myometrium, in: P. J. Lewis, S. Moncada, J. O'Grady (Eds.), Prostacyclin in Pregnancy, Raven Press, New York, 31–36.

Seilhamer, J. J., Pruzanski, W., Vadas, P., Plant, S., Miller, J. A., Kloss, J., & Johnson, L. K., (1989). Cloning and recombinant expression of phospholipase A2 present in rheumatoid arthritic synovial fluid. *J Biol Chem 264*, 5335–5338.

Serhan CN and Levy, B., (2003). Success of prostaglandin E2 in structure-function is a challenge for structure-based therapeutics. Proceedings of the National *Academy of Sciences of the United States of America 100*(15), 8609–8611.

Serhan, C. N., & Prescott, S. M., (2000). The scent of a phagocyte: Advances on leukotriene b(4) receptors. *J Exp Med 192*, F5–8.

Serhan, C. N., Haeggstrom, J. A., & Leslie, C. C., (1996). Lipid mediator networks in cell signaling: Update and imparct of cytokines. *FASEB J 10*, 1147–1158.

Shak, S., & Goldstein, I. M., (1985). The leukotriene B4 omega-hydroxylase in human polymorphonuclear leukocytes is a membrane-associated, NADPH-dependent cytochrome P-450. *Trans Assoc Am Physicians 98*, 352–360.

Sheller, J. R., Mitchell, D., Meyrick, B., Oates, J., & Breyer, R., (2000). EP2 receptor mediates bronchodilation by PGE2 in mice. *Journal of Applied Physiology 88*(6), 2214–2218.

Shenker, A., Goldsmith, P., Unson, C. G., & Spiegel, A. M., (1991). The G protein coupled to the thromboxane A2 receptor in human platelets is a member of the novel Gq family. *J Biol Chem 266*, 9309–9313.

Shichijo, M., Arimura, A., Hirano, Y., Yasui, K., Suzuki, N., Deguchi, M., & Abraham, W. M., (2009). A prostaglandin D2 receptor antagonist modifies experimental asthma in sheep. *Clin Exp Allergy 39*, 1404–1414.

Shimizu, T., Izumi, T., Seyama, Y., Tadokoro, K., Rådmark, O., & Samuelsson, B., (1986). Characterization of leukotriene A4 synthase from murine mast cells: evidence for its identity to arachidonate 5-lipoxygenase. *Proc Natl Acad Sci USA 83(12)*, 4175–4179.

Shirasaki, H., Kikuchi, M., Kanaizumi, E., & Himi, T., (2009). Accumulation of CRTH2-positive leukocytes in human allergic nasal mucosa. *Ann Allergy Asthma Immunol 102*, 110–115.

Shirasaki, H., Kikuchi, M., Seki, N., Kanaizumi, E., Watanabe, K., & Himi, T., (2007). Expression and localization of the thromboxane A2 receptor in human nasal mucosa. *Prostaglandins Leukot Essent Fat Acids 76*, 315–320.

Siggins, G., Hoffer, B., & Bloom, F., (1971). Prostaglandin-norepinephrine interactions in brain: microelectrophoretic and histochemical correlates. *Ann N Y Acad Sci 180*, 302–323.

Singh, E. J., Baccarini, I., & Zuspan, F., (1975). Levels of prostaglandins $F_{2\alpha}$ and E_2 in human endometrium during the menstrual cycle. *Am J Obstet Gynecol 121*, 1003–1006.

Singh, J., Zeller, W., Zhou, N., Hategen, G., Mishra, R., Polozov, A., Yu, P., Onua, E., Zhang, J., Zembower, D., Kiselyov, A., Ramírez, J. L., Sigthorsson, G., Bjornsson, J. M., Thorsteinsdottir, M., Andrésson, T., Bjarnadottir, M., Magnusson, O., Fabre, J. E., Stefansson, K., & Gurney, M. E., (2009). Antagonists of the EP3 receptor for prostaglandin E2 are novel antiplatelet agents that do not prolong bleeding. *ACS Chem Biol 4, 115*–126.

Sinzinger, H., Feigl, W., Silberbauer, K., Oppolzer, R., Winter, M., & Auerswald, W., (1980). Prostacyclin (PG12)-generation by different types of human atherosclerotic lesions. *Exp Pathol 18*, 175–180.

Sirois, P., Borgeat, P., Jeanson, A., Roy, S., & Girard, G., (1980). The action of leukotriene B4 (LTB4) on the lung. *Prostaglandins 5, 429*–444.

Sjostrom, M., Jakobsson, P. J., Juremalm, M., Ahmed, A., Nilsson, G., Macchia, L., & Haeggstrom, J. Z., (2002). Human mast cells express two leukotriene C(4) synthase isoenzymes and the CysLT(1) receptor. *Biochim Biophys Acta* 1583, 1553.

Smith, A. P., Cuthbert, M. F., & Dunlop, L. S., (1975). Effects of inhaled prostaglandins E1, E2 and F2α on the airway resistance of healthy and asthmatic man. *Clin Sci Mol Med 48*, 421–430.

Smith, J. B., (1980). The prostanoids in hemostasls and thrombosis. Am *J Pathol 99*, 743–804.

Smith, W. L., (1992). Prostanoid biosynthesis and mechanisms of action. *Am J Physiol 263*(2 Pt 2), F181–191.

Smith, W. L., & DeWitt, D. L., (1995). Biochemistry of mostadandin endoueroxide H synthase-1 and synthase-2 and their differential susceptibility and nonsteroid anti-inflammatory drugs. *Semin Nephrol 15*(3), 179–194.

Smith, W. L., & Langenbach, R., (2001). Why there are two cyclooxygenases. *J. Clin. Invest. 107*, 1491–1495.

Smith, W. L., & Murphy, R. C., (2002). The eicosanoids: cyclooxygenase, lipoxygenase, and epoxygenase pathways, in Biochemistry of Lipids, Lipoproteins and Membranes (Vance, D. E., Vance, J. E., eds.) 4th edition pp. 341–372, Elsevier, New York.

Smith, W. L., DeWitt, D. L., & Garavito, R. M., (2000). Cyclooxygenases: structural, cellular and molecular biology. *Annu Rev Biochem 69*, 149–182.

Smith, W. L., Marnett, L. J., & DeWitt, D. L., (1991). Prostaglandin and thromboxane biosynthesis. *Pharmacol Ther 49*(3), 153–179.

Smyth, E. M., Grosser, T., Wang, M., Yu, Y., & FitzGerald, G. A., (2009). Prostanoids in health and disease. *J Lipid Res 50*, S423–428.

Snyder, G. D., Capdevila, J., Chacos, N., Manna, S., & Falck, J. R., (1983). *Proc. Natl. Acad. Sci. USA 80*, 3504–3507.

Soberman, R. J., & Christmas, P., (2006). Revisiting prostacyclin: new directions in pulmonary fibrosis and inflammation. *American Journal of Physiology 291(2)*, L142–L143.

Söderström, M., Wigren, J., Surapureddi, S., Glass, & C. K., Hammarström, S., (2003). Novel prostaglandin D2-derived activators of peroxisome proliferator-activated receptor-gamma are formed in macrophage cell cultures. *Biochim Biophys Acta 1631*, 35–41.

Soler, M., Camacho, M., Escudero, J. R., Iniguez, M. A., & Vila, L., (2000). Human vascular smooth muscle cells but not endothelial cells express prostaglandin E synthase. *Circ Res 87*, 504–507.

Sousa, A. R., Parikh, A., Scadding, G., Corrigan, C. J., & Lee, T. H., Leukotriene-receptor expression on nasal mucosal inflammatory cells in aspirin-sensitive rhinosinusitis. *N Engl J Med 347*, 1493–1499.

Sousa, A., Pfister, R., Christie, P. E., Lane, S. J., Nasser, S. M., & Schmitz-Schumann, M., (1997). Enhanced expression of cyclo-oxygenase isoenzyme 2 (COX-2) in asthmatic airways and its cellular distribution in aspirin-sensitive asthma. *Thorax 52*, 940–945.

Spencer, A. G., Woods, J. W., Arakawa, T., Singer, I. I., & Smith, W. L., (1998). Subcellular localization of prostaglandin endoperoxide H synthases-1 and -2 by immunoelectron microscopy. *J Biol Chem 273*, 9886–9893.

Stebbins, K. J., Broadhead, A. R., Baccei, C. S., Scott, J. M., Truong, Y. P., Coate, H., Stock, N. S., Santini, A. M., Fagan, P., Prodanovich, P., Bain, G., Stearns, B. A., King, C. D., Hutchinson, J. H., Prasit, P., Evans, J. F., & Lorrain, D. S., Pharmacological blockade of the DP2 receptor inhibits cigarette smoke-induced inflammation, mucus cell metaplasia, and epithelial hyperplasia in the mouse lung. *J Pharmacol Exp Ther 332*, 764–775.

Steinberg, D., Parthasarathy, S., Carew, T. E., Khoo, J. C., & Witztum, J. L., (1989). Beyond cholesterol. Modifications of low-density lipoprotein that increase its atherogenicity. *N Engl J Med. 320(14)*, 915–924.

Steinhilber, D., Radmark, O., & Samuelsson, B., (1993). Transforming growth factor beta upregulates 5-lipoxygenase activity during myeloid cell maturation. *Proc Natl Acad Sci USA 90*, 5984–5988.

Stichtenoth, D. O., Thoren, S., Bian, H., Peters-Golden, M., Jakobsson, P. J., & Crofford, L. J., (2001). Microsomal prostaglandin E synthase is regulated by proinflammatory cytokines and glucocorticoids in primary rheumatoid synovial cells. *J Immunol 167(1)*, 469–474.

Stjernschantz J & Resul, B., (1992). Phenyl substituted prostaglandin analogs for glaucoma treatment. *Drugs Future 17*, 691–704.

Stoffersen, E., Moncada, S., & Vane, J. R., (1978). Eicosapentaenoic acid and prevention of thrombosis and atherosclerosis? *Lancet II*, 117–119.

Sturino, C. F., O'Neill, G., Lachance, N., Boyd, M., Berthelette, C., Labelle, M., Li, L., Roy, B., Scheigetz, J., Tsou, N., Aubin, Y., Bateman, K. P., Chauret, N., Day, S. H., Lévesque, J. F., Seto, C., Silva, J. H., Trimble, L. A., Carriere, M. C., Denis, D., Greig, G., Kargman, S., Lamontagne, S., Mathieu, M. C., Sawyer, N., Slipetz, D., Abraham, W. M., Jones, T., McAuliffe, M., Piechuta, H., Nicoll-Griffith, D. A., Wang, Z., Zamboni, R., Young, R. N., & Metters, K. M., (2007). Discovery of a potent and selective prostaglandin D2 receptor antagonist, [(3R)-4-(4-chloro-benzyl)-7-fluoro-5- (methylsulfonyl)-1,2,3,4-tetrahydrocyclopenta[b] indol- 3-yl]-acetic acid (MK-0524). *J Med Chem 50*, 794–806.

Sundberg, F., Ingleman-Sundberg, L., & Rydin, G., (1964). The effect of prostaglandin E_2 on the human uterus and the fallopian tubes in vitro. *Acta Obstet Gynecol Scand 43*, 95–102.

Sutyak, J., Austen, K. F., & Soberman, R. J., (1989). Identification of an aldehyde dehydrogenase in the microsomes of human polymorphonuclear leukocytes that metabolizes 20-aldehyde leukotriene B4. *J Biol Chem 264(25)*, 14818–14823.

Suzawa, T., Miyaura, C., Inada, M., Maruyama, T., Sugimoto, Y., Ushikubi, F., Ichikawa, A., Narumiya, S., & Suda, T., (2000). The role of prostaglandin E receptor subtypes (EP1, EP2, EP3, and EP4) in bone resorption; an analysis using specific agonists for the respective EPs. *Endocrinology 141*, 1554–1559.

Swanson, M. L., Lei, Z. M., Swanson, P. H., Rao, C. V., Narumiya, S., & Hirata, M., (1992). The expression of thromboxane A2 synthase and thromboxane A2 receptor gene in human uterus. *47*, 105–117.

Swanston, I. A., McNatty, K. P., & Baird, D. T., (1977). Concentration of prostaglandin $F_{2\alpha}$ and steroids in the human corpus luteum. *J Endocrinol 73*, 115–122.

Szczeklik, A., Gryglewski, R. J., Musiał, J., Grodzińska, L., & Serwońska, M., (1978). Marcinkiewicz, E., Thromboxane generation and platelet aggregation in survivals of myocardial infarction. *Thromb Haemostas 40*, 66–74.

Tada, M., Kuzuya, T., Inoue, M., Kodama, K., Mishima, M., Yamada, M., Inui, M., & Abe, H., (1981). Elevation of thromboxane B2 levels in patients with classic and variant angina pectoris. *Circulation 64*, 107–115.

Tada, M., Kuzuya, T., Inoue, M., Kodama, K., Mishima, M., Yamada, M., Inui, M., & Abe, H., (1981). Elevation of thromboxane 6, levels in patients with classic and variant angina pectoris. *Circulation 64*, 1107–1115.

Tajima, T., Murata, T., Aritake, K., Urade, Y., Hirai, H., Nakamura, M., Ozaki, H., & Hori, M., (2008). Lipopolysaccharide induces macrophage migration via prostaglandin D2 and prostaglandin E2. *J Pharmacol Exp Ther 326*, 493–501.

Takayama, K., Yuhki, K., Ono, K., Fujino, T., Hara, A., Yamada, T., Kuriyama, S., Karibe, H., Okada, Y., Takahata, O., Taniguchi, T., Iijima, T., Iwasaki, H., Narumiya, S., & Ushikubi, F., (2005). Thromboxane A2 and prostaglandin F2α mediate inflammatory tachycardia. *Nat Med 11*, 562–566.

Tanaka, H., Kanako, S., & Abe, S., (2005). Prostaglandin E2 receptor selective agonists E-prostanoid 2 and E-prostanoid 4 may have therapeutic effects on ovalbumin-induced bronchoconstriction. *Chest 128*(5), 3717–3723.

Tanaka, K., Hirai, H., Takano, S., Nakamura, M., & Nagata, K., (2004). Effects of prostaglandin D2 on helper T cell functions. *Biochem Biophys Res Commun 316*, 1009–1014.

Tanaka, K., Roberts, M. H., Yamamoto, N., Sugiura, H., Uehara, M., Mao, X. Q., Shirakawa, T., & Hopkin, J. M., (2002). Genetic variants of the receptors for thromboxane A2 and IL-4 in atopic dermatitis. *Biochem Biophys Res Commun 292*, 776–780.

Tassoni, D., Kaur, G., Weizinger, R. S., & Sinclair, A. J., (2008). The role of eicosanoids in the brain. *Asia Pac J Clin Nutr 17*(1), 220–228.

Terragno, N. A., Terragno, A., Early, J. A., Roberts, M. A., & McGiff, J. C., (1978). Endogenous prostaglandin synthesis inhibitor in the renal cortex. *Clin Sci Mol Med 55*(4), 199–202.

Thuresson, E. D., Lakkides, K. M., Rieke, C. J., Sun, Y., Wingerd, B. A., Micielli, R., Mulichak, A. M., Malkowski, M. G., Garavito, R. M., & Smith, W. L., (2001). Prostaglandin endoperoxide H synthase-l: the functions of cyclooxygenase active site residues in the binding, positioning, and oxygenation of arachidonic acid. *J Biol Chem 276*, 10347–10357.

Tilley, S. L., Coffman, T. M., & Koller, B. H., (2001). Mixed messages: modulation of inflammation and immune responses by prostaglandins and thromboxanes. *J Clin Invest 108*(1), 15–23.

Town, M. H., Casals-Stenzel, J., & Schilinger, E., (1983). Pharmacological and cardiovascular properties of a hydantoin derivative, BW 245 C, with high affinity and selectivity for PGD2 receptors. *Prostaglandins 25*, 13–28.

Tran, P. O., Gleason, C. E., & Robertson, R. P., (2002). Inhibition of interleukin-1 beta-induced COX-2 and EP3 gene expression by sodium salicylate enhances pancreatic islet beta-cell function. *Diabetes, 51*, 1772–1778.

Trist, D. G., Collins, B. A., Wood, J., Kelly, M. G., & Robertson, A. D., The antagonism by BW A868C of PGD$_2$ and BW245C activation of human platelet adenylate cyclase. *Br. J. Pharmacol 96*, 301–306.

Trivedi, S. G., Newson, J., Rajakariar, R., Jacques, T. S., Hannon, R., Kanaoka, Y., Eguchi, N., Colville-Nash, P., & Gilroy, D. W., (2006). Essential role for hematopoietic prostaglandin D 2 synthase in the control of delayed-type hypersensitivity. *Proc Natl Acad Sci USA 103*, 5179–5184.

Tryzelius, Y., Nilsson, N. E., Kotarsky, K., Olde, B., & Owman, C., (2000). Cloning and characterization of cDNA encoding a novel human leukotriene B(4) receptor. *Biochem Biophys Res Commun 274*, 377–382.

Tsafriri, A., Koch, Y., & Lindner, H. R., (1973). Ovulation rate and serum LH levels in rats treated with indomethacin or prostaglandin E2. *Prostaglandins 3*, 461–67.

Uller, L., Mathiesen, J. M., Alenmyr, L., Korsgren, M., Ulven, T., Högberg, T., Andersson, G., Persson, C. G., & Kostenis, E., (2007). Antagonism of the prostaglandin D2 receptor CRTH2 attenuates asthma pathology in mouse eosinophilic airway inflammation. *Respir Res 28*, 16.

Urade, Y., & Hayaishi, O., (2011). Prostaglandin D2 and sleep/wake regulation. *Sleep Med Rev 15(6)*, 411–418.

Ushikubi, F., Aiba, Y. I., Nakamura, K. I., Namba, T., Hirata, M., Mazda, O., Katsura, Y., & Narumiya, S., (1993). Thromboxane A2 receptor is highly expressed in mouse immature thymocytes and mediates DNA fragmentation and apoptosis. *J Exp Med 178*, 1825–1830.

Ushikubi, F., Hirata, M., & Narumiya, S., (1995). Molecular biology of prostanoid receptors; an overview. *J Lipid Mediat Cell Signal 12(2–3)*, 343–359.

Vahouny, G. V., Chanderbhan, R., Hinds, R., Hodges, V. A., & Treadwell, C. R., (1978). ACTH-induced hydrolysis of cholesteryl esters in rat adrenal cells. *J. Lipid Res 19*, 570–577.

Vane, J., & Corin, R. E., (2003). Prostacyclin: a vascular mediator. *European Journal of Vascular and Endovascular Surgery 26(6)*, 571–578.

Vanhoutte, P. M., & Houston, D. S., (1985). Platelets, endothelium, and vasospasm. *Circulation 72*, 728–734.

Vassaux, G., Gaillard, D., Ailhaud, G., & Negrel, R., (1992). Prostacyclin is a specific effector of adipose cell differentiation. Its dual role as a cAMP- and Ca(2+)-elevating agent. *J Biol Chem 267*, 11092–11097.

Vinall, S. L., Townsend, E. R., & Pettipher, R., (2007). A paracrine role for chemoattractant receptor-homologous molecule expressed on T helper type 2 cells (CRTH2) in mediating chemotactic activation of CRTH2(+) CD4(+) T helper type 2 lymphocytes. *Immunology 121*, 577–584.

Walch, L., de Montpreville, V., Brink, C., & Norel, X., (2001). Prostanoid EP1-and TP-receptors involved in the contraction of human pulmonary veins. *Br J Pharmacol 134*, 1671–1678.

Walch, L., Labat, C., Gascard, J. P., de Montpreville, V., Brink, C., & Norel, X., (1999). Prostanoid receptors involved in the relaxation of human pulmonary vessels. *Br J Pharmacol 126*, 859–866.

Walker, R. L., (1970). The fatty acids of adrenal lipids from essential fatty acid-deficient rats. *J Nutr 100*, 355–360.

Wang, M. L., Huang, X. J., Fang, S. H., Yuan, Y. M., Zhang, W. P., Lu, Y. B., Ding, Q., & Wei, E. Q., (2006). Leukotriene D4 induces brain edema and enhances CysLT2 receptor-mediated aquaporin 4 expression. *Biochem Biophys Res Commun 350*, 399–404.

Weems, C. W., Weems, Y. S., & Randel, R. D., (2006). Prostaglandins and reproduction in female farm animals. *Veterinary Journal 171*, 206–228.

Weiss, H. J., & Lages, B. A., (1977). Possible congenital defect in platelet thromboxane synthetase. *Lancet I*, 760–761.

Weiss, J. W., Drazen, J. M., Coles, N., McFadden, E. R. Jr., Weller, P. F., Corey, E. J., Lewis, R. A., & Austen, K. F., (1982). Bronchoconstrictor effects of leukotriene C in humans. Science 216, 196–198.

Welch, K. M. A., Spira, P. J., Knowles, L., & Lance, J. W., (1974). Effects of prostaglandins in the internal and external carotid blood flow in the monkey. *Neurology 24*, 705–10

Wennmalm, A., (1988). Role of eicosanoids in the cardiovascular system. *Acta Physiol Pol 39(4)*, 217–224.

Whalen, J. B., Clancey, C. J., Farley, D. B., & Van Orden, D. E., (1978). Plasma prostaglandins in pregnancy. *Obstet Gynecol 51*, 52–55.

Wheeldon, A., & Vardey, C. J., (1993). Characterization of the inhibitory prostanoid receptors on human neutrophils. *Br J Pharmacol 108*, 1051–1054.

White, R., Terragno, D. A., Terragno, N. A., Hagen, A. A., & Robertson, J. T., (1977). Prostaglandins in porcine cerebral blood vessels. *Stroke 8*, 135.

Whittle, B. J., Hamid, S., Lidbury, P., & Rosam, A. C., (1988). Specificity between the antiaggregatory actions of prostacyclin, prostaglandin E1 and D2 on platelets. *Adv Exp Med Biol 192*, 109–125.

Whittle, B. J., Moncada, S., Mullane, K., & Vane, J. R., (1983). Platelet and cardiovascular activity of the hydantoin BW245C, a potent prostaglandin analog. *Prostaglandins 25*, 205–223.

Widdicombe, J. H., Ueki, I. F., Emery, D., Margolskee, D., Yergey, J., & Nadel, J. A., (1989). Release of cyclooxygenase products from primary cultures of tracheal epithelia of dog and human. *American Journal of Physiology 257(6)*, L361–L365.

Wilborn, J., Bailie, M., Coffey, M., Burdick, M., Strieter, R., & Peters-Golden, M., (1996). Constitutive activation of 5-lipoxygenase in the lungs of patients with idiopathic pulmonary fibrosis. *J Clin Invest 97*, 1827–1836.

Wilhelmsson, L., Wikland, M., & Wiqvist, M., (1981). PGH2, TXA2, and PGI2 have potent and differentiated actions on human uterine contractility, *Prostaglandins 21*, 277–286.

Willerson, J. T., Campbell, W. B., Winniford, M. D., Schmitz, J., Apprill, P., Firth, B. G., Ashton, J., Smitherman, T., Bush, L., & Buja, L. M., (1984). Conversion from chronic to acute coronary artery disease: Speculation regarding mechanisms. *Am J Cardiol 54*, 1349–1354.

Wolfe, L. S., & Coceani, F., (1979). The role of prostaglandins in the central nervous system. *Annu Rev Physiol 41*, 669–684.

Wolfe, L. S., & Marner, O. A., (1974). Measurement of prostaglandin FIG levels in human cerebrospinal fluid in normal and pathological conditions. *Prostaglandins 9*, 183–192.

Wood, C. E., Cudd, T. A., Kane, C., & Engelke, K., (1993). Fetal ACTH and blood pressure responses to thromboxane mimetic U-46619. *Am J Physiol 265*, R858–862.

Woodward, D. F., Lawrence, R. A., Fairbairn, C. E., Shan, T., & Williams, L. S., (1993). Intraocular pressure effects of selective prostanoid receptor agonists involve different receptor subtypes according to radio ligand binding studies. *J Lipid Mediat 6(1–3)*, 545–553.

Woodward, D. F., Pepperl, D. J., Burkley, T. H., & Regan, J. W., (1995). 6-Isopropoxy-9-oxoxanthene-2-carboxylic acid (AH 6809), a human EP2 receptor antagonist. *Biochem Pharmacol 50*, 1731–133.

Xiao, C. Y., Yuhki, K., Hara, A., Fujino, T., Kuriyama, S., Yamada, T., Takayama, K., Takahata, O., Karibe, H., Taniguchi, T., Narumiya, S., & Ushikubi, F., (2004). Prostaglandin E2 protects the heart from ischemia-reperfusion injury via its receptor subtype EP4. *Circulation 109*, 2462–2468.

Xue, L., Gyles, S. L., Wettey, F. R., Gazi, L., Townsend, E., Hunter, M. G., & Pettipher, R., (2005). Prostaglandin D2 causes preferential induction of proinflammatory Th2 cytokine production through an action on chemo attractant receptor-like molecule expressed on Th2 cells. *J Immunol 175*, 6531–6536.

Yamauchi, H., & Katsu, K., (2002). Expression of the EP4 prostaglandin E2 receptor subtype with rat dextran sodium sulfate colitis: colitis suppression by a selectiveagonist, ONO-AEI-329. *Scand J Immunol 56*, 66–75.

Yang, L., Huang, Y., Porta, R., Yanagisawa, K., Gonzalez, A., Segi, E., Johnson, D. H., Narumiya, S., & Carbone, D. P., (2006). Host and direct antitumor effects and profound reduction in tumor metastasis with selective EP4 receptor antagonism.*Cancer Res 66(19)*, 9665–9672.

Yang, Z. M., Das, S. K., Wang, J., Ichikawa, A., & Dey, S. K., (1997). Potential sites of prostaglandin actions in the per implantation mouse uterus: Differential expression and regulation of prostaglandin receptor genes. *Biol Reprod 56*, 368–79.

Ylä-Herttuala, S., Rosenfeld, M. E., Parthasarathy, S., Sigal, E., Särkioja, T., Witztum, J. L., & Steinberg, D., (1991). Gene expression in macrophage-rich human atherosclerotic lesions. 15-lipoxygenase and acetyl low-density lipoprotein receptor messenger RNA colocalize with oxidation specific lipid-protein adducts. *J Clin Invest 87*(4), 1146–1152.

Ylikorkala, O., Makila, U. M., & Viinikka, L., (1981). Amniotic fluid prostacyclin and thromboxane in normal, preeclamptic, and some other complicated pregnancies. Am J Obstet *Gynecol 141*, 487–490.

Yokomizo, T., Kato, K., Hagiya, H., Izumi, T., & Shimizu, T. J., (2001). Hydroxyeicosanoids bind to and activate the low affinity leukotriene B4 receptor, BLT2. *Biol Chem 276*, 12454–1249.

Yokomizo, T., Kato, K., Terawaki, K., Izumi, T., & Shimizu, T. J. (2000). A second leukotriene B(4) receptor, BLT2. A new therapeutic target in inflammation and immunological disorders. *Exp Med 192*, 421–432.

Yue, D. K., McLennan, S., Handelsman, D. J., Delbridge, L., Reeve, T., & Turtle, J. R., (1985). The effects of cyclooxygenase and lipooxygenase inhibitors on the collagen abnormalities on diabetic rats. *Diabetes 34*, 74–78.

Yuhki, K., Kojima, F., Kashiwagi, H., Kawabe, J., Fujino, T., Narumiya, S., & Ushikubi, F., (2011). Roles of prostanoids in the pathogenesis of cardiovascular diseases: Novel insights from knockout mouse studies. *Pharmacol Ther 129(2)*, 195–205.

Zahradnik, H. P., Schäfer, W., Neulen, J., Wetzka, B., Gaillard, T., Tielsch, J., & Casper, F., (1992). The role of eicosanoids in reproduction. *Eicosanoids 5*, S56–59.

Zanagnolo, V., Dharmarajan, A. M., Endo, K., & Wallach, E. E., (1996). Effects of acetylsalicylic acid (aspirin) and naproxen sodium (naproxen) on ovulation, prostaglandin, and progesterone production in the rabbit. *Fertil Steril 65*, 1036–1043.

Zhang, X. Y., Wang, X. R., Xu, D. M., Yu, S. Y., Shi, Q. J., Zhang, L. H., Chen, L., Fang, S. H., Lu, Y. B., Zhang, W. P., & Wei, E. Q., (2013). HAMI3379, a CysLT2 receptor antagonist, attenuates ischemia-like neuronal injury by inhibiting microglial activation. *J Pharmacol Exp Ther 346(2)*, 328–341.

Zor, U., Kaneko, T., Schneider, H. P., McCann, S. M., Lowe, I. P., Bloom, G., Borland, B., & Field, J. B., (1969). Stimulation of anterior pituitary adenyl cyclase activity and adenosine 3',5'-cyclic phosphate by hypothalamic extract and prostaglandin E1. *Proc Natl Acad Sci USA 63(3)*, 918–925.

PART IV

ADVANCES IN
MEDICINAL CHEMISTRY

CHAPTER 8

DEVELOPMENT OF ANTIMALARIAL DRUG ANALOGS TO COMBAT *PLASMODIUM* RESISTANCE

SHIVANI SHARMA,[1] AAKASH DEEP,[2] MANAV MALHOTRA,[3] and BALASUBRAMANIAN NARASIMHAN[4]

[1]*Department of Pharmaceutical Chemistry, Indo-Soviet Friendship (ISF) College of Pharmacy, Ferozepur Road, Moga – 142001, India*

[2]*Department of Pharmaceutical Sciences, Ch. Bansi Lal University, Bhiwani – 127021, India, Mobile: +919896096727; E-mail: aakashdeep82@gmail.com*

[3]*M. K. Drugs, F-10 Industrial Focal Point, Derabassi – 140507, India*

[4]*Faculty of Pharmaceutical Sciences, Maharshi Dayanand University, Rohtak, India*

CONTENTS

ABSTRACT

Malaria infection is a major public health problem worldwide, caused by unicellular protozoan parasites of genus *Plasmodium*: *P. falciparum*, *P. vivax*, *P. malariae*, *P. ovale*, and *P. knowlesi*. Chloroquine was the first synthetic antimalarial agent, introduced in 1944 and became mainstay of therapy and prevention. But later, antimalarial drug treatment has been hampered by the appearance of drug resistance. The emergence of drug resistance parasites led to new approaches involving modification of existing agents, discovery of new natural compounds and identification of new targets. New antimalarial dugs and combinations are being studied but there is not yet sufficient information on their efficacy. This chapter will review potential of newly synthesized compounds against plasmodial resistance for the prevention and treatment of malaria.

8.1   INTRODUCTION

Malaria, a tropical parasitic disease that now remains globally extended to more than 40% population and is one of the major causes of morbidity and death from infectious diseases worldwide (after respiratory infections, HIV/AIDS, diarrheal diseases, and tuberculosis) and the second in Africa, after HIV/AIDS (Breman et al., 2004; Snow et al., 2001). It originated from *Italian* word *"aria male"* meaning bad air, caused by an erythrocytic protozoan first identified by *Alfonse Laveran* in 1880 (White, 2008). The term was shortened to "malaria" in the twentieth century and in 1889, R. Ross discovered that mosquitoes transmitted malaria. There are five identified species of this parasite, namely, *Plasmodium vivax, P. falciparum, P. ovale, P. malariae* and *P. knowlesi*. *Plasmodium falciparum* is more common in sub Saharan Africa and Melanesia (Papua New Guinea, Solomon Islands) (Bozdech et al., 2003); *Plasmodium vivax* in Central and South America, India, North Africa and Middle East; *Plasmodium ovale* in Western Africa and *Plasmodium malariae* is sporadic worldwide (Myrvang, 2010). Of the four common species that cause malaria, the most virulent type is *P. falciparum* that can cause cerebral malaria. However, another relatively new species *P. knowlesi* is also a dangerous species that is typically found in long tailed and pigtailed macaque monkeys. The other three common species of malaria (*P. vivax, P. ovale, P. malariae*) are generally less lethal and are not life threatening. It is possible to be infected with more than one species of plasmodium at the same time (Sibley et al., 2001).

Malaria is transmitted by the female anopheles mosquito that can be treated in just 48 h, if left untreated, they may develop severe complications. According to the WHO, malaria is prevalent in 108 countries of the tropical and semitropical world (Africa; Amazon, central and southern America; central, south and SE Asia; Pacific) that are home to more than half of the world's people. Malaria is prevalent in tropical regions because significant amounts of rainfall, warm temperature and stagnant waters provide habitats ideal for mosquito larve (Rogers and Randolp, 2000; Sutherst, 2004; Wiesner et al., 2003). The number of cases of malaria to be 225 million in 2009 according to World Health Organization and it was estimated 781,000 malarial deaths during 2009 and 89% of them occurred in Africa, 6% in the East Mediterranean region and 5% in South East Asiatic countries. More than 30,000 cases of malaria are reported annually among travelers from developed world visiting malarious areas (Eckstein-Ludwig et al., 2003). Malaria kills one to 3 million peoples annually, many of whom were children under the age of 5 and pregnant women. It was estimated that every 40 seconds a child dies from malaria (Sachs and Malaney, 2002). In 2010, there were 216 million documented cases of malaria that year between 655,000 and 1.2 million people die from this disease in the African region, where the financial cost of malaria is crippling economic development due to high cost of medicines and reduced production.

In the beginning the symptoms of malaria may be unspecific, including joint pain, asthenia and abdominal pain; followed by high fever, shivering, anorexia and vomiting. The most severe form is caused by *P. falciparum*, particularly in children, pregnant women and nonimmune travelers from nonendemic countries (Clark and Schofield, 2000). Currently, about 2 million deaths per year worldwide are due to *Plasmodium* infections and majority occur in children under 5 years of age in sub-Saharan African countries. There are about 400 million new cases per year worldwide. Most people diagnosed in the U.S. obtained their infection outside of the country, usually while living or traveling through an area where malaria is endemic. The morbidity and mortality have risen because of inefficacious first-line oral treatment that increases the proportion of patients, who suffer from severe disease and even the transmission of resistant strains is promoted by unsuccessful treatment. Due to inexorable spread of drug resistance and lack of effective antimalarial vaccine, the search and development of effective, safe and affordable drugs against *falciparum* malaria are of prime importance worldwide (Canfield et al., 1995).

8.2   LIFE CYCLE

The infectious stages of the malaria parasite reside in the salivary glands of female Anopheles mosquitoes that bite humans for a blood meal. During blood extraction, the mosquito injects its saliva into the wound, thereby transferring approximately 15–20 so-called sporozoites into the blood stream. In a matter of minutes, these sporozoites are able to conceal themselves from the host's immune system by entering into the liver cells. Each sporozoite develops into a tissue schizont, containing 10,000–30,000 merozoites (Frudurich et al., 2002). After one to two weeks, the schizont ruptures and releases the merozoites into the blood stream, starting the erythrocytic phase of the parasite's life cycle (Kumar et al., 2002). In the cases of *P. vivax* and *P. ovale*, some sporozoites turn into hypnozoites, a form that can remain dormant in the liver cells, causing relapses months or even years after the initial infection. *P. falciparum* and *P. malariae* lack this liver persistent phase, but *P. malariae* can persist in the blood for many years if inadequately treated (Sachs and Malaney, 2002). Merozoites released into the bloodstream hide again from the host's immune system by invading erythrocytes. In the erythrocyte, the parasite develops from a ring stage via a trophozoite stage into a blood schizont. After a time characteristic for each specific Plasmodium species, the erythrocyte ruptures and releases 16–32 new merozoites into the blood stream which in turn again invade erythrocytes, thereby starting a new erythrocytic cycle. This asexual life cycle, from invasion of the erythrocytes until the schizont ruptures, spans 48 h for *P. falciparum, P. vivax,* and *P. ovale,* and 72 h for *P. malariae* (Makler et al., 1993). After a number of asexual life cycles, some merozoites develop into sexual forms, the gametocytes, which are transferred to a mosquito during another blood meal as shown in Figure 8.1. These gametocytes undergo sexual reproduction within the mosquito mid-gut producing thousands of infective sporozoites, which migrate to the salivary gland where they are ready for a new infection (Ashley et al., 2006; Makler et al., 1993).

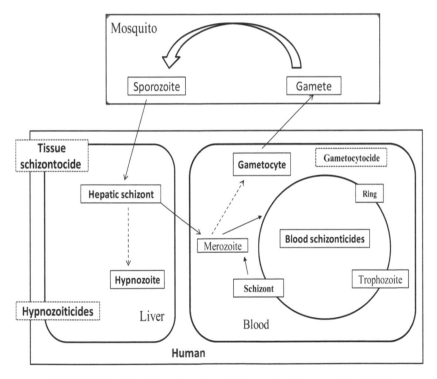

FIGURE 8.1   The life cycle of malarial parasite (Kumar et al., 2002).

With the rupture of the erythrocyte, the parasite's waste and cell debris is released into the blood stream, causing some of the clinical symptoms of malaria. The main symptom is fever, but rarely in the classical tertian (every 48 h) or quartan (every 72 h) patterns. Further symptoms include chill, headache, abdominal and back pain, nausea, and sometimes vomiting. Thus, the early stages of malaria often resemble the onset of an influenza infection. *P. vivax, P. ovale* and *P. malariae* show distinct selectivity toward the age of the infected erythrocytes. For that reason, the degree of total parasitemia is limited. In contrast, *P. falciparum* infects erythrocytes of all ages, leading to high parasitemia. Although the symptoms of *P. vivax, P. ovale* and *P. malariae* infections can be severe in nonimmune persons, these parasites seldom cause fatal disease and malaria caused by these three parasites is often called benign malaria. In contrast, *P. falciparum* malaria (also known as malaria tropica) can progress within a few days from mild to severe disease with a lethal outcome in 10–40% of all cases depending on the time lag between the onset of the symptoms and effective treatment, as well as on the hospital facilities for the management of complications (Wilairatana et al., 2002). Observed complications include coma (cerebral malaria), respiratory distress, renal failure, hypoglycemia, circulatory collapse, acidosis and coagulation failure (Pasvol, 2006).

Traditionally, antimalarial agents are classified by the stages of the malaria life cycle that are targeted by the drug: blood schizonticides acting on the asexual intraerythrocytic stages of the parasites and tissue schizonticides kill hepatic schizonts, and thus prevent the invasion of erythrocytes, acting in a causally prophylactic manner. Hypnozoiticides kill persistent intrahepatic stages of *P. vivax* and *P. ovale*, preventing relapses from these dormant stages. Gametocytocides destroy intraerythrocytic sexual forms of the parasites and prevent transmission from human to mosquito (Esperanc et al., 2010). As there are no dormant liver stages in *P. falciparum* malaria (malaria tropica), blood schizonticidal drugs are sufficient to cure the infection. In cases of *P. vivax* and *P. ovale*, a combination of blood schizonticides and tissue schizonticides is required (Kumar et al., 2003).

8.3 CHEMOPROPHYLAXIS

The chemotherapy of malaria basically involves killing of the asexual parasites and providing supportive therapy to the host to boost its immune system. Prior to the 2nd World War, quinine, pamaquine, chloroquine and mepacrine were developed. These were followed by proguanil and amodiaquine (1940 s), primaquine and pyrimethamine (1950 s), sulfadoxine (1960 s), and artemisinin (1970 s). A number of drugs were introduced in the 1980 s, which include mefloquine, halofantrine, aablaquine, pyronaridine, piperaquine and the artemisinin derivatives, such as artemether, aretsunate and dihydroartemisinin Drug chemoprophylaxis has proved to be an effective preventive strategy in malaria endemic areas (Paredes et al., 2006; Trouiller et al., 2002), both for *P. falciparum* and *nonfalciparum* malaria, despite it does not usually prevent the later relapses that can occur with *P. vivax* and *P. ovale*. Drugs can act on different stages of the Plasmodium biological cycle, on the preerythrocytic liver forms (causal prophylaxis) and on the erythrocytic blood forms (suppressive prophylaxis).

8.4   4-AMINOQUINOLINES

Cinchona bark contains quinoline alkaloids quinine (**1**) and quinidine (**2**), they are chemically known as (1R)-(6-methoxyquinolin-4-yl)[(2S,4S,8R)-8-vinylquinuclidin-2-yl] methanol and (SR)-(6-methoxyquinolin-4-yl) [(2S,4S,8R)-8-vinylquinuclidin-2-yl] methanol respectively, quinine possess plasmodicidal activity and was considered as the first medicine to be used against malaria. In 1856, chemist William Henry Perkins set out to synthesize quinine. Later on, Chloroquine (**3**) was chemically synthesized in 1934, as a substitute for quinine. It acts by selectively deposited in the food vacuole of the parasite, exerting its antimalarial effect by preventing the polymerization of the toxic heme (Foley and Tilley, 1997; Ginsburg et al., 1998; Hoppe et al., 2004; Yayon et al., 1984) and its use became prominent in the early 1950 s, when the World Health Organization (WHO) declared a war on malaria. It has been the milestone of antimalarial drugs up to the 80 s, because of its efficacy, tolerability and safety in pregnancy and childhood. Chloroquine (CQ) is a rapid blood schizonticide recommended in combination with proguanil in areas dominated by *P. vivax* or in areas where *P. falciparum* is still sensitive to chloroquine such as Central America and Middle Eastern countries (Khatoon et al., 2009; Lee et al., 2009). During its prime, CQ was considered a wonder drug, cured billions of clinical episodes of malaria and saved millions of lives all around the world. But soon, CQ-resistant (CQR) parasites started appearing and now the drug is virtually ineffective in most parts of the world. The parasites with greater resistance are generally those harboring Lys76Thr and Ala220Ser mutations in *P. falciparum* chloroquine resistance transporter *(pfcrt)* (Djimde et al., 2001; Fidock et al., 2000) and Asn86tyr mutations in *P. falciparum* multidrug resistant *(pfmdr)* (Cooper et al., 2005; Djimde et al., 2001; Fidock et al., 2000; Valderramos and Fidock, 2006). Pfmdr1 duplication suggests a possible cross-resistance with mefloquine, which needs further studies to be confirmed (Graffing et al., 2010).

(1)   (2)   (3)

One of the strategies to overcome CQ resistance is the attachment of two 4-aminoquinoline moieties by linkers of various length and chemical nature. The activity of such bisquinolines against CQ resistant strains has been explained by their steric bulk, which prevents them from fitting into the substrate binding site of *pfcrt*. The most advanced representative of the bisquinolines, piperaquine (4) was used in China (Davis et al., 2005) but still the resistance has developed in the areas where piperaquine has been used (IC_{50}=240–320 nM).

(4)

The structure activity relationship studies on the 4-amino-quinoline antimalarials suggested that the 7-*chloro* group is essential for the antiplasmodial activity since replacement of 7-*chloro* group either by an electron-donor group like -NH_2, -OCH_3 or by an electron-withdrawing group like -NO_2

resulted in decreased antimalarial activity. The role of carbon chain length in the amino alkyl side chain has also been investigated and the results suggested that both shortening and lengthening of the carbon side chain in CQ led to compounds that retained the antimalarial activity. The replacement of the diethylamino function with a metabolically stable tert-butyl group or heterocyclic ring (piperidyl, pyrrolidinyl, and morpholinyl) in the short chain analogs led to a substantial increase in the antimalarial activity (Egan et al., 2000; Kaschula et al., 2002).

Despite emergence of resistance to CQ and other 4-aminoquinoline drugs in most of the endemic regions, research findings provide considerable support that there is still significant potential to discover new affordable, safe, and efficacious 4-aminoquinoline antimalarials. New side chain modified 4-aminoquinoline derivatives and quinoline–acridine hybrids were synthesized and evaluated in vitro against NF 54 strain of *Plasmodium falciparum*. Out of 25 compounds, compound **(5)** (MIC = 0.125 µg/mL) was equipotent to standard drug CQ (MIC = 0.125 µg/mL) and compound **(6)** (MIC = 0.031 µg/mL) was four times more potent than CQ (Kumar et al., 2010).

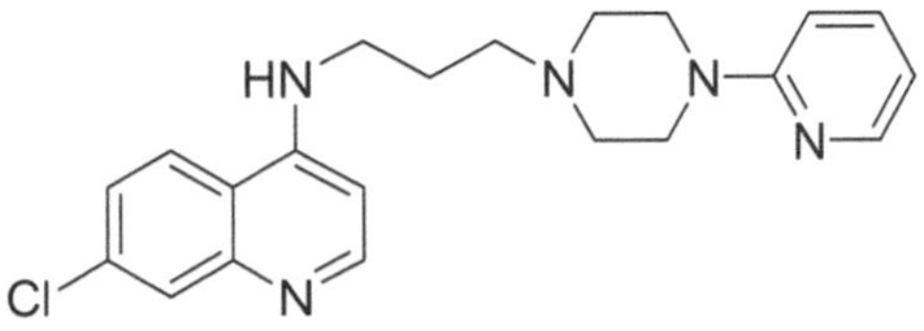

(5)

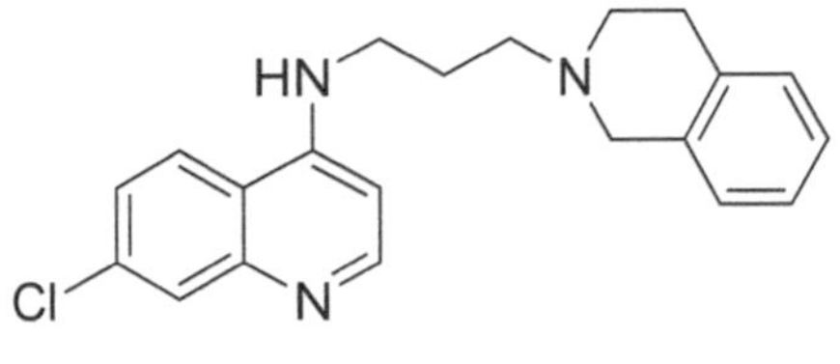

(6)

A series of novel hybrid 4-aminoquinoline 1,3,5-triazine derivatives was synthesized by Bhat et al. and evaluated its *vitro* antimalarial activity against chloroquine-sensitive (3D7) and chloroquine-resistant (RKL-2) strains of Plasmodium falciparum. Among the synthesized derivatives, compound **(7)** was found to be much active with % dead asexual parasites, 25.7 µg/ml and 22.7 µg/ml respectively (Bhat et al., 2015).

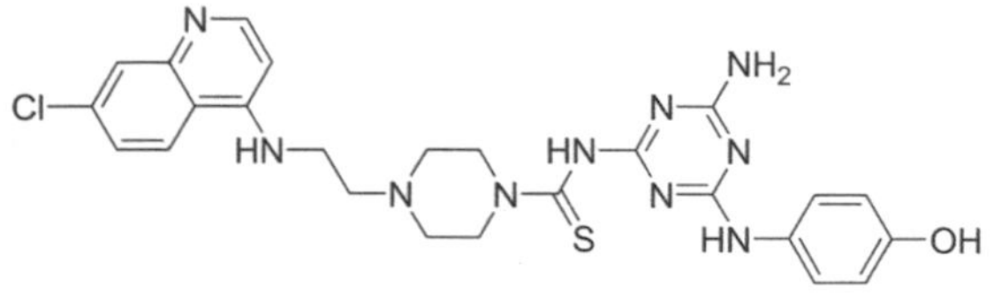

(7)

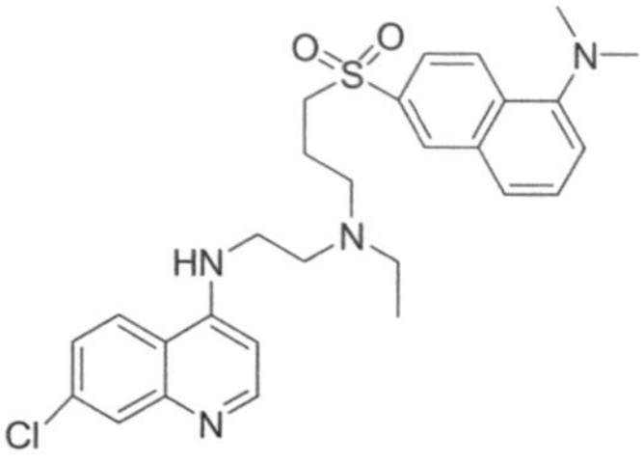

(8)

Wolf et al. (Ekoue Kovi et al., 2009) reported many of the analogs of 4-amino-7-chloro quinolyl derived sulfonamides, ureas, thioureas and amides showed promising antimalarial activity and low resistance indices. Systematic variation of the side chain length and introduction of fluorinated aliphatic and aromatic terminus revealed promising leads that overcame CQ resistance. In particular, sulfonamide (8) containing a short side chain with a terminal dansyl moiety produced high antiplasmodial potency with IC_{50} values of 17.5 and 22.7 nM against HB3 and Dd2 strains, respectively.

The 15 series of 4-aminoquinoline derivatives have been synthesized by Katti et al. (Solomon et al., 2010) and found to be active against both susceptible and resistant strains of *Plasmodium falciparum* in vitro. Compound 1-[3-(7-chloro-quinolin-4-ylamino)-propyl]-3-cyclopropyl-thiourea (9) exhibited superior activity against resistant strains with IC_{50} = 23.9 nM (DQ2) and 14.1 nM (DQ6) as compared to chloroquine (Solomon et al., 2010).

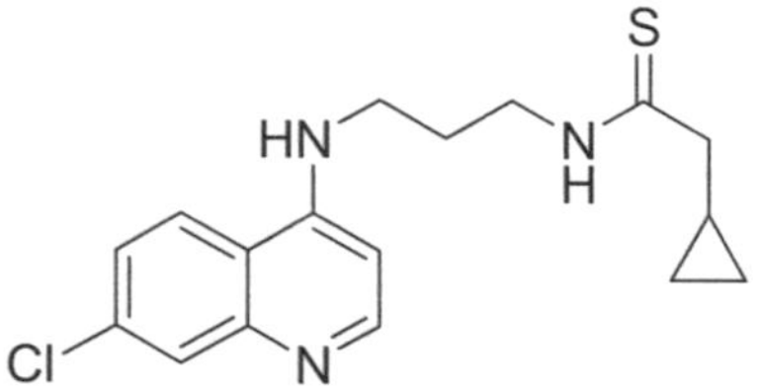

(9)

Amodiaquine (**10**) and mefloquine (**11**) share a common structural feature of basic nitrogen within the hydrogen-bonding proximity to a hydroxyl group. At the physiological pH, the basic nitrogen will be protonated; hence it is possible that the intramolecular hydrogen bonding between the protonated amino and the hydroxyl group may play an important role in increasing activity against CQ resistance *P. falciparum* (Madrid et al., 2005). Upon the basis of these assumptions, analogs containing four different aryl linkers and various aromatic substitutions were synthesized with hydrogen bond accepting capability. Several compounds showed IC_{50} value of less than 5 nM against W2 strain. Analogues containing a *p*-amino cresol motif (**12**) were more potent than those having furan or thiophene ring at the distal basic center (Madrid et al., 2006).

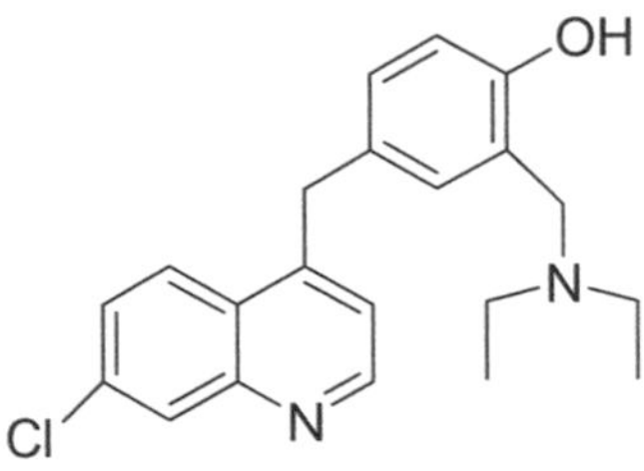

(10) 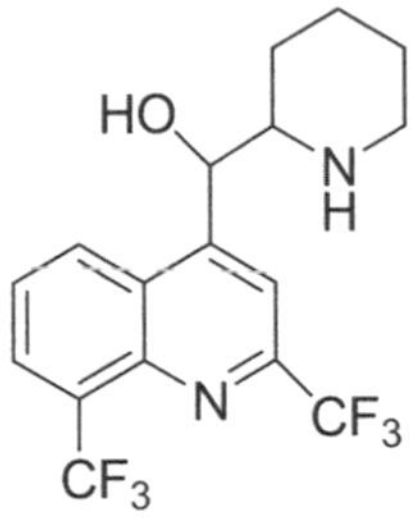   **(11)**

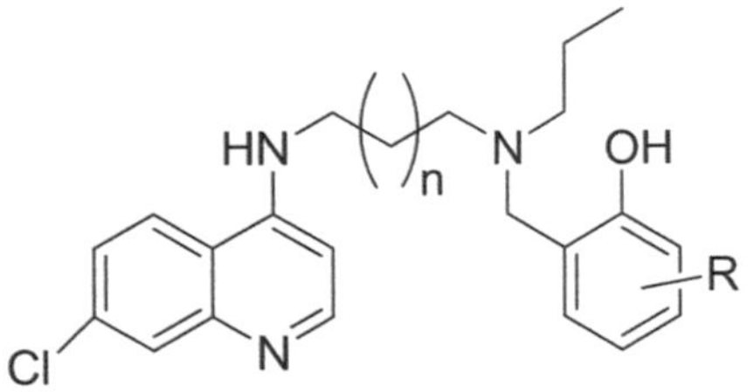

(12)

The mode of action of amodiaquine (**10**) emphasizes on binding to ferriprotoporphyrin IX which is released by degradation of hemoglobin by malarial parasites that is further detoxified by crystallization into hemozoin (Sullivan, 2002).

In order to develop new classes of antimalarial agents, the possibility of replacing the phenolic ring of amodiaquine, tebuquine, and isoquine with other kinds of aromatic nuclei was investigated (Kesten et al., 1987; O'Neill and Posner, 2004; Raynes et al., 1999). Replacement of phenolic ring of amidoquine, tebuquine, and isoquine with a pyrrolic ring associated with a good antimalarial activity. The 66 rare compounds were synthesized and tested against both D10 (CQ-S) and W-2 (CQ-R) strains of *Plasmodium falciparum*. Compound (13) has $IC_{50} = 8.5$ nM (CQ-S) and 13.4 nM (CQ-R) and Compound **14** has $IC_{50} = 23.6$ nM (CQ-S) and 28.8 nM (CQ-R) (Casagrande et al., 2008).

(13)   **(14)**

In order to overcome CQ resistance, later on, compounds were synthesized by O'Neill et al. (Hatton et al., 1986; Neftel et al., 1986; Tingle et al., 1995) having no quinoneimine metabolite which causes drug induced hepatotoxicity and aggranulocytosis by irreversible binding to cellular macromolecules. In series of compounds, GSK369796 **(15)** was much active with better preclinical safety profile and $IC_{50} = 4.8$–6.1 mg/kg (ED_{90}) against chloroquine sensitive *P. falciparum* 3D7 and HB3 strains (O'Neill et al., 2009).

(15)   **(16)**

Further, a series of 4'-fluoro and 4'-chloro analogs of amodiaquine were synthesized by O'Neill et al. (O'Neill et al., 2009) as suitable 'back up' compound for GSK369796. Compound (16) was found to have $IC_{50} = 12.1 \pm 5.3$ nM against chloroquine sensitive *P. falciparum* 3D7 strain and acceptable pharmacokinetic profile.

In order to determine the real significance of the phenolic group in the antimalarial activity and/or cytotoxicity of amodiaquine (AQ) analogs for which this functionality was shifted or modified. The Promising in vitro antimalarial activity was obtained for compounds unable to form intramolecular hydrogen bond. Among the 16 compounds synthesized, a new amino derivative (17) displayed the greatest selectivity index towards the most CQ-resistant strain tested and was active in mice infected by *Plasmodium berghei* (Cochin et al., 2008).

(17)   **(18)**

Based upon the reports that the urea derivatives interact with the dihydrofolate reductase (DHFR) through hydrogen bonding that accumulate in the parasitic food vacuole and inhibit the plasmepsin of the parasite. The antimalarial efficacy and docking studies of 2-methyl-6-ureido-4-quinolinamides demonstrated by Batra et al. (Leban et al., 2004; Madapa et al., 2009; Rastelli et al., 2003) on the *Pf* DHFR and found to be active.

Campiani et al. (Gemma et al., 2009) developed an innovative class of antimalarials based on a polyaromatic pharmacophores which may selectively interact with heme and interfere with the *P. falciparum* heme metabolism that were synthesized by hybridizing the 4-aminoquinoline moiety with the clotrimazole-like pharmacophore characterized by a polyaryl methyl group. The combination of the polyaryl methyl system was able to form and stabilize radical intermediates with the iron-complexing and conjugation-mediated electron transfer properties of the 4-aminoquinoline system led to potent antimalarials. Out of synthesized compounds, compound (18) was most active with IC_{50} value of 22 nM against W2 strain.

Development of resistance to the chloroquine and related drugs is a serious worldwide problem which has been significantly reduced treatment options. One interesting strategy to counteract the chloroquine resistance involves potentiating its effects using compounds with weak antimalarial activity. Search for such agents has led to biological evaluation of various calcium channel blockers (Martin et al., 1987), antidepressants (Bitonti et al., 1988) antihistamines (Peters et al., 1989) and prostaglandin oligomers (Chandra et al., 1993) that have shown to reverse chloroquine resistance in *P. falciparum*, but they could not pursued further since they were effective only in vitro. Moreover, the effective dose of these compounds as resistance reversal agents (or chemosensitizers) is generally close to or higher than therapeutic dose for other clinical applications.

(19)

Bitonti et al. (Gerena et al., 1992) have found that the 9-g-methylaminopropyl-9, 10-dihydro-9, 10-ethanoanthracene (maprotiline) (19) has shown antiMDR activity on *P. falciparum*. This study was further explored to synthesize a series of 9, 10-dihydro-9, 10-ethano or ethenoanthracene (DEEA) derivatives. Compound (3) was more potent than verapamil and promethazine and also less cytotoxic than the reference chemosensitizers. The resistance reversal activity was under the influence of amino group, the nature of the group associated with the amine, the length of the chain carrying the functions and the presence of an ethano bridge versus that of an ethano bridge. It was found that secondary amines had better activity than the primary and quaternary amines. The decrease in potency of some of these derivatives could be explained by their weak lipophilicity. The charged entity of the quaternary ammonium group which could be detrimental to cross membranes lowers the activity. Indeed, these functional groups can be ranked in a decreasing order of activity as: amino > hydroxyl > carbamate

> carboethoxy > carboxylic acid. The double bond in the ethenoanthracene derivatives versus the ethenoanthracene derivatives decreased the potentiating activity (Alibert et al., 2002).

8.5   8-AMINOQUINOLINES

In 1891, Ehrlich and Guttmann cured two malaria patients with methylene blue which became the first synthetic drug ever used in therapy. In 1920, chemists at Bayer in Germany started to modify the structure of methylene blue yielding the first synthetic antimalarial drug pamaquine (**20**) in 1925 and was capable of preventing the relapses in *P. vivax* malaria. However, under clinical trial this drug displayed multiple side effects that limit its use and therefore in 1952, primaquine (**21**) was introduced. The terminal diethylamino moiety of pamaquine was replaced by an unsubstituted primary amine in primaquine (Nodiff et al., 1991).

(20)   **(21)**

The mechanism of action of the 8-aminoquinolines is unclear and there is no firm understanding of the mechanism of primaquine. The antagonistic effect of a quinoneimine metabolite has been suggested, leading to the inhibition of electron transport in the respiratory chain. Furthermore, these metabolites should undergo a redox cycle causing oxidative stress, thereby depleting the glutathione storage in individuals deficient in glucose-6-phosphate dehydrogenase (G6PD). Interestingly, primaquine also exhibits some activity in reversing chloroquine resistance (Bray et al., 2005). Primaquine distinguishes itself from other antimalarials as it shows activity against the liver and the sexual blood stages of different plasmodia while its activity against asexual blood stages is too low to be therapeutically significant. Primaquine is still the only antimalarial drug licensed for the radical cure of *P. vivax* infections. Owing to its short half-life of 4–6 h, primaquine requires daily administration for 14 days to achieve a cure. The most serious side effect of this drug is a potentially life threatening hemolysis in persons deficient in G6PD, a genetic polymorphism particularly abundant in Africa and Asia (Chen and Keystone, 2005) (Table 8.1).

Tefanoquine (**22**), the most lipophilic derivative of primaquine, has trifluoromethylpheoxy substituent in its structure which leads to four fold increase in activity as compared to primaquine (Vennerstrom et al., 1995).

(**22**)

Jain et al. (Vangapandu et al., 2003) synthesized 40 ring substituted 8-aminoquinoline compound. Out of these compounds (**23**) and (**24**) was found to exhibit in vitro and in vivo biological efficacy higher than CQ against both CQ sensitive and CQ resistance strains (IC$_{50}$ of 9.4 and 9.7 ng/ml, respectively).

(**23**)   (**24**)

A series of naphthyridine analogs of primaquine were synthesized by Zhu et al. (Zhu et al., 2007). The (1,5) naphthyridine ring has higher oxidation potential due to the presence of a second nitrogen atom in an aromatic system which may prevent hydroxylation at C-5 position, responsible for chemically reactive and potentially toxic metabolites. A second substitution was introduced at the aliphatic side chain so that oxidative deamination

TABLE 8.1 Antimalarial Site of Action, Drug Resistance, Resistance Markers and Side Effects (Johanna, 2006)

Drug	Site of Action	Drug Resistance	Parasite Gene	t$_{1/2}$ (in days)	Side Effect
Chloroquine phosphate	Hgb metabolism/ food vacuole	Common worldwide	pfcrt (pfmdr1 and others as modifiers)	10	Pruritis
Quinine	Hgb metabolism/ food vacuole	Southeast Asia	Pfmdr1	0.6	Cinchonism
Mefloquine	Hgb metabolism/ food vacuole	Southeast Asia	Pfmdr1	21	Neuropsychiatric
Primaquine phosphate	Unknown	None	–	0.4	Contraindicated G6PD
Sulfadoxine-pyrimethamine	Antifolate	Common:Africa	dhps, dhfr	8/4	Well tolerated
Atovaquone-proguanil	Mitochondria	Rare	Cytochrome b	2.5/0.6	Well tolerated
Artemisinins	SERCA orthologue	Rare	Pfmdr1, pfatp6	0.15	Well tolerated

would be disfavored. The synthesized compounds were less toxic than primaquine among which analog **(25)** was most active ($IC_{50} = 0.045$ μM).

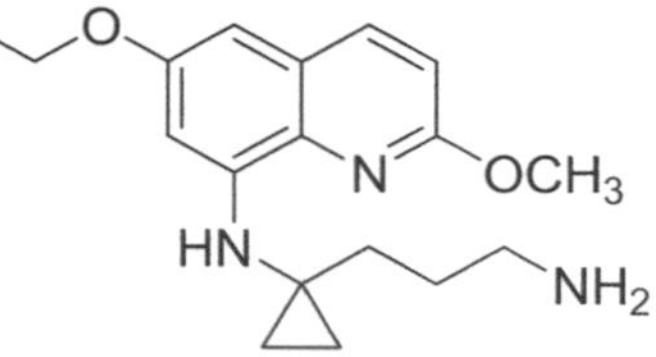

(25)

8.6   ARTEMISININ AND DERIVATIVES

Artemisinin **(26)** was first isolated by Chinese researchers and launched in 1972, to discover new antimalarial drugs with $IC_{50} = 12.1$ nM against *P. falciparum* species (Cummings et al., 1997; Eckstein-Ludwig et al., 2003; Haynes and Krishna, 2004; Jung, 1997; Posner and O'Neill, 2004; Wu et al., 2001).

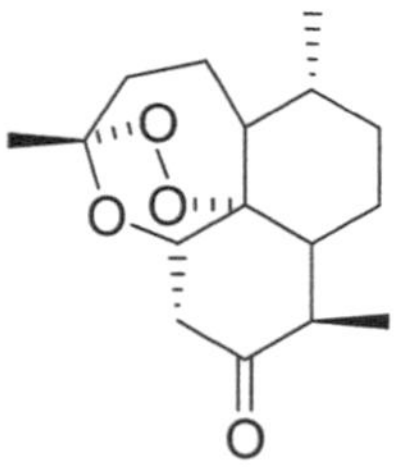

(26)

A key structural feature of all artemisinins is the 1,2,4-trioxane substructure or the endoperoxide ring which is mandatory for antimalarial activity. Their exact mechanism of action is still unresolved and remains a matter of debate. It has been proposed that iron (II)-mediated cleavage of the endoperoxide leads to the formation of different C-centered radicals which react with different protein targets thus preventing heme detoxification and inhibiting a multitude of enzymes. Artemisinins act against all developmental parasite stages including those which do not produce hemozoin. Recently, Krishna and co-workers put forward another theory: endoperoxide cleavage should take place in the cytoplasm catalyzed by a cytoplasmic iron (II) source. The resulting reactive species then very specifically inhibits an ATP-dependent Ca²⁺ pump located on the endoplasmic reticulum. The pump called *Pfatp6* is a homologue of a mammalian sarcoplasmic/endo-plasmic reticulum Ca²⁺ ATPase (SERCA) (Biot and Chibale, 2006; Krishna et al., 2004; Ramharter et al., 2002). As it was possible to create a *Pfatp6* mutant that is resistant to artemisinins by the exchange of just one amino acid (L263E). Indeed, *Pfatp6* mutations were found in isolates displaying significantly decreased susceptibility toward artemether. Furthermore, stable resistance to artemisinins could be introduced in rodent parasites lacking mutations in *atp6* and *mdr1* genes (Uhlemann et al., 2005) as shown in Table 8.1.

As artemisinin is poorly soluble in water and oil, semisynthetic derivatives have been developed. The reduction of the lactone substructure of artemisinin leads to the hemiacetal-containing compound dihydroartemisinin (**27**) and alkylation of hemiacetal yields artemether (**28**) and arteether.

(**27**)   (**28**)

Dihydroartemisinin itself undergoes rapid hydroxylation at positions 5, 7 and 14 and glucuronidation at the hemiacetal OH group to yield highly water-soluble metabolites, resulting in an elimination half life of 40–60 mins. Because of this high clearance rate, artemisinins have to be administered over a period of 5–7 days which leads to poor compliance and ultimately to recrudescence (Haynes, 2001). The modification of dihydroartemisinin into artemether increases lipophilicity which is better absorbed from the GI tract making it suitable for oral administration. Blood levels after i.m. application of an oily solution have shown to be unpredictable and sometimes undetectable. Currently, the application of artemether with lumefantrine is the only artemisinin-based combination therapy available manufactured under GMP standard (Woodrow et al., 2005). Another modification of dihydroartemisinin is artesunate in which the hemiacetal OH group is acylated with succinic acid. Artesunate is an unstable drug; the succinic ester is rapidly cleaved releasing dihydroartemisinin as the active agent. Because of the free carboxylate, artesunate is a water soluble drug that can be administered via the i.v. route. This is of particular importance for the treatment of severe malaria tropica in which the condition of the patients prohibits any other route of administration.

The problem of the first generation semisynthetic artemisinins is their rapid biotransformation that results in a short half life and the formation of the neurotoxic dihydroartemisinin. Much work has been done in the development of second generation artemisinins. Methyl and ethyl residues of the first generation have been replaced by numerous other residues to enhance water solubility.

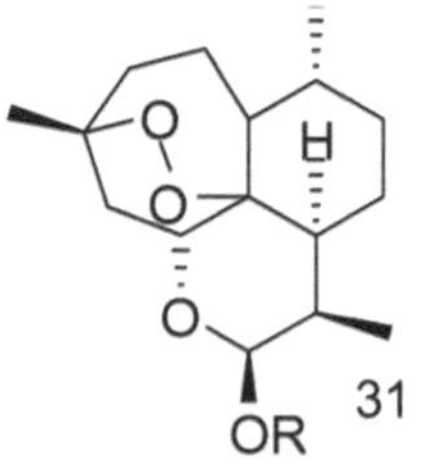

R = -CH$_2$CH$_2$NHCH$_3$ IC$_{50}$ = 2.3nM
R = -C$_6$H$_5$-p-CF$_3$ IC$_{50}$ = 5.4nM

(29)

Most of the variations have been carried out at position 10, where the exocyclic oxygen atom is replaced by carbon substituents (alkyl, aryl and heteroaryl residues) to remove the metabolically sensitive acetal substructure. Some substituents have been used for the formation of dimers.

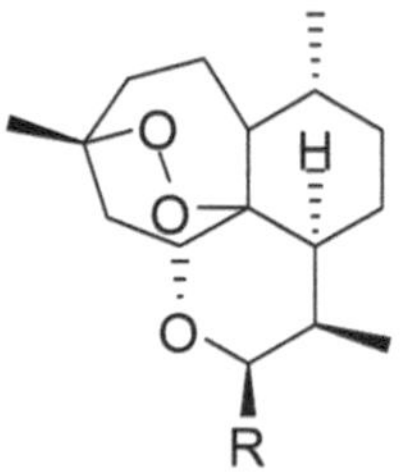

R= -CH$_3$ IC$_{50}$ = 1.0nM
R= -CH$_2$OH   IC$_{50}$ = 1.9nM

(30)

As it became clear that the antimalarial activity of artemisinin derivatives is based on the endoperoxide substructure, several groups having synthetic endoperoxides as antimalarials had been developed, resulting in numerous candidates of varying complexity.

Bachi et al. (Yeates et al., 2008) synthesized 60 compounds of bridged β-sulfonyl-endoperoxides and among them compounds (31) and (32) show IC_{50} value lower than 25 nM against *P. falciparum* NF54 strain and have same potency as that of artemether and arteether.

31 Y= PhSO₂; Z = H; R₁ = OBn
32 Y=H; Z = PhSO₂; R₁ =OBn

(**31**) and (**32**)

In last two decades cyclic peroxides have attracted the attention of chemists and biologists due to their potent activity against malaria. Nowadays, cyclic compounds such as tetroxanes, ozonides and trioxanes are considered as the most promising synthetic peroxides in the antimalarial field. 1,2,4,5-Tetroxanes were found to have a high activity as well as a high stability (Borstnik et al., 2002; Dong, 2002; Dong et al., 1999; Dussault and Davies, 1996; Ellis, 2008; Gelb, 2007; Hamada et al., 2002; Iskra et al., 2003; Jefford, 1997; Jin et al., 2005; Jin et al., 2006; Kim et al., 1999; McCullough et al., 2000; Najjar et al., 2005; Solaja et al., 2002; Vennerstrom et al., 1992; Zmitek et al., 2006). Some of these compounds exhibit impressive antimalarial activity comparable to or higher than that of the widely used natural peroxide artemisinin consisting of a relatively complex tetracyclic endoperoxide skeleton (Amewu et al., 2006; Dong et al., 2006; O'Neill et al., 2010; Singh et al., 2004).

Thirty synthesized tetroxane analogs have been examined for their antimalarial activities against *P. falciparum* K1 by Liebscher et al. (Hamann et al., 2011). Tetroxanes exhibit significant antimalarial activity in vitro with the best IC_{50} = 0.277 μg/mL for compound (33). In contrast, none of the other tetroxanes showed notable antimalarial activity. These results indicate that relatively minor changes in the structure of the tetroxanes have a profound effect to the antimalarial activities.

(33)

Venner-strom et al. worked on 1,2,4-trioxolans, resulted in the first antimalarial endoperoxide OZ-277 (RBx11160) (34) which has recently entered into clinical trials. The 1,2,4-trioxolane system is well known to organic chemists as secondary ozonide, a highly reactive intermediate of the ozonolysis reaction. The key issue in the development was to balance stability against reactivity through the selection of appropriate residues on both sides of the trioxolane system, whereas two cyclohexane rings clearly did not provide enough protection for the sensitive heterocycle, resulting in rapid compound breakdown, two adamantane rings sterically shielded the trioxolane too much resulting in a stable compound albeit one with insufficient activity against Plasmodia. However, by decorating the trioxolane ring with an adamantane residue on one side and a cyclohexane group on the other, the critical balance between stability and reactivity could be obtained. Finally, the addition of an aminoacyl residue provided the correct polarity and solubility resulting in the desired pharmacological properties. OZ-277 displayed high activity against field isolates from Gabon (median IC_{50} = 0.47 nm; range: 0.13–2.23 nm). Its stage specificity is similar to that of artemisinins and activity against *P. vivax* is in the same range as the activity against *P. falciparum*. If the outcome of phase II studies confirm the promising preclinical results, OZ-277 could become an easily accessible alternative to artesunate owing to its short and straightforward synthesis. Development of a pediatric formulation and an intravenous formulation is in progress. The lead for second generation ozonide has an extended half-life and a higher oral bioavailability relative to the parent OZ-277. It displayed a cure rate of 100% at a single dose of 30 mg/kg body weight in a murine model and was as effective as mefloquine, a prophylactic agent (Bathurst and Hentschel, 2006; Dong et al., 2005; Vennerstrom et al., 2004).

OZ-277

(34)

8.7   BIGUANIDES AND PYRIMIDINES

In most species, tetrahydrofolic acid plays a key role in the bio-synthesis of thymine, purine nucleotides and several amino acids (Met, Gly, Ser, Glu, and His) whereas humans depend on dietary intake of preformed dihydro-folic acid as an essential nutrient which is then reduced to tetrahydrofolic acid by the action of enzyme dihydrofolate reductase. Furthermore, *P. falciparum* is able to use exogenous dihydrofolic acid *via* salvage pathway. Inhibitors of two key enzymes of the dihydropteroate synthase (DHPS) and dihydrofolate reductase (DHFR) in folate biosynthetic pathway have long been used in the treatment of protozoal infections. In *P. falciparum*, both enzymes are present not as monofunctional proteins but *DHPS* and *DHFR* activities are present on specific domains of bifunctional proteins. In the case of *DHPS*, the preceding enzymatic activity of hydroxymethyl dihydropterin pyrophosphokinase is located on the same polypeptide. *DHFR*, in turn, is collocated with the subsequent thymidylate synthase activity on a single protein. The use of antifolates against malaria and the possibility of using other enzymes along the folate biosynthetic pathway as drug targets has already been reviewed (Hyde, 2005; Kasekarn et al., 2004). The first antifolate to be used against malaria was the well-known *DHPS* inhibitor sulfachrysoidine (35) and developed in 1932 by Domack as an antibacterial agent. Sulfanilamide (36), an active agent, had arisen from the reductive cleavage of the azo substructure. In 1937, sulfachrys-oidine was successfully used in a trial against malaria but then interest in sulfonamides diminished because of the continuing effectiveness of quinine and the development of other synthetic antimalarials. When sul-fonamides such as sulfadoxine with longer half lives (**Table 8.1**) and improved toxicological profiles were developed in the late 1950 s, sulfon-amides again gained interest especially as combination partners for the *DHFR* inhibitors proguanil (37) and pyrimethamine (38). Proguanil is a prodrug that yields the active metabolite cycloguanil (39) through oxida-tive ring closure. Both *DHFR* inhibitors are structurally closely related; the main difference is the tetrahedral geometry at C-6 of cycloguanil, which removes the heterocyclic planarity of pyrimethamine. Both drugs are highly active inhibitors of *P. falciparum* dihydrofolate reductase (*Pf* DHFR) with K_i values of 1.5 nm and 2.6 nm, respectively (Nzila et al., 2005).

(35)   (36)

(37)   (38)

(39)

A series of 2, 4, 6-trisubstituted pyrimidines was synthesized by Chauhan et al. (Agarwal et al., 2005; Agarwal et al., 2005) and evaluated for their in vitro antimalarial activity against *Plasmodium falciparum*. Out of the 30 compounds synthesized (40) and (41) compounds showed MIC in the range of 0.5 μg/mL. These compounds were several folds more active in vitro than pyrimethamine.

(40)   (41)

8.8   ATOVAQUONE

Baziard-Mouyzett et al. synthesized 18 derivatives of the antimicrobial ato-vaquone which were substituted at the 3-hydroxy group by ester and ether functions. These compounds were evaluated in vitro for their activity against the growth of *P. falciparum*. All the compounds showed potent activity; among them (42) and (43) compounds with IC_{50} = 1.25 nM comparable to those of atovaquone and much higher than chloroquine or quinine (Hage et al., 2009).

(42) R= -C_6H_5
(43) R= -$(CH_2)_4$-CH_3

8.9 4-PYRIDONES

Yeates et al. found introduction of a lipophilic side chain at C-5 on clopidol (44) led to a significant improvement in activity both in vitro against *P. falciparum* and orally in vivo against murine *P. yoelii*, yield a number of derivatives with antimalarial activity in these assays comparable to current drugs. Additional studies on atovaquone resistant strain revealed that GW308678 (45) and GW844520 (46) were potential antimalarials with IC_{50} = 0.003 µm and 0.007 µm (Woodrow et al., 2005)]. They act by inhibiting the plasmodial mitochondrial electron transport chain targeting Cytochrome bc1 (complex III).

GW308678(45) R = 3-CF_3
GW844520(46) R = 4-OCF_3

(44)

The introduction of tertiary amines, esters or polar hydroxyl groups at position C-3 had deleterious effects on the antimalarial activity. With the aim to design novel 4-pyridone derivatives with improved solubility and pharmacokinetic profile, in series of compounds (47) was found to have IC_{50} = 0.002 µm. The increase in solubility of (47) in comparison with its nonhydroxylated analog GW844520 (45) resulted into enhanced oral bioavailability for solid dosage form (suspension in 1% methyl cellulose) both in mice (50% vs 20%) and particularly in dogs (16% vs. 4.4%). The Compound (47) seems to be more sensitive to metabolic clearance as demonstrated by the significant reduction in the half-lives in both species possibly due to the oxidation of the hydroxyl group to carboxylic acid or by direct elimination *via* conjugation. In spite of its shorter half-life, compound (45) had displayed excellent in vivo antimalarial efficacy in our murine model of *P. falciparum* malaria 27 (ED$_{50}$ = 0.6 mg/kg). The -CH$_2$OH group induced a fall in antimalarial activity in vitro when located at position C-2 in comparison with its nonhydroxylated counterpart. It is, therefore, possible to obtain potent derivatives with improved solubility and pharmacokinetic properties by introducing the -CH$_2$OH group at position C-6 of the 4-pyridone ring (Bueno et al., 2011).

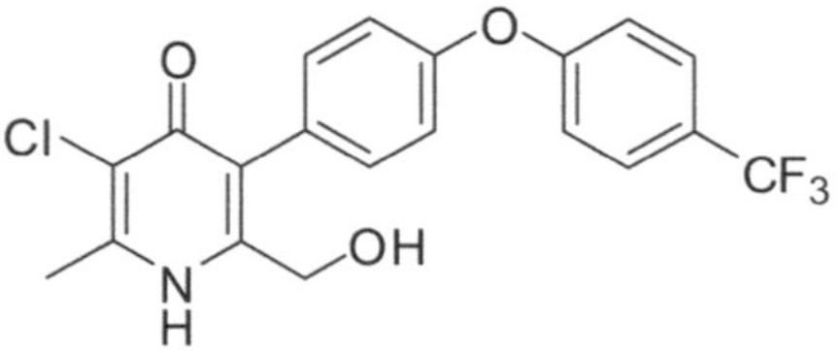

(47)

8.10   1,3,5-TRIAZINES

Gravestock et al. (Gravestock et al., 2011) synthesized 28 compounds of 2,N^6-disubstituted 1,2-dihydro-1,3,5-triazine-4,6-diamines possessing a flexible tether of varying length between the exocyclic nitrogen atom bonded to C-6 of the 1,2-dihydro-1,3,5-triazine-4,6-diamine heterocycle and the distal substituted phenyl ring. They were tested for their antimalarial activity against the cycloguanil resistant FCR-3 *P. falciparum* strain. The 2-furanyl derivative (48) was approximately 5-fold more active (IC$_{50}$ = 0.99 mM) than cycloguanil and the most active compound in series of compounds.

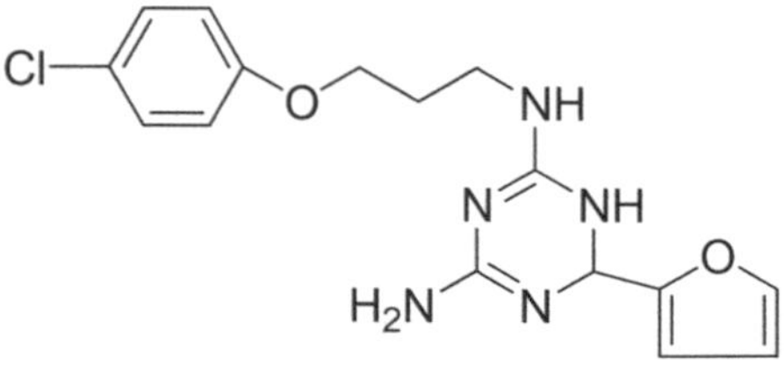

(48)

8.11 1,2,3-TRIAZOLES

Various 1,2,3-triazole derivates are reported in the literature to exhibit several biological activities such as *analgesic*, antibacterial, fungicidal, antiinflammatory, antihypertensive, antiviral and antitumor (Sztanke et al., 2008). 1,2,3-triazole moieties are attractive connecting units since they are stable to metabolic degradation and capable of hydrogen bonding which can be favorable in binding of biomolecular targets as well as for solubility (Horne et al., 2004). The 14 compounds were synthesized and evaluated against *Plasmodium berghei* infected mice model as suggested by the World Health Organization. Compounds **(49)** and **(50)** were more effective than chloroquine at 10 mg/kg on day 9, showing suppression of 82% and 72% (p < 0.05), respectively (Corrales et al., 2011).

(49)   **(50)**

8.12 CHALCONES

Chalcones are aromatic ketones and key biosynthetic intermediates for combinatorial assembly of different heterocyclic scaffolds. They form an important group of natural compounds which are easy to synthesize, and some of them show wide range of biological activities. Licochalcone A, an oxygenated chalcone, isolated from the roots of Chinese licorice was reported to have an antiplasmodial activity (Chen et al., 1994). This encouraged researchers to design and synthesize a variety of chalcone derivative and to evaluate their antimalarial potential (Li et al., 1995; Liu et al., 2003). Firstly the antiplasmodial activity of chalcones with azoles on acetophenone ring was reported. Three of these azole derivatives inhibited the parasite multiplication rate to 50% (IC_{50}) at concentrations of less than 3 mg/ml indicating the potential to be developed as an antimalarial (Mishra et al., 2008). Chalcone derivatives **(51)** and **(52)** each in combination with artemisinin were synthesized. The isobolograms showed that in vitro interaction of the chalcone derivatives with artemisinin was not antagonistic and compound **(51)** in combination with artemisinin showed synergistic antiplasmodial interaction in 3:4 fixed-ratio combinations. Similarly the combination of (52) and artemisinin demonstrated synergistic interaction in two combinations and additive in rest of the two combinations (Bhattacharya et al., 2009).

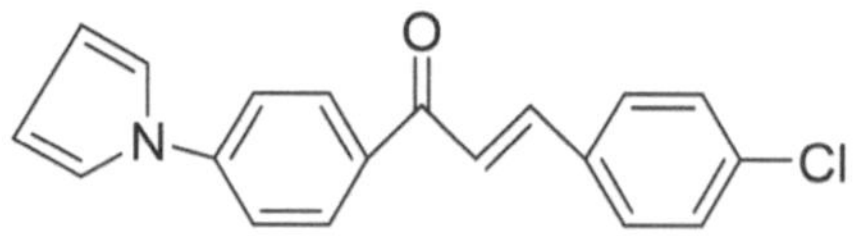

(51)

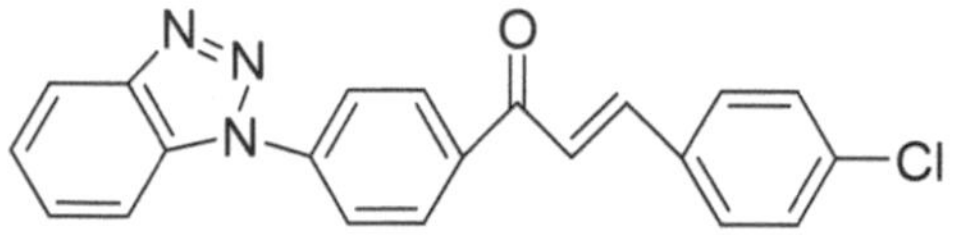

(52)

Bhasin et al. synthesized novel 1, 3-diaryl propenone derivatives were tested for antimalarial activity in vitro against asexual blood stages of human malaria parasite *P. falciparum*. Chalcone derivatives were prepared *via* Claisen-Schmidt condensation of substituted aldehydes with substituted methyl ketones. Antiplasmodial IC_{50} activity of these compounds ranged between 1.5 and 12.3 mg/ml. The chloro series of 1,2,4-triazole substituted chalcone (53) IC_{50} = 1.5 mg/ml was found to be the most effective in inhibiting the growth of *P. falciparum* (Mishra et al., 2008).

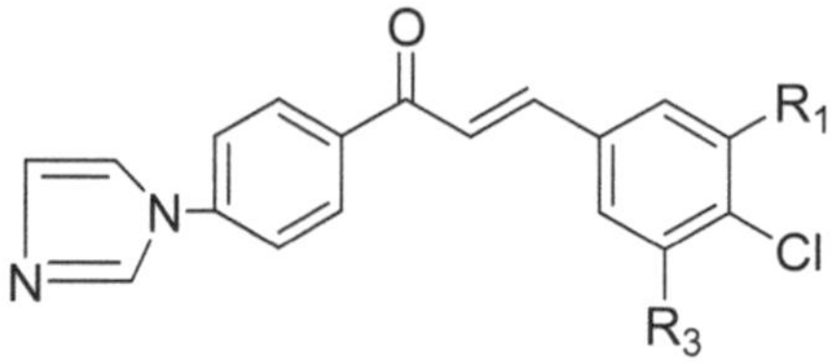

(53)

8.13   1-ARYL-3, 3-DIALKYLTRIAZENES

Till now, 1,3-diaryltriazenes, Isometamidium (Duch et al., 1984) and Berenil (Heischkeil, 1971) are well known compounds having antimalarial activity. Isometamidium is known to inhibit histamine N-methyltransferase and diamine oxidase both in vitro and in vivo which is important for their activity and berenil is 125 times more active than CQ in vivo.

Out of the 14 synthesized compounds, antimalarial activity of 1-aryl-3,3-dialkyltriazenes of 3,3-dimethyl-1-(4-(trifluoromethyl)phenyl)triaz-1-ene (**54**) results with 50% survival rate against chloroquine sensitive *Plasmodium berghei* NK- 65 strain in vivo. These results indicated that the triazenes could be new candidates for antimalarial drugs (Nishiwaki et al., 2007).

(54)   (55)

8.14   2-METHYL-6-UREIDO-4-QUINOLINAMIDES

Eighty new 2-methyl-6-ureido-4-quinolinamides were synthesized and evaluated by Batra et al. for their antimalarial activity. Several analogs elicited the antimalarial effect at MIC of 0.25 mg/mL against the chloroquine sensitive *P. falciparum* strain. Compound (**55**) was found to be most active with $IC_{50} = 2.2$ nM and MIC=0.25 µg/ml (Madapa et al., 2009).

8.15   SULFONYLUREAS

A series of sulfonylureas had been tested for their antimalarial activities against chloroquine resistant strain of *P. falciparum*, in vitro hemoglobin hydrolysis, hemozoin formation and development of *Plasmodium berghei* in murine malaria. The most active antimalarial compound was (E)-1-[40-(3-(2,4-difluorophenyl) acryloyl) phenyl]-3-tosylurea (**56**) with an IC_{50} of 1.2 µM against cultured *P. falciparum* parasites (Leon et al., 2007).

(56)

8.16   4-ANILINOQUINOLINE

Chauhan et al. synthesized 60 compounds having 4-anilinoquinoline moiety, out of these, compounds **(57)** and **(58)** were found to be much active with IC_{50} = 29.74 ng/ml and 23.13 ng/ml, respectively, against CQ-sensitive 3D7 strain of *P. falciparum* and evaluated for their cytotoxicity towards VERO cell line [131].

(57) = R_1 = piperidino, R_2 = N-ethyl piperizino
(58) = R_1 = piperidino, R_2 = 4(3-aminopropyl)morpholino

(59)

Melnyk et al. (Cochin et al., 2008) synthesized 17 compounds of 4-anilinoquinolines bearing an amino side chain linked to the aromatic ring with a carbamate or an amide bond were synthesized. Among them, majority of compounds found to be active in nanomolar range against both chloroquine-sensitive and resistant strains of *Plasmodium falciparum* with relative low cytotoxicity. Compound **(59)** was found to have IC_{50} = 11.2 nM.

8.17   1,3,5-TRISUBSTITUTED PYRAZOLINES

Compounds **(60)** and **(61)** synthesized by Kaushik et al. (Acharya et al., 2010) and were found to be active when evaluated for in vitro antimalarial efficacy against chloroquine sensitive (MRC-02) as well as chloroquine resistant (RKL9) strains of *Plasmodium falciparum* IC_{50} = 0.0532 µg/ml and 0.0314 µg/ml.

(60)   (61)

8.18   6-THIOUREIDO-4-ANILINOQUINAZOLINES

Forty-two new thioureidoquinazoline analogs showed promising antimalarial effect against chloroquine sensitive 3D7 strain of *Plasmodium falciparum*. Compound (62) (IC_{50} = 105.8 nM MIC=1 µg/ml) had 50% curative effect in the mouse model at an oral dose of 100 mg/kg for 4 days against multidrug resistant *Plasmodium yoelii* (Mishra et al., 2009).

(62)

8.19   HISTONE DEACETYLASES INHIBITORS

Acetylation and deacetylation of histones play an important role in transcription regulation of eukaryotic cells. The steady state of histone acetylation is established by the dynamic equilibrium between competing histone deacetylases (HDACs) and histone acetyltransferases (HATs). HATs add acetyl groups to lysine residues while HDACs remove the acetyl groups. In general, hyperacetylation of histones facilitates gene expression whereas histone deacetylation is correlated with transcriptional repression. The HDACs are able to control histone deacetylation which consequently promotes chromatin condensation. HDAC inhibitors (HDACIs) selectively alter gene transcription by permitting chromatin remodeling and changing the composition of multiprotein complexes bound to proximal region of specific gene promoter. Further, the HDACs interact with many nonhistone protein-substrates such as the hormone receptors, chaperone proteins and cytoskeleton proteins which regulate cell proliferation and cell death. In mammalian cells, the HDACs are divided into four classes that depend on their sequence/ structural homology to yeast deacetylases, expression patterns and catalytic mechanisms (Gui et al., 2004; Lehrmann et al., 2002; Minucci and Pelicci, 2006; Marks and Dokmanovic, 2005). Compound (63), vital candidate for drug lead, has HDAC inhibitory and antimalarial activity against D6, W2, TM90C235 and TM90C2A strains with IC_{50} values (nM) of 17, 32, 35, 17, respectively (Chen et al., 2008).

(63)

Kumar et al. synthesized 40 arylidene analogs of Meldrum's acid, among them compound (64) exhibited the most potent antimalarial activity ($IC_{50} = 9.68$ µM) (Sandhu et al., 2010).

(64)   (65) Z = heptyl amine

The 24 compounds of 4-alkylaminoaryl phenyl cyclopropyl methanones were synthesized from 4-fluorochalcones by cyclopropanation of double bond followed by nucleophilic substitution of fluoro with different amines by Tripathi et al. The compounds were screened for antimalarial activities against *Plasmodium falciparum* 3D7 strains in vitro, compound (65) was potent one with $IC_{50} = 0.080$ µg/mL (Ajay et al., 2010).

The antiplasmodial activity was established on new hybrids of 9-anilinoacridine triazines against CQ-sensitive 3D7 strain of *Plasmodium falciparum* and their cytotoxicity was determined on VERO cell line. Out of the evaluated compounds, compounds (66) (IC_{50} = 4.21 nM) and (67) (IC_{50} = 4.27 nM) displayed two times higher potency than CQ (IC_{50} = 8.15 nM) (Kumar et al., 2009).

(66) R_1 = Aniline R_2 = N,N-Dimethylpropylenediamine
(67) R_1 = Aniline R_2 = Hydrazine

Thieno-[3,2-b]benzothiazine S,S-dioxide derivatives were investigated for their abilities to inhibit b-hematin formation, hemoglobin hydrolysis and for their efficacy in rodent infected with *Plasmodium berghei* by Charris et al. (Barazarte et al., 2008) Compound (68) was the most promising inhibitors of hemoglobin hydrolysis but not as efficient as chloroquine.

(68)

A series of new N-alkyl- and N-alkoxy-imidazolidinediones were prepared and assessed for prophylactic and radical curative activities in mouse and Rhesus monkey models by Lin et al. These were found to be metabolically stable with weak activity in vitro against *Plasmodium falciparum* clones (D6 and W2) and in mice infected with *Plasmodium berghei* sporozoites. Out of 80 compounds, compounds (69) and (70) showed good causal prophylactic activity in Rhesus monkeys dosed 30 mg/kg/day for 3 consecutive days by i.m. with delayed patency for 19–21 days and 54–86 days, respectively, as compared to the untreated control.

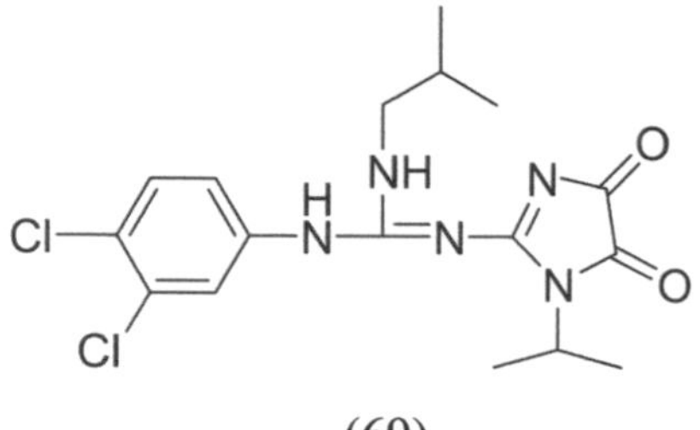

(69) 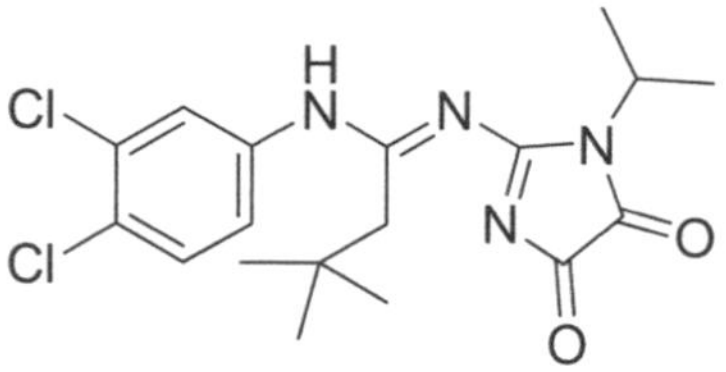   (70)

A series of new 13 compounds of amino functionalized 1, 2, 4-trioxepanes and ester functionalized 1,2,4-trioxepanes have been synthesized by Puri et al. and evaluated against multidrug resistant *Plasmodium yoelii* in Swiss mice. Compound **(71)** was the most active compound of the series (Singh et al., 2008).

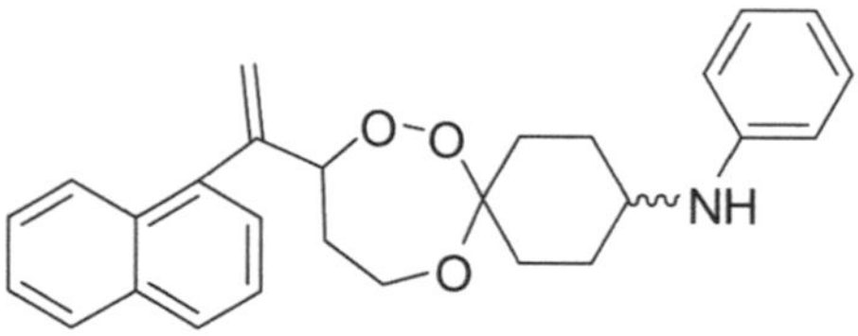

(71)

Compound **(72)** among 20 boron-containing benzoxaborole compound demonstrated the best potency (IC_{50} = 26 nM) against *Plasmodium falciparum* with low molecular weight (206.00), low C logP (0.86) and high water solubility (750 µg/mL at pH 7). A four-step route has been established for the synthesis of 7-(2-carboxyethyl)-1, 3-dihydro-1-hydroxy-2, 1-benzoxaborole **(71)** which is a potent new class boron-containing antimalarial agent in preclinical development with IC_{50} = 26 nM against the malaria parasite *P. falciparum* (Zhang et al., 2011a,b).

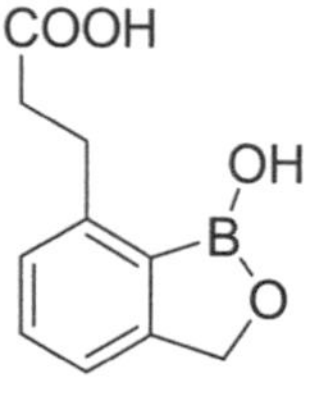

(72)

The new derivatives of 10-allyl-, 10-(3-methyl-2-butenyl) – and 10-(1, 2-propadienyl)-9(10H)-acridinone were assessed in parallel against human MRC-5 cells to determine their selectivity and cytotoxicity. Compound (73) (1-fluoro-10-(3-methyl-2-butenyl)-9(10H)-acridinone) provide a hit for antimalarial drug development exhibiting IC_{50} less than 0.2 mg/mL (Calienes et al., 2011).

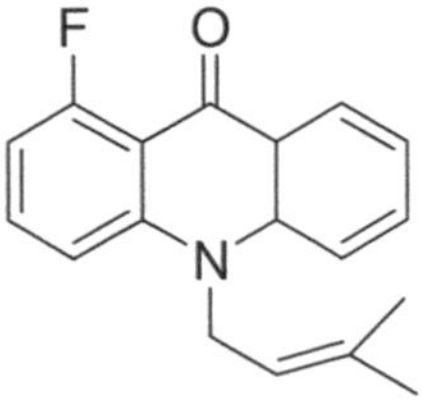

(73)

The natural products like dependensin-1 and it analogs possess antimalarial activity. Therefore, novel benzopyrano[4,3-b]benzopyran derivatives were screened against a chloroquine sensitive P. *falciparum* line (3D7) using standard in vitro growth inhibition assays. Compound (74) was found to be much active with $IC_{50}=1.9$ µM (Devakaram et al., 2011).

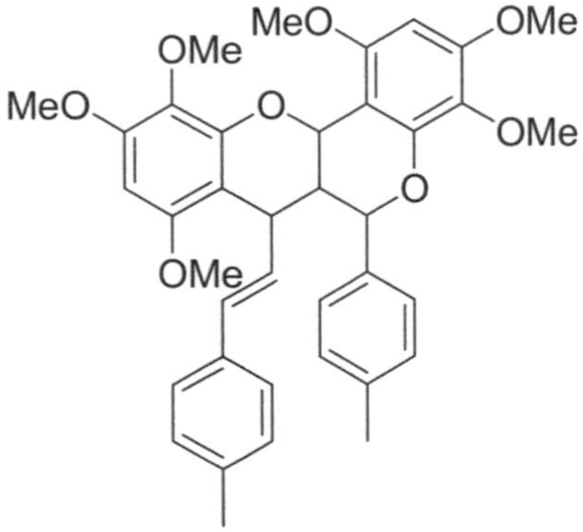

(74)

The semicarbazone and thiosemicarbazone derivatives were prepared by Zani et al. (Oliveira et al., 2008) and tested to evaluate antiplasmodial potential against a chloroquine resistant strain of *P. falciparum* (W2). Three thiosemicarbazones were found to be active against the parasite and non-toxic to human peripheral blood mononuclear cells (PBMC). Among these, compound (75) presented the lowest IC_{50} value and was the least toxic in the PBMC proliferation assay ($IC_{50}= 73.5$ mM).

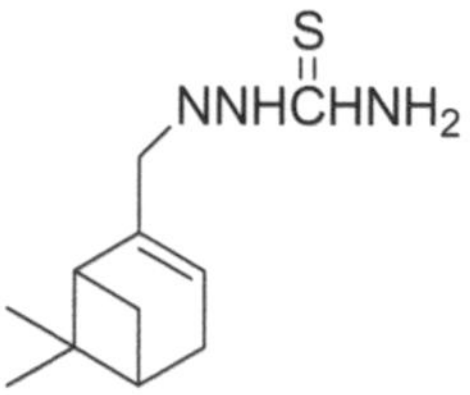

$$\underset{\text{NNHCHNH}_2}{\overset{\overset{\text{S}}{\|}}{}}$$

(75)

8.20 SUMMARY AND CONCLUSION

Public health has long been affected by malaria in the developing world with more than 1 million clinical episodes and 3000 deaths every day. Drug therapy faces major challenges due to development of parasite resistance to first line antimalarials and unavailability of a vaccine. This has stimulated a search of new natural and synthetic antimalarials to counteract resistance. In this review, the main antiplasmodial drugs and new drug candidates have been presented. IC_{50} values obtained against various strains (chloroquine-sensitive and chloroquine-resistant ones) of plasmodium species are mentioned. A large member of newly synthesized compounds displayed moderate to excellent efficacy in pharmacological evaluation. Therefore, the pharmacophores of these drug candidates could further be developed in future to overcome the resistance and cure the disease.

KEYWORDS

- analog
- malaria
- mechanism of action
- molecular target
- plasmodium
- resistance

REFERENCES

Acharya, B. D., Saraswat, D., Tiwari, M., Shrivastava, A. K., Ghopade, R., Bapna, S., & Kaushik, M. P. (2010). Synthesis and antimalarial evaluation of 1,3,5-trisubstituted pyrazolines. *Eur. J. Med. Chem. 45*, 430–438.

Agarwal, A., Srivastava, K., Kumar, S. R., Puri, S. K., & Chauhan, P. M. S. (2005). Antimalarial activity and synthesis of new trisubstituted pyrimidines. *Bio. Med. Chem. 15*, 3130–3132.

Agarwal, A., Srivastava, K., Kumar, S. R., Puri, S. K., & Chauhan, P. M. S. (2005). Synthesis of 2, 4, 6-trisubstituted pyrimidines as antimalarial agents. *Bio. Med. Chem. 13*, 4645–4650.

Ajay, A., Singh, V., Pandey, S., Gunjan, S., Dubey, D., Sinha, S. K., Singh, B. N., Chaturvedi, V., & Tripathi, R. P. (2010). Synthesis and biological evaluation of alkylaminoaryl phenyl cyclopropyl methanones as antitubercular and antimalarial agents. *Bio. Med. Chem. 18*, 8289–8301.

Alibert, S., Santelli-Rouvier, C., Pradines, B., Houdoin, C., Parzy, D., Karolak-Wojciechowska, J., & Barbe, J. (2002). Synthesis and effects on chloroquine susceptibility in Plasmodium falciparum of a series of new dihydroanthracene derivatives. *J. Med. Chem. 45*, 3195–3209.

Amewu, R., Stachulski, A. V., Ward, S. A., Berry, N. G., Bray, P. G., Davies, J., Labat, G., Vivas, L., & O'Neill, P. M. (2006). Design and synthesis of orally active dispiro 1,2,4,5-tetraoxanes; synthetic antimalarials with superior activity to artemisinin. *Org. Biomol. Chem. 4*, 4431.

Ashley, E., McGready, R., Proux, S., & Nosten, F., (2006). *Malaria. Travel Med. Infect. Dis. 4*, 159–173.

Barazarte, A., Camacho, J., Dominguez, J., Lobo, G., Gamboa, N., Rodrigues, J., Capparelli, M. V., Enriz, D., & Charris, J. (2008). Synthesis, antimalarial activity, structure activity relationship analysis of thieno-[3,2-b]benzothiazine S,S-dioxide analogs. *Bio. Med. Chem. 16*, 3661–3674.

Bathurst, I., Hentschel, C. (2006). Medicines for Malaria Venture: sustaining antimalarial drug development. *Trends Parasitol. 22*, 301–307.

Bhat, H. R., Singh, U. P., Thakur, A., Ghosh, S., Gogoi, K., Prakash, A., & Singh, R. K. (2006). Synthesis, antimalarial activity and molecular docking of hybrid 4-aminoquinoline-1,3,5-triazine derivatives. *Exp Parasitol.*, (2015). *157*, 59–67.

Bhattacharya, A., Mishra, L. C., Sharma, M., Awasthi, S. K., & Bhasin, V. K. (2006). Antimalarial pharmacodynamics of chalcone derivatives in combination with artemisinin against Plasmodium falciparum in vitro. *Eur. J. Med. Chem.*, (2009). *44*, 3388–3393.

Biot, C., Chibale, K. (2006). Artemisinin and its Derivatives. Novel approaches to antimalarial drug discovery. *Infect. Disord. Drug Targets.*, *6*, 173–204.

Bitonti, A. J., Sjoerdsma, A., McCann, P. P., Kyle, D. E., Oduola, AMJ., Rossan, R. N., & Davidson, D. E. (1988). Reversal of chloroquine resistance in malaria parasite P. falciparum by desipramine. *Sci., 242*, 1301–1303.

Borstnik, K., Paik, I. H., Shapiro, T. A., & Posner, G. H. (2002). Antimalarial chemotherapeutic peroxides: artemisinin, yingzhaosu A and related compounds. *Int. J. Parasitol., 32*, 1661–1667.

Bozdech, Z., Llinas, M., Pulliam, B. L., Wong, E. D., Zhu, J., & DeRisi, J. L. (2003). The transcriptome of the intraerythrocytic developmental cycle of Plasmodium falciparum. *PLoS Biol., 1*, E5.

Bray, P. G., Deed, S., Fox, E., Kalkanidis, M., Mungthin, M., Deady, L. M., & Tilley, L. (2005). Primaquine synergises the activity of chloroquine against chloroquine-resistant P. falciparum. *Biochem. Pharmacol.*, *70*, 1158–1166.

Breman, J. G., Alilio, M. S., & Mills, A. (2004). Conquering the intolerable burden of malaria: what's new, what's needed: a summary. *Am J Trop Med Hyg.*, *71*, 1–15.

Bueno, J. M., Manzano, P., García, M. C., Chicharro, J., Puente, M., Lore, M., García, A., Ferrer, S., Gómez, R. M., Fraile, M. T., Lavandera, J. L., Fian, J. M., Vidal, J., Herreros, E., & Gargallo-Viola, D. (2011). Potent Antimalarial 4-pyridones with improved physicochemical properties. *Bioorg. Med. Chem. Letters.*, *21*, 5214–5218.

Calienes, A., Pellon, R., Docampo, M., Fascio, M., D'Accorso, N., Maes, L., Mendiola, J., Monzote, L., Gille, L., & Rojas, L. (2011). Antimalarial activity of new acridinone derivatives. *Bio. Med. Chem.*, *65*, 210–214.

Canfield, C. J., Pudney, M., & Gutteridge, W. E. (1995). Interactions of atovaquone with other antimalarial drugs against *Plasmodium falciparum* in vitro. *Exp. Parasito.*, *80*, 373–381.

Casagrande, M., Basilico, N., Parapini, S., Taramelli, D., & Sparatore, A. (2008). Novel amodiaquine congeners as potent antimalarial agents. *Bioorg. Med. Chem.*, *16*, 6813–6823.

Chandra, S., Ohnishi, S. T., & Dhawan, B. N. (1993). Reversal of chloroquine resistance in murine malaria parasites by prostaglandin derivatives. *Am. J. Trop. Med. Hyg.*, *48*, 645–651.

Chen, H., Keystone, J. S. (2005). New strategies for the prevention of malaria in travelers. *Infect. Dis. Clin. N. Am.*, *19*, 185–210.

Chen, M., Theander, T. G., Christensen, S. B., Hviid, L., Zhai, L., Kharazmi, A., & Licochalcone A. (1994). A new antimalarial agent, inhibits invitro growth of the human malaria parasite Plasmodium falciparum and protects mice from P. yoelii infection. *Anti-microb. Agents. Chemother.*, *38*, 1470–1475.

Chen, Y., Lopez-Sanchez, M., Savoy, D. N., Billadeau, D. D., Dow, G. S., & Kozikowski, A. P. (2008). A Series of Potent and Selective, Triazolylphenyl-Based Histone Deacetylases Inhibitors with Activity against Pancreatic Cancer Cells and Plasmodium falciparum. *J. Med. Chem.*, *51*, 3437–3448.

Clark, I. A., Schofield, L. (2000). Pathogenesis of malaria. *Para Tod.*, *16*, 451–454.

Cochin, S., Grellier, P., Maes, L., Mouray, E., Sergheraet, C., & Melnyk, P. (2008). Synthesis and antimalarial activity of carbamate and amide derivatives of 4-anilinoquinoline. *Eur. J. Med. Chem.*, *43*, 2045–2055.

Cochin, S., Grellier, P., Maes, L., Mouray, E., Sergheraet, C., & Melnyk, P. (2008). Synthesis and antimalarial activity of carbamate and amide derivatives of 4-anilinoquinoline. *Eur. J. Med. Chem.*, *43*, 2045–2055.

Cooper, R. A., Hartwig, C. L., & Ferdig, M. T. (2005). pfcrt is more than the Plasmodium falciparum chloroquine resistance gene: a functional and evolutionary perspective. *Acta Trop.*, *94*, 170–180.

Corrales, RCNR., Souza, N. B., Pinheiro, L. S., Abramo, C., Coimbra, E. S., & Silva, A. D. (2011). Thiopurine derivatives containing triazole and steroid: synthesis, antimalarial and antileishmanial activity. *Biomed. Pharmacothe.*, *65*, 198–203.

Cummings, J. N., Ploypradith, P., & Posner, G. H. (1997). Antimalarial activity of artemisnin (qimghaosu) and related trioxanes: Mechanism(s) of action. *Adv. Pharmacol.*, *3*, 253.

Davis, TME., Hung, T., Sim, I., & Karunajeewa, H. A. (2005). Effects of a High-Fat Meal on the Relative Oral Bioavailability of Piperaquine. K. F. *I lett, Drugs.*, *65*, 75–87.

Devakaram, R., Black, D. S., Andrews, K. T., Fisherb, G. M., Davisb, R. A., & Kumar, N. (2011). Synthesis and antimalarial evaluation of novel benzopyrano[4, 3-b]benzopyran Derivatives. *Bio. Med. Chem.*, *19*, 5199–5206.

Djimde, A., Doumbo, O. K., Cortese, J. F., Kayentao, K., Doumbo, S., Diourté, Y., Coulibaly, D., Dicko, A., Su, X. Z., Nomura, T., Fidock, D. A., Wellems, T. E., & Plowe, C. V. (2001). A molecular marker for chloroquine-resistant falciparum malaria. *N Engl J Med.*, *344*, 257–263.

Dong, Y. X., Matile, H., Chollet, J., Kaminsky, R., Wood, J. K., & Vennerstrom, J. L. (1999). Synthesis and Antimalarial Activity of 11 Dispiro-1,2,4,5-tetraoxane Analogues of WR 148999. 7, 8, 15, 16-Tetraoxadispiro[5.2.5.2]hexadecanes Substituted at the 1 and 10 Positions with Unsaturated and Polar Functional Groups. *J. Med. Chem.*, *42*, 1477–1480.

Dong, Y. X., Tang, Y. Q., Chollet, J., Matile, H., Wittlin, S., Charman, S. A., Charman, W. N., Tomas, J. S., Scheurer, C., Snyder, C., Scorneaux, B., Bajpai, S., Alexander, S. A., Wang, X. F., Padmanilayam, M., Cheruku, S. R., Brun, R., & Vennerstrom, J. L. (2006). Effect of functional group polarity on the antimalarial activity of spiro and dispiro-1, 2, 4-trioxolanes. *Bioorg. Med. Chem.*, *14*, 6368.

Dong, Y., Chollet, J., Matile, H., Charman, S. A., Chiu, FCK., Charman, W. N., Scorneaux, B., Urwyler, H., Tomas, J. S., Scheurer, C., Snyder, C., Dorn, A., Wang, X., Karle, J. M., Tang, Y., Wittlin, S., Brun, R., & Vennerstrom, J. L. (2005). Spiro and Dispiro-1, 2, 4-trioxolanes as Antimalarial Peroxides: Charting a Workable Structure–Activity Relationship Using Simple Prototypes. *J. Med. Chem.*, *48*, 4953–4961.

Dong, Y. (2002). Synthesis and antimalarial activity of 1,2,4,5-tetraoxanes. *Mini-Rev. Med. Chem.*, *2*, 113–123.

Duch, D. S., Bacchi, C. J., Edelstein, M. P., & Nichol, C. A. (1984). Inhibitors of histamine metabolism in vitro and in vivo: Correlations with antitrypanosomal activity. *Biochem. Pharmacol.*, *33*, 1547–1553.

Dussault, P. H., Davies, D. R. (1996). Synthesis of 1, 2-dioxanes, 1, 2, 4-trioxanes, and 1, 2, 4-trioxepanes via cyclizations of unsaturated hydroperoxyacetals. *Tetrahedron Lett.*, *37*, 463–466.

Eckstein-Ludwig, U., Webb, R. J., Van Goethem, I. D., East, J. M., Lee, A. G., Kimura, M., O'Neill, P. M., Bray, P. G., Ward, S. A., & Krishna, S. (2003). Artemisinins target the SERCA of Plasmodium falciparum. *Nat.*, *424*, 957–961.

Egan, T. J., Hunter, R., Kaschula, C. H., Marques, H. M., Misplon, A., & Walden, J. C. (2000). Structure–Function Relationships in Aminoquinolines: Effect of Amino and Chloro Groups on Quinoline–Hematin Complex Formation, Inhibition of β-Hematin Formation, and Antiplasmodial Activity. *J. Med. Chem. 43*, 283–291.

Ekoue Kovi, K., Yearick, K., Iwaniuk, D. P., Natarajan, J. K., Alumasa, J., Dios, A. C., Roepe, P. D., & Wolf, C. (2009). Synthesis and antimalarial activity of new 4-amino-7-chloroquinolyl amides, sulfonamides, ureas and thioureas. *Bioorg. Med. Chem.*, *17*, 270–283.

Ellis, G. L., Amewu, R., Hall, C., Rimmer, K., Ward, S. A., & O'Neill, P. M. (2008). An efficient route into synthetically challenging bridged achiral 1,2,4,5-tetraoxanes with antimalarial activity. *Bioorg. Med. Chem. Lett.*, *18*, 1720–1724.

Esperanc, S., Gonza, R., & Menendez, C. (2010). Current knowledge and challenges of antimalarial drugs for treatment and prevention in pregnancy. *Expert Opin. Pharmacother.*, *11*, 1277–1293.

Fidock, D. A., Nomura, T., Talley, A. K., Roland, A., Michael, T., Ferdig, & Lyann, M. B. (2000). Mutations in the *P. falciparum* digestive vacuole transmembrane protein PfCRT and evidence for their role in chloroquine resistance. *Mol Cell.*, *6*, 861–871.

Foley, M., Tilley, L. (1997). Quinoline antimalarials: mechanisms of action and resistance. *Int J Parasitol.*, *27*, 231–240.

Frudurich, M., Dognu, J. M., Angenot, L., & De Mol, P. (2002). New Trends in Anti-Malarial Agents. *Curr. Med. Chem.*, *9*, 1435–1456.

Gelb, M. H. (2007). Drug discovery for malaria: a very challenging and timely endeavor. *Curr. Opin. Chem. Biol.*, *11*, 440.

Gemma, S., Campiani, G., Butini, S., Joshi, B. P., Kukreja, G., Coccone, S. S., Bernetti, M., Persico, M., Nacci, V., Fiorini, I., Novellino, E., Taramelli, D., Basilico, N., Parapini, S., Yardley, V., Croft, S., Keller-Maerki, S., Rottmann, M., Brun, R., Coletta, M., Marini, S., Guiso, G., Caccia, S., & Fattorusso, C. (2009). Combining 4-Aminoquinoline and Clotrimazole-Based Pharmacophores toward Innovative and Potent Hybrid Antimalarials. *J. Med. Chem.*, *52*, 502–513.

Gerena, L., Bass, G. E., Kyle, D. E., Oduola, AMJ., Milhous, W. K., & Martin, R. K. (1992). Fluoxetine hydrochloride enhances in vitro susceptibility to chloroquine in resistant *P. falciparum. Antimicrob Agents Chemother.*, *36*, 2761–2765.

Ginsburg, H., Famin, O., Zhang, J., & Krugliak, M. (1998). Inhibition of glutathione-dependent degradation of heme by chloroquine and amodiaquine as a possible basis for their antimalarial mode of action. *Biochem Pharmacol.*, *56*, 1305–1313.

Graffing, S., Syphard, L., Sridaran, S., McCollum, A. M., Mixson-Hayden, T., Vinayak, S., Villegas, L., Barnwell, J. W., Escalantem, A. A., & Udhayakumar, V. (2010). Pfmdr1 Amplification and fixation of Chloroquine resistan pfcrt alleles in Venezuela. *Antimicrob Agents Chemother*, *54*, 1572–1579.

Gravestock, D., Rousseau, A. L., Lourens, ACU., Moleete, S. S., Zyl, RLV., & Steenkamp, P. A. (2011). Biological Evaluation of novel 2, N⁶-disubstituted 1, 2-dihydro −1,3,5-triazine 4, 6-diamines as potential animalarials. *Eur. J. Med. Chem.*, *46*, 2022–2030.

Gui, C. Y., Ngo, L., Xu, W. S., Richon, V. M., & Marks, P. A. (2004). Histone deacetylase (HDAC) inhibitor activation of p21WAF1 involves changes in promoter-associated proteins, including HDAC1. *Proc. Natl. Acad. Sci.*, *101*, 1241–1246.

Hage, S., Ane, M., Stigliani, J., Marjoire, M., Vial., Bazaird-Mouysset, G., & Payard, M. (2009). Synthesis and antimalarial activity of new atovaquone derivatives. *Eur. J. Med. Chem.*, *44*, 4778–4782.

Hamada, Y., Tokuhara, H., Masuyama, A., Nojima, M., Kim, H. S., Ono, K., Ogura, N., & Wataya, Y. (2002). Synthesis and Notable Antimalarial Activity of Acyclic Peroxides, 1-(Alkyldioxy)-1-(methyldioxy)cyclododecanes. *J. Med. Chem.*, *45*, 1374–1378.

Hamann, H., Hecht, M., Bunge, A., Gogol, M., & Liebscher, J. (2011). Synthesis and antimalarial activity of new 1,2,4,5-tetroxanes and novel alkoxy-substituted 1,2,4,5-tetroxanes derived from gem-dihydroperoxides. *Tetrahedron Letters.*, *52*, 107–111.

Hatton, C. S., Peto, T. E., Bunch, C., Pasvol, G., Russell, S. J., Singer, C. R., Edwards, G., & Winstanley, P. (1986). Frequency of severe neutropenia associated with amodiaquine prophylaxis against malaria. *Lancet.*, *327*, 411–414.

Haynes, R. K. (2001). Artemisnin and derivatives: the future for malaria treatment. *Curr. Opin. Infect. Dis.*, *14*, 719–726.

Haynes, R. K., Krishna, S. (2004). Artemisinins: activities and actions. *Microbes Infect.*, *6*, 1339–1346.

Heischkeil, R. Z. (1971). [Effctiveness of trypanocides berenil and pentamidine in rodent malaria (Plasmodium vinckei)] *Tropen Med. Parasitol., 22*, 243–249.

Hoppe, H. C., Schalkwyk, D. A., Wiehart, U. I., Meredith, S. A., Egan, J., & Weber, B. W. (2004). Antimalarial quinolines and artemisinin inhibit endocytosis in Plasmodium falciparum. *Antimicrob Age Chemother., 48*, 2370–2378.

Horne, W. S., Yadav, M. K., Stout, C. D., & Ghadiri, M. R. (2004). Heterocyclic peptide backbone modifications in an alpha-helical coiled coil. *J Am Chem Soc., 6*, 15366–15367.

Hyde, J. E. (2005). Exploring the folate pathway in Plasmodium falciparum. *Acta Trop. 94*, 191–206.

Iskra, J., Bonnet-Delpon, D., & Begue, J. P. (2003). One-pot synthesis of nonsymmetric tetraoxanes with the H_2O_2/MTO/fluorous alcohol system. *Tetrahedron Lett., 44*, 6309–6314.

Jefford, C. W. (1997). Peroxidic antimalarials. *Adv. Drug Res., 29*, 271–325.

Jin, H. X., Liu, H. H., Zhang, Q., & Wu, Y. K. (2005). Synthesis of 1,6,7-trioxa-spiro[4.5] decanes. *Tetrahedron Lett., 46*, 5767–5777.

Jin, H. X., Zhang, Q., Kim, H. S., Wataya, Y., Liu, H. H., & Wu, Y. (2006). Design, synthesis and in vitro antimalarial activity of spiroperoxides. *Tetrahedron., 62*, 7699–7711.

Johanna, P. (2006). Antimalarial Drug Therapy: The Role of Parasite Biology and Drug Resistance. *J. Clin. Pharmacol., 46*, 1487–1497.

Jung, M. (1997). Synthesis and cytotoxicity of novel artemisinin analogs. *Bio. Med.Chem. Lett., 7*, 1091–1094.

Kaschula, C. H., Egan, T. J., Hunter, R., Basilico, N., Parapini, S., Tarameli, D., Pasini, E., & Monti, D. (2002). Structure–Activity Relationships in 4-Aminoquinoline Antiplasmodials. The Role of the Group at the 7-Position. *J. Med. Chem., 45*, 3531–3539.

Kasekarn, W., Sirawaraporn, R., Chahomchuen, T., Cowman, A. F., & Sira-waraporn, W. (2004). Molecular characterization of bifunctional hydroxymethyldihydropterine pyrophosphokinase-dihydropteroate synthase from Plasmodium falciparum. *Mol. Biochem. Parasitol., 137*, 43–53.

Kesten, S. J., Johnson, J., & Werbel, L. M. (1987). Antimalarial drugs: Synthesis and antimalarial effects of 4-t(7-chloro-4-quinolinyl)amino]-2-[(diethylamino)methyl]-6-alkylphenols and their N.omega.-oxides. *J. Med. Chem., 30*, 906–911.

Khatoon, L., Baliraine, F. N., Bonizzoni, M., Malik, S. A., & Yan, G. (2009). Prevalence of antimalarial drug resistance mutations in Plasmodium vivax and P. falciparum from a malaria-endemic area of Pakistan. *Am J Trop Med Hyg., 81*, 525–528.

Kim, H. S., Shibata, Y., Wataya, Y., Tsuchiya, K., Masuyama, A., & Nojima, M. (1999). Synthesis and Antimalarial Activity of Cyclic Peroxides, 1,2,4,5,7-Pentoxocanes and 1,2,4,5-Tetroxanes. *J. Med. Chem., 42*, 2604–2609.

Krishna, S., Uhlemann, A. C., & Haynes, R. K. (2004). Artemisinins: mechanisms of action and potential for resistance. *Drug Resis Updat., 7*, 233–244.

Kumar, A., Katiyar, S. B., Agawal, A., & Chauhan, PMS (2003). Perspective in Antimalarial Chemotherapy. *Curr. Med. Chem. 10*, 1137–1150.

Kumar, A., Srivastava, K., Kumar, S. R., Puri, S. K., & Chauhan, PMS (2009). Synthesis of 9-anilinoacridine triazines as new class of hybrid antimalarial agents. *Bio. Med. Chem., 19*, 6996–6999.

Kumar, A., Srivastava, K., Kumar, S. R., Puri, S. K., & Chauhan, PMS (2010). Synthesis of new 4-aminoquinolines and quinoline–acridine hybrids as antimalarial agents. *Bio. Med. Chem. Lett. 20*, 7059–7063.

Kumar, A., Srivastava, K., Kumar, S. R., Siddiqi, M. I., Puri, S. K., Sexana, J. K., & Chauhan, PMS (2011). 4-anilinoquinoline triazines; a novel class of hybrid antimalarial agents. *Eur. J. Med. Chem.*, *40*, 676–690.

Kumar, S., Epstein, J. E., Richie, T. L., Nkrumah, F. K., Soisson, L., Carucci, D. J., & Hoffman, S. L. (2002). A multilateral effort to develop DNA vaccines against falciparum malaria. *Trends Parasitol.*, *18*, 129–135.

Leban, J., Pegoraro, S., Dormeyer, M., Lanzer, M., Aschenbrenner, A., & Kramer, B. (2004). Sulfonyl-phenyl-ureido benzamidines: a novel structural class of potent antimalarial agents. *Bioorg. Med. Chem. Lett.*, *14*, 1979–1982.

Lee, S. W., Lee, M., Lee, D. D., Kim, C., Kim, Y. J., Kim, J. Y., Green, M. D., Klein, T. A., Kim, H. C., Nettey, H., Ko, D. H., Kim, H., & Park, I. (2009). Biological resistance of hydroxychloroquine for Plasmodium vivax malaria in the Republic of Korea. *Am J Trop Med Hyg.*, *81*, 600–604.

Lehrmann, H., Pritchard, L. L., & Harel-Bellan, A. (2002). Histone acetyltransferases and deacetylases in the control of cell proliferation and differentiation. *Adv. Cancer Res.*, *86*, 41–65.

Leon, C., Rodrigues, J., Dominguez, N. G., Charris, J., Gut, J., Rosethal, P., & Domineguez, J. N. (2007). Synthesis and evaluation of sulfonylurea derivatives as novel antimalarials. *Eur. J. Med. Chem.*, *42*, 735–742.

Li, R., Kenyon, G. L., Cohen, F. E., Chen, X., Gong, B., Dominguez, J. N., Davidson, E., Kurzban, G., Miller, R. E., Nuzum, E. O., Rosenthal, P. J., & McKerrow, J. H. (1995). In vitro Antimalarial Activity of Chalcones and their Derivatives. *J. Med. Chem.*, *38*, 5031–5037.

Liu, M., Wilairat, P., Croft, S. L., Tan, A. L., & Go, M. L. (2003). Structure–Activity Relationships of Antileishmanial and Antimalarial Chalcones. *Bioorg. Med. Chem.*, *11*, 2729–2738.

Madapa, S., Tusi, Z., Puri, S. K., Shukla, P. K., & Batra, S. (2009). Search for new pharmacophores for antimalarial activity: synthesis and antimalarial activity of new 2-methyl-6-ureido-4-quinolinamides. *Bio Med. Chem.*, *17*, 203–221.

Madrid, P. B., Liou, A. P., DeRisi, J. L., & Guy, R. K. (2006). Incorporation of an Intramolecular Hydrogen-Bonding Motif in the Side Chain of 4-Aminoquinolines Enhances Activity against Drug-Resistant *P. falciparum*. *J. Med. Chem.*, *49*, 4535–4543.

Madrid, P. B., Sherrill, J., Liou, A. P., Weisman, J. L., DeRisi, J. L., & Guy, R. K. (2005). Synthesis of ring-substituted 4-aminoquinolines and evaluation of their antimalarial activities. *Bio. Med. Chem. Lett.*, *15*, 1015–1018.

Makler, M. T., Ries, J. M., Williams, J. A., Bancroft, J. E., Piper, R. C., Gibbins, B. L., & Hinrichs, D. J. (1993). Parasite lactate dehydrogenase as an assay for Plasmodium falciparum drug sensitivity. *Am. J. Trop. Med. Hyg.*, *48*, 739–741.

Marks, P. A., Dokmanovic, M. (2005). Histone deacetylase inhibitors:discovery and development as anticancer agents. *Expert Opin. InVest. Drugs.*, *14*, 1497–1511.

Martin, S. K., Oduola, A. M., & Milhous, W. K. (1987). Reversal of chloroquine resistance in Plasmodium falciparum by verapamil. *Sci.*, *235*, 899–901.

McCullough, K. J., Wood, J. K., Bhattacharjee, A. K., Dong, Y. X., Kyle, D. E., Milhous, W. K., & Vennerstrom, J. L. (2000). Methyl-substituted dispiro-1,2,4,5-tetraoxanes: Correlations of structural studies with antimalarial activity. *J. Med Chem.*, *43*, 1246–1249.

Minucci, S., Pelicci, P. G. (2006). Histone deacetylase inhibitors and the promise of epigenetic (and more) treatments for cancer. *Nat. Re V. Can.*, *6*, 38–51.

Mishra, A., Srivastava, K., Puri, S. K., & Batra, S. (2009). Search for new pharmacophores for antimalarial activity: synthesis and antimalarial activity of new 2-methyl-6-ureido-4-quinolinamides. *Bio Med. Chem.*, *44*, 4404–4412.

Mishra, N., Arora, P., Kumar, B., Mishra, L. C., Bhattachrya, A., Awasthi, S. K., & Bhasin, V. K. (2008). Synthesis of novel substituted 1, 3-diaryl propene derivatives and their antimalarial activity in vitro. *Eur. J. Med. Chem.*, *43*, 1530–1535.

Myrvang, B. (2010). A fifth plasmodium that can cause malaria. *Tidsskr. Nor. Laegeforen.*, *130*, 282–283.

Najjar, F., Gorrichon, L., Baltas, M., Andre-Barres, C., & Vial, H. (2005). Alkylation of natural endoperoxides G3-factor synthesis and antimalarial activity studies. *Org. Biomol. Chem.*, *3*, 1612–1614.

Neftel, K. A., Woodtly, W., Schmid, M., Frick, P. G., & Fehr, J.(1986). Amodiaquine induced agranulocytosis and liver damage. *Br. Med. J.*, *292*; 721–723.

Nishiwaki, K., Okamoto, A., Matsuo, Kawaguchi., Y, Hayase, Y., & Ohba, K. (2007). Antimalarial activity of 1-aryl-3,3-dialkyltriazenes. *Bioorg. Med. Chem.*, *15*, 2856–2859.

Nodiff, E. A., Chatterje, S., & Musallam, H. A (1991). Antimalarial activity of the 8-aminoquinolines. *Prog. Med. Chem.*, *28*, 1–40.

Nzila, A., Ward, S. A., Marsh, K., Sims, P. F., & Hyde, J. E. (2005). Comparative folate metabolism in humans and malaria parasites (part II): activities as yet untargeted or specific to Plasmodium. *Trends Parasitol.*, *21*, 334–339.

O'Neill, P. M., Amewu, R. K., Nixon, G. L., ElGarah, F. B., Mungthin, M., Chadwick, J., Shone, A. E., Vivas, L., Lander, H., Barton, V., Muangnoicharoen, S., Bray, P. G., Davies, J., Park, B. K., Wittlin, S., Brun, R., Preschel, M., Zhang, K., & Ward, S. A. (2010). Identification of a 1,2,4,5-tetraoxane antimalarial drug-development candidate (RKA 182) with superior properties to the semisynthetic artemisinins. *Angew. Chem., Int. Ed.*, *49*, 5693–5697.

O'Neill, P. M., Posner, G. H. (2004). A Medicinal Chemistry Perspective on Artemisinin and Related Endoperoxides. *J. Med. Chem.*, *47*, 2945–2964.

O'Neill, P. M., Shone, A. E., Stanford, D., Nixon, G., Asadollahy, E., Park, B. K., Maggs, J. L., Roberts, P., Stocks, P. A., Biagini, G., Bray, P. G., Davies, J., Berry, N., Hall, C., Rimmer, K., Winstanley, P. A., Hindley, S., Bambal, R. B., Davis, C. B., Bates, M., Gresham, S. L., Brigandi, R. A., Gargallo, D. V., Parapini, S., Vivas, L., Lander, H., & Taramelli, DWard (2009). Synthesis, Antimalarial Activity, and Preclinical Pharmacology of a Novel Series of 4'-Fluoro and 4'-Chloro Analogues of Amodiaquine. Identification of a Suitable "Back-Up" Compound for N-tert-Butyl Isoquine. *J. Med. Chem.*, *52*, 1828–1844.

Oliveira, R. B., Souza-Fagundes, E. M., Soares, R., Andrade, A. A., Krettli, A. U., & Zani, C. L. (2008). Synthesis and antimalarial activity of semicarbazone and thiosemicarbazone derivatives. *Eur. J. Med. Chem.*, *43*, 1983–1988.

Paredes, C. F., José, I., Santos-Preciado, J. I. (2006). Problem pathogens: prevention of malaria in travelers. *Lancet Infect. Dis.*, *6*, 139–149.

Pasvol, G. (2006). The treatment of complicated and severe malaria. *Br. Med. Bull.*, *75*, 29–47.

Peters, W., Ekong, R., Robinson, B. L., Warhurst, D. C., & Pan, X. Q. (1989). Antihistaminic drugs that reverse chloroquine resistance in P. falciparum. *Lanc.*, *2*, 334–335.

Posner, G. H., O'Neill, P. M. (2004). Knowledge of the proposed chemical mechanism of action and cytochrome p450 metabolism of antimalarial trioxanes like artemisinin allows rational design of new antimalarial. *Acc. Chem. Res.*, *37*, 397–404.

Ramharter, M., Noedl, H., Thimarsan, K., Wiedermann, G., Wernsdorfer, G., & Wernsdorfer, W. H. (2004). In vitro activity of tafenoquine alone and in combination with artemisinin against Plasmodium falciparum. *Am. J. Med. Hyg.*, (2002). *67*, 39–43.

Rastelli, G., Pacchioni, S., Sirawaraporn, W., R. Sirawaraporn, R., Parenti, M. D., & Ferrari, A. M. (2003). Docking and Database Screening Reveal New Classes of *Plasmodium falciparum* Dihydrofolate Reductase Inhibitors. *J. Med. Chem.*, *46*, 2834–2845.

Raynes, K. J., Stocks, P. A., O'Neill, P. M., Park, B. K., & Ward, S. A. (1999). New 4-Aminoquinoline Mannich Base Antimalarials. 1. Effect of an Alkyl Substituent in the 5'-Position of the 4'-Hydroxyanilino Side Chain. *J. Med. Chem.*, *42*, 2747–2756.

Rogers, D. J., Randolp, S. E. (2000). The global spread of malaria in a future warmer world. *Sci.*, *289*, 1763–1766.

Sachs, J., Malaney, P. (2002). The economic and social burden of malaria. *Nat.*, *415*, 680–685.

Sandhu, H. S., Sapra, S., Gupta, M., Nepali, K., Gautam, R., Yadav, S., Kumar, R., Chugh, M., Gupta, M. K., Suri, O. P., & Dhar, K. L. (2010). Synthesis and biological evaluation of arylidene analogs of meldrum's acid as a new class of antimalarial and antioxidant agents. *Bio. Med. Chem.*, *18*, 5626–5633.

Sibley, C. H., Hyde, J. E., Sims, P. F., Plowe, C. V., Kublin, J. G., Mberu, E. K., Cowman, A. F., Winstanley, P. A., Watkins, W. M., & Nzila, A. M. (2001). Pyrimethamine-sulfadoxine resistance in Plasmodium falciparum: what next? *Trends Parasitol.*, *17*, 582–588.

Singh, C., Malik, H., & Puri, S. K. (2004). Orally active amino functionalized antimalarial 1, 2, 4-trioxanes. *Bioorg. Med. Chem. Lett.*, *14*, 459–462.

Singh, C., Pandey, S., Kushwaha, A., & Puri, K. (2008). New functionalized 1, 2, 4-trioxepanes: synthesis and activity against multidrug resistant P. *yoeii* in mice. *Bio. Med. Chem.*, *18*, 5190–5193.

Snow, R. W., Trape, J. F., & Marsh, K. (2001). The past, present and future of childhood malaria mortality in Africa. *Trends Parasitol.*, *17*, 593–597.

Solaja, B. A., Terzic, N., Pocsfalvi, G., Gerena, L., Tinant, B., Opsenica, D., & Milhous, W. K. (2002). Mixed Steroidal 1,2,4,5-Tetraoxanes: Antimalarial and Antimycobacterial Activity. *J. Med. Chem.*, *45*, 3331–3336.

Solomon, V. R., Haq, W., Smilkestein, M., Srivastava, K., Puri, S. K., & Katti, S. B. (2010). 4-aminoquinoline derived antimalarials: synthesis, antiplasmodial activity and heme polymerization inhibition studies. *Eur J. Med. Chem.*, *45*, 4990–4996.

Sullivan, D. J. (2002). Theories on malarial pigment formation and quinoline action. *Int. J. Parasitol. 32*, 1645–1653.

Sutherst, R. W. (2004). Global change and human vulnerability to vector-borne diseases. *Clin. Micro. Rev.*, *17*, 136–173.

Sztanke, K., Tuzimski, T., Rzymowska, J., Pasternak, K., & Szerszen, M. K. (2008). Synthesis, structure elucidation and identification of antitumoral properties of novel fused 1, 2, 4-triazine aryl derivatives. *Eur. J. Med. Chem.*, *43*, 1085–1094.

Taunton, J., Hassig, C. A., & Schreiber, S. L. (1996). A mammalian histone deacetylase related to the yeast transcriptional regulator Rpd3p. *Sci.*, *272*, 408–411.

Tingle, M. D., Jewell, H., Maggs, J. L., O'Neill, P. M., & Park, B. K.(1995). The bioactivation of amodiaquine by human polymorphonuclear leucocytes in vitro: chemical mechanisms and the effects of fluorine substitution. *Biochem. Pharmacol.*, *50*, 1113–1139.

Trouiller, P., Olliaro, P., Torreele, E., Orbinski, J., Laing, R., & Ford, N.(2002). Drug development for neglected diseases: a deficient market and a public-health policy failure. *Lancet.*, *359*, 2188–2194.

Uhlemann, A. C., Cameron, A., Eckstein-Ludwig, U., Fischbarg, J., Iserovich, P., Zuniga, F. A., East, M., Lee, A., Brady, L., Haynes, R. K., & Krishna, S. (2005). *Nat. Struct. Mol. Biol.*, *12*, 628–629

Valderramos, S. G., Fidock, D. A. (2006). Transporters involved in resistance to antimalarial drugs. *Trends Pharmacol Sci.*, *27*, 594–601.

Vangapandu, S., Sachdeva, S., Jain, M., Singh, S., Singh, P. P., Kaul, C. L., & Jain, R. (2003). 8-Quinolinamines and their Prodrug Conjugates as Potent Blood-Schizontocidal Antimalarial Agents. *Bioorg. Med. Chem.*, *11*, 4557–4568.

Vennerstrom, J. L., Arbe-Barnes, S., Brun, R., Charman, S. A., Chlu, F. C. K., Chollet, J., Dong, Y., Dorn, A., Hunziker, D., Matile, H., McIntosh, M., Pad-manilayam, M., Tomas, J. S., Scheurer, C., Scorneaux, B., Tang, Y., Urwyler, H., Wittlin, S., & Charman, W. N. (2004). Identification of an antimalarial synthetic trioxolane drug development candidate. *Nat*, *430*, 900–904.

Vennerstrom, J. L., Fu, H. N., Ellis, W. Y., Ager, A. L., Wood, J. K., Andersen, S. L., Gerena, L., & Milhous, W. K. (1992). Dispiro-1,2,4,5-tetraoxanes: a new class of antimalarial peroxides. *J. Med. Chem.*, *35*, 3023–3027.

Vennerstrom, J. L., Nuzum, E. O., Miller, R. E., Dorn, A., Gerena, L., Dande, P. A., Ellis, W. E., Ridley, R. G., & Milhous, W. K. (1995). Prophylaxis of *Plasmodium falciparum* Infection in a Human Challenge Model with WR 238605, a New 8-Aminoquinoline. *Antimal. Antimicrob. Agen Chemother.*, *43*, 598–602.

White, N. J. (2008). Qinghaosu (Artemisinin): The Price of Success. *Science, 320*, 330–334.

Wiesner, J., Ortmann, R., Jomaa, H., & Schlitzer, M. (2003). New antimalarial drugs. *Angew Chem Int Engl.*, *42*, 5274–5293.

Wilairatana, P., Krudsood, S., Reeprasertsuk, S., Chalermrut, K., & Looare Suwan, S. (2002). The Future Outlook of Antimalarial Drugs and Recent Work on the Treatment of Malaria. *Arch. Med. Res.*, *33*, 416–421.

Woerdenbag, H. J., Moskal, T. A., Pras, N., Maringle, T. M., ElFeraly, F. S., Kampinga, H. H., & Konings, AWT (1993). Cytotoxicity of Artemisinin-Related Endoperoxides to Ehrlich Ascites Tumor Cells. *J. Nat. Prod.*, *56*, 849–856.

Woodrow, C. J., Haynes, R. K., & Krishna, S. (2005). Artemisinins. Postgrad. Med. J., *81*, 71–78.

Wu, J. M., Shan, F., Wu, G. S., Li, Y., Ding, J., Xiao, D., Han, J. X., Atassi, G., Leonce, S., Caignard, D. H., & Renard, P. (2001). Synthesis and cytotoxicity of artemisinin derivatives containing cyanoarylmethyl group. *Eur. J. Med. Chem.*, *36*, 469–479.

Wu, Y. L., Li, Y. (1995). Study on the chemistry of artemisinin. *Med. Chem. Res.*, *5*, 569–586.

Yayon, A., Cabantchik, Z. I., & Ginsburg, H. (1984). Identification of the acidic compartment of Plasmodium falciparum-infected human erythrocytes as the target of the antimalarial drug chloroquine. *Embo J. 3*, 2695–2700.

Yeates, C. L., Batchelor, J. F., Capon, E. C., Cheesman, N. J., Fry, M., Hudson, A. T., Pudney, M., Trimming, H., Woolven, J., Bueno, J. M., Chicharro, J., Fernández, E., Fiandor, J. M., Gargallo-Viola, D., Heras, F., Herreros, E., & León, M. L. (2008). Synthesis and Structure–Activity Relationships of 4-Pyridones as Potential Antimalarials. *J. Med. Chem.*, *51*, 2845–2852.

Zhang, Y., Plattner, J. J., Easom, E. E., Waterson, D., Ge, M., Li, Z., Li, L., & Jian, Y. (2011). An efficient synthesis for a new class antimalarial agent 7-(2-carboxyethyl)-1, 3-dihydro-1-hydroxy- 2, 1-benzoxaborole. *Tetrahedron Lett.*, *52*, 3909–3911.

Zhang, Y., Plattner, J. J., Easom, E. E., Waterson, D., Ge, M., Li, Z., Li, L., & Jian, Y. (2011). Synthesis and structure activity relationships of novel benzoxaborales as a new class of antimalarial agents. *Bio. Med. Chem.*, *21*, 644–651.

Zheng, G. Q. (1994). Cytotoxic terpenoids and flavonoids from Artemisia annua. *Planta Med.*, *60*, 54–57.

Zhu, S., Zhang, Q., Gudise, C., Meng, L., Wei, L., Smith, E., & Kong, Y. (2007). Synthesis and evaluation of naphthyridine compounds as antimalarial agents. *Bioorg. Med. Chem. Lett.*, *17*, 6101–6106.

Zmitek, K., Stavber, S., Zupan, M., Bonnet-Delpon, D., Charneau, S., Grellier, P., & Iskra, J. (2006). Synthesis and antimalarial activities of novel 3,3,6,6-tetraalkyl-1,2,4,5-tetraoxanes. *Bio. Med. Chem.*, *14*, 7790–7795.

CHAPTER 9

CHEMISTRY AND PHARMACOLOGY OF β-LACTAM ANALOGS

ANKUR VAIDYA[1] and SHWETA JAIN[2]

[1]Pharmacy College, Uttar Pradesh University of Medical Sciences, Saifai, Etawah 206130, Uttar Pradesh, India, Tel: +91-7582-266171; E-mail: ankur_vaidya2000@yahoo.co.in

[2]ADINA College of Pharmacy, Sagar, M.P., 470003, India

CONTENTS

9.1   INTRODUCTION

Antimicrobial or antibacterials, drugs are the greatest contribution of the twentieth century to therapeutics. Their advent changed the outlook of the physician about the powerful drugs can have on diseases. The word antibiotics derived from the two Greek words *anti* mean "against", and *bios* means "life" (bacteria are life forms), so the antibiotics are the substances (both natural and synthetics) used to control the growth of or kill bacteria.

The ancient Egyptians, Chinese, Indians and of central Americans all used molds to treat infected wounds, still they did not appreciate the connection of the antibacterial properties of molds and the treatment of diseases. In late 1880 scientist introduced 'germ theory of disease', which results scientists began to devote time to searching for drugs that would kill these diseases-causing bacteria (Wainwright et al., 1986). In 1871 British surgeon Joseph Lister, reported that urine contaminated with mold not allow the successful growth of bacteria. In 1890 German doctors, Rudolf Emmerich and Oscar Low reported the so-called first antibiotics pyocyanase from microbes, which was often did not work against bacteria. On the morning of September 3rd, 1928, a British Professor Alexander Fleming found a ring shape mold over a *Staphylococcus* pre coated glass plate. The area around the ring seemed to be free of the bacteria *Staphylococcus*. Curious, Alexander Fleming grew the mold of *Penicillium Notatum* in a pure culture and found that it produced substance that killed a number of disease-causing bacteria. Later he proposed the name of this chemical is penicillin, the first official antibiotic of medicine. With the success of penicillin, the race to produce other antibiotics began and today, physicians can choose from over 100 of antibiotics now on the market, and they're being prescribed in very high numbers. These available antibiotics work on variety of microorganism either by killing (bactericidal) or by inhibiting their growth (bacteriostatic) (Curtis et al., 2007). Theses antibiotics classified by their ability to treat infections as broad spectrum and narrow spectrum antibiotics:

9.1.1 BROAD SPECTRUM ANTIBIOTICS

Theses antibiotics are effective against a broad range or variety of bacteria and are useful to treat many infections together. For example: Cephalosporins are broad spectrum antibiotics used for treating pneumonia, meningitis and blood infections (Martin, 2003).

9.1.2 NARROW SPECTRUM ANTIBIOTICS

These are effective against a limited group of organisms or a single bacterium that causes multiple infections. For example: Azithromycin is a macrolide antibiotic specific for bacteria causing throat infections, laryngitis, tonsillitis etc.

In 1884 Hans Christian Gram, characterized bacteria based on the structural characteristics of their cell walls. The thick layers of peptidoglycan in bacteria termed as "gram-positive" (stain purple), while the thin layer called "gram-negative" (appears pink).

9.2 β-LACTAM ANTIBIOTICS

The modern era of β-lactam antibiotics began in 1928 with the discovery of penicillin by Professor Alexander Fleming. In 1940, Florey and Chain extracted penicillin from *Penicillium Notatum* and showed that it possessed powerful chemotherapeutic values in infected mice and that it was nontoxic. Penicillin antibacterial effects in man were confirmed in 1941 by the treatment of a policeman who was infected with *Staphylococcal* and *Streptococcal septicemia* with multiple absences. Ten years later *Penicilium Notatum* becomes rich source of penicillin G for clinical use. After the discovery of penicillin numerous penicillin derivatives were developed that are stable at acid *p*H, resistant to β-lactamase, and active against both gram-positive and gram-negative bacteria.

Later penicillin and penicillin derivatives were classified as β-lactam antibiotics because of their unique four-member β-lactam ring. This β-lactam ring consists of three carbon atoms and one nitrogen atom in a closed four membered cyclic ring system. It is named so, because the nitrogen atom is attached to the β-carbon relative to the carbonyl (C=O) group. The β-lactam antibiotics are a large class of diverse compounds used clinically in both the oral and parenteral forms. Todays β-lactam antibiotics have become the most widely used therapeutic antimicrobials because of their broad antibacterial spectrum and excellent safety profile. This includes penicillins, cephalosporins, monobactams and carbapenems. Most β-lactam antibiotics are primary bactericidal agents and work by inhibiting bacterial cell wall biosynthesis by interference with the synthesis of peptidoglycan (Cohen, 1983).

These β-lactam antibiotics are classified according to their ring fused:
- β-lactam fused to saturated five-membered ring
  - β-lactam attached to thiazolidine rings are named penams

- – β-lactam attached to oxazolidine rings are named oxapenams or clavams
- – β-lactam attached to pyrrolidine rings are named carbapenams
- • β-lactam fused to unsaturated five-membered ring
  - – β-lactam having 2,3-dihydro-1H-pyrrole rings are named carbapenems
  - – β-lactam having 2,3-dihydrothiazole rings are named penems
- • β-lactam fused to unsaturated six-membered ring
  - – β-lactam attached to 3,6-dihydro-2H-1,3-thiazine rings are named cephems
  - – β-lactam attached to 1,2,3,4-tetrahydropyridine rings are named carbacephems
  - – β-lactam attached to 3,6-dihydro-2H-1,3-oxazine rings are named oxacephems
- • β-lactam not fused to any other ring is called monobactam (other ring is missing or absent)

9.3   PENICILLINS

The penicillins constitute one of the most important groups of β-lactam antibiotic to be used clinically since 1941. The success of penicillin led to the discovery and development of various β-lactam antibiotics: penicillins, cephalosporins, monobactams and carbapenems which all contain the four membered β-lactam ring. Penicillin (or penicillin G) was initially obtained from the fungus *Penicillium notatum*, but the present source is a high yielding mutant of *P. chrysogenum*. Natural penicillin or penicillin G has several limitations, that is, unstable in gastric pH, therefore not given orally; susceptible to destruction by β-lactamase; relatively inactive against gram negative bacteria; narrow spectrum of action and short duration of action. To avoid these limitations new or modified penicillins are investigated and used. These include penicillin V, ampicillin, amoxicillin, methicillin, carbenicillin, cloxacillin etc. (Dürckheimer et al., 1985).

9.3.1   PREPARATION

Natural penicillin is obtained by aerobic fermentation in fed batch ferments made with some high yield molds of penicillium strains, usually *Penicillium notatum* or *Penicillium chrysogenum* that transforms substrates rich in carbohydrates into penicillin. As with other antibiotic production processes, the penicillin process operated at antibiotics involves four stages. The incubation of the culture strain provides the seed that grows in seed ferments until a stage of maturity is reached. Then, the seed is transferred to a final-stage ferment. These ferments are operated in a fed – batch mode under standard conditions in order to optimize the synthesis of penicillin. After that, the product is withdrawn by solvent extraction in the downstream. Commercially *Penicillium chrysogenum* ATCC 16520 (obtained from American Type Culture Collection) is used for the production of penicillin (Hamed et al., 2013).

Penicillin obtained by this method is secondary metabolite and is produced when growth of the fungus is inhibited by stress. Production is also limited by feedback in the synthesis pathway of penicillin.

$$\alpha\text{-ketoglutarate} + \text{AcCoA} \rightarrow \text{homocitrate} \rightarrow \text{L-}\alpha\text{-aminoadipic acid} \rightarrow$$
$$\text{L-lysine} + \beta\text{-lactam}$$

The by-product, L-lysine, inhibits the production of homocitrate, so the presence of exogenous lysine should be avoided in penicillin production.

The available carbon sources are also important: glucose inhibits penicillin production, whereas lactose does not. The *p*H and the levels of nitrogen, lysine, phosphate and oxygen of the batches must also be carefully controlled.

9.3.2 RING NOMENCLATURE

The basic structure of all penicillins has sulfur containing thiazolidine ring fused to β-lactam ring, forming 6-aminopenicillanic acid (6-APA).

Thiazolidine ring————————————————————————Beta lactam ring

The nomenclature of penicillins, as with most antibiotics, is complex and very cumbersome. Two numbering systems for the fused bicyclic heterocyclic system are proposed.

The chemical abstract system is perfect and unambiguous but too complex for ordinary use. In chemical abstract system the numbering initiates with sulfur atom and assigns the ring nitrogen as the 4-position. The chemical name of benzyl penicillin or penicillin G is 4-thia-1-azabicyclo [3,2,0] heptanes. Thus according to Chemical Abstract System penicillin designated as 6-acylamino-2,2-dimethylpenam-3-carboxylic acids. The atoms comprising the 6-aminopenicillanic acid (6-APA) derived biosynthetically from two amino acids, L-cysteine (S-1, C-5, C-6, C-7 and 6-amino) and L-valine (2,2-dimethyl, C-2, C-3, N-4 and 3-carboxyl) (Dalhoff et al., 2006).

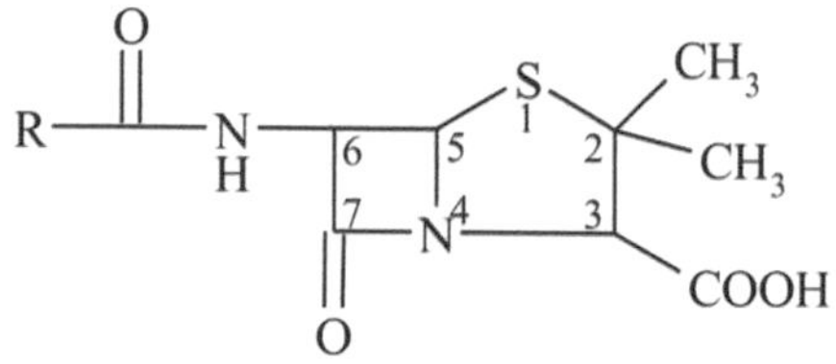

Chemical Abstract

United State Pharmacopeia uses a different system which results in the atoms differently, assigning the nitrogen atom 1 and the sulfur atom as number 4.

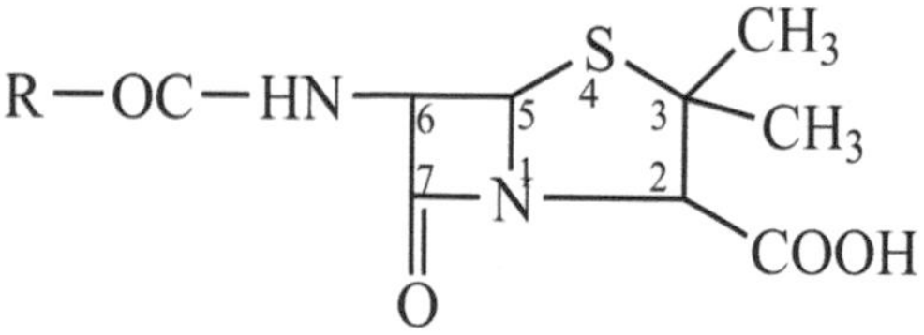

USP System

All naturally, synthetic and semi synthetic penicillins contain three asymmetric carbon atoms (C-3, C-5 and C-6). The carbon atom attached to carboxyl group (C-3) has D-configuration where as carbon atom bearing the acylamino group (C-6) has L-configuration. Thus the carboxyl group and acylamino are trans to each other, with the former in the β- and the latter in the α-orientation relative to the penam ring system. The absolute stereochemistry of penicillin is designated as 3S:5R:6R.

9.3.3 CLASSIFICATION OF PENICILLINS

Penicillins can be classified into the following groups:
1. *Natural Penicillin*: Penicillin G, Penicillin V.
2. *Penicillinase Resistant Penicillin*: Methicillin, Oxacillin, Cloxacillin, Dicloxacillin, Naficillin.
3. *Aminopenicillins*: These are extended spectrum penicillin. Ampicillin, Amoxycillin.
4. *Carboxy-Penicillin*: These are also extended spectrum penicillin. Carbenicillin, Ticarcellin.
5. *Ureidopenicillins*: These are also extended spectrum penicillin. Mezlocillin, Piperacillin.

Carboxypenicillins and Ureidopenicillins are also called as antipseudomonal penicillins due to their activity against pseudomonas eruginosa (Darouiche et al., 1994) (Table 9.1).

9.3.4 SPECTRUM OF ACTION

PnG is a narrow spectrum antibiotic; and mainly used in the treatment of gram-positive bacteria infections and few others. PnG is susceptible to hydrolysis by β-lactamases enzymes produced by especially *staphylococcal*. Antistaphylococcal penicillins or penicillinase resistant penicillins (e.g.,

TABLE 9.1   Structure of Penicillin

Generic Name	Chemical Name	R Group
Natural Penicillin		
Penicillin G	Benzyl Penicillin	
Penicillin V	Phenoxymethylpenicillin	
Penicillinase Resistant Penicillin		
Methicillin	2,6-Dimethoxyphenyl-penicillin	
Oxacillin	5-Methyl-3-phenyl-4-isoxazoylpenicillin	
Cloxacillin	5-Methyl-3-(2-chlorophenyl)-4-isoxazoylpenicillin	

Generic Name	Chemical Name	R Group
Dicloxacillin	5-Methyl-3-(2,6-dichlorophenyl)-4-isoxazoylpenicillin	
Nafcillin	2-Ethoxy-1-naphthyl-penicillin	
Aminopenicillins		
Ampicillin	D-α-Aminobenzyl-penicillin	
Amoxicillin	D-α-Amino-*p*-hydroxybenzyl-penicillin	
Carboxy-Penicillin		
Carbenicillin	α-Carboxybenzylpenicillin	
Ticarcillin	α-Carboxy-3-thienyl-penicillin	

Generic Name	Chemical Name	R Group
Ureidopenicillins		
Mezlocillin	α-(1-methanesulfonyl-2-oxoimidazolidino-carbonylamino)benzyl penicillin	
Piperacillin	α-(4-Ethyl-2,3-dioxo-1-piperazinylcarbonyl-amino) benzylpenicillin	

methicillin, oxacillin, cloxacillin and nafcillin) are resistant to *staphylococcal* lactamases or β-lactamases and thus shown activity against *staphylococci* and *streptococci* bacteria but inactive against enterococci, anaerobic bacteria, and gram-negative cocci and rods. Extended-spectrum penicillins (ampicillin and the antipseudomonal penicillins) retain the antibacterial spectrum of penicillin and have improved activity against gram-negative organisms, but they are destroyed by β-lactamases (Sykes et al., 1981).

9.3.5 MECHANISM OF ACTION

β-lactam antibiotics plays a center role in chemotherapy because: a potent and rapid –cidal action against bacteria in the growth phase and a very low frequency of toxic and other adverse reactions in the host. All β-lactam antibiotics act by inhibiting bacterial cell wall synthesis (Park et al., 2008). Penicillins act by binding to proteins called penicillin binding proteins (PBPs). Penicillin binding proteins (PBPs) are usually seven different functional proteins:

PBPs 1$_a$ and **1**$_b$ are transpeptidase engaged in peptidoglycan synthesis associated with cell elongation. Inhibition results in spheroplast formation and rapid cell lysis caused by autolysins.

PBP 2 is also transpeptidase engaged in maintaining the rod shape of bacteria. Inhibition results in ovoid forms that endure delay lysis.

PBP 3 is also transpeptidase entailed for septum formation during cell division. Inhibition results in the formation of filamentous forms containing rod-shaped units that cannot separate.

PBP 4–6 are carboxypeptidases responsible for the hydrolysis of D-terminal-D-terminal peptide bonds of the cross linking peptides.

Penicillins differ in their affinities for PBPs, but penicillin G binds preferentially to PBP 3, while first generation cephalosporin binds with high affinity to PBP 1$_a$. Amdinocillin or mecillinam binds to PBP 2, and imipenem binds to PBP 1$_b$ and PBP 2 only.

The cell wall of bacteria composed of peptidoglycan a heteropolymeric part which provides rigid mechanical stability by virtue of its highly crosslinked latticework structure. The peptidoglycan composed of two alternative amino sugars N-acetylglucosamine and N-acetylmuramic acid that are cross-linked by peptide chains (Zapun et al., 2008). Gram-positive cell wall is 50–100 molecules thick while in gram-negative it is only 1 or 2 molecule thick. The biosynthesis of peptidoglycan involves about 30 bacterial enzymes. Penicillin inhibits number of theses enzymes. Penicillins acylate D-transpeptidase enzyme a specific bacterial enzyme, responsible for peptide cross-linking of two linear peptidoglycan strands by transpeptidation and loss of D-alanine. This results in inhibition of dipeptidoglycan synthesis and ultimately loss of strength and rigidity of bacterial cell wall.

9.3.6 DEVELOPMENT OF RESISTANCE

The use of penicillin is limited by the fact that, many microorganisms have developed resistance to the penicillins, and serious hospital epidemics involving infants and surgical patients have been caused by penicillin-resistant bacterias. Some organisms are resistant because PnG is unable to penetrate the bacterial cell, alterations of penicillin target enzymes the penicillin binding proteins (PBPs) and produce an enzyme, penicillinase, that destroys the antibiotic (Fisher et al., 2005).

The primary bacterial resistance to the β-lactam antibiotics is caused by the inability of the agent to penetrate to its site of action. In gram-positive bacteria, the peptidoglycan polymer are very near the cell surface while in gram-negative bacteria, their surface structure is more complex, and their inner membrane, which is analogous to the cytoplasmic membrane of gram-positive bacteria, is covered by the outer membrane, lipopolysaccharide, and capsule. The outer membrane functions as an impenetrable barrier for some antibiotics, for example, penicillin G. Thus penicillin G is highly sensitive for gram positive bacteria and insensitive for gram negative. Broader-spectrum penicillins (such as ampicillin and amoxicillin) are small hydrophilic antibiotics, however, diffuse through aqueous channels in the outer membrane of gram negative bacilli that are formed by proteins called *porins* and inhibit its cell wall.

Penicillin-binding proteins, PBPs, are crucial enzymes represent the targets for β-lactam antibiotics and are thus involved in the evolution of penicillin resistance in *Streptococcus pneumoniae*. Whereas distinct point mutations in individual PBP genes occur mutants selected for resistance, gene transfer events play an additional role in the emergence and spread in clinical isolates. Active efflux pumps serve as another mechanism of resistance, removing the antibiotic from its site of action before it can act. This is an important mechanism of β-lactam resistance in *P. aeruginosa, E. coli,* and *Neisseria gonorrheae* (Kong et al., 2009).

The most important biochemical mechanism of penicillin resistant is the bacterial elaboration of enzymes called penicillinase that destroys the antimicrobial action of the drug. Commercially many penicillinases are known, and are produced by a wide variety of bacteria. These penicillinase are classified into two category on the basis of their reaction catalyze.

The most important class of penicillinase is β-lactamase that breaks the β-lactam ring, and deactivating the molecule's antibacterial properties.

β-lactamase produced by some bacteria especially gram-negative organisms (e.g., *Staphylococcus aureous*). These bacteria are usually secreted β-lactamase enzyme, when antibiotics are present in the environment. Synthesis of bacterial β-lactamase may be under chromosomal and plasmid R factor control and may be either constitutive or inducible. Some β-lactamase enzymes are either cytoplasmic that remains in the bacterial cells (e.g., in gram negative) or those which synthesized by bacterial cell wall and released extracellular (e.g., in *Staph. aureus*). Now-a-days, variety of β-lactamase have been discovered by their bacterial source, by their structure, their substrate, and there inhibitory specificities. Some of them are summarized in the following subsections.

9.3.6.1 TEM β-Lactamases

Based upon different combinations of amino acid substitutions, currently 140 TEM-type enzymes have been described. Although TEM-type β-lactamases are most often found in *E. coli* and *K. pneumoniae*, they are also found in other species of gram-negative bacteria with increasing frequency. TEM-1 is the most prevalent β-lactamases in gram-negative bacteria; they hydrolyze the β-lactam bond in susceptible β-lactam antibiotics, thus conferring resistance to penicillins and cephalosporins. TEM-3, TEM-4 and TEM-5 are capable of hydrolyzing cefotaxime, ceftazidime and ceftazidime respectively. TEM-6 is capable of hydrolyzing ceftazidime and aztreonam. TEM-8/CAZ-2, TEM-16/CAZ-7 and TEM-24/CAZ-6 are markedly active against ceftazidime.

9.3.6.2 SHV β-Lactamases

SHV β-lactamases enzymes belong to the molecular class A of serine and share extensive functional and structural similarity to TEM β-lactamases. The group of plasmid-mediated SHV β-lactamases includes SHV-1 and at least 20-three variants, most of which possess extended-spectrum (ES) activity. Their likely ancestor is a chromosomal penicillinase of *Klebsiella pneumoniae*. Foremost β-lactamases produced by the *Enterobacteriaceae* is the SHV family. SHV-1 shares 68 percent of its amino acids with TEM-1 and have a similar overall structure.

9.3.6.3   CTX-M β-Lactamases

CTX-M β-lactamases were named for their greater activity against cefo-taxime than other oxyimino-β-lactam substrates (e.g., ceftazidime, ceftri-axone, or cefepime). Currently more than 80 CTX-M enzymes are known. Few CTX-M β-lactamases are more active on ceftazidime than cefotaxime. These enzymes are not very closely related to TEM or SHV β-lactamases in that they show only approximately 40% identity with these two commonly isolated β-lactamases.

9.3.6.4   OXA β-Lactamases

OXA β-lactamases are less common β-lactamase variety that could charac-terized by their high hydrolytic activity against oxacillin and cloxacillin and that they are poorly inhibited by clavulanic acid. These β-lactamases differ from the TEM and SHV enzymes in that they belong to molecular class D and functional group 2d.

Second important penicillinase enzyme is acylase that are capable to hydrolyze the acylamino side chain of penicillin. These acylase enzymes have been obtained from several species of gram-negative bacteria, but their possible role in bacterial resistance has not been well defined.

9.3.7   ADVERSE DRUG REACTIONS

As with most antibiotics, penicillin exhibits common side effects and adverse reactions. Common adverse drug reactions (≥1% of patients) associ-ated with use of the penicillins include allergic reactions or hypersensitivity, neurotoxicity, urticaria, rash, diarrhea, nausea, and superinfection (includ-ing candidiasis). Infrequent adverse effects (0.1–1% of patients) include fever, vomiting, erythema, dermatitis, angioedema, seizures (especially in people with epilepsy), and pseudomembranous colitis. Serious allergies are the most common adverse reactions to penicillins, with about 10% of people reporting an allergy. All forms of natural and synthetic penicillins can cause allergy, but it is more commonly seen after parenteral than oral administration.

Two kinds of allergic reactions, that is, immediate and delayed are noted clinically to penicillin. Immediate reactions response within 20 min of administration and is very rare and usually occurs after parenteral therapy. These immediate reactions range in severity from urticaria and pruritus to angloneurotic edema, laryngospasm, bronchospasm, hypotension, vascular collapse and death. While delayed reactions to penicillin therapy starts within 1 to 2 weeks after initiation of therapy. Manifestations include serum sickness-like symptoms, i.e., fever, malaise, urticaria, myalgia, arthralgia, abdominal pain and various skin rashes, ranging from maculopapular eruptions to exfoliative dermatitis. These adverse reactions may be controlled with antihistamines and, if necessary, systemic corticosteroids. Whenever such reactions occur, the drug should be discontinued unless, in the opinion of the physician, the condition being treated is life-threatening and amenable only to penicillin therapy (Hautekeete, 1995).

Penicillin G shows highest incidence of drug allergy. Incidence is further increased with procaine penicillin, as procaine itself is allergenic. The penicillin hypersensitivity is unpredictable, that is, an individual who tolerated penicillin earlier may show allergy on subsequent administration and *vice versa*. However, the majority of people who believe they are allergic can take penicillin without a problem, either because they were never truly allergic or because their allergy to penicillin has resolved over time. The proposed mechanism involved in penicillin allergy is there rearrangement product formed in vivo, that is, penicillenic acid which react with lysine-ξ-amino acids of proteins to form penicilloyl proteins, which are antigenic determinants. However, true allergy, namely acute and subacute reactions mediated by IgE and IgG antibodies respectively. The acute allergic reaction outcome from reaction of IgE to penicillin instantly or rapidly within minutes to an hour and includes sudden anaphylaxis with hypotension, bronchospasm, angioedema and urticaria. This results release of histamine and other mediators from mast cells produce the signs and symptoms typical of a true anaphylactic reaction. The subacute reaction caused by preformed IgG to penicillin (prepared previous penicillin treatment) results in the activation of the complement reactions producing inflammation resulting in the symptoms include urticaria, fever and arthralgias or arthritis. Skin rashes including scarlatiniform, morbilliform, vesicular and bullous eruption may be caused by allergy to penicillin. Purpuric lesions are uncommon and usually are the result of a vasculitis; thrombocytopenic purpura may occur very rarely. More severe skin reactions are exfoliative dermatatitis and exudative

erythema multiforme. The incidence of skin rashes seems to be highest following the use of ampicillin (Fossieck and Parker, 1974). Polymeric impurities in ampicillin dosage forms have been implicated as possible dosage forms
have been implicated as possible antigenic determinants and shown highest
incidence of allergic incidence especially with semisynthetic penicillin.

9.3.8 STRUCTURE ACTIVITY RELATIONSHIP (SAR)

A large number of penicillin analogs have been synthesized and studied. The
results of these studies led to the following conclusions:

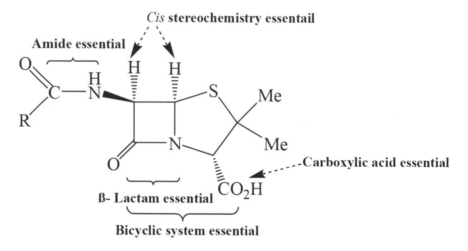

1. β-lactam ring or 6-aminopenicillanic acid is essential for biological
    activity.
2. Presence of H atom at C-5 and C-6 must be present in Cis form for
    biological activity.
3. Sulphur at position 1 of penicillin responsible for antibacterial activ-
    ity of penicillin.
4. The germinal dimethyl group at C-2 is characteristic for activity.
5. The design and development of C-6 substituents has been aimed at
    strengthening various weaknesses which have traditionally hampered
    penicillins in terms of activity, stability, resistance and absorption/
    distribution.

6. The C-6 amine moiety is necessary for appreciable antibacterial activity, but substitution of the amine via monoacylation can offer much more potent congeners.

7. Only carboxamido derived at C-6 position are tolerated; sulphonation or phosphoramide containing substituent are devoid of antibacterial activity. Similarly imide or carbamate containing at C-6 are inactive.

8. Introducing the bulk substitution around β-lactam ring make the penicillin more stable towards penicillinase enzyme.

9. Attachment of more hydrophilic substitution near β-lactam ring enhance the activity against gram negative bacteria.

10. Spectrum of activity further increase on adding strong acidic groups at the α-carbonyl center of the side chain.

11. The derivatization of carboxylic group of C-3 is not tolerated unless the free penicillin carboxylic acid can be generated in vivo. Derivatization of carboxylic to ester such as alkanoyloxyalkyl congeners undergo rapid cleavage in vivo to generate active penicillin, for example, pivampicillin and becampicillin.

12. The nitrogen atom at the ring junction is imperative for antibacterial activity. The nitrogen atom contributes to the reactivity of the β-lactam carbonyl center.

9.3.9   PENICILLIN G

The term "penicillin" is often used generically to benzylpenicillin or penicillin G. Penicillin G is a narrow spectrum penicillin that is given intravenously or intramuscularly as a treatment for syphilis, meningitis, endocarditis, pneumonia, lung abscesses and septicaemia in children. It is in the list of World Health Organization's Essential Medicines, a list of the most important medications needed in a basic health system. Penicillin G obtained from *P.chrysogenum* (previously known as *Penicillium notatum*) cultures. In normal conditions, the rate of penicillin production is low, but in terms of controlled substances in the culture media and good physical condition, fungus can grow more and the rate of penicillin production will increase, consequently.

Benzylpenicillin is acid labile and thus avoid oral route and can therefore only be given parenteral. The alkali salts of benzylpenicillin are rapidly and completely absorbed after IM injection. Peak plasma levels of 150 to

200 IU/ml are attained 15 to 30 min after IM injections of 10 million IU of penicillin G sodium. After attaining plasma level benzylpenicillin poorly accessible tissues, for example, cardiac valves, bone, CSF and empyema, etc., while inflamed tissue is more readily penetrated. It appears in pleural, pericardial, peritoneal, and synovial fluids but diffuses less readily into the eye and only to a small extent into abscess cavities and avascular areas. Benzylpenicillin diffuses across the placenta into the fetal circulation and found in breast milk are relatively low and very little passes into the CSF unless the meninges are inflamed. The volume of distribution is about 0.3 to 0.4 l/kg in adults and approaches 0.75 l/kg in children. Plasma protein binding is approximately 55%. In adults with normal kidney function plasma half-life is approximately 30 min. Almost administered dose (50 to 80%) is eliminated along renal pathways in an unchanged form (85 to 95%). Tubular excretion is inhibited by probenecid which is sometimes given to increase plasma-penicillin concentrations.

As with other β-lactam antibiotics, it may be expected that untoward reactions will occur in individuals including hypersensitivity like angioedema and anaphylaxis may occur. The jarisch-herxheimer a systemic reaction, may occur after the initiation of penicillin G therapy in patients with syphilis or other spirochetal infections (i.e., lyme disease and relapsing fever). The release from the spirochaete of host stable pyrogen is the proposed pathogenesis of the herxheimer reaction in patients. Hypersensitivity especially Anaphylaxis is the gravest potential adverse reaction to a penicillin G. It is usually associated with the administration of parenteral rather than oral dosage forms. Gastrointestinal reactions including pseudomembranous colitis have been reported with the onset occurring during or after penicillin G treatment. Nausea, vomiting, stomatitis, black or hairy tongue, and other symptoms of gastrointestinal irritation may occur, especially during oral therapy. Sometimes hematological reactions including hemolytic anemia, neutropenia, thrombocytopenia, thrombocytopenic purpura, eosinophilia, leukopenia, and agranulocytosis have also been observed in patients receiving prolonged high doses of penicillin G (e.g., bacterial endocarditis). Penicillin G sodium may cause serious and even fatal electrolyte disturbances when given intravenously in large doses.

9.3.10 PENICILLIN V

Phenoxymethylpenicillin, commonly known as penicillin V was reported by Behrens et al. as a biosynthetic product. Penicillin v is less active than

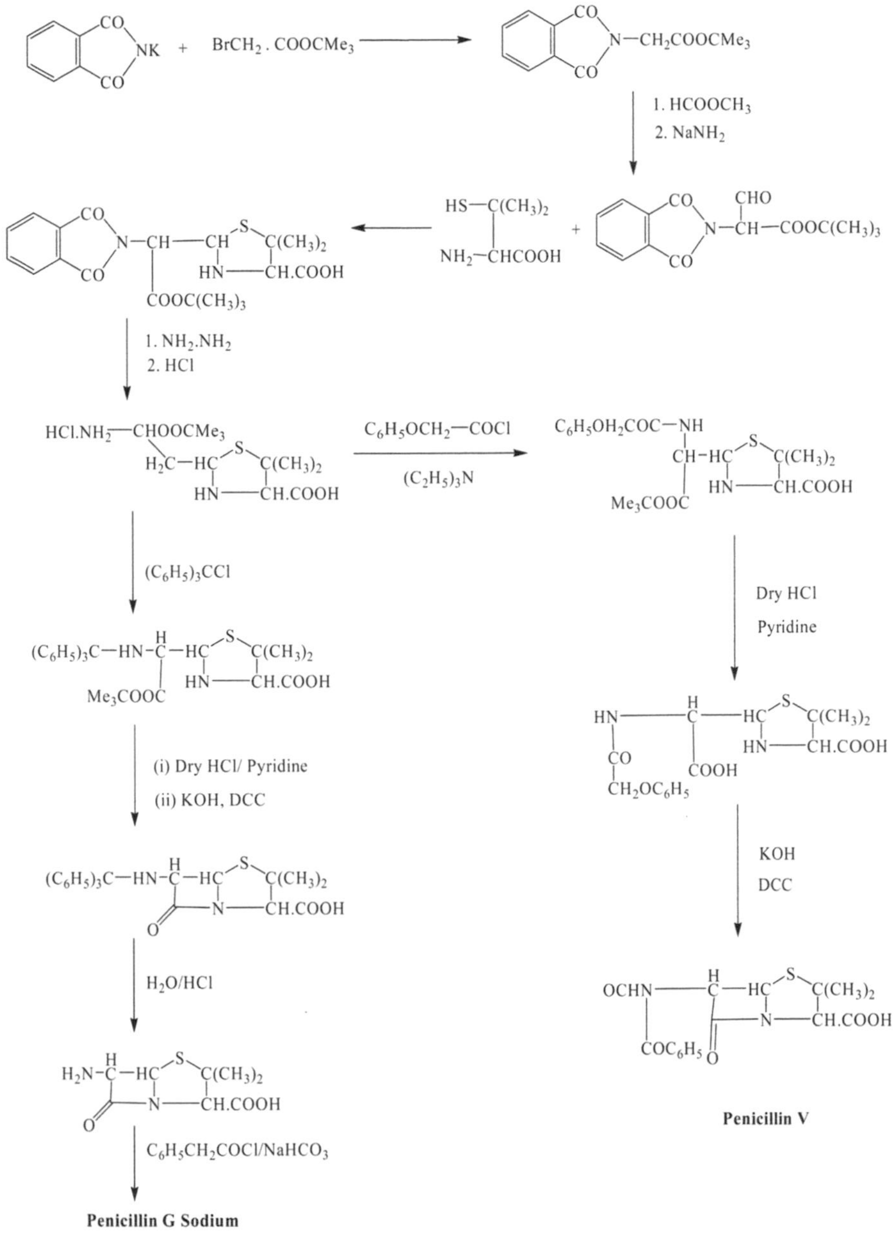

SCHEME 9.1 Synthesis of Penicillin G.

benzylpenicillin (penicillin G) against gram-negative bacteria but it is more acid-stable than benzylpenicillin, which allows it to be given orally. The acid stability of penicillin V is caused by the electronegative oxygen atom present in C-7 amide side chain inhibiting participation in the β-lactam bond hydroly-sis. Penicillin V absorbed better by oral route and produce uniform concentra-tion in blood peak blood level is reached in 1 h and plasma $t_{1/2}$ is 30–60 min. Approximately 25–73% of an oral dose of penicillin V absorbed from the GI tract in healthy, fasting adults. Penicillin V is widely distributed into body tissues and highest concentrations in kidneys; lower concentrations in liver, skin, intestines, bile, tonsils and pericardial fluids. Small amounts of drug observe in CSF. It crosses the human placenta and is distributed into human milk. The plasma protein binding of penicillin V is found to be approximate 75–89%. Penicillin V is less potent and is not, as a rule, used for acutely seri-ous infection as compare to benzyl penicillin. The antimicrobial profile and clinical spectrum of phenoxymethyl penicillin is almost same as penicillin G and is used only for streptococcal pharyngitis, sinusitis, otitis media, prophy-laxis of rheumatic fever (when an oral drug has to be selected), less serious pneumococcal infections and trench mouth. The potassium salt of penicillin V is highly soluble in water and used for parenteral preparations. Potassium salt of penicillin V is also used for oral preparation provides rapid, effec-tive plasma concentration. N,N'-bis(dehydroabietyl)ethylenediamine salt of penicillin V (hydrabamine, Compocillin-V) a water insoluble preparation is use for long-acting form of this compound (for synthesis refer Scheme 9.1).

9.3.11   METHICILLIN

Methicillin is a semisynthetic derivative of penicillin was first developed by Beecham in 1959 as a type of antibiotic called a penicillinase-resistant penicillin. Methicillin contained a modification to the original penicillin structure that made it resistant to a bacterial enzyme called penicillinase (β-lactamase). The steric protection afforded by the 2- and 6-methoxy group makes the methicillin particularly resistant to penicillinase catalyzed hydro-lysis. It was previously used to treat infections caused by susceptible gram-positive bacteria, in particular, penicillinase-producing organisms such as *Staphylococcus aureus* that would otherwise be resistant to most penicillins, but it is no longer clinically used today. Methicillin is highly penicillinase resistant but not acid resistant and thus must be administered parentrally. Currently methicillin role in therapy has been largely replaced by flucloxa-cillin and dicloxacillin, but the term meticillin-resistant *Staphylococcus aureus* (MRSA) continues to be used to describe *S. aureus* strains resistant to all penicillins. The MRSA altered PBPs and thus do not bind penicillins.

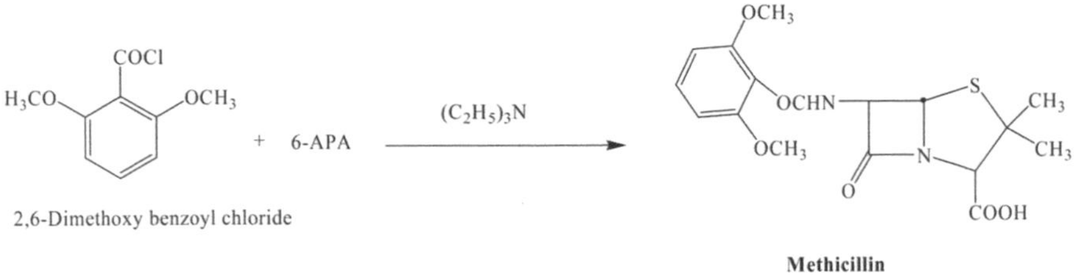

SCHEME 9.2   Synthesis of Methicillin.

The drug of choice for these organisms is vancomycin/linezolid, but cipro-floxacin can also be used. The sodium salt of methicillin is a white, crystal-line solid that is extremely soluble in water forming clear neutral solutions. Methicillin is highly sensitive to moisture, losing about half of its activity in 5 days at room temperature. The loss of activity can be prevented upto 20% by refrigeration at 5°C. Side effects associated with its use included intesti-nal nephritis, a probably hypersensitivity reaction is reported higher as com-pare to other penicillins. Methicillin is primary metabolized into liver (20 to 40%) and mainly excreted through kidney (62 to 80%). The total protein binding of methicillin is 28 to 49%; and volume of distribution is 22 L/kg.

9.3.12   OXACILLIN

Oxacillin sodium is (5-methyl-3-phenyl-4-isoxazolyl)penicillin sodium mono-hydrate and is highly penicillinase as well as acid resistant. Oxacillin is similar to methicillin, and has replaced methicillin in clinical use. Since oxacillin is resistant to penicillinase enzymes, produced by *Staphylococcus aureus*, it is widely used clinically to treat penicillin-resistant *Staphylococcus aureus*. Like methicillin oxacillin is less potent than benzylpenicillin against gram-positive microorganism (generally *Staphylococci* and *Strepetococci*) that do not pro-duce a β-lactamase but retain their potency against those that do. Oxacillin is somewhat more acid stable and thus may be taken orally. Microorganism resistant to methicillin generally also is resistant to oxacillin. With the intro-duction and widespread use of both oxacillin and methicillin, antibiotic-resis-tant strains called methicillin-resistant and oxacillin-resistant *Staphylococcus aureus* (MRSA/ORSA) have become increasingly prevalent worldwide. MRSA/ORSA is treated using vancomycin. Oxacillin sodium is resistant to gastric acid inactivation, but is only partially absorbed after oral administra-tion. The oral bioavailability of oxacillin in humans has been reported to range from 30–35%, and, if given with food, both the rate and extent of absorption is decreased. After IM administration, oxacillin is rapidly absorbed and peak plasma levels generally occur within 30 min. The drug is distributed to the lungs, kidneys, bone, bile, pleural fluid, synovial fluid and ascitic fluid. The volume of distribution is reportedly 0.4 L/kg in human adults. As with the other penicillins, only minimal amounts are distributed into the CSF, but levels are increased with meningeal inflammation. In humans, approximately 89–94% of the drug is bound to plasma proteins. Oxacillin is partially metabolized to both active and inactive metabolites. These metabolites and the parent compound are rapidly excreted in the urine via both glomerular filtration and tubular secretion mechanisms. A small amount of the drug is also excreted in the feces via biliary elimination. The serum half-life in humans with normal renal function ranges from about 18–48 min.

9.3.13   CLOXACILLIN

Cloxacillin is [3-(o-chlorophenyl)-5-methyl-4-isoxazolyl]penicillin sodium monohydrate discovered and developed by Beecham. Cloxacillin has a weaker antibacterial activity than benzylpenicillin, and is devoid of serious toxicity except for allergic reactions. It is active against *Staphylococci* that produce β-lactamase, due to its large side chain, which does not allow the β-lactamases to act on β lactam ring. An added bonus exist in that cloxacillin is somewhat more acid stable acid stable and thus may be taken orally, and it is more potent as well. It has similar pharmacokinetic profile as oxacillin (for synthesis refer Scheme 9.3).

9.3.14   DICLOXACILLIN

Dicloxacillin   is   [3-(2,6-dichlorophenyl)-5-methyl-4-isoxazolyl]penicillin sodium monohydrate a narrow-spectrum β-lactam antibiotic of the penicillin class and used to treat infections caused by susceptible gram-positive bacteria and active against β-lactamase-producing organisms such as *Staphylococcus aureus*, which would otherwise be resistant to most penicillins. The presence of the isoxazolyl group similar to oxacillin and cloxacillin on the side chain of it's and facilitates the β-lactamase resistance, since they are relatively intolerant of side-chain steric hindrance. Dicloxacillin is more acid-stable than many other penicillins and can be given orally, in addition to parenteral routes.

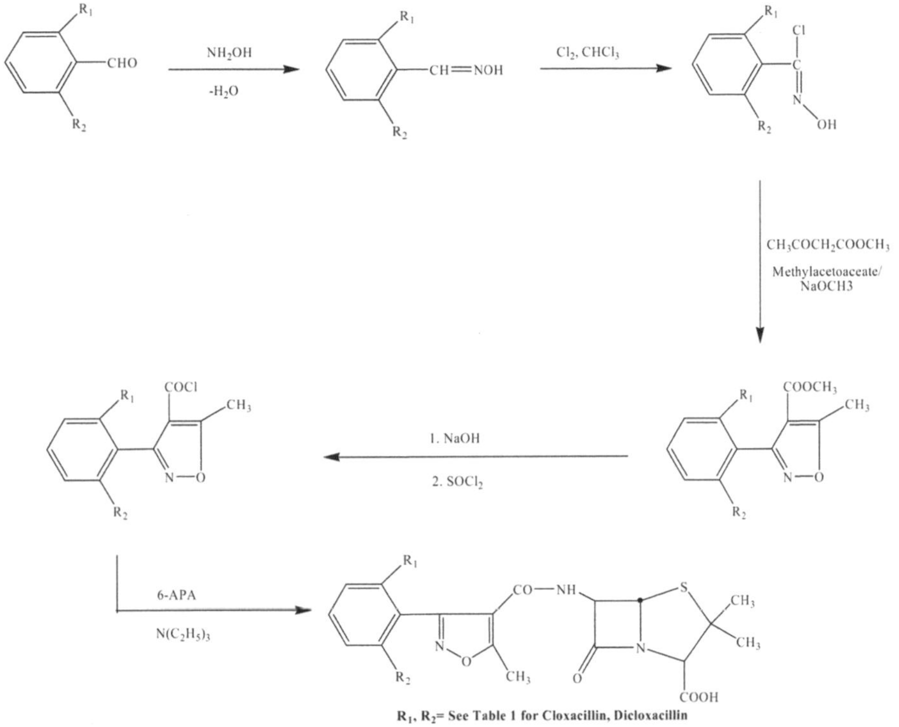

SCHEME 9.3   Synthesis of Oxacillin.

However, like methicillin, it is less potent than benzylpenicillin against non-β-lactamase-producing gram-positive bacteria (for synthesis refer Scheme 9.3).

9.3.15   AMPICILLIN

Ampicillin is 6-[D-α-aminophenylacetamido]penicillianic acid introduced by the British company Beecham. Ampicillin has been used extensively to treat bacterial infection since 1961. Instead of penicillin G, ampicillin demonstrated activity against gram-negative organisms such as *H. influenzae*, Coliforms and *Proteus* spp. Ampicillin was the first of a number of so-called broad spectrum penicillins specially used to treat certain infections caused by bacteria such as pneumonia; bronchitis; and infections in ear, lung, skin, and urinary tract. The presence of ionized or polar group (i.e., amine moiety) into the α-position of the side chain benzyl carbon atom of penicillin G confers activity against gram-negative bacterias. Ampicillin is particularly

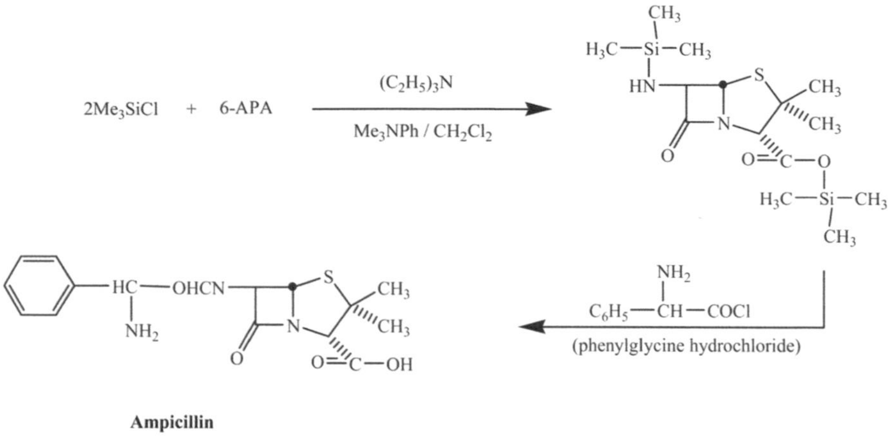

SCHEME 9.4   Synthesis of Ampicillin

useful for the treatment of urinary tract infection caused by *E. coli* or *Proteus mirabilis* and is the agent of choice against *Haemophilus influenza* infections. Probencid inhibit the tubular excretion of ampicillin and thus become a treatment of choice for gonorrhea in recent years. However, ampicillin is active against both gram-positive and gram-negative bacteria but it is not resistant to penicillinase and thus ineffective against penicillinase producing strains of *Staphylococcus aureous*. Ampicillin is water soluble and stable to acid and thus administered orally. It is absorbed from intestinal tract and shown peak plasma concentration in about 2 hr. Ampicillin is rapidly excreted and unchanged through the kidney thus oral dose must be repeated every 6 hr. Ampicillin is available as white, crystalline, anhydrous powder, sparingly soluble in water or as the colorless or slightly buff-colored crystalline trihydrate that is soluble in water. As with other penicillin drugs, ampicillin is relatively nontoxic and adverse effects of a serious nature are encountered only infrequently.

9.3.16   AMOXICILLIN

Amoxicillin is 6[D-(–)-α-Amino-*p*–hydroxyphenylacetamido]penicillanic acid a *p*-hydroxy analog of ampicillin. It is a close congener of ampicillin having moderate-spectrum, bacteriolytic, β-lactam antibiotic in the aminopenicillin family used to treat susceptible gram-positive and gram-negative bacteria. Amoxicillin is resistant to gastric acid and usually the drug of

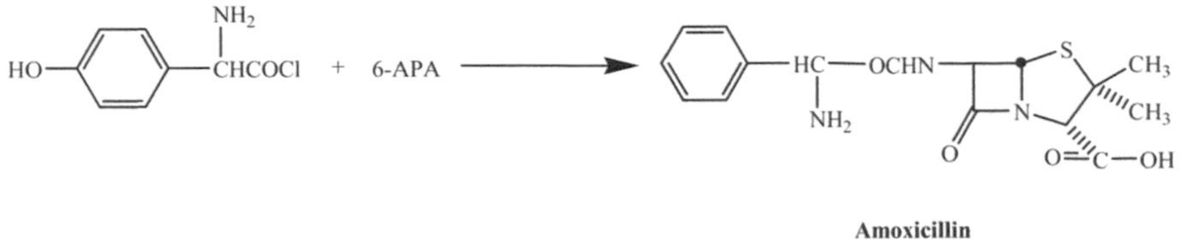

Amoxicillin

SCHEME 9.5 Synthesis of Amoxicillin.

choice within the class because it is better-absorbed, following oral administration, than other β-lactam antibiotics. Oral absorption is better; food does not interfere with its absorption; higher and more sustained blood levels are produced. Amoxicillin shows low incidence of diarrhea. Amoxicillin is susceptible to degradation by β-lactamase-producing bacteria, and thus it is often combined with clavulanic acid, a β-lactamase inhibitor. This drug combination is commonly called Co-amoxiclav. Combining the drugs increases effectiveness by reducing susceptibility to β-lactamase resistance. Amoxicillin-clavulanate is recommended as the first-choice drug for bacterial sinusitis, but most sinusitis is caused by viruses, for which amoxicillin and amoxicillin-clavulanate are ineffective. Amoxicillin is used in the treatment of a number of infections, including acute otitis media, streptococcal pharyngitis, pneumonia, skin infections, urinary tract infections, *Salmonella* infections, Lyme disease, and chlamydia infections. It is also used to prevent bacterial endocarditis in high-risk people having dental work done, to prevent *Streptococcus pneumonia* and other encapsulated bacterial infections in those without spleens, such as people with sickle-cell disease, and for both the prevention and the treatment of anthrax.

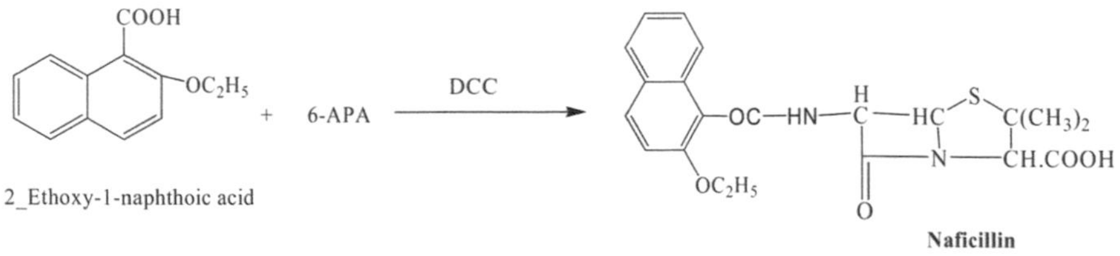

2_Ethoxy-1-naphthoic acid

Naficillin

SCHEME 9.6 Synthesis of Nafcillin.

9.3.17 NAFCILLIN

Nafcillin sodium is a narrow-spectrum β-lactam antibiotic having a fused naphthalene ring on one flank and an ethoxy moiety on the other of the side-chain amide linkage. The presence of bulky substituted naphthalene moiety play steric role in stabilizing naficillin against penicillinase. Naficillin is used in the treatment of infections caused by gram-positive bacteria, in particular, species of *Staphylococci* that are resistant to other penicillins. Nafcillin is widely distributed in various body fluids (e.g., bile, pleural, amniotic, syno-vial) and has high CSF penetration in presence of inflamed meninges and having protein binding is 89.9%, mainly albumin. Nafcillin is metabolized mainly in the liver and excreted in urine (30% as unchanged), and primarily eliminated in bile.

9.3.18 PENICILLIN G PROCAIN

Penicillin G procaine is a first semisynthetic amine salt of penicillin G pre-pared by combining penicillin G with procaine. Chemically penicillin G procaine is (2S,5R,6R)-3,3-Dimethyl-7-oxo-6-(2-phenylacetamido)-4-thia-1-azabicyclo[3.2.0]heptane-2-carboxylic acid compound with 2-(diethyl-amino)ethyl p-aminobenzoate (1:1) monohydrate. It is a white crystalline powder, with slightly soluble in water and used in the treatment of subacute bacterial endocarditis and venereal disease during pregnancy.

9.3.19 PENICILLIN G BENZATHINE

Benzathine benzylpenicillin is a diamine salt of penicillin G, in which two moles of penicillin are available from each molecule of penicillin G benza-thine. It is slowly absorbed into the circulation, after intramuscular injection, and hydrolyzed to benzylpenicillin in vivo. It is the drug-of-choice when prolonged low concentrations of benzylpenicillin are required (known as a long-acting natural penicillin) and appropriate, allowing prolonged antibi-otic action over 2–4 weeks after a single IM dose.

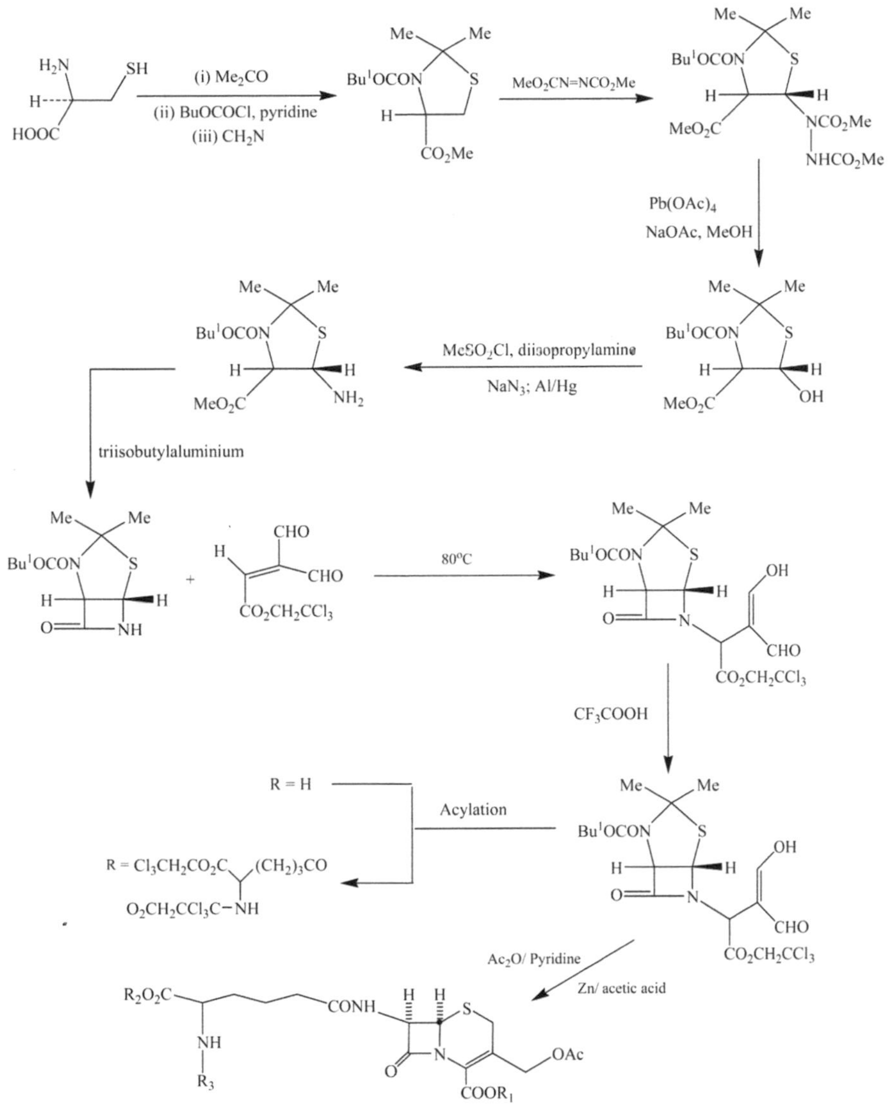

SCHEME 9.7 Synthesis of Carbenicillin.

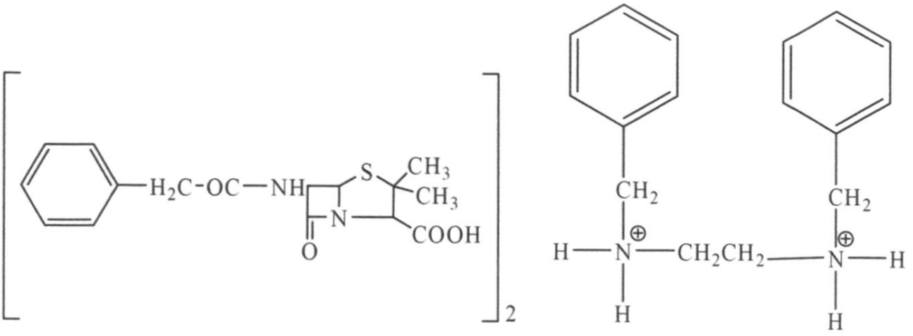

9.3.20 CARBENICILLIN

Carbenicillin is another semisynthetic analog of the naturally occurring benzyl-penicillin belonging to the carboxypenicillin subgroup. It was discovered by scientists of Beecham and marketed as Pyopen. Carbenicillin is not absorbed orally because it readily decarboxylated to benzylpenicillin in the presence of gastric acid and therefore must be administered parentrally. It is susceptible to degradation by β-lactamase enzymes, although it is more resistant than ampicillinto degradation. Carbenicillin is primary active against gram-negative bacteria including *Pseudomonas aeruginosa* but limited to gram-positive bacteria. Carbenicillin is highly soluble in water and its sodium salt is used in a dose of 1–2 g i.m. or 1–5 g i.v. every 4–6 h. At the higher doses, enough Na ion may be administered

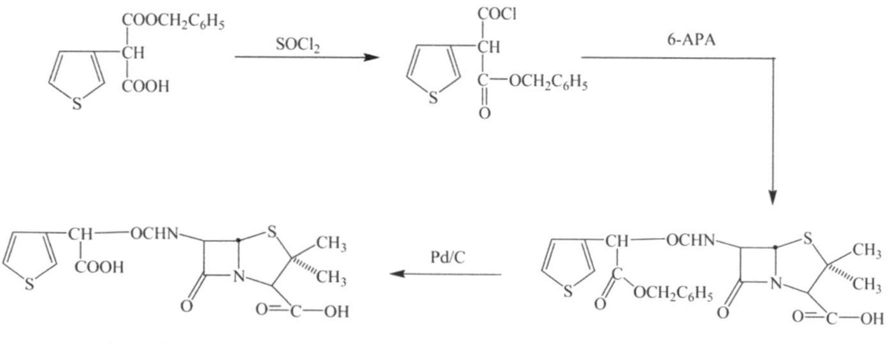

SCHEME 9.8   Synthesis of Ticarcillin.

to cause fluid retention and CHF in patients with borderline renal or cardiac function. Use of carbenicillin can cause hypokalemia by promoting potassium loss at the distal convoluted tubules of the kidney (Shattil et al., 1980).

9.3.21 TICARCILLIN

Ticarcillin is another carboxypenicillin, but it cannot decarboxylate as the carboxyl group of carbenicillin does. Because it is penicillin, thus it has similar mechanism as other penicillins. It is almost invariably sold and used in combination with clavulanate as timentin since the combination has enhanced antipsedomonad activity due to stability of lactamase. Its main clinical use is as an injectable antibiotic for the treatment of gram-negative bacteria, particularly *Pseudomonas aeruginosa* and also one of the few antibiotics capable of treating *Stenotrophomonas maltophilia* infections (Boon et al., 1986).

9.3.22 UREIDOPENICILLINS

They are acylated derivatives of ampicillin in which the D-side amino chain has been converted to a variety of ureas. The added side chain mimics a longer segment of the peptidoglycan chain, more than ampicillin, and thus would bind more easily to the penicillin-binding proteins. Examples of ureidopenicillins include azlocillin, mezlocillin and piperacillin. They are gastric acid labile and not resistant to β-lactamases. Thus they are used parenteral, and are particularly used in the infections caused by gram-negative bacteria especially *Pseudomonas aeruginosa*. Ureidopenicillins have greater penetration through the cell membrane of bacterial species especially *Klebsiella, Enterobacter* and *Pseudomonas aeruginosa*, thus prefer to carbenicillin (Fu and Neu, 1978).

9.3.23 AZLOCILLIN

It demonstrates antibacterial activity against a broad spectrum of bacteria, including *Pseudomonas aeruginosa* and, in contrast to most cephalosporins, exhibits activity against *Enterococci*. Azlocillin and piperacillin are more active than carbenicillin or ticarcillin against *Pseudomonas aeruginosa* (Eliopoulos and Moellering, 1983).

9.3.24 MEZLOCILLIN

Mezlocillin and piperacillin demonstrate significant activity against the *Enterobacteriaceae*, including many strains of *Klebsiella pneumoniae* against which the older penicillins carbenicillin, ticarcillin, and ampicillin are ineffective. As other ureidopenicillins, mezlocilin is not effective against β-lactamases producing bacteria, nor is it active orally. Mezlocillin is white, crystalline, water soluble sodium salt for injection. Mezlocillin injection is available as ready to use in market (Bergan, 1983).

9.3.25 PIPERACILLIN

Piperacillin is an extended-spectrum most commonly used ureidopenicillin. It is more active than mezlocillin against gram-negative aerobic bacilli, like *Serratia marcescens, Proteus, Enterobacter, Citrobacter* and *Ps. Aeruginosa*. It is not absorbed orally and destroys by gastric acid, and must therefore be given by intravenous or intramuscular injection. Piperacillin is destroyed by β-lactamase enzyme and thus must be used together with a β-lactamase inhibitor, notably in the combination piperacillin/tazobactam (Charbonneau, 1994).

9.3.26   β-LACTAMASE INHIBITORS

A β-lactamase inhibitor block the activity of β-lactamase enzymes and used in conjunction with a β-lactam antibiotic to extend its spectrum of activity. Although β-lactamase inhibitors have little or no antibacterial activity of their own, they instead inhibit the activity of β-lactamases enzyme that break the β-lactam ring that allows penicillin-like antibiotics to work, thereby conferring bacterial resistance. Several β-lactam derivatives have been reported as β-Lactamase inhibitors; however, only clavulanic acid, sulbactam and tazobactam have reached clinical importance. Different β-lactamases inhibitors differ in their substrate affinities. On the basis of chemistry β-lactamases inhibitors are classified into two categories: Class I inhibitors, have a hetero atom leaving group at position 1, that is, clavulanic acid and sulbactam and Class II inhibitors that do not have, that is, carbapenems. Class I inhibitors can cause prolong inactivation of certain β-lactamases thus they can use particularly useful in congener with extended-spectrum β-lactamase-sensitive penicillins to treat infections caused by β-lactamases producing bacteria. In spite of class I, class II inhibitor, the carbapenem derivative imipenem has potent antibacterial activity in addition to its potent inhibition of β-lactamases (Buynak, 2013).

9.3.27   CLAVULANIC ACID

Clavulanic acid was discovered around 1974/75 by British scientists working at the drug company Beecham. It was obtained from *Strentomuces claauligerus*, having β-lactam ring but no antibacterial activity of its own. Clavulanic acid functions as a mechanism-based β-lactamase inhibitor while not effective by itself as an antibiotic. When it combines with penicillin-group antibiotics, it can overcome antibiotic resistance in bacteria that secrete β-lactamase, which otherwise inactivates most penicillins (Finlay, 2003). Clavulanic acid inhibits a wide variety (class II to class V) of β-lactamases (but not class I cephalosporinase) produced by both gram-positive and gram-negative bacteria and a 'progressive' inhibitor: binding with β-lactamase is reversible initially, but becomes covalent later-inhibition increasing with time. It is also called a 'suicide, inhibitor, because it gets inactivated after binding to the enzyme. Clavulanic acid is biosynthesized by the bacterium *Streptomyces clavuligerus*, using glyceraldehyde-3-phosphate, L-arginine as the starting materials and clavaminate synthase, β-lactam synthetase and N^2-(2-carboxyethyl)-L-arginine synthase(CEA) as catalyzing enzyme. Although all of the intermediates of the pathway are known, the exact mechanism of how each enzymatic reaction is not fully understood. Clavaminate synthase is a nonheme iron α-keto-glutarate dependent oxygenase that regulates 3 steps in the overall synthesis of clavulanic acid. All three steps occur in the same region of the catalytic iron center, yet do not occur in-sequence and affect different areas of the clavulanic acid structure. β-lactam synthetase involve in synthesize of β-lactam is a 54.5 kDa protein shows similarity to asparagine synthase – Class B enzymes. The exact mechanism on how this enzyme works to synthesize the β-lactam is not proven, but is believed to occur in coordination with a CEA synthase and ATP. CEA synthase is a 60.9 kDA protein, couple together glyceraldehyde-3-phosphate with L-arginine in the presence of thiamine diphosphate (TDP or thiamine pyrophosphate), which is the first step of the clavualnic acid biosynthesis. Clavulanic acid has rapid oral absorption and shows 60% bioavailability. It can also be injected depend upon the severity of infection. It is eliminated mainly by glomerular filtration and is largely hydrolyzed and decarboxylated before excretion, while amoxicillin is mainly excreted unchanged by tubular secretion. Clavulanic acid reestablishes the activity of amoxicillin (called Coamoxiclav) against β-lactamase producing resistant *Staph. aureus* (but not MRSA that have altered PBPs), *H. influenzae, N. gonorrheae,*

E. coli, Proteus, Klebsiella, Salmonella and *Shigella. Bact. fragilis* and *Branhamelln catarrhalis* are not responsive to amoxicillin alone, but are inhibited by the combination. Amoxicillin sensitive strains are not affected by the addition of clavulanic acid. Adverse effects of clavulanic acid with penicillins have been associated with an increased incidence of cholestatic jaundice and acute hepatitis during therapy or shortly after. Clavulainc acid tolerance is poorer-especially in children. Other side effects are candida stomatifis/vaginitis and rashes.

9.3.28 SULBACTAM

Sulbmctan is another semisynthetic β-lactamase inihibitor, related chemically as well as in activity to clavulanic acid. Strucurally it is penicillanic acid sulfone or 1,1-dioxopenicillanic acid. Sulbactam is an irreversible inhibitor of β-lactamase, highly active against class II to class V β-lactamase but poorly active against class I. Similar to clavulanic acid it is given in combination with β-lactam antibiotics to inhibit β-lactamase, but on weight basis, it is 2–3 times less potent than clavulanic acid for most types of the enzyme, but the same level of inhibition can be obtained at the higher concentrations achieved clinically. It does possess some antibacterial activity when administered alone, but it is too weak to have any clinical importance. As compare to clavulanic acid sulbactam does not induce chromosomal β-lactamases. Oral absorption of sulbactam is inconsistent. Therefore, it is preferably given parenteral especially in combination with ampicillin for use against β-lactamase producing resistant strains. Fixed dose combinations of sulbactam sodium and ampicillin sodium, marketed under the trade name Unasyn. This combination is used for the treatment of infection caused by β-lactamases producing strains of *Staph. Aureus, E. coli, Klebsiella, Pr. Mirabilis, B. fragilis, Enterobacter,* and *Acinetobacter* (Akova, 2008).

9.3.29 TAZOBACTAM

Tazcbactam is also β-lactamase inhibitor similar to sulbactam. Tazobactam is more potent β-lactamase inhibitor than sulbactam, and has a slightly broader spectrum of activity than clavulanic acid. It is commonly used as its sodium salt, tazobactam sodium and combined with the extended spectrum β-lactam antibiotic piperacillin in the drug piperacillin/tazobactam. Its pharmacokinetics matches with piperacillin and their combination (marketed as Zosyn, 8:1 ratio) use in severe infections like peritonitis, pelvic/urinary/respiratory infections caused by β-lactamase producing bacilli (Khan, 2005). However, the combination is not active against piperacillin-resistant *Pseudomonas*, because tazobactam (like clavulanic acid and sulbactam) does not inhibit inducible chromosomal β-lactamase produced by *Enterobacteriaceae*. Tazobactam is derived from the penicillin nucleus and is a penicillinic acid sulfone. Tazobactam has short half-life (1 hr), minimal protein binding, modest metabolism, and excreted as active form in the urine in high concentration.

9.3.30   CARBAPENEMS

The carbapenems are β-lactam antimicrobial agents plays a critical role in our antibiotics with a broad spectrum of antibacterial activity. Of the many hundreds of different β-lactams, carbapenems possess the broadest spectrum of activity and greatest potency against both gram-positive and gram-negative bacteria and anaerobes. Structurally carbapenems are very similar to the penicillins (penams), but the sulfur atom in position 1 of the structure has been replaced with a carbon atom, and an unsaturation has been introduced—hence the name of the group, the carbapenems. Olivanic acids possess a "carbapenem backbone" was the first discovered carbapenems and act as broad-spectrum β-lactams. Due to chemical instability and poor penetration into the bacterial cell, the olivanic acids were not further pursued. Shortly thereafter, thienamycin from *Streptomyces cattleya* the first "carbapenem" were noticed and would eventually serve as the parent or model compound for all carbapenems. A series of other carbapenems were also identified; however, the discovery of thienamycin was paramount. Carbapenems also are thus far the only β-lactams capable of inhibiting L,D-transpeptidases. As a result, they are often used as "last-line agents" or "antibiotics of last resort" when patients with infections become gravely ill or are suspected of harboring resistant bacteria such as resistant strains of *Escherichia coli* (*E. coli*) and *Klebsiella pneumoniae*. They have a structure that renders them highly resistant to most β-lactamases. Carbapenem antibiotics were originally developed from the carbapenem thienamycin, a naturally derived product of *Streptomyces cattleya*. The carbapenems are thought to share their early biosynthetic steps in which the core ring system is formed. Malonyl-CoA is condensed with gluta-mate-5-semialdehyde with concurrent formation of the five-membered ring. Next, a β-lactam synthetase uses ATP to form the β-lactam and the saturated carbapenam core. Further oxidation and ring inversion provides the basic carbapenem. Carbapenems have poor oral bioavailability and are thus adminis-tered intravenously specially in hospital settings for more serious infections. However, research is underway to develop an effective oral carbapenem. To date, more than 80 carbapenems with mostly improved antimicrobial proper-ties has been described in the literature. The carbapenems that are presently in clinical use includes imipenem-cilastatin, meropenem, ertapenem, doripe-nem, panipenem-βmipron and biapenem (Shah, 2008).

9.3.31   IMIPENEM (PRIMAXIN)

Imipenem is an intravenous carabapenem discovered by Merck scientists Burton Christensen, William Leanza, and Kenneth Wildonger in 1980. It was the first member of the carbapenem antibiotics and plays a key role in the treatment of infections not readily treated with other antibiotics. Imipenem act by inhibiting cell wall synthesis of various gram-positive and gram-negative bacteria and remains very stable in the presence of β-lactamase (both penicillinase and cephalosporinase) produced by some bacteria, and is a strong inhibitor of β-lactamases from some gram-negative bacteria that are resistant to most β-lactam antibiotics (Clissold, 1987). A limited feature of imipenem is its rapid hydrolysis by the enzyme dihydropepetidase I located on the brush border of tubular cells. A pioneering solution of this problem is its combination with cilastatin, a reversible inhibitor of dihydropepetidase I. Common adverse drug reactions of imipenem are nausea and vomiting. People who are allergic to penicillin and other β-lactam antibiotics should take caution if taking imipenem, as cross-reactivity rates is low.

9.3.32   MEROPENEM

Meropenem is an injectable β-lactam carbapenem similar to imipenem used to treat a wide variety of infections. Meropenem was originally developed by Dainippon Sumitomo Pharma, was initially marketed by Astra Zeneca under the trade name Merrem and gained US FDA approval in July 1996. It penetrates well into many tissues and body fluids, including cerebrospinal fluid, bile, heart valve, lung, and peritoneal fluid. It inhibits bacterial wall synthesis like other β-lactam antibiotics. Similar to imipenem, meropenem is highly resistant to degradation by β-lactamases or cephalosporinases. The possible mechanism of resistance is due to mutations in penicillin-binding proteins, production of metallo-β-lactamases, or resistance to diffusion across the bacterial outer membrane. Unlike imipenem, it is stable to dehydropeptidase-1, so can be given without cilastatin (Goyal and Rajput, 2014).

9.3.33   ERTAPENEM

Similar to Imipenem and meropenem ertapenem is another carbapenem antibiotic marketed by Merck as Invanz. Its structur is very similar to meropenem having a 1-β-methyl group. In compare to other members of the carbapenem group (imipenem, doripenem, and meropenem) having broad spectrum antibacterials, ertapenem having a somewhat less broad spectrum of activity (not against Pseudomonas aeruginosa), thus not used for infections caused by difficult to treat or multidrug-resistant bacteria (such as ESBL expressing Klebsiella pneumonia) (Parakh et al., 2009). Ertapenem possess extended serum half-life and allows it to be administered once every 24 h. Similar to other β-lactam antibiotics ertapenem also works by stopping the growth of bacteria. Ertapenem is effective against both gram-negative and gram-positive bacteria but not active against MRSA, ampicillin-resistant enterococci, *Pseudomonas aeruginosa*, or *Acinetobacter* species. It also has clinically useful activity against anaerobic bacteria. Akinto other carbapenems ertapenem is also administer parentrally usually 1 g given by intravenous injection over 30 min, or 1 g diluted with 3.2 mL of 1% lidocaine given intramuscularly. Ertapenem is excreted primarily (80%) by the kidneys and metabolized by the liver. Patients on haemodialysis should be given ertapenem at least 6 h before dialysis. The common side effects of ertapenem are swelling, redness, pain, or soreness at the injection site may occur. Thou a very serious allergic reaction to this drug is rare but a seek of immediate medical attention require if notice any symptoms of a serious allergic reaction, including: rash, itching/swelling (especially of the face/tongue/throat), severe dizziness, trouble breathing.

9.3.34 DORIPENEM

Doripenem monohydrate is an ultra-broad-spectrum injectable β-lactam antibiotic and belongs to the subgroup of carbapenems. It was introduced by Shionogi Co. of Japan under the brand name Finibax in 2005 and is being marketed outside Japan by Johnson & Johnson. Doripenem act by decreases the process of cell wall growth, which eventually leads to elimination of the infectious cell bacteria together. Doripenem have a broad spectrum of bacterial activity including both gram-positive and gram-negative bacteria but it is not active against MRSA. It is stable against β-lactamases including those with extended spectrum, but it is susceptible to the action of carbapenemases (Mandell, 2009). Thus it can be used in the treatment of infections such as: complex abdominal infections, pneumonia within the setting of a hospital, and complicated infections of the urinary tract including kidney infections with septicemia. Doripenem is also more active against *Pseudomonas aeruginosa* then other carbapenems. Doripenem white to somewhat yellowish crystalline powder which is moderately soluble in water, slightly soluble in methanol, and virtually insoluble in ethanol. Doripenem is also solution in N,N-dimethylformamide. The chemical configuration of doripenem's has 6 asymmetrical carbon atoms (6 stereocenters) and is most commonly supplied as one pure isomer. In terms of doripenem for injection, the crystallized powered drug can form a monohydrate when mixed with water. However, doripenem has not been proven to possess polymorphism. Doripenem is metabolized in the liver by the enzyme dehydropeptidase-I into an inactive ring-opened metabolite and excreted by the kidney.

9.3.35 PANIPENEM/β-MIPRON

Panipenem is a parenteral carbapenem antibacterial agent marketed by Daiichi Sankyo Co. of Japan. It is a broad spectrum antibacterial agent covering a wide range of gram-negative and gram-positive aerobic and anaerobic bacteria, including *Streptococcus pneumoniae* and species producing β-lactamases. Panipenem is coadministered with βmipron to inhibit panipenem renal uptake into the renal tubule and also prevent nephrotoxicity (much like the imipenem/cilastatin combination) (Kohno et al., 1998). Panipenem in combination with βmipron demonstrated good clinical and bacteriological efficacy (similar to that of imipenem/cilastatin) in adults with respiratory tract or urinary tract infections and also effective in adults with surgical or gynecological infections, and in pediatric patients with respiratory tract and urinary tract infections in noncomparative trials. Panipenem/βmipron is well tolerated with few adverse events reported in clinical trials, most commonly elevated serum levels of hepatic transaminases and eosinophils, rash and diarrhea.

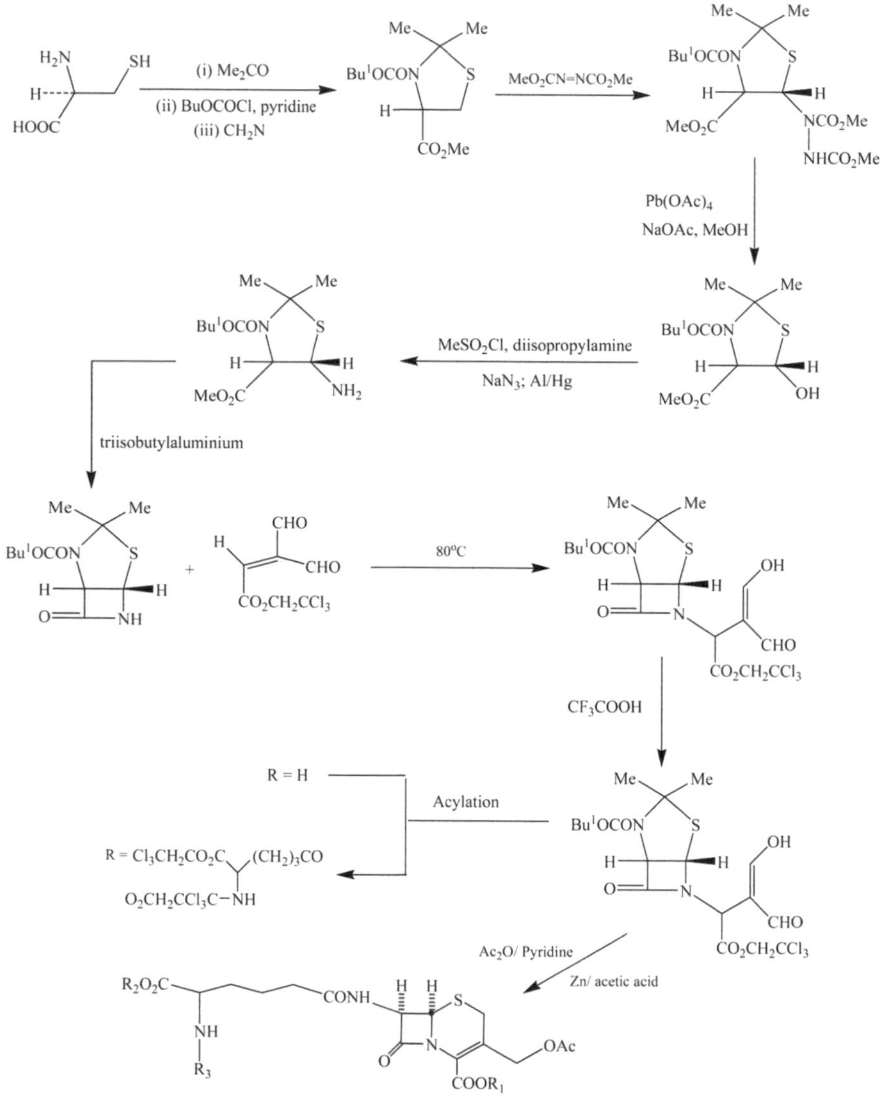

SCHEME 9.9   The synthetic route for cephalosporins.

9.4 CEPHALOSPORINS

The cephalosporins are the most diverse family of β-lactam antibiotics having structurally and pharmacologically related to the penicillins. Cephalosporins are a group of broad spectrum, semisynthetic β-lactam antibiotics derived from the mold *Cephalosporium*. The first cephalosporin was isolated from cultures of bacteria *Cephalosporium acremonium* in 1948 by Italian scientist Giuseppe Brotzu. The first marketed cephalosporin for parenteral use, cephalothin (cefalotin) was launched by Eli Lilly in 1964. The second cephalosporin for parenteral use became available little later under the name cephaloridine. The clinical successes of these two cephalosporins urged researchers to improve the pharmacological properties and develop more agents (Sweetman, 2011). Today we have thousands of semisynthesized analogs of natural cephalosporin compounds based on the knowledge gained by intensive research on the chemistry of those two starting materials. Chemically cephalosporins are divided into three groups: cephalosporin N and C are chemically related to penicillins and cephalosporin P a steroid antibiotic resembles fusidic acid. They have a β-lactam ring fused to a 6-membered dihydrothiazine ring to form 7-aminocephalospornic acid. By addition of a side chain at position 7 of β-lactam ring and position 3 of dihydrothiazine ring a large number of semisynthetic compounds can be produced.

9.4.1 PREPARATION

The most common synthesis of cephalosporin is Woodword synthesis method (Aharonowitz et al., 1992).

9.4.2 MECHANISM OF ACTION

Cephalosporins are bactericidal and have similar mechanism of action as other β-lactam antibiotics (such as penicillins), but are less susceptible to β-lactamases. Cephalosporins disrupt the synthesis of the peptidoglycan layer of bacterial cell walls. The peptidoglycan layer is important for cell wall structural integrity. The final transpeptidation step in the synthesis of the peptidoglycan is facilitated by transpeptidases known as penicillin-binding proteins (PBPs). PBPs bind to the D-Ala-D-Ala at the end of mucopeptides (peptidoglycan precursors) to crosslink the peptidoglycan. B-lactam antibiotics mimic the D-Ala-D-Ala site, thereby irreversibly inhibiting PBP crosslinking of peptidoglycan and ultimate inhibit cell wall synthesis (Martens, 1989).

9.4.3 NOMENCLATURE OF CEPHALOSPORIN

The nomenclature of cephalosporin is complex as penicillin nomenclature. According to Chemical Abstract system, the fused ring system is designated as 5-thia-1-azabicyclo[4,2,0]oct-2-ene. These available cephalosporins are named as 3-cephems or Δ^3-cephems represents the position of double bond. In spite of 3-cephems, 2-cephems are inactive presumably because the β-lactam lack the necessary ring strainto be sufficiently reactive.

9.4.4 STRUCTURE ACTIVITY RELATIONSHIP

The structure–activity relationship (SAR) of cephalosporins forecast the relationship between the chemical or 3D structure of cephalosporins to its biological activity. The analysis of SAR enables the determination of the chemical groups responsible for evoking a target biological effect in the organism. This allows modification of the effect or the potency of a bioactive compound (typically a drug) by changing its chemical structure (Goldberg, 1987).

1. B-lactam ring is required for PBP reactivity and antibacterial activity.
2. The derivatives where Y= S exhibits greatest activity than if Y = O, but the reverse is true when stability toward β-lactamse is considered.
3. The 7β amino group is essential for antibacterial activity.

4. Replacement of the hydrogen at C-7 (X=H) with an alkoxy (X=OR) results in improvement of the antimicrobial activity of cephalosporins.
5. The addition of a 7α methoxy also improves cephalosporins stability towards β-lactamse.
6. The 6α hydrogen is essential for biological activity.
7. Activity is further increase when Z is a 5-membered heterocyclic as compare to 6-membered heterocyclic.

9.4.5 SPECTRUM OF ACTION

The cephalosporins have different spectrum of activities according to their generations.

The first generation cephalosporinps possess good gram-positive cocci coverage including *Streptococci, Staphylococci, Enterococci,* but not effective against methicillin-resistant *Staph. aureus,* penicillin-resistant *Strep. pneumoniae.* They also show modest gram-negative bacteria coverage like *Escherichia coli, Proteus mirabilis,* and *Klebsiella pneumoniae,* though susceptibilities may vary but poor activity against *Moraxella catarrhalis* and *Hemophilus influenzae.*

The second generation cephalosporin also shows activity against gram-positive cocci: "True" 2nd generation cephalosporins are almost comparable to 1st generation agents against *Streptococci.* Slight loss of activity against *Staphylococci* (NOT active against methicillin-resistant strains). Cephamycins are less active against gram-positive cocci than 1st generation agents. The members of second generation cephalosporin, have activity against gram-negative aerobes include *Hemophilus influenzae, Moraxella catarrhalis, Proteus mirabilis, E. Coli, Klebsiella, Neisseria gonorrheae.* They are not active against *Pseudomonas* and *Enterococci.*

The third generation cephalosporins show limited activity against gram-positive cocci (particularly agents available in an oral formulation). Cefotaxime, ceftriaxone, and ceftizoxime have the best gram-positive coverage of the third-generation agents: methicillin-susceptible *Staphylococcus aureus* (though less than 1st and some 2nd generation agents), very active against Groups A and B *Streptococci,* and *Viridans streptococci.* Cefotaxime and ceftriaxone are more active than ceftizoxime against *Streptococcus pneumoniae.* None of them are active against methicillin-resistant *Staphylococci, Enterococci* and *Listeria.* The third generation cephalosporins are very active against gram-negative bacteria including *Hemophilus influenzae, Moraxella catarrhalis, Neisseria meningitidis, Enterobacteriaceae* (*Escherichia coli, Klebsiella*species, *Proteus* (including strains resistant to aminoglycosides), *Providencia, Citrobacter* and *Serratia*. Cefotaxime, ceftriaxone, and ceftizoxime have adequate activity against oral anaerobes. They have no pseudomonal activity.

The fourth generation cephalosporins shows good activity against both gram-positive cocci including *Streptococcus pneumonia*, groups A and B *Streptococci*, methicillin-susceptible *Staphylococcus aureus* (less potent than the first and second generation agents) and gram-negative bacteria. They have excellent activity against *Enterobacteriaceae* and *Pseudomonas aeruginosa*, but minimal anaerobic coverage (Williams, 1987).

In general, first generation cephalosporins have better activity against gram-positive bacteria and less gram-negative activity, while third generation agents, with a few exceptions, have better gram-negative activity and less gram-positive activity. The only fourth generation agent has both gram-positive and gram-negative activity.

9.4.6 DEVELOPMENT OF RESISTANCE

Resistance to cephalosporin antibiotics can involve several mechanisms includes (Minami et al., 1980, Livermore, 1987):
1. Resistance to cephalosporin antibiotics can involve either reduced affinity of existing PBP components or the acquisition of a supplementary β-lactam-insensitive PBP.
2. Destruction of β-lactam ring by β-lactamases; an intact β-lactam ring is essential for antibacterial activity. This β-lactamases is produced by resistant bacteria
3. Decreased penetration of antibiotic to the target site or cell membrane. This is only applicable to gram-negative bacteria because gram-positive bacteria lack an outer cell membrane, and therefore penetration to the target site is not a problem.

9.4.7 QUANTITATIVE STRUCTURE ACTIVITY RELATIONSHIP

With advances in computational chemistry and the rapid accumulation of experimental data, these ADME properties of cephalosporins can now be predicted by computationally method. Screening of newly synthesized molecules for ADME properties has become a very important issue in drug development. The method has the advantage that it can take place even before molecules are synthesized. These methods try to build relationships between a dataset consisting of known values for the property of interest and some calculated theoretical descriptors. Theoretical descriptors are derived from molecular representations of the molecules thus no synthetization or experimental setups are required. These relationships are called quantitative structure – activity relationships (QSAR). A quantitative structure-activity relationship study of the binding of 28 commercially available cephalosporins to human serum proteins shows that polar interactions and the forming of hydrogen-bonding properties of cephalosporins are two major factors affecting the binding ability and ultimate biological activity.

9.4.8 ADVERSE DRUG REACTIONS

Cephalosporins are generally well tolerated, but are more toxic than penicillin. The most common side effects of cephalosporins associated with allergic hypersensitivity. Hence, it was commonly stated that they are contraindicated in patients with a history of severe, immediate allergic reactions (urticaria, anaphylaxis, interstitial nephritis, etc.) to cephalosporins. Skin tests for sensitivity to cephalosporins are unreliable. A positive coomb's test occurs in many, but haemolysis is rare. Several cephalosporins are associated with hypoprothrombinemia and a disulfiram-like reaction with ethanol. These include latamoxef, cefmenoxime, moxalactam, cefoperazone, cefamandole, cefmetazole, and cefotetan. This is thought to be due to the N-methylthiotetrazole side-chain of these cephalosporins, which blocks the enzyme vitamin K epoxide reductase (likely causing hypothrombinemia) and aldehyde dehydrogenase (causing alcohol intolerance). Highest incidence of nephrotoxicity with cephaloridine, which consequently has been withdrawn. Cephalothin and a few others have low-grade nephrotoxicity which may be accentuated by preexisting renal disease or concurrent administration of an aminoglycoside or loop diuretic. Sometimes diarrhea is more common with oral cephradine and parenteral cefoperazone due to alteration of gut ecology or irritative effect. Neutropenia and thrombocytopenia are rare adverse effects reported with ceftazidime and some others. Pain after i.m. injection occurs with many which are as severe with cephalothin as to interdict i.m. route, but many others can be injected i.m. (see individual compounds). Thrombophlebitis of injected vein can also occur (Norrby, 1987; Quin, 1989).

9.4.9 CLASSIFICATION OF CEPHALOSPORIN

The cephalosporin class is very wide so a fine classification system is necessary to distinguish different cephalosporins from each other. There are few chemical and activity features that could be used for classification, for example chemical structure, side chain properties, pharmacokinetic, spectrum of activity or clinical properties. Cephalosporins can also be classified based upon: spectrum, generation, chemical structure, resistance to β-lactamases and clinical pharmacology.

Despite these variable features the most common classification system for cephalosporins is to divide them into generations. Cephalosporin drugs are divided into different generations depending upon their microbial spectrum. The first discovered cephalosporins were classified into first-generation cephalosporins, whereas, later, more extended-spectrum cephalosporins were classified as second-generation cephalosporins. Each newer generation has significantly greater gram-negative antimicrobial properties than the preceding generation, in most cases with decreased activity against gram-positive organisms. Fourth-generation cephalosporins, however, have true broad-spectrum activity (Tables 9.2–9.4).

9.4.9.1 First Generation Cephalosporins

The first generation cephalosporins were originally spelled "ceph-" in english-speaking countries and were the first cephalosporins in the market. They have good antimicrobial activity against gram-positive bacteria but limited activity against gram-negative species. The chemical structures of the first generation cephalosporins are fairly simple. The examples of first generation cephalosporin are cephalexin, cephradine and cefadroxil have a single methyl group at position C-3. The methyl group at position C-3 gives low affinity for common PBP which can in part explain the relatively low activity of these first generation drugs. Cefaclor however has a chlorine (-Cl) group at position C-3 which gives it better binding to PBP and thus better antimicrobial

TABLE 9.2 Structures of First and Second Generation Cephalosporins

Compound	R$_1$	R$_2$
Cefazolin		
Cephalothin	-CH$_2$OAc	
Cephapirin	-CH$_2$OAc	
Cephalexin	-CH$_3$	
Cefadroxil	-CH$_3$	
Cefaclor	-Cl	
Cefotetan		
Cefprozil	-CH=CH-CH$_3$	

Compound	R$_1$	R$_2$
Cefoxitin	-CH$_2$OCONH$_2$ (-OCH$_3$ at C7 also)	
Cefuroxime	-CH$_2$OCONH$_2$	

TABLE 9.3   Structures of Third Generation Cephalosporins
(i) With 2-Aminotrhiazolyloximino moiety

Compound	R$_1$	R$_2$
Cefotaxime	-CH$_2$OAc	H
Ceftriaxone		H
Ceftizoxime	H	H
Cefixime	CH$_2$=CH-	-CH$_2$COOH
Ceftazime		

(ii) With different acyl residues

Compound	R$_1$	R$_2$
Cefoperazone	N—N (triazole with N–N, N·N–CH$_3$) S—CH$_2$—	HO—⟨C$_6$H$_4$⟩—CH(HN—CO—N piperazine-2,3-dione N—C$_2$H$_5$)
Cefpiramide	N—N (triazole, N·N–CH$_3$) S—CH$_2$—	HO—⟨C$_6$H$_4$⟩—CH(HN—CO—pyridine with CH$_3$, HO)
Cefivitril	N—N (triazole, N·N–CH$_3$) S—CH$_2$—	N—N (triazole, N·N–CH$_3$) S—CH$_2$—

TABLE 9.4 Fourth Generation Cephalosporin

Structure: aminothiazole ring (H$_2$N–C$_3$N S) —C(=NOCH$_3$)—OCHN— attached to cephem nucleus with S, N, =O (β-lactam), R, COOH.

Compound	R$_1$
Cefepime	pyrrolidinium N$^{\oplus}$ with $\overline{C}H_2$ and CH$_3$
Cefpirome	cyclopenta-fused pyridinium N^{+}—$\overline{C}H_2$

activity, but it is often classified as such because of its C-7 side chain which is more related to the first generation than the second. All of the first generation cephalosporins have an α-amino group at position C-7. This structure makes them vulnerable to hydrolysis by β-lactamases (McEniry, 1987).

9.4.9.1.1   Cefazolin Sodium

Cefazolin or cephazolin is a first-generation cephalosporin antibiotic useful for the treatment of a number of bacterial infections. As other cephalosporins, it also falls into the category of β-Lactam (β-lactam) antibiotics which

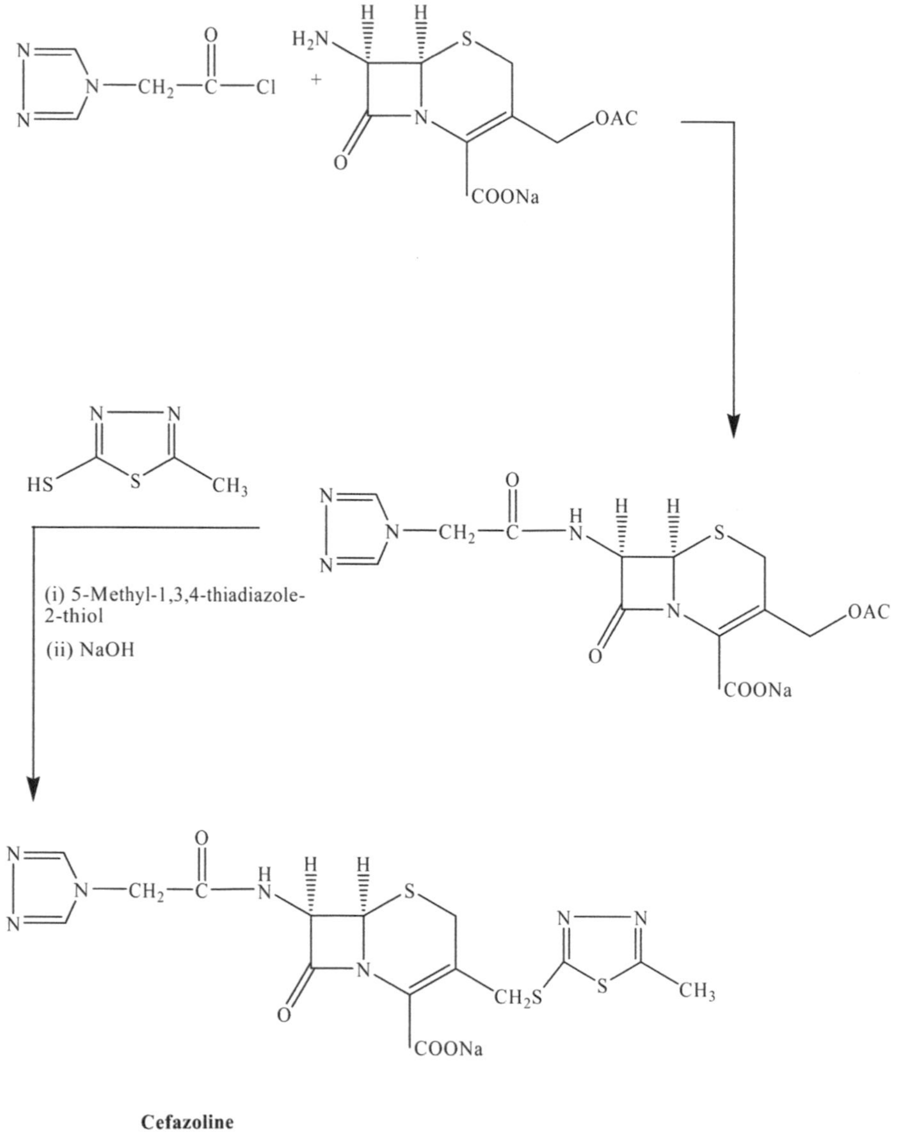

Cefazoline

SCHEME 9.10   Synthesis of Cefazolin Sodium.

works by inhibiting cell wall synthesis of the bacteria. The IUPAC name of Cefazolin is (6*R*,7*R*)-3-{[(5-methyl-1,3,4-thiadiazol-2-yl)thio]methyl}-8-oxo-7-[(1*H*-tetrazol-1-ylacetyl)amino]-5-thia-1-azabicyclo[4.2.0]oct-2-ene-2-carboxylic acid. In Cefazolin the C-3 acetoxy functional group has been replaced by a thiol-containing heterocyclic: 5-methyl-2-thio-1,3,4-thiaziazole group. It also contains unwanted tetrazolyl acetyl acylating group. Cefazolin is a bactericidal agent used to treat many kinds of bacterial infections, including severe or life-threatening forms. Cefazolin is primary active against some gram-negative bacteria such as *Escherichia coli* and *Klebsiella pneumoniae* and also clinically effective against infections caused by *Staphylococci* and *Streptococci* which are gram-positive bacteria commonly found on human skin. Cefazolin has been shown to be very effective in treating methicillin-susceptible *staphylococcus aureus* (MSSA) but does not work in cases of methicillin-resistant *staphylococcus aureus* (MRSA). In many instances of MSSA, the use of cefazolin is preferred over the use of other antibiotics such as vancomycin (Deguchi, 1988). It is use to treat severe bacterial infections involving the lung, bone, joint, stomach, blood, heart valve, and urinary tract. Cefazolin is ineffective against infection caused by *Enterococcus*, anaerobic bacteria or atypical bacteria among others. It is a safe drug in pregnancy and can also be use in breastfeed. Cefazolin is often used as a prophylactic antibiotic before a wide range of surgical operations due to being one of the most widely studied antibiotics with proven efficacy. The drug is usually administered by either intramuscular injection (injection into a large muscle) or intravenous infusion (intravenous fluid into a vein). The common possible side effects of cefazolin include diarrhea, stomach pain or upset stomach, vomiting, and rash. Patients with penicillin allergies may also have a reaction to cefazolin. Cefazolin is not metabolized inside the body and excreted unchanged in urine. Approximately 60% excreted in urine in 6 h and 70% to 80% within 24 h. Peak urine concentrations after IM doses of 500 mg and 1 g are approximately 2,400 and 4,000 μg/mL, respectively.

9.4.9.1.2 Cefalotin Sodium

Cefalotin or cephalothin (USAN) is a first-generation cephalosporin antibiotic. It was the first cephalosporin marketed (1964) and continues to be widely used. It is white to off-white crystalline powder which

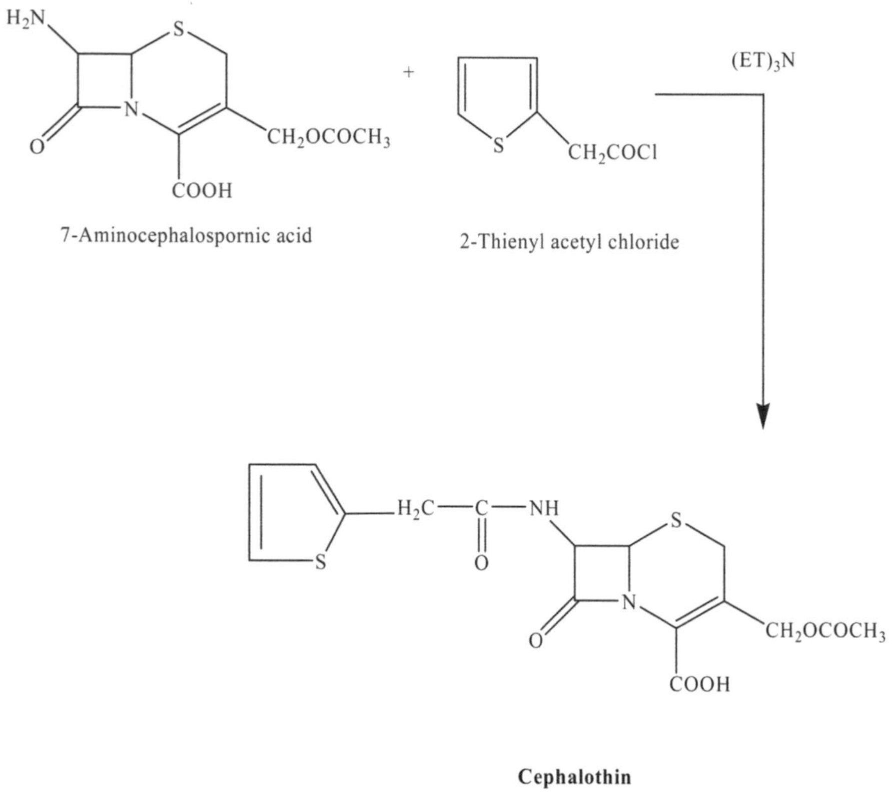

7-Aminocephalospornic acid

2-Thienyl acetyl chloride

Cephalothin

SCHEME 9.11   Synthesis of Cefalotin Sodium.

is soluble in water and insoluble in most organic solvents. The IUPAC name of cefalotin is (6R,7R)-3-[(acetoxy)methyl]-8-oxo-7-[(2-thienylace-tyl)amino]-5-thia-1-azabicyclo[4.2.0]oct-2-ene-2-carboxylic acid. It is active against gram-positive (less active against gram negative) *Staphylococci* (except MRSA), *Strep. pneumoniae, Haemophilus influ-enzae, E. coli, Klebsiella* and *Proteus mirabilus*. Cefalotin is absorbed poorly from the gastrointestinal tract and must be given by either intra-venous or intramuscular injection with a similar antimicrobial spectrum to cefazolin and the oral agent cefalexin. Cefalotin sodium is marketed as Keflin (Lilly) and under other trade names. Cefalotin is rapidly excreted through the kidney and about 60% being lost within 6hr of administra-tion. Common side effects of cefalotin include hypersensitivity reactions (up to 15% of patients with a history of penicillin allergy). Manifestations may include urticarial or maculopapular rash, bronchospasm, and drug fever. Anaphylaxis, including severe hypotension and cardiac arrest, is also reported. Rare cases of renal insufficiency associated with cephalothin may be hypersensitivity-mediated since fever, eosinophilia, and rash are often also present (Hameed, 2002).

9.4.9.1.3 Cefapirin Sodium

Cefapirin (also spelled cephapirin) is an injectable, first-generation cephalosporin antibiotic. It closely resembles cephalothin in chemical and pharmacokinetic properties. Cephapirin, have cephalosporanic acid core with the acetyloxymethyl group at the 3rd position and having IUPAC name (6R,7R)-3-(Acetoxymethyl)-8-oxo-7-{[(pyridin-4-ylsulfanyl)acetyl]amino}-5-thia-1-azabicyclo [4.2.0] oct-2-ene-2-carboxylic acid. It is unstable in acid and must be administered parenteral. It has similar mechanism as other cephalosporins. It is marketed under the trade name Cefadyl. It is effective against a wide variety of gram-positive and gram-negative bacteria; used as the sodium salt. Among the most serious adverse reactions of cefapirin, that is, neutropenia, leukopenia, anemia, bone marrowdepression, and allergic reactions, it has been discontinued in the United States (Wiesner, 1972).

9.4.9.1.4 Cephalexin

Cephalexin or cefalexin is a first-generation cephalosporin designed especially as an orally active antibiotic. It is white crystalline monohydrate, freely soluble in water, acid and used to treat bone, ear, skin, and urinary tract infections. It was developed in 1967 and first marketed in 1969 and 1970 by a number of companies, including Glaxo Wellcome and Eli Lilly and company under the names Keflex and Ceporex. It's IUPAc name is (7R)-3-Methyl-7-(α-D-phenylglycylamino)-3-cephem-4-carboxylic   acid monohydrate. The oral inactivation of cephalexin has been attributed in the presence of α-amino group which makes it acid stable. It is taken by mouth and is active against gram-positive bacteria and some gram-negative bacteria. It has similar activity as other agents within this group, including the intravenous agent cefazolin. Cephalexin may be used in those patients who have mild or moderate allergies to penicillin, but is not recommended in those with severe allergies. It has no effect against viral infections, such as the common cold or acute bronchitis. Cefalexin is used to treat a number of

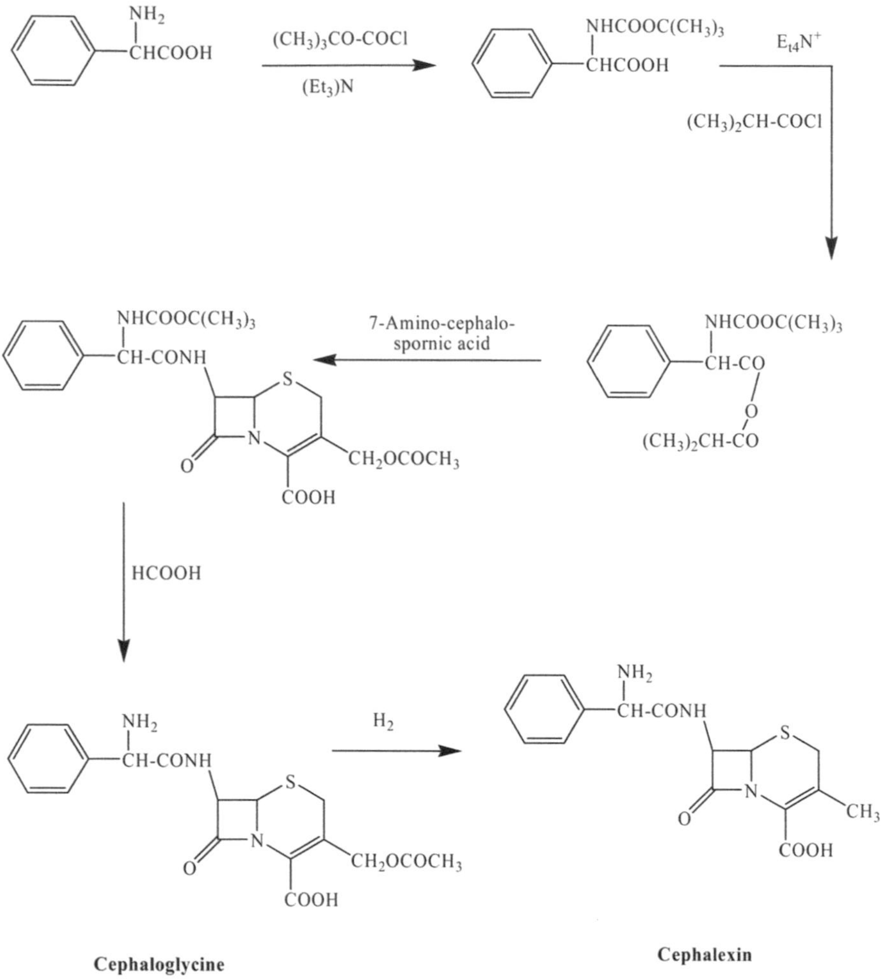

SCHEME 9.12   Synthesis of Cephalexin.

infections including: otitis media, streptococcal pharyngitis, bone and joint infections, pneumonia, cellulitis, and urinary tract infections. It may be also be used to prevent bacterial endocarditis and can be used for the prevention of recurrent urinary-tract infections. Cefalexin does not treat methicillin-resistant *Staphylococcus aureus* infections. Cefalexin is a useful alternative to penicillins in patients with penicillin intolerance. Caution must be exercised when administering cephalosporin antibiotics to penicillin-sensitive patients, because cross sensitivity with β-lactam antibiotics has been documented in up to 10% of patients with a documented penicillin allergy. Cephalexin is rapidly absorbed in upper intestine and distributes to the tissues, other than the spinal fluid and aqueous humor. Cephalexin does not penetrate into the host tissue cells, which probably accounts for its low incidence of side effects. Binding to human serum proteins is low, and there is no measurable destruction or metabolism of cephalexin during its sojourn in the body fluids. Cephalexin is rapidly cleared from the body by the kidneys. Seventy to 100% of the dose is found in the urine within 6–8 hr after each dose. The most common adverse effects of cefalexin are gastrointestinal disturbances and hypersensitivity reactions. Gastrointestinal disturbances include nausea, vomiting, and diarrhea. Hypersensitivity reactions include

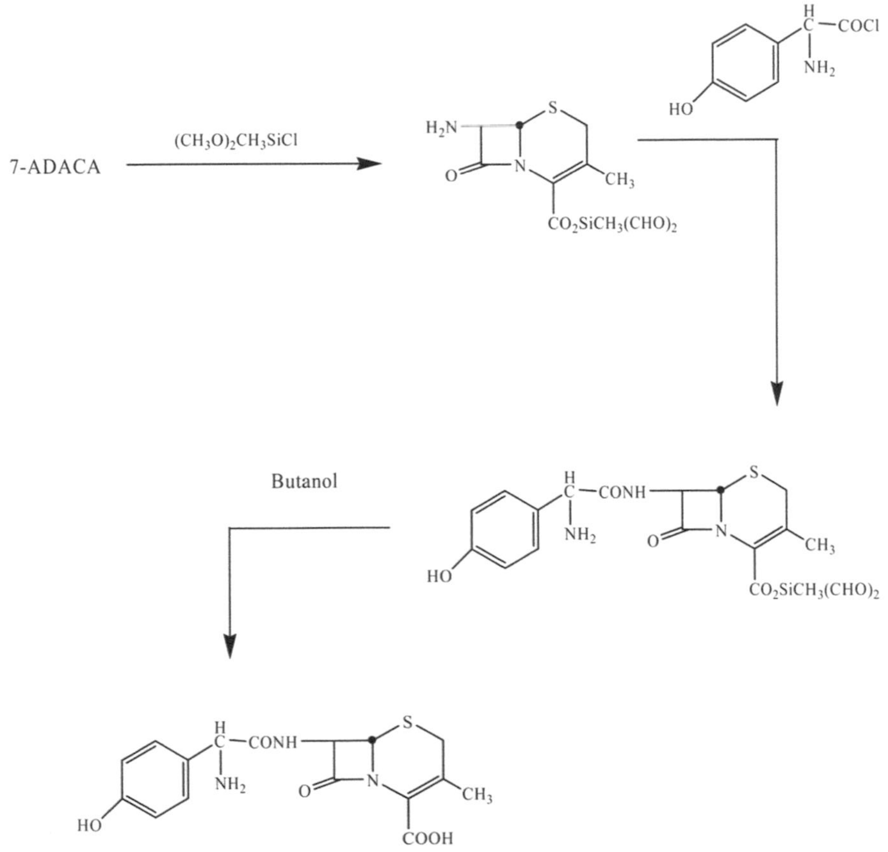

Cefadroxil

SCHEME 9.13 Synthesis of Cefadroxil.

skin rashes, urticaria, fever, and anaphylaxis. Pseudomembranous colitis has also been reported with use of cephalexin. Overall, cefalexin allergy occurs in less than 0.1% of patients, but it is seen in 1% to 10% of patients with a penicillin allergy (Speight, 1972).

9.4.9.1.5   Cefadroxil

Cefadroxil is an oral cephalosporin antibiotic used to treat infections of skin, throat, and urinary tract infections. It is the para-hydroxy deriva-tive of cefalexin, having (6R,7R)-7-{[(2R)-2-amino-2-(4-hydroxyphenyl) acetyl]amino}-3-methyl-8-oxo-5-thia-1-azabicyclo[4.2.0]oct-2-ene-2-car-boxylic acid IUPAC name. Cefadroxil stop bacteria from multiplying by preventing bacteria from forming the walls that surround them. Cefadroxil is active against many gram-positive and gram-negative bacterial, including *Staphylococcus aureus, Streptococcus pneumoniae, Streptococcus pyogenes, Moraxella catarrhalis, E. coli, Klebsiella,* and *Proteus mirabili*s. Cefadroxil is almost completely absorbed from the gastrointestinal tract, widely distrib-uted to body tissues and fluids and excreted unchanged in the urine within 24 h by glomerular filtration and tubular secretion (Buck, 1977). The slow excretion of cefadroxil lead to produce prolongs duration of action of this compound and thus permits once-a-day dosing. It crosses the placenta and appears in breast milk. The most common side effects of cefadroxil are diar-rhea, nausea, upset stomach, and vomiting. Other side effects include rashes, hives, and itching. The D-*p*-hydroxyphenylglycyl isomer is more active than the L-isomer.

9.4.9.1.6.   Cephaloridine (Cefaloridine)

Cephaloridine (or cefaloridine) is a first generation semisynthetic deriva-tive of cephalosporin C with pyridinium-1-ylmethyl and 2-thienylacet-amido side groups. It is unique among cephalosporins in that it exists as a zwitterion. It's IUPAC name is (6R,7R)-8-oxo-3-(pyridin-1-ium-1-ylmethyl)-7-[(2-thiophen-2-ylacetyl)amino]-5-thia-1-azabicyclo   [4.2.0] oct-2-ene-2-carboxylate. Cephaloridine is a white crystalline powder that discolurs when expose to light. Cephaloridine is highly soluble in water but worsen rapidly in aqueous solutions. Conformations around the β-lactam rings are quite similar to the molecular nucleus of penicillin, while those at the carboxyl group exocyclic to the dihydrothiazine and thiazolidine rings respectively are different (Larry, 1975). Cephaloridine is easy absorbed after intramuscular injection and poorly absorbed from the gastrointestinal tract. Cephaloridine is distributed well into the liver, stomach wall, lung and spleen and is also found in fresh wounds one hour after injection. However, the drug is poorly penetrated into the cerebrospinal fluid and is found in a

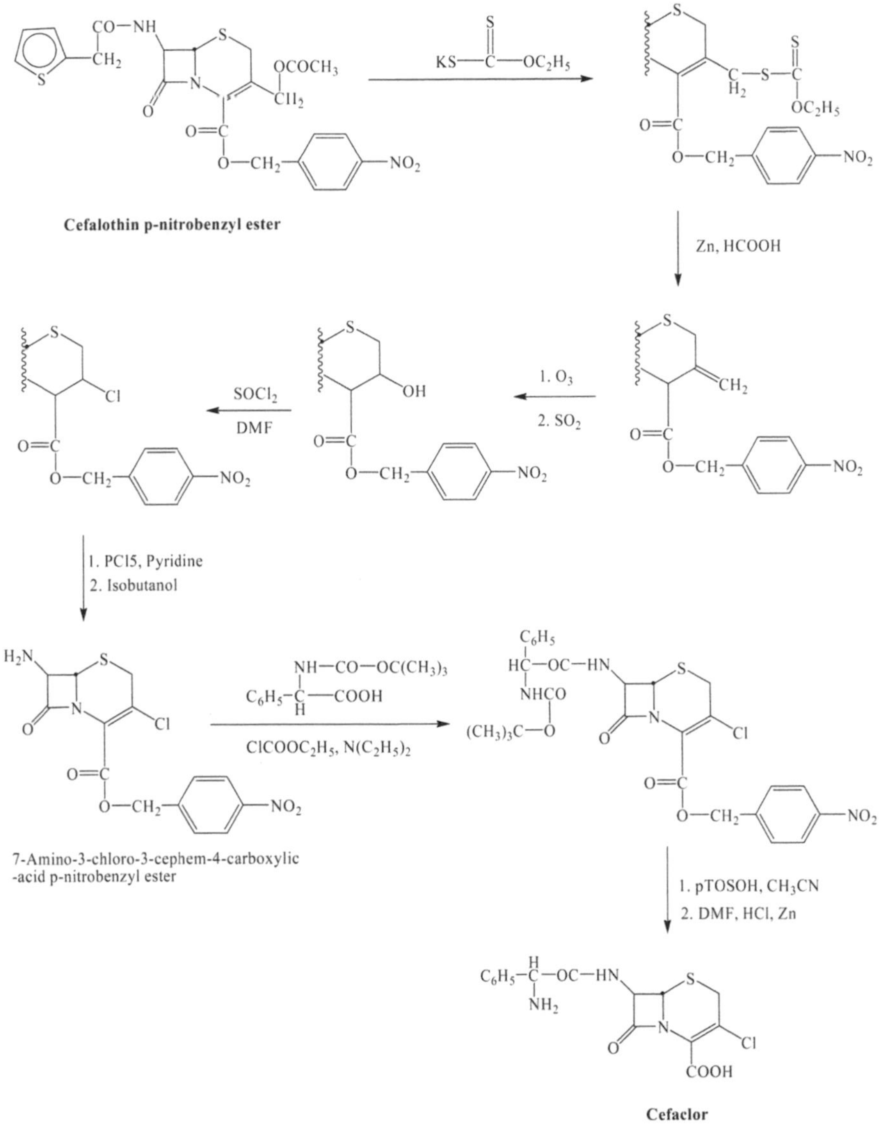

SCHEME 9.14 Synthesis of Cefaclor.

much smaller amount in the cerebral cortex. The minor pathway of elimination is biliary excretion. Renal clearances were reported to be 146–280 mL/min, a plasma clearance of 167 mL/min/1,73 m² and a renal clearance of 125 mL/min/1,73 m². A serum half-life of 1–2 h and a volume of distribution of 16 L were reported.

9.4.9.2 Second Generation Cephalosporins

Second generation cephalosporins followed the first generation cephalosporins. The second generation cephalosporins have a variable activity against gram-positive bacteria but have a greater activity against gram-negative bacteria. They are more resistant to β-lactamase and especially useful for treating upper and lower respiratory tract infections, sinusitis and otitis media. Cefaclor, cefotetan, cefoxitin, cefprozil, ceftin, cefuroxime and cefzil are classed as second-generation cephalosporins (Riaz, 2013).

9.4.9.2.1   Cefaclor

Cefaclor or cefachlor is a semisynthetic second-generation cephalosporin antibiotic for oral administration. It is chemically designated as 3-chloro-7-D-(2-phenylglycinamido)-3-cephem-4-carboxylic acid monohydrate. It acts by interfering with formation of the bacteria's cell wall while it is growing. It is used to treat some infections caused by bacteria such as pneumonia sand infections of the ear, lung, skin, throat, and urinary tract. Cefaclor is active against many bacteria, including both gram-negative and gram-positive organisms. Cefaclor should be avoided if there is a history of immediate hypersensitivity reaction. Allergic reactions includes rashes, pruritus (itching), urticaria, serum sickness-like reactions with rashes, fever and arthralgia, and anaphylaxis. Other side effects include gastrointestinal disturbances (e.g., diarrhea, nausea and vomiting, abdominal discomfort, disturbances in liver enzymes, transient hepatitis and cholestatic jaundice), headache, and Stevens–Johnson syndrome. Cefaclor is a well-absorbed by oral route and peak concentrations in serum are attained within 30–60 min. Food intake reduces the rate, but not the extent of absorption of cefaclor. Cefaclor is not metabolized to a significant degree, but it degrades chemically in the body with an approximate half-life of 2 h. Most of the drug is excreted unchanged in the urine, the serum half-life after oral administration is 0.5–0.7 h (Mangla, 2012).

9.4.9.2.2 Cefotetan Disodium

Cefotetan developed by Yamanouchi and marketed by Astra Zeneca with the brand names Apatef and Cefotan. It is of cephamycin type for prophylaxis and use in the treatment of bacterial infections. It is often grouped together with second-generation cephalosporins and has a similar antibacterial spectrum, but with additional antianaerobe coverage (Ronald, 1988). Cefotetan is broad-spectrum, β-lactamase resistant cephalosporin antibiotics for parenteral administration. It is active against a wide range of both aerobic and anaerobic gram-positive and gram-negative microorganisms. It

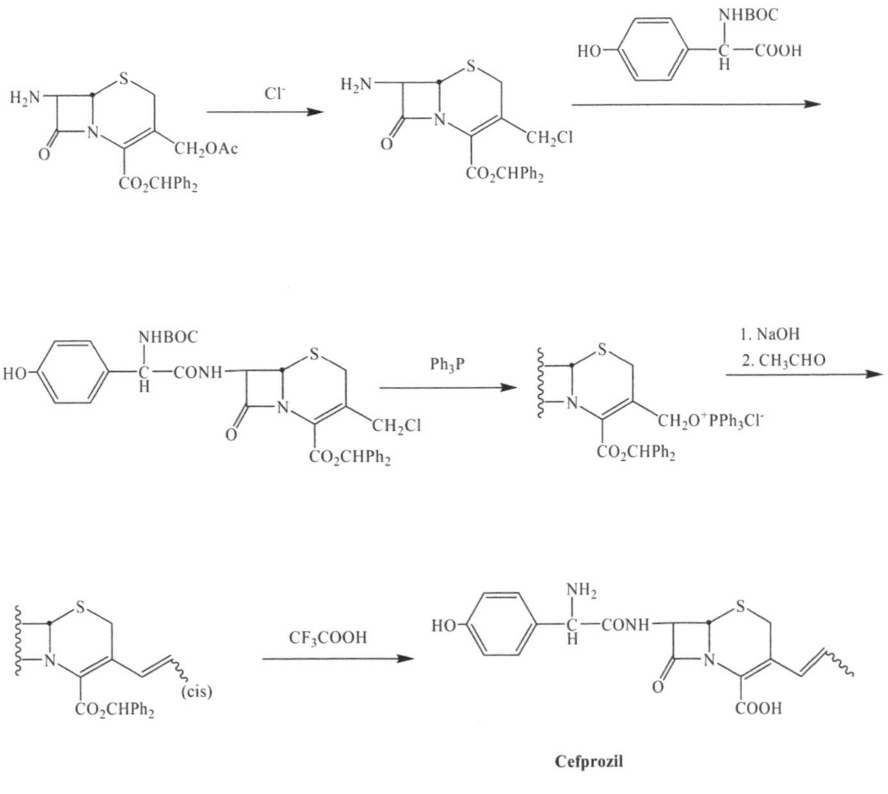

Cefprozil

SCHEME 9.15 Synthesis of Cefprozil.

is the disodium salt chemically [6R-(6α,7α)]-7- [[[4-(2-amino-1-carboxy-2-oxoethylidene)-1,3- dithietan-2-yl]carbonyl]amino]-7- methoxy-3-[[(1-methyl-1*H*-tetrazol-5-yl) thio]methyl]-8-oxo-5-thia-1-azabicyclo [4.2.0] oct-2-ene-2- carboxylic acid. Its molecular formula is $C_{17}H_{15}N_7Na_2O_8S_4$ with a molecular weight of 619.57. Cefotetan contains the MTT group that has been associated with hypoprothrombinemia and alcohol intolerance. Cefotetan is administered intravenously or intramuscularly and shows about 88% plasma protein bound. No active metabolites of cefotetan have been detected; and it excrete unchanged by the kidneys over a 24 h period, which results in high and prolonged urinary concentrations (Ward, 1985).

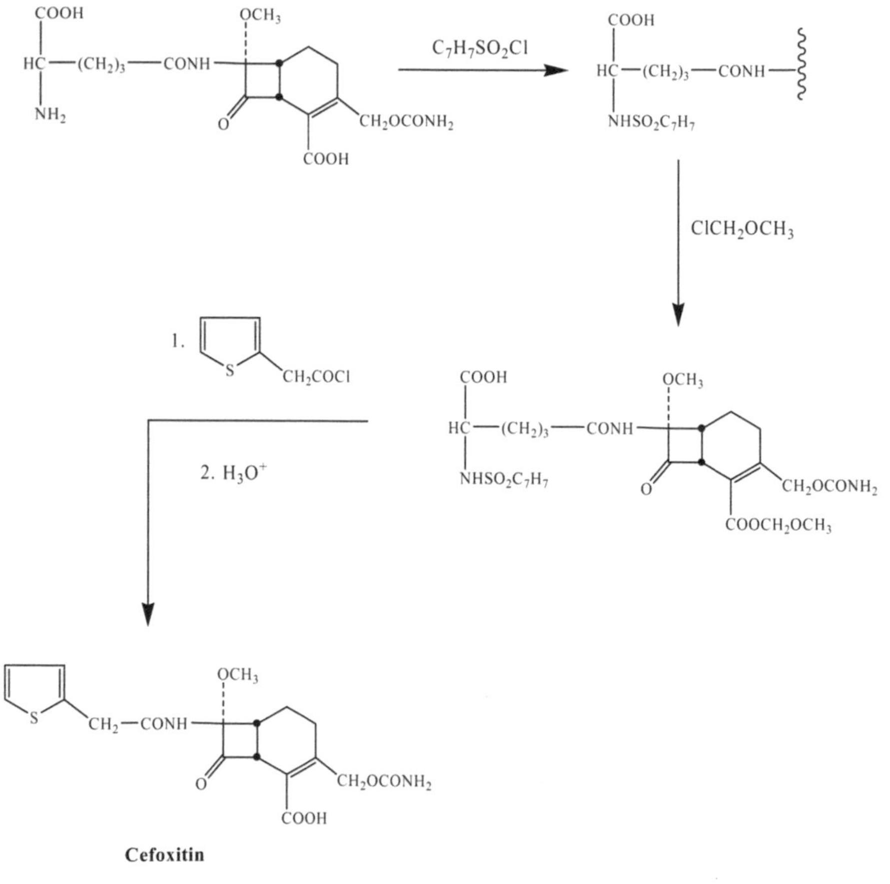

Cefoxitin

SCHEME 9.16   Synthesis of Cefoxitin Sodium.

9.4.9.2.3 Cefprozil

Cefprozil, sometimes spelled cefproxil is a white to yellowish powder with a molecular formula for the monohydrate is $C_{18}H_{19}N_3O_5S \cdot H_2O$. It is a second-generation cephalosporin type antibiotic. It is a cis and trans isomeric mixture, having chemical name (6R,7R)-7-[(R)-2-Amino-2-(p-hydroxyphenyl)acetamido]-8-oxo-3-propenyl-5-thia-1-azabicyclo[4.2.0] oct-2-ene-2-carboxylic acid monohydrate. It is orally active cephalosporin, used to treat bronchitis, ear infections, skin infections, and other

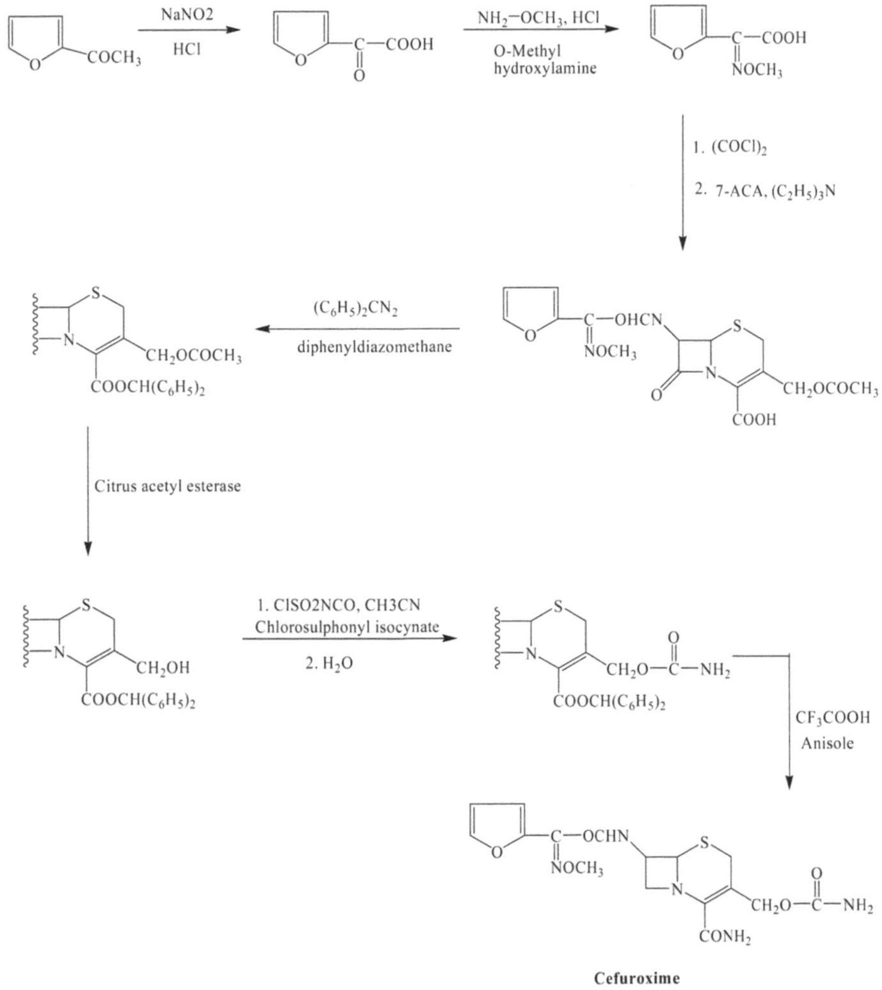

SCHEME 9.17 Synthesis of Cefuroxime Sodium.

bacterial infections (Bhargava, 2003). Cefprozil exhibits greater in vitro activity against *Streptococci, Neisseria spp.*, and *Staph. aureus* than does cefadroxil. The plasma half-life is 1.2 to 1.4 hr and thus permits twice-a-day dosing for the treatment of most infections caused by susceptible organism.

9.4.9.2.4   Cefoxitin Sodium

Cefoxitin is a semisynthetic broad spectrum antibiotics derived from cephamycin C, which is produced by *Streptomyces lactamdurans*. It's molecular formula is $C_{16}H_{16}N_3NaO_7S_2$ and IUPAC name (6*S*,7*R*)-4-(carbamoyloxymethyl)-7-methoxy-8-oxo-7-[(2-thiophen-2-ylacetyl) amino]-5-thia- 1-azabicyclo[4.2.0]oct-2-ene-2-carboxylic acid. Cefoxitin has been used in the treatment of skin, bone, respiratory, urinary tract infections and used in the infection caused by the bacteria include some *Staphylococci, Enterococci, Streptococci*, and others. It is a potent competitive inhibitor of many β-lactamase and active against both gram-positive and gram-negative bacteria including anaerobes (Brogden, 1979). The half-life of cefoxitin is relatively short and thus administered three to four times daily. It is also used to treat gonorrhea caused by β-lactamase producing strains.

9.4.9.2.5   Cefuroxime Sodium

Cefuroxamie is a second-generation oral or parenteral cephalosporin antibiotic, discovered by the Glaxo company (now GlaxoSmithKline) and marketed in 1978 as Zinacef and received approval from the U. S. Food and Drug Administration in October of 1983. It is α-methoximinoacyl-substituted cephalosporin, having IUPAC name (6*R*,7*R*)-3-{[(aminocarbonyl)oxy]methyl}-7-{[(2*Z*)-2-(2-furyl)-2-(methoxyimino)   acetyl]amino}-8-oxo-5-thia-1-azabicyclo[4.2.0]oct-2-ene-2-carboxylic acid. It is active against β-lactamase producing strains such as *E. Coli, K. pneumonia, N. gonorhhoeae* and *H. influenza*. It has biological half-life 1.4 to 1.8 hrs with 33% plasma protein binding. It is well tolerated by i.m. route and attains relatively higher CSF levels, but has been superseded by 3rd generation cephalosporins in the treatment of meningitis (Brogden, 1979).

9.4.9.2.6 Cefuroxime Axetil

Cefuroxime axetil is a broad-spectrum acetoxyetyl-ester-prodrug of cefuroxime. Its molecular formula is $C_{20}H_{22}N_4O_{10}S$, and having IUPAC name 1-Acetoxyethyl (6R,7R)-3-[(carbamoyloxy)methyl]-7-{[(2Z)-2-(2-furyl)-2-(methoxyimino) acetyl] amino}-8-oxo-5-thia-1-azabicyclo[4.2.0]oct-2-ene-2-carboxylate. Cefuroxime axetil is an amorphous powder which is effective orally. After oral administration, cefuroxime axetil is absorbed from the gastrointestinal tract and rapidly hydrolyzed by nonspecific esterases in the intestinal mucosa and blood to cefuroxime. Cefuroxime is subsequently distributed throughout the extracellular fluids. The axetil moiety is metabolized to acetaldehyde and acetic acid. Approximately 50% of serum cefuroxime is bound to protein (Leder, 1997).

9.4.9.3 Third-Generation Cephalosporins

Third-generation cephalosporins are broad-spectrum antimicrobial agents useful in a variety of clinical situations and having similar mechanism of action to that of other β-lactam antibiotics. Six third-generation cephalosporins are cefotaxime, moxalactam, cefoperazone, ceftizoxime, ceftriaxone, and cefmenoxime. These agents are particularly active against gram-positive bacteria, including methicillin-resistant *Staphylococcus aureus*; and the drugs are also effective against gonococci, haemophilus influenzae, and neisseria meningitidis. Several common gram-negative pathogens are susceptible to the third-generation cephalosporins, including *Escherichia coli*, *Klebsiella*, *Citrobacter diversus*, *Proteus*, and *Morganella*. Adverse effects associated with use of the third-generation cephalosporins are generally similar to those that occur with other β-lactam antibiotics with the exception of coagulopathies and the disulfiram reaction seen with moxalactam and cefoperazone. Despite the relatively high cost of the third-generation cephalosporins, they are often cost effective because of their reduced dosing frequencies, broad spectra of activity, and effectiveness in serious infections for which more toxic antibiotics have been required in the past (Adu, 1995).

Third-generation cephalosporins differ from earlier generations in the presence of a C=N-OCH₃ group in their chemical structure. This group provides improved stability against certain β-lactamase enzymes produced by gram-negative bacteria. These bacterial enzymes rapidly destroy

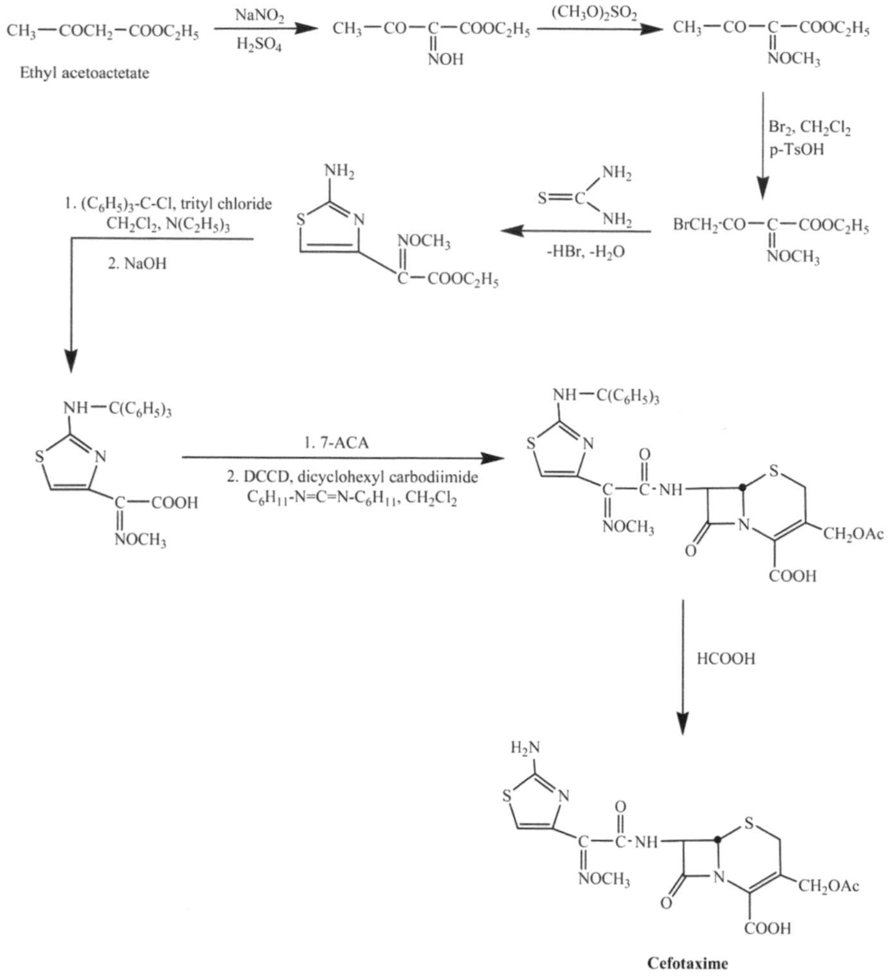

SCHEME 9.18 Synthesis of Cefotaxime Sodium.

earlier-generation cephalosporins by breaking open the drug's β-lactam chemical ring, leading to antibiotic resistance. Though initially active against these bacteria, with widespread use of third-generation cephalosporins, some gram-negative bacteria known as extended-spectrum β-lactamases (ESBLs) are even able to inactivate the third-generation cephalosporins. Infections caused by ESBL-producing gram-negative bacteria are of particular concern in hospitals and other healthcare facilities.

9.4.9.3.1 Cefotaxime Sodium

Cefotaxime sodium is a semisynthetic, broad spectrum cephalosporin anti-biotic for parenteral administration. It is the sodium salt of 7-[2-(2-amino-4-thiazolyl) glyoxylamido]-3-(hydroxymethyl)-8-oxo-5-thia-1-azabicyclo [4.2.0] oct-2-ene-2-carboxylate 72 (Z)-(o-methyloxime), acetate (ester). Cefotaxime is used for a variety of infections, including, lower respiratory tract infections, genitourinary system, gynecologic infections, bacteremia/ septicemia, intraabdominal infections, bone and join and CNS infections. Cefotaxime is active against numerous gram-positive and gram-negative bacteria, including several with resistance to classic β-lactamse such as penicillin (Jones, 1982). The *syn*-isomer of cefotaxime is more active than *anti*isomer against β-lactamse-producing bacteria. Cefotaxime has modest activity against the anaerobic *Bacteroides fragilis* and not active include *Pseudomonas* and *Enterococcus*. Cefotaxime is administered by intramus-cular injection or intravenous infusion. As cefotaxime is metabolized to both active and inactive metabolites by the liver and largely excreted in the urine, dose adjustments may be appropriate in people with renal or hepatic impair-ment. Cefotaxime reaches the cerebrospinal fluid in sufficient concentration for the treatment of meningitis.

9.4.9.3.2 Ceftriaxone Disodium

Ceftriaxone is a white crystalline cephalosporin used to prevent or treat infections caused by bacteria. The IUPAC name of ceftriazone is (6R,7R)-7-{[(2Z)-2-(2-amino-1,3-thiazol-4-yl)->2-(methoxyimino)acetyl] amino}-3-{[(2-methyl-5,6-dioxo-1,2,5,6-tetrahydro-1,2,4-triazin-3-yl) thio]methyl}-8-oxo-5-thia-1-azabicyclo[4.2.0]oct-2-ene-2-carboxylic acid. Ceftriaxone is given by injection only into a muscle or vein. It is not available orally. Like other third-generation cephalosporins, it has broad-spectrum activity against gram-positive bacteria and expanded gram-neg-ative coverage compared to second-generation agents. Ceftriaxone may cause allergic reactions similar to those caused by penicillin. Other com-mon side effects include local irritation at the injections site, rash, and diarrhea. Hypoprothrombinaemia and bleeding are specific side-effects. Haemolysis is reported. It has also been reported to cause post renal fail-ure in children. Like other third-generation cephalosporins, ceftriaxone

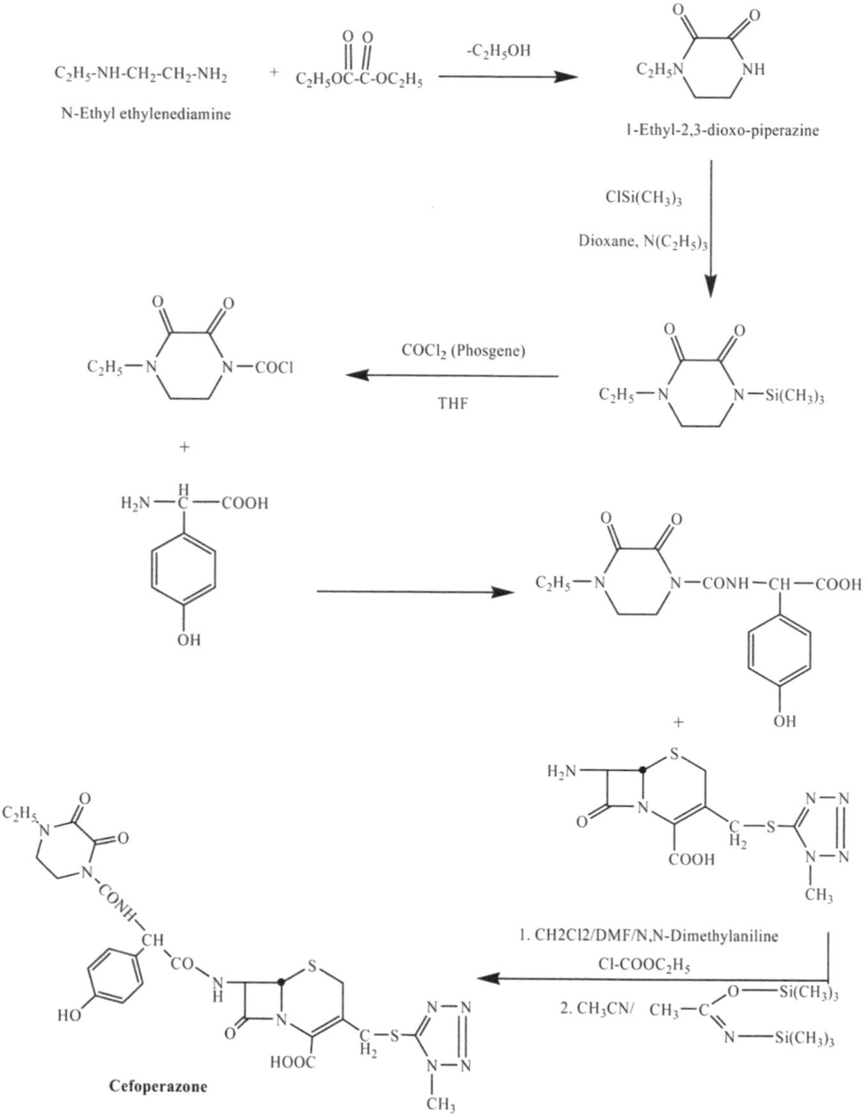

SCHEME 9.19   Synthesis Cefoperazone Sodium.

is active against *Citrobacter, S. marcenscens,* and β-lactamase-producing strains of haemophilus and neisseria. However, unlike ceftazidime and cefoperazone, ceftriaxone does not have useful activity against *Pseudomonas aeruginosa.* Like other third-generation cephalosporins, ceftriaxone penetrates body fluids and tissues well, and can achieve levels in the cerebrospinal fluid sufficient to inhibit most pathogens. To reduce the pain of intramuscular injection, ceftriaxone may be reconstituted with 1% lidocaine. Ceftriaxone posses prolong duration of action due to high protein binding and slow urinary excretion. It is excreted both in the bile and in the urine and show low volume of distribution (Guglielmo, 2000).

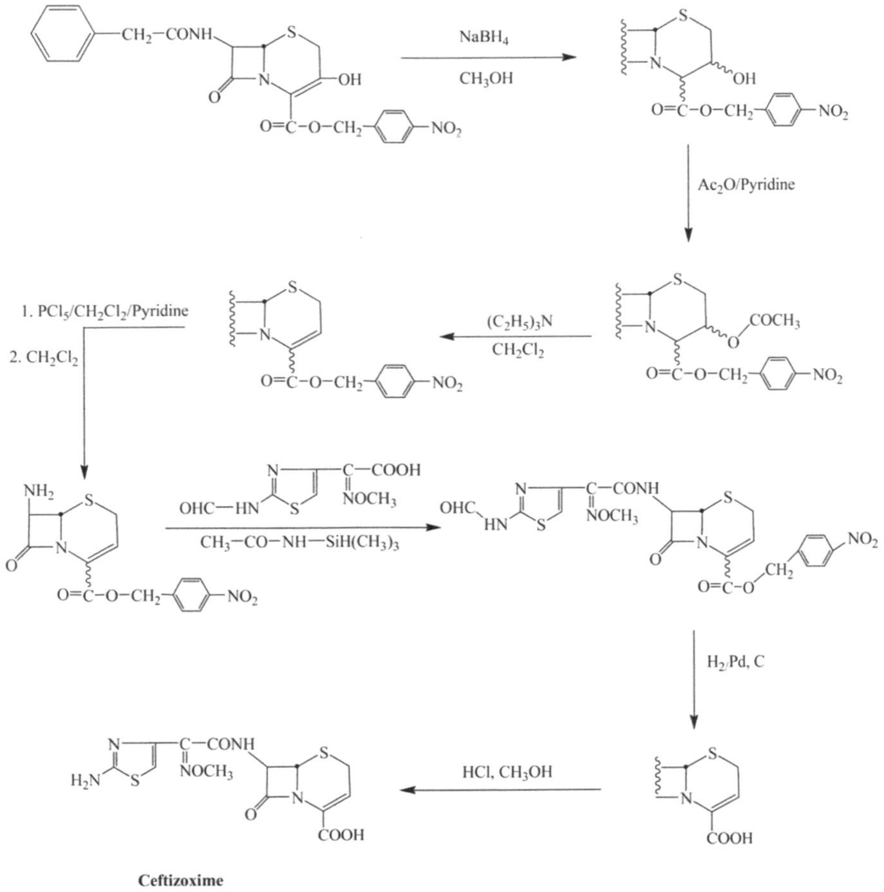

Ceftizoxime

SCHEME 9.20 Synthesis of Ceftizoxime Sodium.

9.4.9.3.3 Cefoperazone Sodium

Cefoperazone is a semisynthetic third-generation broad-spectrum ceph-alosporin with a tetrazolyl moiety that is resistant to β-lactamase. It is one of few cephalosporin antibiotics effective in treating *Pseudomonas* bacterial infections which are otherwise resistant to these antibiotics. Cefoperazone exerts its bactericidal effect similar to other cephalospo-rins, that is, by inhibiting the bacterial cell wall synthesis, and use in combination with sulbactam (a β-lactamase inhibitor), to increase the antibacterial activity of cefoperazone against β-lactamase-producing organisms. Cefoperazone has a broad spectrum of activity and has been used to target bacteria responsible for causing infections of the respira-tory, urinary tract, skin, and the female genital tract system. As other cephalosporins, cefoperazone also exerts hypersensitvities. Cefoperazone contains an *N*-methylthiotetrazole (NMTT or 1-MTT) side chain which is broken down in the body, and releases free NMTT, which can cause hypoprothrombinemia (likely due to inhibition of the enzyme vitamin K epoxide reductase) and a reaction with ethanol similar to that produced by disulfiram (Antabuse), due to inhibition of aldehyde dehydrogenase. It is primary excreted in bile and only about 25% free antibiotic is recovered in the urine. The half-life of cefoperazone is about 2 hr allows for dosing twice a day (Brogden, 1981).

9.4.9.3.4 Ceftizoxime Sodium

Ceftizoxime is a third-generation iminomethoxy aminothiazolyl cephalo-sporin available for parenteral administration. Unlike other third-generation cephalosporins, the whole C-3 side chain in ceftizoxime has been removed to prevent deactivation by hydrolytic enzymes. The IUPAC name of ceftizoxime is (6*R*,7*R*)-7-{[(2*Z*)-2-(2-amino-1,3-thiazol-4-yl)-2-methoxyiminoacetyl] amino}-8-oxo-5-thia-1-azabicyclo[4.2.0]oct-2-ene-2-carboxylic acid. Ceftizoxime is overall similar in antibacterial activity to cefotaxime and moxalactam and inhibits a wide variety of aerobic, anaerobic gram-posi-tive and gram-negative bacteria. Ceftizoxime is not hydrolyzed by common plasmid and chromosomal β-lactamases. Serum levels of ceftizoxime after intramuscular and intravenous injection are similar to those of cefotaxime and moxalactam. The half-life is 1.6 to 1.9 h in normal individuals. The

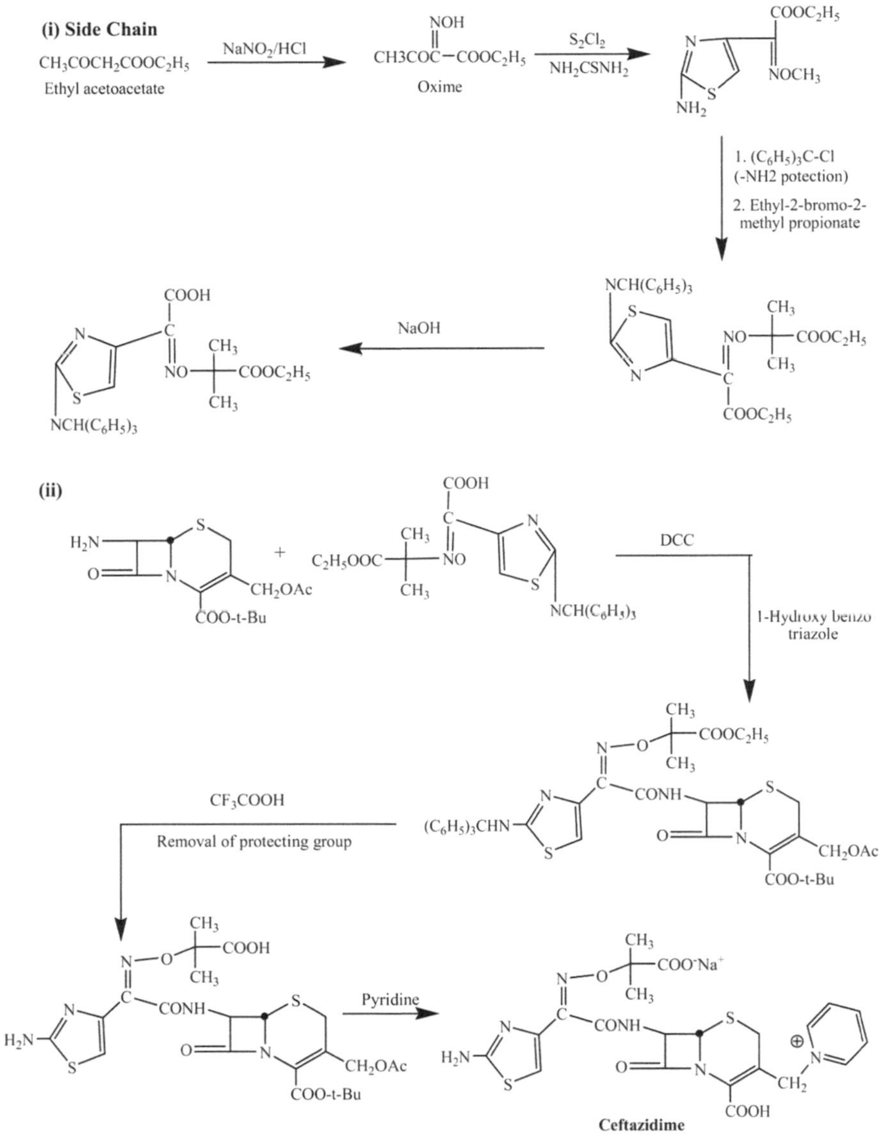

(i) Side Chain

(ii)

SCHEME 9.21 Synthesis of Ceftazidime Sodium.

compound is not metabolized and is cleared from the body by glomerular filtration. Ceftizoxime enters most body fluids, including the cerebrospinal fluid, to produce therapeutic concentrations against clinically important bacteria. Ceftizoxime has proved to be an effective chemotherapeutic agent

when used in the treatment for pneumonia, urinary tract infections, osteomyelitis, septic arthritis, meningitis, peritonitis, gonorrhea, including penicillinase-producing isolates, and gynecological infections. No major adverse
reactions have been associated with the use of ceftizoxime and it has produced neither disulfram -like reactions nor bleeding (Richards, 1985).

9.4.9.3.5   Ceftazidime Sodium

Ceftazidime is a broad-spectrum, parenteral third-generation cephalosporin-
type antibiotic. Chemically it is (6R,7R,Z)-7-(2-(2-aminothiazol-4-yl)-2-(2-
carboxypropan-2-yloxyimino)acetamido)-8-oxo-3-(pyridinium-1-ylmethyl)-
5-thia- 1-aza-bicyclo [4.2.0]oct-2-ene-2-carboxylate. As other third generation cephalosporins ceftazidime is also active against both gram-positive
and gram-negative bacteria. Ceftazidime is one of the few in this class with
activity against *Pseudomonas*. It is used in the treatment of infections of the
lower respiratory tract, urinary tract, skin, abdomen, blood, bones and joints,
and central nervous system. It is not active against methicillin-resistant
Staphylococcus aureus. The drug is given intravenous (IV) or intramuscular (IM) every 8–12 h (two or three times a day), with dose and frequenc-
ing varying by the type of infection, severity, and/or renal function of the
patient. Ceftazidime is generally well-tolerated. Still the most commonly
local side effects from the intravenous line site include allergic reactions,
and gastrointestinal symptoms. The allergic reactions including itching,
rash, and fever, happened in fewer than 2% of patients. Rare but more serious allergic reactions, such as toxic epidermal necrosis, Stevens-Johnson
syndrome, and erythema multiform, have also been reported with this class
of antibiotics, including ceftazidime. Gastrointestinal symptoms, including
diarrhea, nausea, vomiting, and abdominal pain, were reported in fewer than
2% of patients (Wiens, 2014).

9.4.9.3.6   Cefixime

Cefixime is one of the most popular third generation cephalosporin antibiotic. It is used to treat many different types of infections caused by bacteria.
Cefixime is a broad spectrum cephalosporin antibiotic and is commonly used
to treat bacterial infections of the ear, urinary tract, and upper respiratory tract.

The bactericidal action of cephalosporin is due to the inhibition of cell wall synthesis. It binds to one of the penicillin binding proteins (PBPs) which inhibit the final transpeptidation step of the peptidoglycan synthesis in the bacterial cell wall, thus inhibiting biosynthesis and arresting cell wall assembly resulting in bacterial cell death. Only about 40–50% is absorbed from the GI tract (oral), which may be further decreased if taken with food. Absorption from oral suspensions than tablet is more. Adverse drug reactions include diarrhea, dyspepsia, nausea and vomiting. Hypersensitivity reactions like skin rashes, urticaria and Stevens-Johnson syndrome have been reported. There is no specific antidote for cefixime overdosage. Gastric lavage may perform. Dialysis will not remove cefixime in significant quantities (Naqvi, 2011).

9.4.9.4 Fourth Generation

Fourth generation cephalosporins followed the third generation cephalosporins. The fourth generation cephalosporins have activity against grampositive cocci and a broad array of gram-negative bacteria, including *P. aeruginosa* and many of the *Enterobacteriaceae* with inducible chromosomal β-lactamases. Fourth-generation cephalosporins are zwitterions that can penetrate the outer membrane of gram-negative bacteria and have a greater resistance to β-lactamases than the third-generation cephalosporins. Many can cross the blood–brain barrier and are effective in meningitis. They are also used against *Pseudomonas aeruginosa*." Despite the fact that a 4th generation cephalosporin is well-suited for the treatment of polymicrobial infections, the following should be kept in mind: (I) MRSA strains and *Bacteroides fragilis* group are not included in their spectrum of activity. (II) Cefpirome is the only cephalosporin with in vitro activity against *Enterococci*. (III) Severe surgical infections of nosocomial origin, and particularly in settings where *Enterobacter* spp predominate, represent the major indication for empirical use of a 4th generation cephalosporin in combination with a nitroimidazole (Garau, 1997).

9.4.9.4.1 Cefepime

Cefepime or maxpipime is a fourth-generation cephalosporin developed in 1994. It is (6*R*,7*R*,*Z*)-7-(2-(2-aminothiazol-4-yl)-2-(methoxyimino) acetamido)-3-((1-methylpyrrolidinium-1-yl) methyl)-8-oxo-5-thia-1-aza-

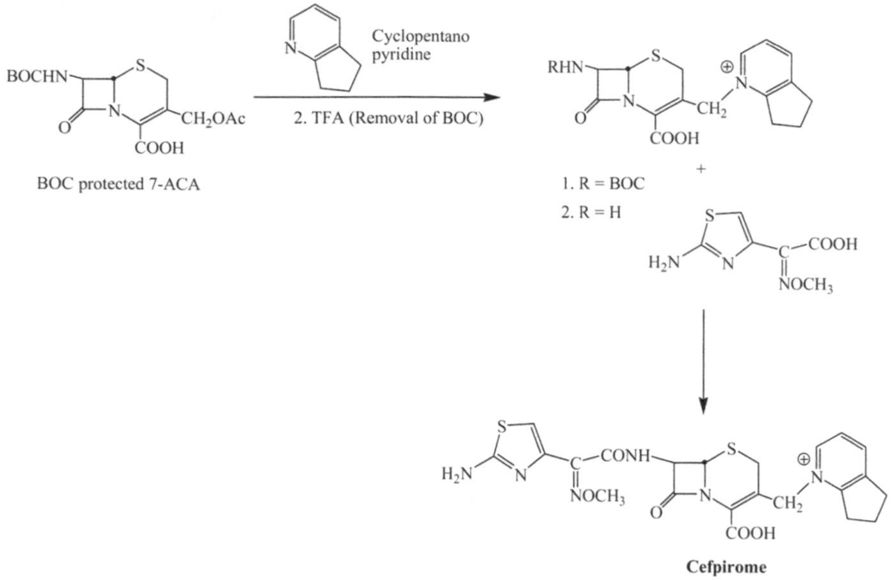

SCHEME 9.22   Synthesis of Cefpirome.

bicyclo[4.2.0]oct-2-ene-2-carboxylate. Cefepime has an extended spectrum of activity against both gram-positive and gram-negative bacteria,
as compare to third-generation agents. Cefepime is usually reserved to
treat moderate to severe nosocomial pneumonia, infections caused by
multiple drug-resistant microorganisms (e.g., *Pseudomonas aeruginosa*) and empirical treatment of febrile neutropenia. It has good activity against important pathogens including *Pseudomonas aeruginosa*,
Staphylococcus aureus, and multiple drug-resistant *Streptococcus pneumoniae*. A particular strength is its activity against *Enterobacteriaceae*.
Whereas other cephalosporins are degraded by many plasmid- and
chromosome-mediated-β-lactamases, cefepime is stable and is a front-
line agent when infection with *Enterobacteriaceae* is known or suspected. Cefepime is injected by intravenous route and it was excreted
primarily unchanged in urine. The recovery of intact cefepime in urine
was invariant with respect to the dose and accounted for over 80% of the
dose (Shahid, 2010).

9.4.9.4.2   Cefpirome

Cefpirome is a fourth-generation cephalosporin for parenteral use as sulfate salt. Its IUPAC name is 1-{[(6R,7R)-7-[(2E)-2-(2-amino-1,3-thiazol-4-yl)-2-(methoxyimino)acetamido]-2-carboxylato-8-oxo-5-thia-1-azabicyclo[4.2.0] oct-2-en-3-yl]methyl}-5H,6H,7H-cyclopenta [b] pyridin-1-ium. Cefpirome is effective against numerous clinically significant gram-positive and gram-negative bacteria, including both aerobes and anaerobes. It is stable against the action of most β-lactamases. Cefpirome is rapidly and widely distributed in body fluids and achieves excellent tissue penetration, reaching concentrations which exceed the minimum inhibitory concentrations for most pathogens. Mean peak serum concentrations of 80 to 90 µg/ml are attained after a single intravenous 1 g dose. The elimination half-life is about 2 h and is prolonged in patients with renal impairment. Cefpirome is less than 10% bound to plasma proteins. Cefpirome is widely distributed into body tissues and fluids and appears in breast milk. It is mainly excreted by the kidneys and 80 to 90% of a dose is recovered unchanged in urine. Significant amounts are removed by haemodialysis (Wiseman, 1997).

KEYWORDS

- **cephalosporin**
- **penicillin**
- **pharmacokinetics**
- **pharmacology**
- **synthesis**
- **β-lactam**

REFERENCES

Adu, A., & Armour, C. L., (1995). Drug Utilization Review (DUR) of the Third Generation Cephalosporins. *Drugs 50*(3), 423–439.

Aharonowitz, Y., Cohen, G., & Martin, J. F., (1992). Penicillin and Cephalosporin Biosynthetic Genes: Structure, Organization, Regulation, and Evolution. *Annu Rev Microbi* *46*, 461–495.

Akova, M., (2008). Sulbactam-containing b-lactamase inhibitor combinations. *Eur Soci Clinil Microbio Infect Dis 14*, 185–188.

Bergan, T., (1983). Review of the pharmacokinetics of mezlocillin. *J Antimicrob Chemother 11*, 1–16.

Bhargava, S., Lodha, R., & Kabra, S. K., (2003). Cefprozil: A Review. *Int. J Paedria 70*(5), 395–400.

Boon, R. J., Beale, A. S., & Sutherland, R., (1986). Bactericidal effects of ticarcillin-clavulanic acid against beta-lactamase-producing bacteria in vivo. *Antimicro Agen Chemother 29*(5), 838–844.

Brogden, R. N., Carmine, A., Heel, R. C., Morley, P. A., Speight, T. M., & Avery, G. S., (1981). Cefoperazone: A Review of its *In Vitro* Antimicrobial Activity, Pharmacological Properties and Therapeutic Efficacy. *Drugs 22*(6), 423–460.

Brogden, R. N., Heel, R. C., Speight, T. M., & Avery, G. S., (1979). Cefoxitin: A Review of its Antibacterial Activity, Pharmacological Properties and Therapeutic Use. *Drugs 17*(1), 1–37.

Brogden, R. N., Heel, R. C., Speight, T. M., & Avery, G. S., (1979). Cefuroxime: A Review of its Antibacterial Activity, Pharmacological Properties and Therapeutic Use. *Drugs 17*(4), 233–266.

Buck R E, Price, K. E., & Cefadroxil, (1977). A New Broad-Spectrum Cephalosporin Antimicrob Agents. *Chemother 11*(2), 324–330.

Buynak John, D., (2013). β-Lactamase inhibitors: a review of the patent literature (2010–2013). *Expert Opinion on Therapeutic Patents 23*(11), 1469–1481.

Charbonneau, P., (1994). Review of piperacillin/tazobactam in the treatment of bacteremic infections and summary of clinical efficacy. *Intensive Care Medi 20*(3), S43–S48.

Clissold, S. P., Todd, P. A., & Deborah, M., (1987). Richards Campoli. Imipenem/Cilastatin. *Drugs 33*(3), 183–241.

Cohen, N. C., (1983). β-lactam antibiotics: Geometrical requirements for antibacterial activities. *J Med Chem 26*(2), 259–264.

Curtis, R., & Jones, J., (2007). Robert Robinson and penicillin: an unnoticed document in the saga of its structure. *J Peptide Sci 13*, 769–775.

Dalhoff, A., Janjic, N., & Echols, R., (2006). Redefining penems. *Biochem Pharmacol 71*(7), 1085–1095.

Darouiche, R. O., & Hamill, R. J., (1994). Antibiotic penetration of and bactericidal activity within endothelial cells. *Antimicrob Agents Chemother 38*(5), 1059–1064.

Deguchi, Y., Koshida, R., Nakashima, E., Watanabe, R., Taniguchi, N., Ichimura, F., & Tsuji, A., (1988). Interindividual changes in volume of distribution of cefazolin in newborn infants and its prediction based on physiological pharmacokinetic concepts. *J. Pharm. Sci 77*, 674–678.

Dürckheimer, W., Blumbach, J., Lattrell, R., & Scheunemann, K. H., (1985). Recent Developments in the Field of β-Lactam Antibiotics. *Angewandte Chem Intern Edit Eng 24*(3), 180–202.

Eliopoulos, G. M., & Moellering, R. C., (1982). Azlocillin, Mezlocillin, and Piperacillin: New Broad-Spectrum Penicillins. *Ann Intern Med 97*(5), 755–760.

Finlay, J., Miller, L., & Poupard, J. A., (2003). A review of the antimicrobial activity of clavulanate. *J Antimicrob Chemother 52*, 18–23.

Fisher, J. F., Meroueh, S. O., & Mobashery, S., (2005). Bacterial Resistance to β-Lactam Antibiotics: Compelling Opportunism, Compelling Opportunity. *Chem Rev 105*(2), 395–424.

Fossieck, B. J., & Parker, R. H., (1974). Neurotoxicity during intravenous infusion of penicillin. *J Clin Pharmacol 14*(10), 504–512.

Fu, K. P., & Neu, H. C., (1978). Azlocillin and Mezlocillin: New Ureido Penicillins. *Antimicro Agen Chem 13*(6), 930–938.

Garau, J., Wilson, W. W., Wood, M., & Carlet, J., (1997). Fourth-generation cephalosporins: a review of in vitro activity, pharmacokinetics, pharmacodynamics and clinical utility. *Clin Microbio Infec 3*(1), 87–101.

Goldberg, D. M., (1987). The cephalosporins [review]. *Med Clin North Am 71*, 1113–1133.

Goyal, V. K., Rajput, S. S., (2014). Meropenem: Current perspective. *Int. J Med Sci Res Practice 1*(1), 03–05.

Guglielmo, B. J., Andrew, D. L., Paletta, D. Jr., & Richard, A. J., (2000). Ceftriaxone Therapy for Staphylococcal Osteomyelitis: A Review. *Clin Infec Dis 30*, 205–7.

Hamed, R. B., Gomez, C., Ruben, J., Henry, L., Ducho, C., Donough, M., Michael, A., & Schofield, C. J., (2013). The enzymes of β-lactam biosynthesis. *Natu Prod Repo 30*(1), 21–107.

Hameed, T. K., & Robinson, J. L., (2002). Review of the use of cephalosporins in children with anaphylactic reactions from penicillins. *Cana J Infec Dis 13*(4), 253–8.

Hautekeete, M. L., (1995). Hepatotoxicity of antibiotics. *Acta Gastroenterol Belg 58*(3–4), 290–6.

Jones Ronald, N., & Thornsberry, C., (1982). Cefotaxime: A Review of in Vitro Antimicrobial Properties and Spectrum of Activity. *Clin Infec Dis 4*(2), S300–S315.

Khan, F. Y., (2005). Severe neutropenia secondary to piperacillin/tazobactam Therapy. *Indian J Pharmacol 37*(3), 192–193.

Kohno, S., Tomono, K., Maesaki, S., Hirakata, Y., & Hara, K., (1998). Comparison of Four Carbapenems; Imipenem-Cilastatin, Panipenem-Betamipron, Meropenem, and Biapenem with Review of Clinical Trials in Japan. *Acta Med Nagasaki 43*, 12–18.

Kong, K. F., Schneper, L., & Mathe, K., (2009). Beta-Lactam Antibiotics: From Antibiosis To Resistance And Bacteriology. *J Compil 118*, 1–36.

Larry, S., Fisher, M. D., Anthony, W., Chow, M. D., Thomas, T., Yoshikawa, M. D., and Lucien, B., & Guze, M. D., (1975). Cephalothin and Cephaloridine Therapy for Bacterial Meningitis: An Evaluation. *Ann Intern Med 82*(5), 689–693.

Leder, R. D., & Carson, D. S., (1997). Cefuroxime Axetil (Ceftin(R)) A Brief Review. *Infec Dis Obs Gynec 5*, 211–214.

Livermore, D. M., (1987). Mechanisms of resistance to cephalosporin antibiotics [review]. *Drugs 34*(2), 64–88.

Mandell, L., (2009). Doripenem: A New Carbapenem in the Treatment of Nosocomial Infection. *Clini Infec Dise 49*, S1–3.

Mangla, P., & Aggarwal, K. K., (2012). Review of Antibiotics in the Management of Respiratory Infections: Cefaclor vs Amoxicillin Clavulanate. *Int. J Clin Prac 22*(11), 571–575.

Martens, M. G., (1989). Cephalosporins. *Obstet Gynecol Clin North Am 16*, 291–304.

Martin, E. A., (2003). *Oxford Concise Medical Dictionary 6*, 45–55.

McEniry, D. W., & Gorbach, S. L., (1987). Cephalosporins in surgery Prophylaxis and therapy [review]. *Drugs 34*(2), 39–216.

Minami, S., Yotsuji, A., Inoue, M., & Mitsuhashi, S., (1980). Induction of Plactamase by various β-lactam antibiotics in Enzterobacter cloacae. *Antimicrob Agen Chemother 18*, 382–385.

Naqvi, I., Saleemi, A. R., & Naveed, S., (2011). Cefixime: A drug as Efficient Corrosion Inhibitor for Mild Steel in Acidic Media.Electrochemical and Thermodynamic Studies. *Int. J. Electrochem. Sci 6*, 146–161.

Norrby, S. R., (1987). Side effects of cephalosporins [review]. *Drugs 34*(2), 20–105.

Parakh, A., Krishnamurthy, S., & Bhattacharya, M., (2009). Ertapenem. *Kathmandu Unive Med J 7*(4), 454–460.

Park, J. T., & Uehara, T., (2008). How bacteria consume their own exoskeletons (turnover and recycling of cell wall peptidoglycan). *Microbiol Mol Biol Rev 72*, 27–211.

Quin, J. D., (1989). The nephrotoxicity of cephalosporins [review]. *Adverse Drug React Acute Poisoning Rev 8*, 63–72.

Riaz, B., & Khatoon, H., (2013). Evaluation of the use of cephalosporin antibiotics in pediatrics. *J App Pharmaceu. Sci. 3*(4), 63–66.

Richards, D. M., & Heel, R. C., (1985). Ceftizoxime. *Drugs 29*(4), 281–329.

Ronald, M. D., & Jones, N., (1988). Cefotetan: A review of the microbiologic properties and antimicrobial spectrum. *Ame J Surgeo 155*(5), 16–23.

Shah, P. M., (2008). Parenteral carbapenems. *Eur Soci Clini Microbio Infect Dis 14*(1), 175–180.

Shahid, S. K., (2010). Cefepime: A Review of Its Use in the Treatment of Serious Bacterial Infections. *2*, 1–10.

Shattil, S. J., Bennett, J. S., McDonough, M., & Turnbull, J., (1980). Carbenicillin and Penicillin G Inhibit Platelet Function In Vitro by Impairing the Interaction of Agonists with the Platelet Surface. *J Clin Invest 65*(2), 329–337.

Speight, T. M., Brogden, R. N., & Avery, G. S., (1972). Cephalexin: A Review of its Antibacterial, Pharmacological and Therapeutic Properties. *Drugs 3*(1), 9–78.

Sweetman, S., (2011). Martindale: The Complete Drug Reference. Pharm Press.

Sykes, R. B., Cimarusti, C. M., Bonner, D. P., Bush, K., Floyd, D. M., & Georgopapadakou, N. H., (1981). Monocyclic beta-lactam antibiotics produced by bacteria. *Nature 291*, 489–491.

Wainwright, M., & Swan, H. T., (1986). C. G. Paine and the earliest surviving clinical records of penicillin therapy. *Med Hist 30*(1), 42–56.

Ward, A., & Richards, D. M., (1985). *Cefotetan. Drugs 30*(5), 382–426.

Wiens, P. L., Walkty, A., & Karlowsky, J. A., (2014). Ceftazidime–avibactam: an evidence-based review of its pharmacology and potential use in the treatment of Gram-negative bacterial infections. *Core Evid 9*, 13–25.

Wiesner, P., Mac Gregor, R., Bear, D., Berman, S., Holmes, K., & Turck, M., (1972). Evaluation of a New Cephalosporin Antibiotic, Cephapirin Antimicrob Agents. *Chemother 1*(4), 303–309.

Williams, J. D., & Moosdeen, F., (1987). *In vitro* antibacterial effects of cephalosporins [review]. *Drugs 34*(2), 44–63.

Wiseman, L. R., & Harriet, M. L., (1997). Cefpirome. *Drugs 54*(1), 117–140.

Zapun, A., Contreras-Martel, C., & Vernet, T., (2008). Penicillin-binding proteins and beta-lactam resistance. *FEMS Microbiol Rev 32*, 361–385.

CHAPTER 10

GENERAL ANESTHETICS

T. N. V. GANESH KUMAR

Chebrolu Hanumaiah Institute of Pharmaceutical Sciences, Chandramoulipuram, Chowdavaram, Guntur, Andhra Pradesh – 522019, India, Tel: +91-8985283248, E-mail: ganeshtnv@gmail.com

CONTENTS

10.1 INTRODUCTION

Anaesthesia (Greek *an-*, "without"; and *aisthēsis*, "sensation") is a temporary state with unconsciousness, loss of memory, lack of pain, and muscle relaxation (Snow and Joubert, 1847). One of the greatest discoveries in medical field is overcoming the pain associated with surgeries, wounds and disorders by the help of small molecules. One of the truly great moments in the long history of medicine occurred on a tense fall morning in the surgical amphitheater of Boston's Massachusetts General Hospital. Morton named his "creation" Letheon, after the Lethe River of Greek mythology, which erased painful memories (Snow and Joubert, 1847). Hardly such an exotic elixir, Morton's stuff was actually sulfuric ether. Yet while the discovery of anesthesia was a bonafide blessing for humankind, it hardly turned out to be that great for its "discoverer," William T. G. Morton. Most of the General Anesthetics (GA) was in the form of gases.

10.1.1 PURPOSE OF GA

1. To relieve anxiety, apprehension and to make the patient more comfortable.
2. Relief in salivary and mucous secretion.
3. To prevent undesirable side effects (bradycardia and muscle spasms).
4. To be used as an adjunct with the general anesthetic and make the induction more smooth.
5. Relief from pain.

10.2 STAGES OF ANESTHESIA

In 1937, Arthur Ernest Guedel was first-classified the stages of anesthesia. The introduction of neuromuscular blocking agents like succinyl choline and tubocurarine changed the concept of general anesthesia by producing temporary paralysis without deep anesthesia. Depth of general anesthesia can now be estimated using devices such as the BIS monitor (Artusio, 1954; Laycock, 1953; McCulloch, 2005).

Stage I: It is the stage of analgesia or disorientation. It is from beginning of induction of general anesthesia to loss of consciousness.

Stage II: It is the stage of excitement or delirium. It is from loss of consciousness to onset of automatic breathing. Eyelash reflex disappear but other reflexes remain intact. Coughing, vomiting, struggling, irregular respiration and breath-holding may occur.

Stage III: It is the stage of surgical anesthesia. It is from onset of automatic respiration to respiratory paralysis. It is divided into four planes:

Plane I: It is from onset of automatic respiration to cessation of eyeball movements. Eyelid reflex is lost, swallowing reflex disappears, marked eyeball movement may occur.

Plane II: It is from cessation of eyeball movements to beginning of paralysis of intercostal muscles. Laryngeal reflex is lost, corneal reflex disappears, tears secretion increases, respiration is regular, movement as a response to skin stimulation disappears.

Plane III: It is from beginning to completion of intercostal muscle paralysis. Diaphragmatic respiration persists but there is progressive intercostal paralysis, pupil dilates and light reflex is abolished. This was the desired plane for surgery when muscle relaxants were not used.

Plane IV: It is from complete intercostal paralysis to diaphragmatic paralysis.

Stage IV: It is the stage of medullary paralysis which is from stoppage of respiration to till death. Anesthetic overdose cause medullary paralysis with respiratory arrest and vasomotor collapse. Pupils are widely dilated and muscles are relaxed.

10.3   MINIMUM ALVEOLAR CONCENTRATION (MAC)

MAC is a measure of anesthetic potency (Aranake et al., 2013; MAC, 2003; Nickalls and Mapleson, 2003; Nicholas, 2008). It is defined as "the minimum alveolar concentration of anesthetic at 1 atmosphere (atm), which produces immobility in 50% of subjects exposed to a noxious stimulus". The concentration should be specified, which means the atmospheric pressure should be quoted when giving a MAC value for precision. If the pressure departs significantly from 1 atm, then this is essential. The difficulty is avoided by expressing MAC as a partial pressure (MAP). The alveolar concentration of the agent is assumed to be in equilibrium with that in the brain. For this to be a valid assumption, sufficient time must be allowed for the brain concentration to come into equilibrium with that of the lung alveolus before a MAC determination is made. MAC is accepted as a valid measure of potency of inhalational general anesthetics because it remains fairly constant for a given species even under varying conditions.

MAC can be estimated from the following equation:

$$MAC \times \lambda \approx 1.82 \text{ atmospheres}$$

where, λ =olive oil/gas partition coefficient. The product of MAC and lipid solubility is a constant; as lipid solubility decreases, MAC increases. Thus, if a new volatile anesthetic was introduced with a MAC of 9%, we could conclude that it was less potent (given its higher MAC) and less soluble than desflurane (which has a MAC of 6%). The lesser the MAC, greater the potency of agent. Halothane (MAC 0.75) is much more potent than nitrous oxide (MAC 104). Values are known to decrease with age and were tabulated based on a 40 year old (MAC_{40}) (Table 10.1).

10.4   THEORIES OF ANESTHETIC ACTION

10.4.1   LIPID SOLUBILITY-ANESTHETIC POTENCY CORRELATION (THE MEYER-OVERTON CORRELATION) (2014)

Von Bibraand Harless in 1847 suggested that GA act by dissolving in the fatty fraction of brain cells and removing fatty constituents from them, thus changing activity of brain cells and inducing anesthesia. Experimental evidence by Meyer and Overton independently published the fact that anesthetic potency is related to lipid. The anesthetic concentration required to induce anesthesia in 50% of a population of animals (the EC_{50}) was

TABLE 10.1   General Anesthetics and Their Non-Logarithmic MAC_{40} Values (Approximately)

General anesthetics	MAC Value
Nitrous oxide	104
Xenon	72
Desflurane	6.6
Ethyl Ether	3.2
Sevoflurane	1.8
Enflurane	1.63
Isoflurane	1.17
Halothane	0.75
Chloroform	0.5
Methoxyflurane	0.16

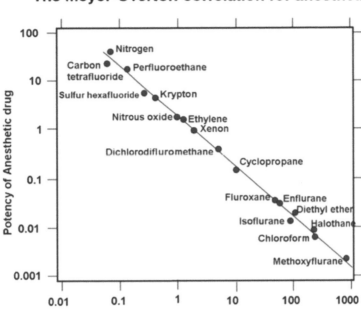

FIGURE 10.1 The Meyer-Overton correlation for anesthetics.

independent of the means by which the anesthetic was delivered, i.e., the gas or aqueous phase.

Meyer and Overton had discovered the striking correlation between the physical properties of general anesthetic molecules and their potency: The greater is the lipid solubility of the compound, greater is its anesthetic potency. The correlation is true for a wide range of anesthetics with lipid solubility over 4–5 orders of magnitude if olive oil is used as the oil phase (Figure 10.1). As the oil: gas solubility increases, the drug potency increases and the dose required to produce anesthesia reduces.

10.4.1.1 Objections to the Outdated Lipid Hypotheses

a. Stereoisomers of an anesthetic drug (Nicholas, 2008)

Enantiomers or optical isomers which are the mirror images representing stereoisomers. For example, Isomers of R-(+)- and S-(−)-etomidate. Physicochemical effects of enantiomers aresame in an achiral environment like in the lipid bilayer. However, in the biological environment enantiomers of general anesthetics, for example, isoflurane, thiopental, etomidate can differ greatly in their anesthetic potency despite the similar oil/gas partition coefficients. For example, the R-(+) isomer of etomidate is 10 times more potent anesthetic than its S-(−) isomer. This objection provides a compelling evidence that the primary target for anesthetics is not the achiral lipid bilayer itself but rather stereoselective binding sites on membrane proteins that provide chiral environment for better interactions.

b. Nonimmobilizers (The Meyer-Overton Theory of Anesthesia, 2014)

General anesthetics induce immobilization (absence of movement) through depression of spinal cord functions, but their amnesic actions are exerted within the brain. The anesthetic potency of the drug is directly proportional to its lipid solubility, but there are many compounds that do not follow this rule. These drugs are similar to general anesthetics and are predicted to be potent anesthetics based on their lipid solubility, but they exert only one constituent of the anesthetic action (amnesia) and do not suppress movement (i.e., do not depress spinal cord functions) as all anesthetics do. These drugs are referred to as nonimmobilizers. Halogenated alkanes that are very hydrophobic belong to this class.

c. Effect vanishes beyond a certain chain length (The Meyer-Overton Theory of Anesthesia, 2014)

According to the Meyer-Overton correlation, in any general anesthetic (e.g., n-alcohols, or alkanes), increasing the chain length increases the lipid solubility, and thereby should produce a corresponding increase in anesthetic potency. However, beyond a certain chain length the anesthetic effect disappears. For example, n-alcohols cutoff occurs at a carbon chain length of about 13 and for the n-alkanes at a chain length of between 6 and 10, depending on the species.

A plot of chain length vs. logarithm of the lipid bilayer/buffer partition coefficient K is linear, with the addition of each methylene group causing a change in the Gibbs free energy of −3.63 kJ/mol.

The cutoff effect was first interpreted as evidence that anesthetics exert their effect not by acting globally on membrane lipids but rather by binding directly to hydrophobic pockets of well-defined volumes in proteins. Thus the volume of the n-alkanol chain at the cutoff length provides an estimate of the binding site volume. This objection provided the basis for protein hypothesis of anesthetic effect.

10.4.2 MODERN LIPID HYPOTHESIS (LIU ET AL., 1993; THE MEYER-OVERTON THEORY OF ANESTHESIA, 2014; UEDA, 1999)

The lipid hypothesis states that anesthetic effect comes if solubilization of general anesthetic in the bilayer causes a redistribution of membrane lateral pressures. The membrane proteins like ion channels are sensitive to changes in this lateral pressure distribution. According to the lipid hypothesis a change in the membrane lateral pressure profile shifts the conformational equilibrium of certain membrane proteins. This mechanism is also nonspecific because the potency of the anesthetic is determined not by its actual chemical structure, but by the positional and orientation distribution of its segments and bonds within the bilayer. However, it is still not obvious what the exact molecular mechanism is. A detailed mechanism of general anesthesia was suggested and investigated using lattice statistical thermodynamics (Figure 10.2). It was proposed that incorporation of amphiphilic and other interfacially active solutes (e.g., general anesthetics) into the bilayer increases the lateral pressure selectively near the aqueous interfaces, which is compensated by a decrease in lateral pressure toward the center of the bilayer. General anesthesia likely involves inhibition of the opening of the ion channel in a postsynaptic ligand-gated membrane protein by the following mechanism:

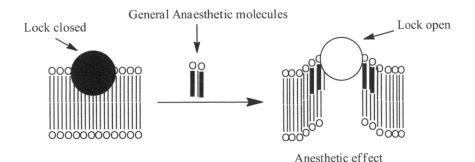

FIGURE 10.2 General anesthetic changes membrane pressure profile and determines conformation of membrane ion channel shown as round lock.

- A nerve impulse opens the channel thus by increasing the cross-sectional area of the protein more near the aqueous interface than in the middle of the bilayer;
- The anesthetic induced increase in lateral pressure near the interface shifts the protein conformational equilibrium back to the closed state, since channel opening will require greater work against the higher pressure at interface.

Thus, according to the modern lipid hypothesis anesthetics do not act directly on their membrane protein targets, but rather perturb specialized lipid matrices at the protein-lipid interface, which act as mediators. This is a new kind of transduction mechanism. Oleamide (fatty acid amide of oleic acid) is an endogenous anesthetic found in cat's brain and it is known to potentiate sleep and lower the temperature of the body by closing the gap junction channel connexon.

10.4.3 MEMBRANE PROTEIN HYPOTHESIS OF GENERAL ANESTHETIC ACTION (FRANKS, 2006; LIU ET AL., 1993)

General anesthetics may also interact with hydrophobic protein sites of certain proteins, rather than affect membrane proteins indirectly through nonspecific interactions with lipid bilayer as mediator. It was shown that anesthetics alter the functions of many cytoplasmic signaling proteins, including protein kinase C; however, the proteins considered the most likely molecular targets of anesthetics are ion channels. According to this theory general anesthetics are much more selective than in the frame of lipid hypothesis and they bind directly only to small number of targets in CNS mostly ligand gated ion channels in synapse and G-protein coupled receptors altering their ion flux. Cys-loop receptors are plausible targets for general anesthetics that bind at the interface between the subunits. The Cys-loop receptor super family includes inhibitory receptors (GABA A, GABA C, glycine receptors) and excitatory receptors (acetylcholine receptor and 5-HT3 serotonin receptor) (Figure 10.3).

Various studies have shown that low affinity drugs including inhaled general anesthetics do not usually interact with their target proteins via specific lock-and-key binding mechanism because they do not change molecular structures of transmembrane receptors, ion channels and globular proteins. Proteins of four-α-helix bundle structural motif served as models of monomer of pentameric Cys-loop receptor because binding pockets of inhaled

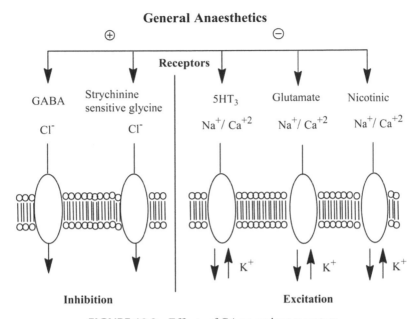

FIGURE 10.3   Effects of GA on various receptors.

anesthetics are believed to be within transmembrane four-α-helix bundles of Cys-loop receptors (Figure 10.4). Inhaled general anesthetics do not change protein structure but they may exert their effect on proteins by modulating protein dynamics in a slow microsecond-millisecond timescale and/or by disrupting the modes of motion essential for function of this protein.

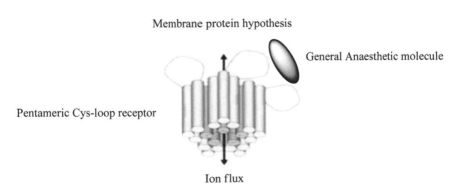

FIGURE 10.4   Inhaled general anesthetics frequently do not change structure of their target protein (of Cys-loop receptor here) but change its dynamics especially dynamics in the flexible loops that connect α-helices in a bundle thus disrupting modes of motion essential for the protein function.

10.5   CLASSIFICATION OF DRUGS (FRANKS AND LIEB, 1994; JANOFF ET AL., 1981; LERNER, 1997; MEYER, 1899, 1937; MIHIC ET AL., 1997; MOHR ET AL., 2005; OVERTON, 1901)

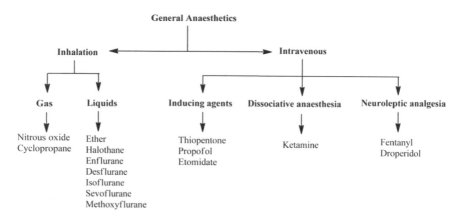

10.5.1   INHALATION ANESTHETICS

10.5.1.1   Volatile Liquids

a. Ethers

i. Diethyl ether: CH3-CH2-O-CH2-CH3

It is not used nowadays mainly because of its flammability.

Advantages:

- Ether produces good muscle relaxation.
- Ether is a safe anesthetic.
- It does not depress cardiac contractility and respiration markedly (mechanical ventilation is not required).

Disadvantages:

- Light induces peroxidation of ether; ether peroxides are toxic
- Causes strong excitation during the stage of delirium.
- Irritates the airways and induces hyper secretion.

ii. Divinyl ether (Vinydan): $CH_2=CH-O-CH=CH_2$

It is also rarely used due to its flammability. It is more potent than diethyl ether, and is less irritative to the airways. It is used in minor surgeries like removal of pharyngeal tonsils.

b. Halogenated hydrocarbons

i. Chloroform (CHCl₃): It is introduced in 1847 by James Simpson. Its sweetish odor made chloroform popular, but is no longer used because of its hepatotoxicity and cardiotoxicity. Its hepatotoxicity causing metabolite is phosgene, an electrophilic acyl chloride, formed by oxidative dehalogenation.

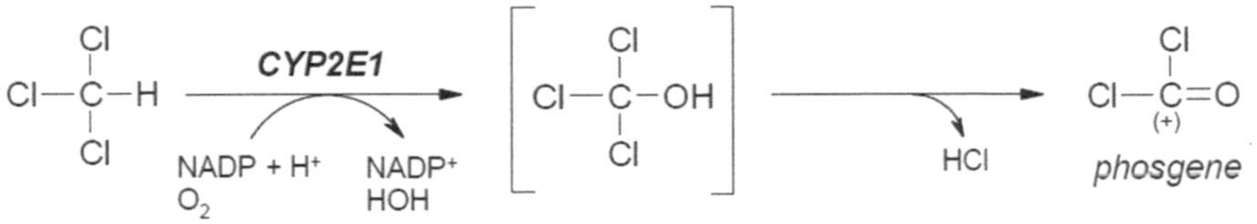

ii. Halothane (Narcotan): It is the only inhalational anesthetic agent containing a bromine atom. Its use has been drastically curtailed due to its idiosyncratic hepatotoxicity.

Advantages:
- It is nonflammable.
- It causes smooth induction with virtually no stage of delirium.
- It rarely induces postoperative nausea and vomiting.
- It has bronchodilatory effect.

Disadvantages:
- Insufficient decrease of muscle tone.
- Cardio depressive effect and arrythmogenic effect.
- Hypotension.
- Slightly sensitizes the heart to catecholamines.
- Respiratory depressive effect and depression of mucociliary function.

Rare adverse effects:

a. Halothane hepatitis, an immunologically determined reaction and symptoms like fever, eosinophilia start 2–5 days after anesthesia and hepatic failure rate may be as high as 50%.

   The mechanism involves oxidative debromination (20% of dose) by CYP2El to reactive trifluoro-acetylchloride (a reactive electrophilic acyl chloride) to neoantigene formation to immune hepatitis.

   Disulfiram pretreatment has been recommended before halothane anesthesia to prevent hepatitis.

trifluoro-acetylchloride
(binds covalently to hepatic proteins)

Disulfiram is a potent inhibitor of CYP2E1 (not only of aldehyde dehy-drogenase) and thus decreases formation of trifluoro-acetylchloride.

b. Malignant hyperthermia (MHT) – a genetically determined reaction, and caused by any halogenated inhalation anesthetic and also by suc-cinylcholine. Hyperthermia, muscle rigidity, hyperkalemia, rhabdo-myolysis and myoglobinuria are the common symptoms.

Therapy includes stopping the delivery of the inhalation anesthetic. Infuse $NaHCO_3$ to correct acidosis and prevent myoglobin precipitation in renal tubules. Decrease the body temperature. Inject dantrolene to relax the skeletal muscle.

c. Halogenated ethers

i. Enflurane (Ethrane): It combines some of the properties of diethyl ether and some of the properties of halothane. It is not in use at present. 2–5% of the dose undergoes biotransformation by dehalogenation.

Advantages:
• It has a good muscle relaxant effect.
• It does not sensibilize the heart toward catecholamine's.
• It is not flammable and smoothly induces anesthesia.

Disadvantages:
• Mildly stimulates the trachea bronchial secretions.
• Causes postoperative nausea and vomiting.
• It decreases cardiac contractility (but not the heart rate).
• It depresses the respiration (assisted ventilation is needed).
• Rare adverse effect: tonic-clonic seizures (specific for Enflurane).
• It causes hypocarbia, which sensitizes the CNS to Enflurane-induced seizures.

ii. Isoflurane (Forane; an isomer of enflurane): Most commonly used volatile anesthetic.

Its properties are very similar to those of enflurane, except for five main differences:
- Isoflurane's dehalogenation in the body is negligible (0.2%).
- Isoflurane does not cause cardiac depression.
- Decreases BP by reducing peripheral vascular resistance.
- Isoflurane does not induce seizures.
- Isoflurane is a preferred anesthetic for brain surgery.

iii. Methoxyflurane (Penthrane)

It used for inducing analgesia, but not for general anesthesia. It has good lipid-solubility as well as potency (MAC = 0.2%). However, its use as a GA has been discontinued due mainly to its nephrotoxicity.

It causes "High-output renal failure", a concentrating defect (refractory polyuria). 50–70% of methoxyflurane undergoes biotransformation via CYP-catalyzed O-demethylation and subsequent oxidative dehalogenation. Its extensive biotransformation is mainly due its accumulation in fatty tissues. Methoxy fluoroacetic acid, dichloroacetic acid, oxalic acid, and fluoride ion are the metabolites which cause nephrotoxicity. However, it is used in subanesthetic dose for emergency services.

iv. Desflurane (suprane):

This contains only fluorine substitutions and has lower lipid solubility and blood solubility than other ether derivatives. It induces and recovers from anesthesia more rapidly. It do not cause cardiac depression, but hypotension may occur due to vasodilatation. Its Dehalogenation is negligible (0.02%). The adverse effect involved irritancy in the airways and causes coughing, breath holding, secretions and laryngospasm.

v. Sevoflurane (Ultane):

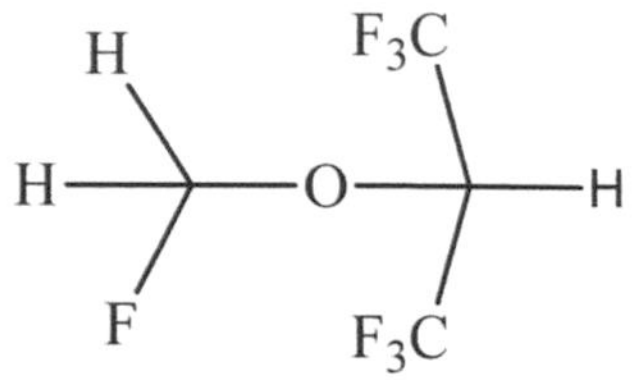

It is most often used inhalation anesthetic containing only fluorine substitution. Its properties are similar to Desflurane except that its dehalogenation is minimal (3%). It has a pleasant odor

Sevoflurane can react with the CO_2 absorberbarium hydroxidein the anesthesia machine, which causes two adverse outcomes:

a. Sevoflurane undergoes an exothermic reaction with the CO_2 absorber, which is overheated, if desiccated can induce ignition and explosion.

b. In the basic CO_2 absorber, sevoflurane is dehydrofluorinated to "Compound A", which causes nephrotoxicity.

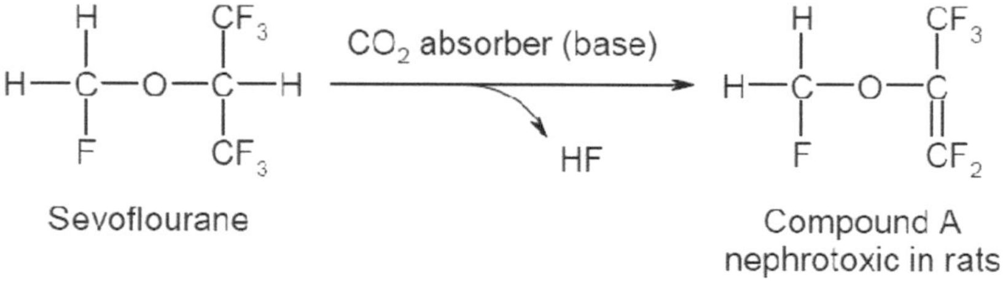

10.5.1.2 Anesthetic Gases

i. Cyclopropane: It is not in use because of its explosive nature. It sensitizes the heart to catecholamines, and can induce arrhythmias, including ventricular fibrillation.

ii. Nitrous oxide (N_2O): Also called as "laughing gas" due to its euphoric nature. It is the least lipid soluble and least potent inhalation anesthetic (MAC = 105%) and used only under hyperbaric conditions.

Advantages:
- Non flammable.
- Lacks harmful effects.
- Does not induce malignant hyperthermia.
- It is the least likely to increase cerebral blood flow and intracranial pressure.

Disadvantages:
- Used only in combination with other general anesthetics.
- Has no muscle relaxant activity.
- Cause adverse effects like "diffusional hypoxia", megaloblastic anemia.

10.5.2   INTRAVENOUS ANESTHETICS

These are relatively hydrophobic and lipophilic compounds. These include: Barbiturates (thiopental and methohexital), Others (propofol, etomidate and ketamine).

10.5.2.1   Mechanism of Action

Except Ketamine, all other agents activate GABA-Areceptor to exert their action. Barbituratesalso inhibit the neuronal N-type acetylcholine receptor. Ketamineinhibits the NMDA-type glutamate receptor.

10.5.2.2   Pharmacokinetics

Immediately after IV injection, they distribute rapidly to the well-perfused tissues, including the brain. General anesthesia occurs rapidly within 1 min of administration. They redistribute to the less well-perfused tissues, such as the muscle, skin and the adipose tissue. The anesthetic action of IV anesthetics is terminated by redistribution, not by elimination. The rate of redistribution is same for all IV anesthetics. Therefore, after a single dose, their duration of action is also similar (5–10 min). These are eliminated by biotransformation (by CYP, glucuronidation, hydrolysis).

Among all the IV anesthetics except Thiopental ($t_{1/2}$= 12 hrs, othersare eliminated relatively rapidly ($t_{1/2}$= 1–4 hrs). Propofol is eliminated most rapidly (by glucuronidation) ($t_{1/2}$= 1–2 hrs).

10.5.2.3   Specific Properties of IV Anesthetics

i.   Barbiturates:
a.   Thiopental (a thiobarbiturate)

Dose: 3–5 mg/kg IV. Its onset of action is nearly 20 seconds. Its induction is rapid and smooth with duration of action nearly 5–10 min. It is eliminated by CYP mediated oxidative desulfuration. The metabolite product is pentobarbital, a hypnotic. Its $t_{1/2}$ is 12 h.

b.   Methohexital (an oxobarbiturate)

Dose: 1–1.5 mg/kg IV. Its onset of action, induction and duration is same as thiopental. It is eliminated by CYP mediated N-demethylation and hydroxylation. Its $t_{1/2}$ is 4 h.

Advantages:
• The cerebral blood flow and intracranial pressure were maintained.
• These exert anticonvulsive effect.

Disadvantages:
- Incompatibility with co administrated muscle relaxants due to their precipitation during injection.
- Barbiturates exert vascular irritative effect (IV thiopental may induce pain and thrombophlebitis.
- Barbiturates induce respiratory depression and apnea at high dose.
- Causes negative inotropic effect.
- Barbiturates induce ALAsynthetase, whichcauses accumulation of porphyrins (neurotoxic).

ii. Others

a. Propofol (Diprivan): 2,6-disopropylphenol.

propofol-glucuronide (inactive)

Dose: 1.5–2.5 mg/kg. Its onset of action, induction and duration is same as barbiturates. Recovery is 10–40 min. It is eliminated rapidly by glucuronidation and sulfation at the hydroxyl group. Its $t_{1/2}$ is 2 hrs.

Propofol is a water immiscible oily substance, hence used as emulsion and its solvent contains soybean oil, egg phospholipids and glycerol, which supports bacterial growth. To overcome this problem, the water soluble prodrug of propofol, fospropofol is now marketed as Lusedra in the USA. Fospropofol is the phosphate ester of propofol, which is hydrolyzed by alkaline phosphatase (AP) in the body, thus releasing propofol.

Fospropofol   Propofol

Advantages:
- Propofol has antiemetic effect.
- Rapidly metabolized.

Disadvantages:
- Its formulation easily gets contaminated due to fatty acids.
- It causes veno irritative effect.
- Causes pain.
- Causes respiratory depressive effect (stronger than barbiturates).
- Decreases blood pressure because of vasodilatation and negative inotropic effect.
- Inhibits mitochondrial electron transport.

b.   Etomidate (Amidate)

Dose: 0.2–0.4 mg/kg. It is an ultra short acting anesthetic (4–8 min). It induces anesthesia rapidly. It is eliminated by ester hydrolysis in the liver. Its $t_{1/2}$ is 3 hrs. Prolonged infusion is contraindicated because it inhibits cortisol synthesis and blunts the stress response and may cause death.

Advantages:
- Less respiratory depression.
- No effect on blood pressure.
- Etomidate is usually reserved for patients at risk for hypotension and myocardial ischemia.

Disadvantages:
- Often induces nausea and vomiting.
- Causes pain on injection.
- Appears to have proconvulsive effect (contraindicated in seizures).

c.   Ketamine (Ketalar): NMDA receptor antagonist

Dose: 0.5–1.5 mg/kg. Its onset of action is relatively slow (1–2 min). Its duration of anesthesia is 10–15 min. Its elimination is by CYP-catalyzed N-demethylation. Its $t_{1/2}$ is 3 hrs.

Ketamine has three peculiar effects:

1. It has *analgesic effect* that outlasts the anesthetic effect.
2. Ketamine may induce *disagreeable dreams* and hallucinations, when emerging from the anesthesia. This occurs seldom in children.
3. Ketamine has *indirect sympathomimetic effect* because it increases neuronal reuptake of catecholamines both peripherally and centrally (like cocaine); the resultant effects are: increase in blood pressure, heart rate, cardiac output, myocardial O_2 consumption, cerebral blood flow, intracranial pressure, pupillary dilation, lacrimation and bronchodilation.

10.6 PRE-ANESTHETIC MEDICATION

Pre anesthetic medication means the delivery of drugs before the general anesthetics are given.

Drugs which are given for Pre-anesthetic medication are listed as below:

a. **Opioids**: Morphine 10 mg or Pethidine 50–100 mg IM, is given prior to anesthetic to relieve anxiety and pain control.
b. **Antianxiety drugs**: Benzodiazepines like Diazepam 5–10 mg oral, Lorazepam 2 mg IM, is given to produce tranquility and smooth induction. Midazolam is given to produce amnesia.
c. **Sedative-hypnotics:** Barbiturates like Pentobarbitone, Secobarbitone or Butabarbitone 100 mg oral have been used night before to ensure sleep and in the morning to calm the patient.
d. **Anticholinergics:** Atropine or hyoscine 0.6 mg IM/IV to reduce salivary and bronchial secretions.
e. **H2 blockers:** Patients undergoing prolonged operations, caesarian section, and obese patients are at increased risk of gastric regurgitation and aspiration pneumonia. Ranitidine 150 mg or famotidine 20 mg is given at night before and in the morning.
f. **Antiemetic:** Metoclopramide 10–20 mg IM, preoperatively is effective in reducing postoperative vomiting.

10.7 COMPLICATIONS OF GENERAL ANESTHESIA

The general anesthetics have wide importance in relieving pain in many cases of medication and surgeries. However, their use also gives some

TABLE 10.2   Complications of General Anesthesia

During Anesthesia	After Anesthesia
1. Respiratory depression and hypercarbia	1. Nausea and vomiting
2. Salivation, respiratory secretions	2. Persisting sedation
3. Cardiac arrhythmias	3. Pneumonia
4. Fall in BP	4. Organ toxicities
5. Aspiration of gastric contents	5. Nerve palsies
6. Laryngospasm	6. Emergence delirium
7. Delirium, convulsions	
8. Fire and explosion	

complications. These can be characterized as during its action and after its use. These effects were tabulated in Table 10.2.

KEYWORDS

- **anesthetic**
- **drug action**
- **pre-anesthetic**
- **safety**
- **theories**
- **toxicity**

REFERENCES

Aranake, A., Mashour, G. A., & Avidan, M. S. (2013). Minimum alveolar concentration: ongoing relevance and clinical utility. *Anaesthesia, 68*, 512–522.

Artusio, J. F. (1954). Di-ethyl ether analgesia: a detailed description of the first stage of ether analgesia in man. *J Pharmacol Exp Ther 111*, 343–334.

David White Uses of MAC, *British Journal of Anesthesia. 91*(2), August 2003.

Franks, N. P. (2006). "Molecular Targets Underlying General Anaesthesia". *Br. J. Pharmacol. 147*(1), 72–81.

Franks, N. P., & Lieb, W. R. (February 1994). "Molecular and Cellular Mechanisms of General Anesthesia". *Nature. 367* (17), 607–14

Nicholas P. Franks, (2008). "General anesthesia: from molecular targets to neuronal pathways of sleep and arousal". *Nature Reviews Neuroscience. 9*(5), 370–386. doi: 10.1038/nrn2372.

Janoff, A. S., Pringle, M. J., & Miller, K. W. (1981). "Correlation of general anesthetic potency with solubility in membranes". *Biochim. Biophys. Acta. 649*(1), 125–128.

John Snow & Meyer Joubert. Five stages of narcotism; on the inhalation of ether in surgical operation, London, 1847.

Laycock, J. D. (1953). "Signs and stages of anesthesia; a restatement". *Anaesthesia. 8*(1), 15–20.

Lerner, R. A. (1997). "A hypothesis about the endogenous analog of general anesthesia". *Proc. Natl. Acad. Sci. USA94*(25), 13375–13377.

Liu, J., Laster, M. J., Taheri, S., Eger, E. I., Koblin, D. D., & Halsey, M. J. (1993). "Is There a Cutoff in Anesthetic Potency for the Normal Alkanes?". *M Anesth Analg 77*(1), 12–18.

McCulloch, T. J. (2005). "Use of BIS Monitoring Was Not Associated with a Reduced Incidence of Awareness". *Anesthesia & Analgesia. 100*(4), 1221.

Mechanism of action of inhaled anesthetic agents. *Anaesthesia UK.* 2005.

Meyer, H. H. (1899). "ZurTheorie der Alkoholnarkose". *Arch. Exp. Pathol. Pharmacol. 42* (2–4), 109–118.

Meyer, K. H. (1937). "Contributions to the theory of narcosis". *Trans Faraday Soc. 33*, 1062–1068.

Mihic, S. J., Ye, Q., Wick, M. J., Koltchine, V. V., Krasowski, M. D., Finn, S. E., Mascia, M. P., Valenzuela, C. F., Hanson, K. K., Greenblatt, E. P., Harris, R. A., & Harrison, N. L. (1997). "Sites of alcohol and volatile anesthetic action on GABA(A) and glycine receptors". *Nature. 389*(6649), 385–389.

Mohr, J. T., Gribble, G. W., Lin, S. S., Eckenhoff, R. G., & Cantor, R. S. (2005). "Anesthetic Potency of Two Novel Synthetic Polyhydric Alkanols Longer than the n-Alkanol Cutoff: Evidence for a Bilayer-Mediated Mechanism of Anesthesia?" *J. Med. Chem. 48*(12), 4172–4176.

Nickalls, R. W. D., & Mapleson, W. W. (August 2003). "Age-related iso-MAC charts for isoflurane, sevoflurane, and desflurane in man". *British Journal of Anaesthesia. 91*(2), 4–174.

Overton, C. E. (1901). "Studienüber die Narkosezugleichein Beitragzurallgemeinen Pharmakologie". *Gustav Fischer,* Jena, Switzerland.

The Meyer-Overton Theory of Anesthesia. *The Pauling Blog* (blog). 2009. Retrieved 18 March 2014.

Ueda, I. (1999). "The window that is opened by optical isomers". *Anesthesiology. 90* (1), 336. doi: 10.1097/00000542-199901000-00068. PMID9915358.

CHAPTER 11

STEREOCHEMICAL ASPECTS IN MEDICINAL CHEMISTRY

SIDHARTHA S. KAR

Department of Pharmaceutical Chemistry, IGIPS, Bhuabneswar – 751015, India

CONTENTS

11.1 INTRODUCTION

Implementation of stereochemistry in pharmaceuticals dates back to the late 1850's, when Pasteur reported that most natural organic products, the vital products of life, are asymmetric and possess such asymmetry that they are not superimposable on their images. In 1874, Van't Hoff and La Bel independently reported the relationship between three-dimensional molecular structure and optical activity and the concept of chiral carbon atom. Subsequently, the physiological and toxicological significance of chiral compounds was chiral molecules was explored. Ariens, in the late 1980's raised the question "why we in some cases have to give doses to the patient where half of the content has no effect or the opposite effect?" After this revival of stereochemistry, the regulatory authorities defined more strict requirements on drug discovery and chiral compounds. Besides the ethical reasons, the therapeutic benefit including safety and, in several instances, extensions of the life cycle of drugs have been essential for developing single enantiomers (Shafaati, 2007). Chiral drugs are made up of molecules with the same chemical structure, but different three-dimensional arrangements in the space. The pharmacological and toxicological effects of the chiral drugs are considered instrumental in drug discovery because of the different activities shown by different enantiomers. Almost all of the biological macromolecules, such as DNA, RNA, protein, polynucleotides, and even the amino acids, the basic structural units of life, are chiral. Most of the natural drugs are chiral, but in nature only the biologically active enantiomer is synthesized. For example, *Papaver somniferum* only synthesizes (−)-(5R,6S,9R,13S,14R)-morphine which is the active enantiomer used as powerful analgesic. Currently, 56% of the marketed drugs are chiral products and 8.8% of them are marketed as racemic mixtures consisting of an equimolar quantity of two enantiomers.

Chiralityis generally defined as the geometric property of a molecule which is not being superimposable with its mirror image. Molecules that can be superimposed on their mirror images are not chiral (achiral). Chirality is a property of matter found throughout biological systems, from the basic building blocks of life such as amino acids, carbohydrates, and lipids to the layout of the human body. The two nonidentical mirror images of a chiral molecule are termed *enantiomers*. Both molecules of an enantiomer pair have the same chemical composition and can be drawn the same way in two dimensions, but in chiral environments such as the receptors and enzymes in the body, they can behave differently. A racemate is a mixture of equal amounts of both enantiomers of a chiral drug. Single enantiomers are sometimes referred to as *single isomers* or stereoisomers. For example, molecules that are isomers of each other share the same stoichiometric molecular formula but may have very different structures. However, many discussions of chiral drugs use the terms enantiomer, single isomer, and/or single stereoisomer interchangeably. The two enantiomers of a chiral drug are best identified on the basis of their absolute configuration or their optical rotation. Other designations such as D and L are used for sugars and amino acids but are specific to these molecules and are not generally applicable to other compounds. The terms *d*, or *dextro,* and *l*, or *levo*, are considered obsolete and should be avoided. Instead, the *R/S* system for absolute configuration and the +/– system for optical rotation should be used. The absolute configuration at a chiral center is designated as *R* or *S* to unambiguously describe the 3-dimensional structure of the molecule. *R* is from the Latin *rectus* and means to the right or clockwise, and *S* is from the Latin *sinister* for to the left or counterclockwise. There are precise rules based on atomic number and mass for determining whether a particular chiral center has an *R* or *S* configuration. A chiral drug may have more than one chiral center, and in such cases it is necessary to assign an absolute configuration to them. Scientists have started using prefixes such as lev- (e.g., levibuprofen), dex- (e.g., dexmethylphenidate, dexibuprofen), ar- (e.g., arbupivacaine) and es- (e.g., esomeprazole) to describe enantiomerically pure active pharmaceutical ingredients (API). Salbutamol is available as a single isomeric preparation of R-isomer as levalbuterol, The recent trend in industry is to market the drugs in a pure enantiomeric form to give new life to old drugs by patenting the pure enantiomer (Gohel, 2003).

11.2 STEREOSELECTIVITY IN BIOLOGICALLY ACTIVE COMPOUNDS

The importance of stereochemistry in drug action is gaining greater attention in medical practice, and a basic knowledge of the subject will be necessary for clinicians to make informed decisions regarding the use of single-enantiomer drugs. Toward a biological target, the potency of two enantiomers can sometimes differ considerably, sometimes be very similar (Table 11.1). Often the activity is concentrated in only one enantiomer. When such a high stereoselectivity arises, it is admitted that the mechanism of action at the molecular level involves a highly specific interaction between the ligand, a chiral molecule and the recognition site, a chiral environment. It is to be expected that the most active isomer, in terms of affinity, achieves a better steric complementarity to the receptor than the less active one. When considering in vivo activities, the difference in activity observed for the two enantiomers is neither always and nor exclusively the result of the quality of the ligand–receptor fit. It must be kept in mind that in vivo the pharmacokinetic processes (ADME) may account for the observed difference in activity. The interpretation of pharmacological data obtained from in vivo assays should thus be questioned and does not allow anticipating the quality of the ligand–receptor interaction. For some therapeutics, single-enantiomer formulations can provide greater selectivities for their biological targets, improved therapeutic indices, and/or better pharmacokinetics than a mixture of enantiomers (McConathy and Owens, 2003). There are obvious benefits to studying the properties of the enantiomers of a chiral drug molecule with respect to therapeutic efficacy and safety.

11.2.1 THE THREE-POINT CONTACT MODEL

The biological activity of chiral substances often depends upon their stereochemistry, since the living body is a highly chiral environment (proximity). The enantiomers of a chiral drug may vary in their interactions with chiral environments such as enzymes, proteins, receptors, etc. of the body. These variations may lead to differences in biological activities such as pharmacology, pharmacokinetics, metabolism, toxicity, immune response etc. Indeed, biological systems can recognize the two enantiomers as two different substances, and their interaction each other will therefore elicit different responses. But, why do enantiomers have different biological activities? (Nguyen et al., 2006). When, in a compound exhibiting stereoselectivity, only one asymmetric center is present in the molecule, it is thought that the substituents on the chiral carbon atom make a three point contact with the receptor. Such a fit ensures a very specific molecular orientation which can only be obtained for one of the two isomers. A three-point fit of this type was first suggested by Easson and Stedman, and the corresponding model proposed by Beckett in the case of R(–)-adrenaline (R(–)-epinephrine). The more active natural R (–)-adrenaline establishes contacts with its receptor through the following three interactions (Figure 1a):

1. Acceptor–donor or hydrophobic interaction between the aromatic ring of adrenaline and an aromatic ring of the receptor protein.
2. A hydrogen bond at the alcoholic hydroxyl group.
3. An ionic bond between the protonated amino group and an aspartic or glutamic carboxylic group of the receptor.

The biologically weak optical isomer, S(–)-adrenaline, can make contact through only two groups.

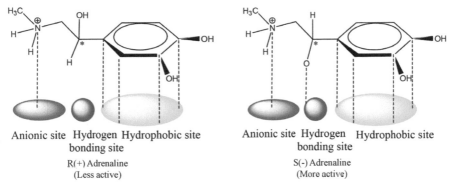

Anionic site Hydrogen   Hydrophobic site
            bonding site

R(+) Adrenaline
(Less active)

Anionic site Hydrogen   Hydrophobic site
            bonding site

S(-) Adrenaline
(More active)

FIGURE 11.1A   Interaction of Adrenaline enantiomers at the receptor binding site.

11.2.2   STEREOSELECTIVITY RATIOS

Rauws stated that "Stereoselectivity is the extent to which a macromolecule (target) shows affinity towards a pair of isomers of one molecule. The enantiomer having desired pharmacological activity (better affinity) is called *eutomer* (Greek, "eu" – good) and enantiomer which may have undesired pharmacological activity or inactivity (lesser affinity) is called *distomer* (Greek, "dys" – bad). The presence of the distomer in the chiral drug can have a number of consequences on biological system (Ariens, 1986). Lehmann mathematically expressed stereoselectivity as the ratio of activity of eutomer and distomer. It is also stated as eudismic ratio.

Eudismic ratio = Activity of eutomer/Activity of distomer

The eudismic ratio determines the therapeutic effect of the enantiomers. ER can be calculated by dividing the EC_{50} or IC_{50} of the distomer by the eutomer. The eudismic ratio for Propranolol (antihypertensive agent) is 130. It signifies that the active (S)-Propranolol is 130 times more potent than (R)-Propranolol as beta adrenoceptor antagonist (Table 11.1) (Peepliwal, et al., 2010). In some cases eudismic ratio (ER) is so high that it is required to separate out the two enantiomers instead of using it as a racemic mixture.

In a series of agonists or antagonists, Eudismic Index (EI) and eudismic affinity quotient (QEA) can be derived from the following equation:

$$EI = a.b \text{ Log Affin. } Eu$$

where a is a constant, b is the quotient of eudismic affinity (QEA) which attributes to the stereoselectivity. When the activity of the eutomer "Eu"

TABLE 11.1 Drugs and Their Eudismic Ratio

Drug	Eutomer	Receptor	Eudismic ratio
Propranolol	(S)-Propranolol	β-adrenoceptor	130
Dexetimide	(S)-(+)-Dexetimide	Muscarinic	10,000
Butaclomol	(+)-Butaclomol	D_2-dopaminergic	1250
Baclofen	(–)-Baclofen	GABA	800
Cyclazosin	(+)-Cyclazosin	$5\text{-}HT_{1A}$	19,000
Ibuprofen	(S)-Ibuprofen	COX I	100
Methadone	(R)-Methadone	μ opioid	20

is compared to that of the racemic mixture "*Rac*," four possibilities can arise:

i. The activity ratio= 2: Eu/Rac = 2/1. In this case the activity is only concentrated in the eutomer and the distomer does not contribute significantly to the observed activity. The chiral compound shows stereoselectivity.

ii. The activity ration > 2: Eu/Rac = 2 (Eu/Rac = 2/0.3). In this case the distomer represents a competitive antagonist of the eutomer.

iii. If the activity ratio < 2: Eu/Rac = 2 (Eu/ Rac = 2/1.6), then both the isomers are active. The distomer strengthens the activity of the eutomer. The eutomer has low affinity for the receptor (poor molecular complementarily).

iv. If the activity ratio = Eu/Rac = 1; then both isomers are equipotent and stereoselectivity is not observed. This can be explained by the assumption that:

a. the compounds act through a nonspecific mechanism,

b. the active compound makes a two-point contact with the receptor (Figure 11.1b),

c. the chiral center is not involved in the contact (is located in a "silent region").

11.2.3 PFEIFFER'S RULE

Pfeiffer in 1956 stated that the eudismic quotient (isomeric activity ratio) of a highly active couple of isomers is always superior to that of a less active couple. The greater the difference between the pharmacological effect of the Rand the S isomers, the greater is the potency of the active isomer."

In other word the 'Pfeiffer's encompasses that lower the effective dose of a drug, the greater the different in the biological activity of optical isomers

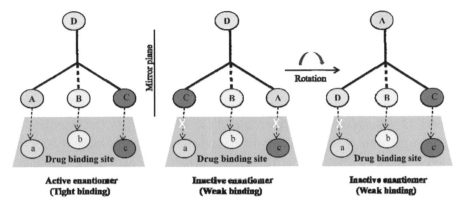

FIGURE 11.1B Interaction of active and inactive enantiomers at the drug binding site (the active enantiomer has a 3 dimensional structure that allows drug domain A interact with binding site a, B with binding site b, C with binding site c whereas inactive enantiomer cannot be aligned to bind the same three sites simultaneously).

(Pfeiffer, 1956). Pfeiffer's rule also states that the eudismic index should increase linearly with increasing efficacy or potency of the eutomer.

11.2.3.1 Exceptions to Pfeiffer's Rule

Due to the conformational flexibility of the ligands others reside in an improper selection of "homologous" sets of compounds as illustrated with muscarinic agonists and antagonists. Quantitative analyzes of the correlations between biological activity and the structure of stereoisomeric compounds is difficult.

11.3 PHARMACODYNAMIC EFFECT OF ENANTIOMERS

Enantiomers are stereoisomers whose mirror images cannot be superimposed. Enantiomers have identical physical and chemical properties except that they rotate the plane of polarized light in opposite directions and behave differently in a chiral environment. Mixtures of equimolar amounts of enantiomers are called racemates. They may exist as racemic mixtures (conglomerates) or racemic compounds (true racemates). In the solid state, the physical properties such as melting points, solubilities and heats of fusion of the racemates may differ from those of the individual enantiomers. They interact at different rates with other chiral compounds including many biological macromolecules (Table 11.2).

The biological response induced by a pair of enantiomers can differ in potency (quantitative difference) or in nature (qualitative difference) (Table 11.3). In the latter case, it is assumed that one enantiomer acts at one receptor site, whereas its antipode is recognized by other sites and possesses a different activity and toxicity profile. Two stereoisomers can compete for binding to same receptors like S-methadone antagonizes respiratory depression action of R-methadone. If the two isomers are of agonist and antagonist type, then racemic mixture acts as partial agonist like picendol and sulfinpyrazone inhibits the metabolism of S-Warfarin significantly but not of R-Warfarin (Chhabra et al., 2013). No generalizations can be made concerning enantiomers since they exhibit a wide variation in effects. Examples of these effects consist of the following (Davies and Teng, 2003):

TABLE 11.2   Comparison of isomer potency of chiral drugs (Nguyen et al., 2006)

Category	Drugs	Isomer potency
Adrenoreceptor blocking drugs (β-blockers)	Propranolol, acebutolol, atenolol, alprenolol, betaxolol, carvedilol, metoprolol, labetalol, pindolol, sotalol	(−) > (+) Ex: S(−)-propranolol>R(+)-propranolol
Calcium channel antagonists	Verapamil, nicardipine, nimodipine, nisoldipine, felodipine, mandipine	(−) > (+) Ex: S(−)-verapamil > R(+)-verapamil
β2-Adrenoceptor agonists: Bronchodilators	Albuterol (salbutamol), salmeterol, terbutaline	(−) > (+) Ex: R(−)-albuterol>S(+)-albuterol
Hypnotics, Sedatives	Hexobarbital, secobarbital, mephobarbital, pentobarbital, thiopental, thiohexital	(−) > (+) Ex: S(−)-secobarbital>R-(+) secobarbital
Anesthetics	Ketamine, isoflurane	(+)>(−) Ex: S(+)-ketamine>R(−)-ketamine
Analgesics, Anti-inflammatory: (NSAID)	Ibuprofen, ketoprofen, benoxaprophen, fenprofen	(+)>(−) Ex: S(+)-ibuprofen>R(−)-ibuprofen
Tranquilizers	Oxazepam, lorazepam, temazepam	(+)>(−) Ex: S(+)-oxazepam>R(−)-oxazepam

(−): levorotary; (+): dextrorotary.

TABLE 11.3 Enantiomers and Their Pharmacological Activity

Drug	Enantiomers	Activity
Propoxyphene	(+) 2R,3S-propoxyphene	Analgesic
	(−) 2S,3R-propoxyphene	Antitussive
Methorphan	Levomethorphan	Opiod analgesic
	Dextromethorphan	Cough suppressant
Thalidomide	R-Thalidomide	Sedative
	S-Thalidomide	Teratogenic
Naproxen	R-Naproxen	Arthralgesic
	S-Naproxen	Teratogenic
Dopa	L-dopa	Antiparkinson
	D-dopa	Deficiency of WBCs
Indacrinone	R(−)Indacrinone	Diuretic
	S(+)Indacrinone	Induces uric acid secretion
Penicillamine	(S)-Penicillamine	Antiarthritic activity
	(R)-Penicillamine	Extremely toxic

i. Equipotent enantiomers (Ex. cyclophosphamide, flecainide).
ii. One enantiomer with all or most of the activity (Ex. NSAIDs, β-blockers).
iii. Both enantiomers active with similar therapeutic and toxic effects but differ in magnitude (Ex. warfarin).
iv. Both enantiomers active but with quantitatively different therapeutic and toxic effects (Ex. verapamil).

Many calcium channel antagonists are used under racemic form such as verapamil, nicardipine, nimodipine, nisoldipine, felodipine, mandipine, etc.

11.4 PHARMACODYNAMIC EFFECT OF DIASTEREOISOMERS

The term stereoisomer comprises of diastereoisomers (including *cis-trans* isomers) and enantiomers. When more than one asymmetric center is involved the complexity of the problem increases rapidly. Diastereoisomers are chemically distinct and often pharmacologically different compounds (Table 11.4), and can generally be separated by achiral analytical methods.

TABLE 11.4   Active Enantiomers of Some Diastereomers with Their Activity

Drug	Active enantiomer	Activity
Chloramphenicol	(–)-Threo-chloramphenicol	Anti-infective
Ephedrine	(–)-(1R,2S)-ephedrine	Bronchodilator
	(+)-(1S,2S)-pseudoephedrine	Decongestant
Labetalol	(S, R)-labetalol	Alpha-blocker
	(R, R)-labetalol	Beta-blocker
Ethambutol	(S,S)-ethambutol	Tuberculostatic
	(R,R)- ethambutol	Optical neuritis
Morphine	(–)-(5R,6S,9R,13S,14R)-morphine	Analgesic

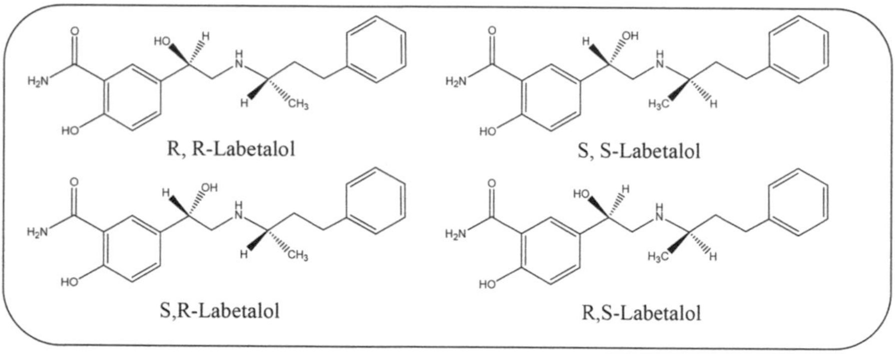

FIGURE 11.2   Structure of labetalol with all its isomers.

Labetalol (Figure 11.2) has four optical isomers, two pairs of enantiomers (R,S & S,R and R,R & S,S). Each possesses different pharmacological properties. Labetalol is formulated as a racemic mixture of four isomers, two of these isomers—the (S, S) and (R, S) isomers—are relatively inactive, a third (S, R) is a potent alpha-blocker and the fourth one (R, R) is a potent beta-blocker. Labetalol has a 3:1 ratio of beta: Alpha antagonism after oral administration. The most active RR enantiomer was developed some years ago as dilevalol, but had to be withdrawn after some months due to a slightly higher than average degree of hepatic toxicity. The stereoisomers of some oxotremorine analogs containing two chiral centers and acting as oxotremorine antagonists to show in vivo tremorolytic activity. They show stereoselectivity ratios as high as 1 to 200.

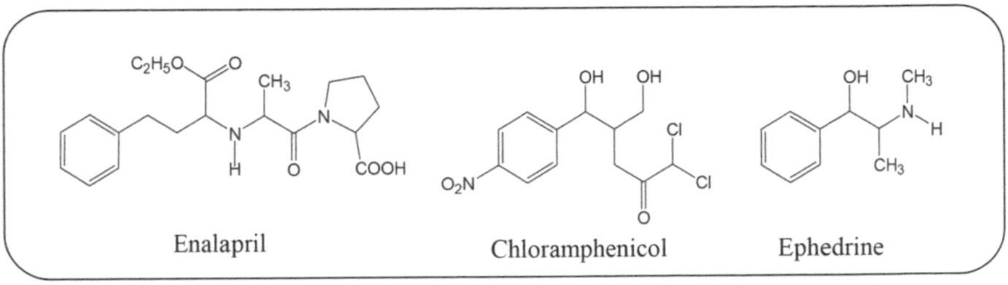

FIGURE 11.3 Structures of some drugs with two or more than two chiral centers.

11.4.1 PHARMACODYNAMIC EFFECT OF CIS-TRANS ISOMERS

Stereoisomerism as a result of the rigid configuration due to double bonds or cyclicity is known as geometrical isomerism. Cis/trans isomerism is a well known term used to represent geometrical isomerism. Another system to describe stereochemistry about the double bond is based on the Cahn-Ingold-Perlog (CIP) convention used to rank atoms and groups. The Z descriptor (from zusammen, German for "together") is used when both the substituents with the highest priority are arranged on the same side of the double bond. The E descriptor (from entgegen, German for "opposing") is applied when they are on opposite sides. In inorganic transition metal complexes, cis/trans-sisomers are the result of arrangements of ligands around a metal ion.

These geometrical isomers differ in their physical, chemical as well as biological properties. The trans and the cis form of a drug can exhibit different properties. One geometric isomer may have the desired pharmacological effect, while the other may be less effective or have adverse side effects.

Example: Cisplatin (Figure 11.4) is a drug used to treat testicular and ovarian cancer. The medicine exhibits geometric isomerism. However, only the cis form of the drug is effective—the trans form has no effects whatsoever. Only cisplatin which is having a square planar structure can react with two guanine molecules in the DNA.

Commercial preparations of the tricyclic antidepressant doxepin contain 15% of the more active cis-doxepin and 85% of the transisomer. In most 'in vivo' and ' in vitro' tests, the cis-isomer is the more potent of the two geometric forms. Doxepin is metabolized to a variety of phase I and phase II metabolites, in which the N-desmethyl metabolite is thought to make an important contribution to therapeutic activity and it has been suggested that

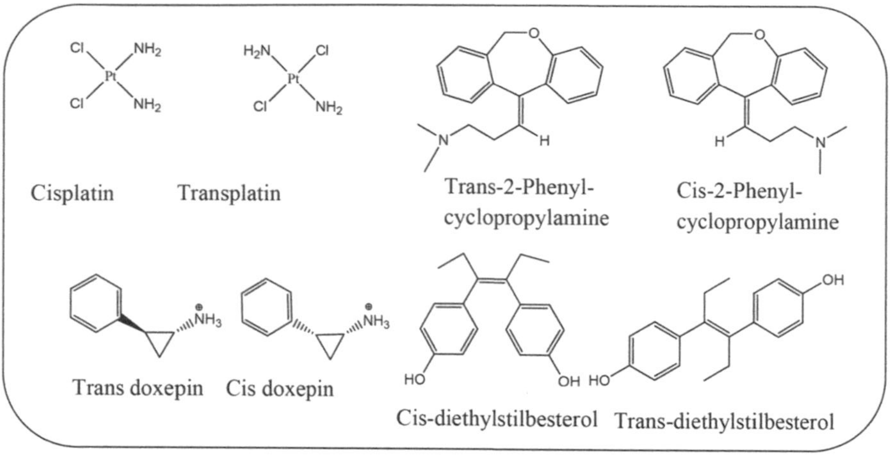

FIGURE 11.4 Structures of some cis-trans isomers used as drugs.

plasma concentrations of (cis plus trans) doxepin and (cis plus trans) des-methyldoxepin show better correlation with antidepressant activity than (cis plus trans) concentrations of the parent drug alone.

Antiestrogen drug diethystilbesterol exhibits geometrical isomerism (Figure 11.4). Cis-diethylstilbesterol is unstable and has one tenth the activity of transdiethylstilbesterol. Trans-diethylstilbesterol has structural resemblance with estradiol. In case of transdiethylstilbesterol, the resonance interactions and minimum steric hindrance facilitate the aromatic rings and the ethylene bridge to stay in the same plane. This orientation reinforces the antiestrogenic effect by blocking the binding of estrogen at the estrogen receptor.

Another type of geometrical isomerism is observed in ring compounds where the ring provides the rigidity to forbid the rotation. For example, trans2-cyclopropylamine is more stable and potent MAO inhibitor than the cis isomer.

11.5   PHARMACOKINETIC EFFECT OF CHIRAL DRUGS

Absorption, distribution, metabolism and elimination of drugs from the body also involves with their interactions with various proteins. Hence then the pharmacokinetics of enantiomers can also be different. For example:

   i.   Bioavailability of (R)-verapamil is twofold more than that of (S)-verapamil due to reduced hepatic first-pass metabolism

ii. Absorption of L-Methotrexate is better absorbed than D-Methotrexate, hence its more bioavailable.

iii. Volume of distribution of (R)-methadone is double that of (S)-methadone due to lower plasma binding and increased tissue binding. S-Warfarin is more extensively bound to albumin than R-Warfarin, hence it has lower volume of distribution. Levocetrizine has smaller volume of distribution than its dextroisomer hence demonstrate lesser side effect. (+)-Propranolol is more extensively bound to proteins than (−)-Propranolol.

iv. Clearance of (R)-fluoxetine is about fourfold higher than that of (S)-fluoxetine due to a higher rate of biotransformations. Clearance and volume of distribution directly influence half-life a drug. Thus the half-life of (S)-fluoxetine is one quarter that of (R)-fluoxetine.

v. Renal clearance of (R)-pindolol is one quarter less than (S)-pindolol due to reduced renal tubular secretion.

In addition, these pharmacokinetic properties can be modified in a stereoselective manner by disease, genetics, ethnicity, age and other drugs.

11.5.1 SOME EXAMPLES FOR STEREOSELECTIVE BIOTRANSFORMATION OF DRUGS

Both enantiomers of omeprazole are equipotent, however, their metabolism is different. (R)-omeprazole is mainly metabolized by the polymorphic CYP2C19 enzyme. There is a 7.5-fold difference in the systemic exposure to (R)-omeprazole in patients who are poor metabolizers compared to extensive metabolizers. With (S)-omeprazole this difference is reduced to about three-fold so it was argued that use of esomeprazole would be associated with less interindividual variability in efficacy. S-Warfarin is more potent and metabolized by ring oxidation while R-Warfarin is less potent and metabolized by side chain reduction, half life of S-Warfarin is 32 h while it is 54 h for R-Warfarin.

Racemization may also occur in physiological fluids, such as the exposure to the acidic environment of the stomach. Enzymatic inversion is concerned with inversion under physiological conditions. Under these conditions, enantiomers may be inverted with the outcome of a racemate enriched in one of the antipodes. Demonstration of metabolic chiral inversion may have profound consequences for the development of a new pharmaceutical entity. A better understanding of the factors facilitating such interconversion may greatly aid drug development by identifying this feature at an early stage, thereby reducing bioanalytical and toxicology workload.

11.5.2 CHIRALITY AND DRUG HAZARDS

Recent interest has focused on the role of the different properties of individual drug enantiomers in causing drug toxicity. For drugs with a single chiral center, both enantiomers may be therapeutically active. However, if the main therapeutic benefit is in only one enantiomer, several possibilities exist for the other enantiomers to be toxic (Table 11.5).

In 1957, a pharmaceutical company in West Germany introduced and marketed thalidomide as a racemic mixture of the (+)(R)-thalidomide and (−)(S)-thalidomide (Figure 11.5). Tragically, thalidomide was found to have

TABLE 11.5 Toxicity of Some Enantiomers

Enantiomer	Toxicity
(R)-bupivacaine	Cardiotoxicity
(−)-ephedrine	Neurotoxicity
(−)(S)-thalidomide	Teratogenic
(R,R)-ethambutol	Blindness
D-levodopa	Granulocytopenia
(R)-penicillamine	Hepatotoxicity
S-Naproxen	Teratogenic

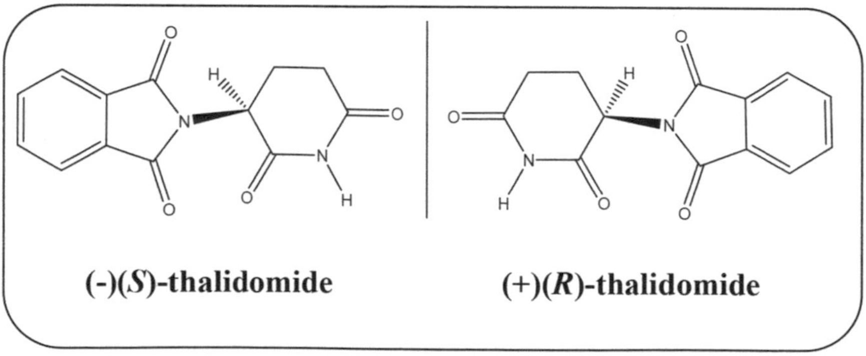

(-)(S)-thalidomide   **(+)(R)-thalidomide**

FIGURE 11.5 Enantiomers of thalidomide.

serious side-effects; thousands of babies were born with missing or abnormal arms, hands, legs, or feet (Phocomelia, Figure 11.6). It was banned by many countries in 1961. Now scientists know that it is the (–)(S)-thalidomide that caused the severe teratogenic effect. Human liver contains enzyme that can convert (+) (R)-thalidomide to (–)(S)-thalidomide. Therefore, even administration of enantiomerically pure (+)(R)-thalidomide results in a racemic mixture.

In view of this, since 1992 the FDA and the European Committee for Proprietary Medicinal Products have required that the properties of each enantiomer be studied separately before decisions are taken to market the drug as one of the enantiomers or as a racemate (Announcement, 1992).

11.6   CHIRAL SWITCH

The recent trend in industry is to market the drugs in a pure enantiomeric form to give new life to old drugs by patenting the pure enantiomer. Chiral switch or racemic switch means that racemic mixtures are redeveloped as single enantiomers (Stephen, 2001). Getting marketing approval for a chiral switch usually requires relatively few new studies to be conducted if the racemate is already marketed (Table 11.6). The single enantiomer can be ready for launch before the patent for the racemate expires and before the marketing of any generics (which tend to substantially drive down the cost of the racemate) (Somogyi and Bochner, 2004). One prominent example of a chiral switch occurred in 2001, when the US Food and Drug Administration (FDA)

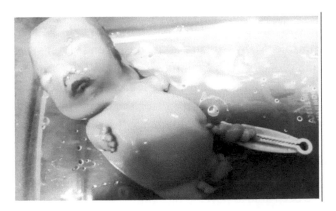

FIGURE 11.6   Fetus with phocomelia (Agrawal and Parulekar, 2015).

TABLE 11.6 Marketed Chiral Switches and Their Racemic Precursors (Gellad et al., 2014)

Racemic Precursor	Year Approved	Chiral switch	Year Approved	Approved Indication(s)
Methylphenidate	1955	Dexmethyl-phenidate	2001	Attention deficit/hyperactivity disorder
Omeprazole	1989	Esomeprazole	2001	Gastroesophageal reflux disease and erosive esophagitis
Citalopram	1998	Escitalopram	2002	Major depression and depression maintenance in adults
Zopiclone	1986	Eszopiclone	2004	Insomnia
Formoterol	2001	Arformoterol	2006	Chronic obstructive pulmonary disease
Cetirizine	1995	Levocetirizine	2007	Seasonal allergic rhinitis, perennial allergic rhinitis, chronic idiopathic urticaria
Lansoprazole	1995	Dexlansoprazole	2009	Gastroesophageal reflux disease and erosive esophagitis

approved AstraZeneca's esomeprazole (Nexium), a proton pump inhibitor (PPI) used to treat gastroesophageal reflux disease and erosive esophagitis. Omeprazole is a racemic mixture of R-omeprazole and S-omeprazole, while esomeprazole, as its name implies, is isolated S-omeprazole. The S-omeprazole enantiomer was responsible for the drug's clinical properties while the R-omeprazole enantiomer was inactive. After FDA approval, esomeprazole was marketed as "the new purple pill" to be used in place of generic omeprazole. Many existing drugs have gone chiral switch. Cetrizine to levocetrizine is one of such examples. The Food and Drug Administration recently approved a chiral switch drug, levalbuterol (the pure l-isomer of albuterol) as a preservative-free nebulizer solution. Single enantiomers of popular drugs like bupivacaine, citalopram, ofloxacin, salbutamol, ketamine, methylphenidate, and oxybutinin have recently been developed through chiral switch for market approval. When the original developer of a block-buster did not patent individual enantiomers, third-party companies, such as Sepracor, have been able to develop the single isomer and enter into licensing agreements with the company that marketed the racemic mixture.

11.6.1 ADVANTAGES OF CHIRAL SWITCHING

There are several possible health benefits to chiral switching. They include:

i.   an improved safety margin (therapeutic index) through increased receptor selectivity and potency, and reduced adverse effects.

ii.   a longer or shorter duration of action due to pharmacokinetic considerations (e.g., half-life) resulting in a more appropriate dosing frequency.

iii. decreased interindividual variability in response commonly due to polymorphic metabolism.

iv.   decreased potential for drug-drug interactions.

11.6.2 DISADVANTAGES

In some cases, chiral switching has been of no benefit. For example, the clinical development of (R)-fluoxetine for depression (based on a more acceptable half-life and less propensity for significant drug-drug interactions) was stopped because of a small but statistically significant prolongation of the QT interval with high doses. Dilevalol was thought to have advantages over labetalol, but was removed from the Japanese market because of hepatotoxicity. Chiral switching is a controversial practice. Some claim that single enantiomerdrugs offer little clinical advantage and are used by pharmaceutical manufacturers to perpetuate revenues as the original racemic pill approaches the end of its market exclusivity. The case of esomeprazole, in fact, has been cited as an example of "corporate waste" of healthcare resources (Goozner, 2004). From 2001 to 2011, the US Food and Drug Administration approved 9 single-enantiomer drugs with racemic precursors. Only 3 had preapproval studies that compared the single-enantiomer with its precursor, and none showed evidence of improved patient outcomes.

11.7 GUIDELINES ON STEREOCHEMICAL ISSUES IN CHIRAL DRUG DEVELOPMENT

FDA initial guidance on chiral drugs was set forth in 1992 to give direction to the discovery and development of safe and efficacious drugs (Announcement, 1992). Identified chiral-specific issues included proper manufacturing controls, product stability, pharmacokinetic evaluations and estimation that accounted for chiral differences and proper interpretation of the pharmacokinetic data of isomers in animals and human. The guideline suggests that the knowledge of composition of a chiral drug has to be known before subjected to preclinical and clinical studies.

The decision whether to develop a single enantiomer, racemate, or non-racemic mixture (enantiomeric mixture other than racemate) rests with the sponsor and should be based on scientific data relating to quality, safety and efficacy and ultimately to the risk/benefit assessment of the drug under the proposed conditions of use. Cases where the development of a racemate may be justified include, but are not limited to, the following:

i.   The enantiomers are configurationally unstable in vitro or undergo racemization in vivo.
ii. The enantiomers have similar pharmacokinetic, pharmacodynamic and toxicological properties.
iii. It is not technically feasible to separate the enantiomers in sufficient quantity and/or with sufficient quality.

The cases where development of a nonracemic mixture may be justified include those where a specific enantiomeric ratio is expected to improve the therapeutic profile. Enantioselective assays should be developed and validated at an early stage of drug development, and used wherever relevant unless it has been clearly demonstrated that a nonenantioselective assay provides results equivalent to those obtained with the enantioselective assay.

The stereoisomeric composition of a drug with a chiral center should be known and the quantitative isomeric composition of the material used in pharmacologic, toxicological, and clinical studies known. Specifications for the final product should assure identity; strength, quality, and purity from a stereochemical viewpoint.

FDA invites discussion with sponsors concerning whether to pursue development of the racemate or the individual enantiomer. All information developed by the sponsor or available from the literature that is relevant to the chemistry, pharmacology, toxicology, or clinical actions of the stereoisomers should be included in the IND and NDA submissions.

11.7.1   CHEMISTRY, MANUFACTURING AND CONTROLS

The chemistry section of the application should contain the requisite information to assure the identity, quality, purity and strength of the drug substance and drug product. In addition, the following considerations should be taken into account when dealing with chiral drug substances and drug products.

11.7.1.1 Methods and Specifications

11.7.1.1.1 Drug Substance

Applications for enantiomeric and racemic drug substances should include a stereochemically specific identity test and/or a stereochemically selective assay method. The choice of the controls should be based upon the substance's method of manufacture and stability characteristics.

11.7.1.1.2 Drug Product

Applications for drug products that contain an enantiomer or racemic drug substance should include a stereochemically specific identity test and/or a stereochemically selective assay method. The choice of the controls should be based upon the product's composition, method of manufacture and stability characteristics.

11.7.1.1.3 Stability

The stability protocol for enantiomeric drug substances and drug products should include a method or methods capable of assessing the stereochemical integrity of the drug substance and drug product. However, once it has been demonstrated that stereochemical conversion does not occur, stereoselective tests might not be needed.

11.7.1.1.4 Labeling

The labeling should include a unique established name and a chemical name with the appropriate stereochemical descriptors.

11.7.2 PHARMACOLOGY

Unless it proves particularly difficult, the main pharmacologic activities of the isomers should be compared in in vitro systems, in animals and/or in humans. The pharmacologic activity of the individual enantiomers should be characterized for the principal pharmacologic effect and any other important pharmacological effect, with respect to potency, specificity, maximum effect, etc.

11.7.3 PHARMACOKINETIC PROFILE

To evaluate the pharmacokinetics of a single enantiomer or mixture of enantiomers, manufacturers should develop quantitative assays for individual
enantiomers in in vivo samples early in drug development. This will allow
assessment of the potential for interconversion and the absorption, distribution, metabolism, and excretion (ADME) profile of the individual isomers.
When the drug product is a racemate and the pharmacokinetic profiles of the
isomers are different, manufacturers should monitor the enantiomers individually to determine such properties as dose linearity and the effects of altered
metabolic or excretory function and drug-drug interactions. If the pharmacokinetic profile is the same for both isomers or a fixed ratio between the plasma
levels of enantiomers is demonstrated in the target population, an achiral assay
or an assay that monitors one of the stereoisomers should suffice for later evaluation. In vivo measurement of individual enantiomers should be available to
help assess toxicological findings, but if this cannot be achieved, it would be
sufficient in some cases to establish the kinetics of the isomers in humans.

11.7.4 TOXICOLOGY

It is ordinarily sufficient to carry out toxicity studies on the racemate. If toxicity other than that predicted from the pharmacologic properties of the drug
occurs at relatively low multiples of the exposure planned for clinical trials, the
toxicity study where the unexpected toxicity occurred should be repeated with
the individual isomers to ascertain whether only one enantiomer was responsible for the toxicity. If toxicity of significant concern can be eliminated by
development of single isomer with the desired pharmacologic effect, it would
in general be desirable to do so. The agency would be pleased to discuss any
cases where questions exist regarding the definition of "significant toxicity."

11.7.5 IMPURITY LIMITS

It is essential to determine the concentration of each isomer and define limits for
all isomeric components, impurities, and contaminants on the compound tested
preclinically that is intended for use in clinical trials. The maximum allowable
level of impurity in a stereoisomeric product employed in clinical trials should
not exceed that present in the material evaluated in nonclinical toxicity studies.

11.7.6   CLINICAL AND BIOPHARMACEUTICAL

Where little difference is observed in activity and disposition of the enantiomers, racemates may be developed. In some situations, development of a single enantiomer is particularly desirable (Ex. where one enantiomer has a toxic or undesirable pharmacologic effect and the other does not). A signal that should trigger further investigation of the properties of the individual enantiomers and their active metabolites is the occurrence at clinical doses of toxicity with the racemate that is not clearly expected from the pharmacology of the drug or the occurrence of any other unexpected pharmacologic effect with the racemate. These signals might be explored in animals but human testing may be essential.

It should be appreciated that toxicity or unusual pharmacologic properties might reside not in the parent isomer, but in an isomer-specific metabolite. In general, it is more important to evaluate both enantiomers clinically and consider developing only one when both enantiomers are pharmacologically active but differ significantly in potency, specificity, or maximum effect, than when one isomer is essentially inert. Where both enantiomers are fortuitously found to carry desirable but different properties, development of a mixture of the two, not necessarily the racemate, as a fixed combination might be reasonable.

If a racemate is studied, the pharmacokinetics of the two isomers should be studied in Phase 1. Potential interconversion should also be examined. Based on Phase 1 or 2 pharmacokinetic data in the target population, it should be possible to determine whether an achiral assay or monitoring of just one enantiomer where a fixed ratio is confirmed will be sufficient for pharmacokinetic evaluation.

If a racemate has been marketed and the sponsor wishes to develop the single enantiomer, evaluation should include determination of whether there is significant conversion to the other isomer, and whether the pharmacokinetics of the single isomer are the same as they were for that isomer as part of the racemate.

11.8   CONCLUSION

Chiral drugs represent a significant portion of the therapy in the market. There are no simple answers to the enantiopure drug versus racemate debate and each illustration must be examined thoroughly on a case-by-case basis. However, it is unethical to start witch trial for racemic drugs already in market having proven safety and efficacy. The pharmaceutical industry will continue to have a niche in chiral compounds, because of the efforts to improve drug ability and to cut down the costs of drug development. Medicinal chemist's everlasting quest for macromolecular targets in the biological system boosts the growth of chiral drugs in the discovery pipeline. A lion's view on the stereochemical issues of chiral drugs and an insight into their pharmacokinetic and pharmacodynamic effect at the molecular level will assist their clinical use.

KEYWORDS

- **chiral switch**
- **diastereomer**
- **enantiomer**
- **stereochemistry**
- **stereoselectivity**
- **toxicity**

REFERENCES

Agrawal, S., & Parulekar, S. V. (2015). Phocomelia, *Journal of Postgraduate Gynecology & Obstetrics, 2* (3).

Announcement: FDA's policy statement for the development of new stereoisomeric drugs. Chirality (1992). *4*(5), 338–340.

Ariens, E. J. (1986). *Chirality in Bioactive Agents and Its Pitfalls, 7*, 200–205.

Chhabra, N., Aseri, M. L., & Padmanabhan, D. (2013). A review of drug isomerism and its significance, *Int J Appl Basic Med Res. 3*(1), 16–18.

Davies, N. M., & Teng, X. W. (2003). Importance of Chirality in Drug Therapy and Pharmacy Practice: Implications for Psychiatry, *Advances in Pharmacy, 1* (3), 242–252.

Gellad, W. F., Choi, P., Mizah, M., Good, C. B., & Kesselheim, A. S. (2014). Assessing the Chiral Switch: Approval and Use of Single-Enantiomer Drugs, 2001 to 2011. *Am J Manag Care. 20* (3), e90–e97.

Gohel, M. C. (2003). Overview on Chirality and Applications of Stereo-selective Dissolution testing in the Formulation and Development Work, Dissolution Technologies, 16–20.

Goozner, M. (2004). The $800 Million Pill: The Truth behind the Cost of New Drugs. Berkeley: University of California Press.

McConathy, J., & Owens, M. J. (2003). Stereochemistry in Drug Action. *Primary Care Companion J Clin Psychiatry, 5,* 70–73

Nguyen, L. A., He, H., & Pham-Huy, C. (2006). Chiral Drugs: An Overview, *International Journal of Biomedical Science,* 85–100.

Peepliwal, A. K., Bagade, S. B., & Bonde, C. G. (2010). A review: Stereochemical consideration and eudismic ratio in chiral drug development, *J Biomed Sci and Res., 12* (1), 29–45.

Pfeiffer, C. C. (1956). Optical isomerism and pharmacological action, a generalization. *Science, 124,* 29–31.

Shafaati, A. (2007). Chiral Drugs: Current status of the industry and the market. *Iranian Journal of Pharmaceutical Research, 6* (2), 73–74.

Somogyi, A., & Bochner, F. (2004). David Foster Inside the isomers: the tale of chiral switches, *Aust Prescr. 27,* 47–49.

Stephen, C. S. (2001). "Chiral Pharmaceuticals." *Chemical and Engineering News, 79* (40), 79–97.

CHAPTER 12

PHARMACEUTICAL INTERACTIONS IN DRUG PRACTICES

SACHINKUMAR PATIL, SHITALKUMAR PATIL, SUDHA KHARADE, and DIPALI KAMBLE

Ashokrao Mane College of Pharmacy, Peth-Vadagaon, District Kolhapur, Maharashtra, India

CONTENTS

12.1 INTRODUCTION

Medications, both prescription and over the counter, are used every day to treat acute and chronic illness. Research and technology constantly improve the drugs that have available and introduce new ones. Medications can help people live healthy lives for a prolonged period. Although medicines are prescribed often, it is important to realize that they must still be used with caution. Drug therapy becomes more complex when two drugs are administered in close sequence to each other. They may be interacting to enhance or diminish the intended effect of one or both the drug. A drug interaction may result into beneficial or desired effect or into adverse or undesired effect. Many studied demonstrated that patient receive multiple drug therapy to increase effectiveness, for example, potassium sparing diuretics in digitalis therapy. As the number of drug regimen increases, the risk of occurrence of toxicity becomes greater and less effective. Unfortunately drug interaction may be recognized only when sever toxicity occurs, for example, a hypertensive crisis may occurs when mono-amino oxidase inhibitor like tranycypromine, and methamphetamine are used together.

However, recently a lot of attention has been focused on the phenomenon of drug interaction and information pertaining to its occurrence is so widely published. The concept of drug interaction is also including drug-drug interaction, drug-food interaction, drug-herb interaction and drug-laboratory interaction.

The term drug interaction is defined as an alteration in duration or magnitude (or both) of pharmacological effect of one drug produced by another drug.

12.2 FACTORS CONTRIBUTING TO OCCURRENCE OF DRUG INTERACTION

Many adverse drug interactions can be avoided if the following predisposing factors are kept in mind.

- **Non prescribed medicine:**
  Self-medication, wherein the patient start treating himself and lends into problems, also insufficient knowledge of effective combination of drug therapy, patient may taking non prescribed drugs which physician may not be aware of, similarly number of over the counter preparations can interact prescribed drug, for example, salicylates increase bleeding with anticoagulants.
- **Drug abuse:**
  Drug abuse or drug misuse also leads to drug-drug interaction. Many patients suffering from psychiatric disorders continue to use drugs of abuse like barbiturates, narcotics, and amphetamine. Excessive CNS depression may result if physician prescribes some antipsychotic

agents prior to surgery while due to fear or anxiety some CNS depressants may be also be consumed by the patient without physicians' knowledge. These may seriously interfere with anesthetics and analgesics administered during surgery.

- **Patient non Compliance:**
  Poor compliance by the patient, that is, fails to take the medication as per direction. This may be due to lack of instructions or confusion which results in excessive dose of drug that can increase risk of drug interaction. Addiction, like Smoking increase metabolism of many drugs: alcohol consumption altered drug metabolism and potentiating action of CNS depressants.

- **Multiple drugs at a time:**
  Inpolypharmacy approach, multiple drugs are prescribed for individual by physician. It's true, all drugs are not clinically significant, nor are all of them unintended. The fact that two drugs interact does not mean they cannot be administered together. In certain situations a drug interaction may be useful to produce indented therapeutic effect (antibiotics with vitamin supplements). Although, a drug is contraindicated when the other drug is used, however, it can be given concurrently as long as precautions are taken like oral anticoagulants and sedatives.

  Over the counter drug like analgesics, laxatives, vitamins forms so much a part of daily life that they no longer are regarded as drugs and the physician may not even aware of the simultaneous ingestion of these drugs.

  Also most of drugs do not possess only one specific type of activity but influence many physiological system of body, which result in secondary effect. Ascorbic acid excretion is increased by such drugs as aspirin, atropine, barbiturates. Absorption of fat soluble vitamins is decreased by intake of even 20 ml of mineral oils.

- **Multiple prescribers:**
  The common practice of consulting more than one physician at a time can increase the risk of drug interaction. The patient may visit a specialist, in addition to his family physician, for example, an ophthalmologist may prescribe pilocarpine eye drops for a patient who is taken an anticholinergic preparation prescribed by another physician for a GI condition. This might alter intraocular pressure to some extent, leading to complication.

- **Dietary factor:**
  Constituents of individual's diet can interact with certain drugs, for example, MAOI interacts with tyramine containing food stuff inducing hypertensive crises. Antibacterial drug (Tetracycline) may interfere with dairy product.

12.3   FACTOR INFLUENCING THE RESPONSE TO A DRUG IN PATIENT

- **Age:**
  The patient characteristic which has the most bearing on drug interactions is age. Certain patient groups, for example, the elderly may have an increased risk of suffering a clinically significant drug interaction due to polypharmacy. It is estimated that for patients taking 2–5 drugs daily the incidence of a potential drug interaction is 19%. This rises to over 80% for those taking six or more drugs. Renal or in particular, hepatic impairment, either age-related or otherwise may affect the ability to metabolize drugs. Patients with severe underlying disease may be less tolerant of changes in plasma concentration of their therapy. The disease being treated and any concomitant diseases may also influence drug interactions as can the patients preexisting clinical status. Genetic characteristics relating to approximately 10% of the population, may affect some drug interactions, for example, grapefruit juice and terfenadine resulting in an increased risk of cardiotoxicity. This appears only to be important in the small number of patients who are poor metabolizes of terfenadine. This risk factor may also explain the propensity of warfarin and tricyclic antidepressants to cause problems.
- **Diet:**
  Drug-food interactions can lead to a loss of therapeutic efficacy or toxic effects of drug therapy. Generally, the effect of food on drugs results in a reduction in the drug's bioavailability; however, food can also alter drug clearance. Some foods greatly affect drug therapy, resulting in serious side effects, toxicity, or therapeutic failure. In some instances, the interaction may have a beneficial effect by increasing drug efficacy or diminishing potential side effects (refer Table 12.1).

TABLE 12.1 Effect of Drug-Food Interactions

Diet	Drug	Interaction
Tyramine containing food, for example, cheese, red wines, ripped bananas, yogurt, shrimp paste, salaami	Monoamino oxidase inhibitor	Hypertensive crises
Spinach, broccoli containing vitamin K	Warfarin	Antagonism of drug effect
Milk	Tetracycline	Complex with calcium
Normal food	Griseofulvine, Metoprolol, Propranonol, Phenytoin, Dextropropoxyphene, Dicumarol	Increase the bioavailability
Normal food	NSAIDs, Teracycline, Ethanol	Delays or reduces the bioavailability
Grape fruit and oranges	Indinavir, Saquinavir, Midazolam, Nimodipine, Nifedipine, Lovastatine, Carbamazepine, Verapamil	Inhibit CYP3A4 isoenzyme of system, increasing the bioavailability of drugs
Rich protein diet	Acidic urine (pH 5.9)	Promotes excretion of basic drugs
Low protein diet	Alkaline urine pH 7.5	Promotes excretion of Acidic drugs
Orange juice, coffee, tea	Alendronate	Reduce bioavailability
Food	Carbamazepine	Increased bile production, enhanced dissolution and absorption.
Food	Dicumerol	Increased bile flow, delayed gastric emptying permits dissolution and absorption
Food	Isoniazide	Rise in pH, preventing dissolution and absorption
High protein diet	Levodopa	Drug compete with amino acid for absorption

- **Alcohol and Smoking:**
  Heavy drinking can speed up the metabolism of drugs such as warfarin, phenytoin, tolbutamide, probably by increasing the activity of liver enzyme. Alcohol with sedatives and other depressant drugs could result in an excessive depressant response due to synergistic

effects. A major clinical problem develop with alcohol is that psychological tolerance and drug interaction.

Smoking increases the activity of drug metabolizing enzymes in liver with the result that certain therapeutic agents are metabolize more rapidly and their effect is decreased, for example, effect of CNS depressant and Chlordiazepoxide was less in heavy smokers.

- **Diseased State:**
  Several diseases may coexist, especially elderly and each requires treatment, for example, association of hypertension, coronary artery disease, diabetes mellitus and arthritis. The benefit to risk ratio is of important in such cases. Treatment of various disease at a time altered enzyme system of body which may affect handling of drug by body, leads to adverse drug interaction.

- **Organ Function Abnormities:**
  Pathological conditions like liver disease, kidney damage, or alter enzyme system may affect the handling of drug by the body and this can lead drug interactions. If there is renal impairment, there can be a prolonged effect on such drugs as are excreted unchanged through kidney. As additional doses are given, blood levels will build up. It may result intoxicity of that drug. Hepatic damage affect metabolism rate of drug, either slower rate of metabolism and exhibit a prolonged effect of drug or vice-versa.

- **Individual Factors:**
  Inter- and intrapatient variations in disposition of a drug must be taken into consideration in choosing a treatment regime, such as the prescribed dose and its compliance. No of factor likes the actually administered dose, drug concentration at the site of action, genetic variations and the effect of the drug at the receptor, and also Body weight, administration time, tolerance, body temperature, and pathological condition of individual are other factors of concern.

12.4   MECHANISM OF DRUG INTERACTION

There are several mechanisms by which drug may interact but most can be classified as:

- **Pharmaceuticaldrug–drug interactions** (in vitro) occur when the formulation of one drug is altered by another before it is administered.

For example, precipitation of sodium thiopentone and vecuronium within an intravenous giving set.

- **Pharmacokinetic drug–drug interactions** (in vivo) occur when one drug changes the systemic concentration of another drug, altering 'how much' and for 'how long' it is present at the site of action.
- **Pharmacodynamicdrug–drug interactions** (in vivo) occur when interacting drugs have either additive effects, in which case the overall effect is increased, or opposing effects, in which case the overall effect is decreased or even 'canceled out'.

12.4.1 PHARMACEUTICAL DRUG–DRUG INTERACTIONS

Drug interaction may occur outside the body, that is, in vitro. It may be physical or chemical drug interaction. These interactions may occur during formulation and mixing of drug and the term incompatibility is often used to designate these in vitro interactions.

Physical term applies when physical state of either drug is altered when chemicals are mixed, for example, amphotericin gets precipitated if mixed with normal saline instead of 5% dextrose.

Chemical interaction applies when components of a drug mixture interact to form chemically altered products;, for example, methicillin and kanamycine, dopamine and sodium bicarbonate. In most of the cases chemical incompatibilities are manifested by precipitation or color change.

12.4.2 PHARMACOKINETIC DRUG–DRUG INTERACTIONS

Pharmacokinetic interactions occur when the absorption, distribution, metabolism or elimination processes of the object drug is altered by the precipitant drug.

In general, pharmacokinetic interactions are considered clinically significant when at least a 30% change is seen in C_{max}, T_{max} and AUC.

a) **Absorption:**
   Many factors influence drug absorption from the digestive tract, which include dissolution rate of the ingested drug, gastric emptying time, gut motility, blood flow. Gastric emptying time and gut motility determines rate and extent of absorption of drug. Gastrointestinal absorption is accelerated by drug that hasten gastric emptying, for

example, metoclopramide cisapride, domperidone. Decreased gut motility by anticholinergics like atropine increases total absorption of drug which are otherwise slowly incompletely absorbed. Milk, antacids containing calcium magnesium, iron preparations from chelate/complexes with tetracycline and hence decrease their absorption.

i) **Changes in gastrointestinal pH:** Absorption in the gut is governed by the gut pH, lipid solubility and pKa of the drug, and action of the P-glycoprotein. While changes in gastric pH induced by H_2 and proton pump blockers and antacids containing Al/Mg formulations have been shown to significantly reduce drug bioavailability; the alteration in pH has certain clinical implications as it can result in a significant reduction in the absorption of ketoconazole and itraconazole which are insoluble in water and are only ionized at low pH, hence gastric acidity plays an important part in this interaction. Likewise salicylic acid absorption is greater at low pH. The absorption of quinolone is also reduced when given along with antacids. Other drugs that are influenced by changes in pH are glipizide, glyburide, cefuroxime, and cefpodoxime.

ii) **Interactions affecting intestinal active transport:** The absorption of the drug is altered due to inhibition or induction of transport proteins (transporters). Inhibition or absence of an intestinal uptake transporter can result in decreased systemic drug exposure and/or lower C_{max}. Inhibition of an intestinal efflux transporter may result in increased systemic drug exposure and/or increased C_{max} either due to a primary increase in absorption and/or, secondarily, due to decreased availability of drug to intestinal drug metabolizing enzymes (e.g., CYP3A).

iii) **Inhibition of GI Enzymes:** The absorption of some drugs depend on their metabolism by the enzymes in gut, if these enzymes are inhibited then absorption of drug also decreases. Folic acids in dietary sources are poorly absorbed polyglutamate. Folic acid deficiency anemia is caused in-patient receiving phenytoin because ability to phenytointo inhibit intestinal conjugates enzyme. Oral contraceptives may interfere with deconjugation of polyglutamate form of folic acid resulting in folic acid deficiency.

    iv) Malabsorption states: Certain drugs, such as neomycin, laxative, colchicines, cholestyramine and aminosalicylic acid causes malabsorption problems that result in decreased absorption of vitamins and nutrients from GIT.

b) Distribution:

    i) Plasma protein binding: When drug absorbed into the blood stream, most of their molecule attach to plasma protein. Many drugs interact by displacement of each other's binding to plasma proteins. Acidic drugs are known to have an affinity to bind to plasma proteins, hence when two or more are given concomitantly, competitive binding for the same site or receptor may displace one drug from the protein binding site increasing the amount of the displaced free drug in plasma and various tissues setting up an interaction leading to an enhanced toxicity. The risk of an interaction occurring is greatest with drug that are highly protein bound and also have small apparent volume of distribution.

    ii) Warfarin and phenylbutazone are highly protein bound; phenylbutazone has a greater affinity for the binding sites, resulting in displacement of warfarin. The free drug quantities of warfarin are thus available in larger amount and activity of anticoagulant is increased which leads hemorrhage. Bilirubin is bound to albumin binding site and drugs like sulphonamides are capable of displacement of bilirubin from albumin. Salicylate and sulphonamides are capable of displacing methotrexate from its binding site leads to methotrexate toxicity agranulocytosis.

c) Metabolism:

The processes by which the enzymes alter an active drug inside the body to an inactive one or into active or toxic metabolites are referred to as drug metabolism or biotransformation.

    Most drugs need to reach a receptor site in order to exert their systemic effect and need to be lipid soluble so as to be able to penetrate the lipid plasma membrane barrier. The lipid soluble drugs further need to be converted into a water-soluble form to be excreted chiefly by the renal route and the chief role of metabolism is to enable these processes in two phases.

    In phase I, oxidation/reduction reactions convert the drugs into a more polar form, while phase II reactions provide another set of mechanisms involving conjugation/hydrolysis with substances like glucuronic acid and glucuronyl transferase for modifying drugs into inactive compounds to enable their excretion. Phase I reactions are catalyzed by a family of mixed function oxygenases called the "Cytochrome P450" class, expressed chiefly within the microsomal smooth endoplasmic reticulum hepatocytes and to a lesser extent in other cells. Drug interactions mostly occur by enzyme induction or enzyme inhibition.

i)   **Enzyme induction:** Liver microsomal enzyme involved in drug metabolism can be stimulated by drug including barbiturate, hydantoins, griseofulvin, chlorinated hydrocarbon insecticides and many others. Enzyme induction result in increased metabolism and excretion and reduced effect of drug. The stimulation of the hepatic enzyme activity is a factor in the development of drug interaction and also responsible for the development of tolerance to certain drugs that are given for prolonged period.

Meprobamate and glutethimide develop tolerance as they have the ability to stimulate their own metabolism. Phenobarbitone increase rate of metabolism of warfarin and also digoxin, this result in decrease anticoagulant activity of warfarin and plasma level of digoxin as it is being metabolized and excreted rapidly.

Rifampicin and barbiturates, with oral contraceptives, administered concurrently, reduction in blood level of the contraceptive is observed. Also rifampicin induce metabolism of corticosteroids which result in increased risk of graft rejection.

Smoking and alcohol increases rate of metabolism of certain drugs, polycyclic hydrocarbons which are present in cigarette smoke, tend to increase activity of hepatic enzyme.

Enzyme induction can be used for its beneficial effects also, for example, Phenobarbitone is used in the treatment of neonatal hyperbilirubinaemia, because it accelerates the metabolism of bilirubinto more readily excreted compounds.

ii) **Enzyme inhibition:** Compound that interferes with the activity of inactivating enzyme can potentiate the action of other drugs, for example, MAOIs inhibits the normal functioning of mono-amine oxidase, elevate level of biogenic amines and produce hypertensive crises in the presence of presser amines. Xanthine oxidase inhibitor, allopurinol, increases the plasma level of mercaptopurine by blocking its breakdown. Physostigmine and neostigmine block the degradation of choline esters, and can enhance the effect of acetylcholine.

Anticonvulsant phenytoin is metabolized by hepatic microsomal enzyme inhibitors like p-aminosalicylic acid, disulfiram, isoniazid and methylphenidate due to accumulation of unmetabolized drug. These increases level of toxicity.

d) **Excretion:**

i) **Renal excretion:** Renal filtration accounts for most drug excretion. Drugs that are chiefly excreted by the kidneys can get involved in drug interactions by different mechanisms such as competition at active transport sites, or alterations in glomerular filtration, passive renal tubular reabsorption or active secretion and urinary pH.

- **Glomerular filtration:** About one-fifth of the plasma reaching the glomerulus is filtered through pores in the glomerular endothelium. The rate of excretion of a drug or its metabolites can be influenced by other drugs that increase or decrease glomerular filtration due to changes in renal blood flow. For drugs with a low therapeutic index like digoxin, phenytoin and warfarin, any increase in renal clearance decreases their steady state plasma concentrations and conversely any reduction in their renal excretion increases circulating levels of the drugs resulting intoxicity.

- **Tubular reabsorption:** Nearly all water and most electrolytes are passively and actively reabsorbed from the renal tubules back into the circulation. However, polar compounds, which account for most drug metabolites, cannot diffuse back into the circulation and are excreted unless a specific transport mechanism exists for their reabsorption (e.g., as for glucose, ascorbic acid, and B vitamins). Small dose of aspirin impair the uricosuric action of probenecid by interfering with active secretion of uric acid into the renal tubules. Excretion of lithium gets altered by diuretics and NSAIDs and inhibits renal tubular reabsorption, leading to compensatory reabsorption of lithium and lithium toxicity.

- **Change in fluid and electrolyte:** Changes in electrolytes level and fluid level induced by drugs can affect therapeutic effectiveness and toxicity of other drugs, especially those acting on heart, kidney, and skeletal muscles. Hypokalaemia produced by diuretics and corticosteroids increases digitalis toxicity and can antagonize the antiarrhythmic activity of quinidine, lidocaine, procainamide, phenytoin and disoyramide.

- **Urine pH and Ionization:** Urine pH, which varies from 4.5 to 8.0, may markedly affect drug reabsorption and excretion because urine pH determines the ionization state of a weak acid or base. Acidification of urine increases reabsorption and decreases excretion of weak acids, and, in contrast, decreases reabsorption of weak bases. Alkalization of urine has the opposite effect. In some cases of overdose, these principles are used to enhance the excretion of weak bases or acids; e.g., urine is alkalized to enhance excretion of acetylsalicylic acid. The extent to which changes in urinary pH alter the rate of drug elimination depends on the contribution of the renal route to total elimination, the polarity of the un-ionized form, and the molecule's degree of ionization. Anions and cations are handled by separate transport mechanisms. Normally, the anion secretory system eliminates metabolites conjugated with glycine, sulfate, or glucuronic acid. Anions compete with each other for secretion. This competition can be used therapeutically; e.g., probenecid blocks the normally rapid tubular secretion of penicillin, resulting in higher plasma penicillin concentrations for a longer time. In the cation transport system, cations or organic bases (e.g., pramipexole, dofetilide) are secreted by the renal tubules; this process can be inhibited by cimetidine, trimethoprim, prochlorperazine, megestrol, or ketoconazole. Likewise, cimetidine competes with metformin (both being cationic drugs) for elimination by renal tubular secretion.

ii) **Biliary excretion:** Some drugs and their metabolites are extensively excreted in bile. Because they are transported across the biliary epithelium against a concentration gradient, active secretary transport is required. When plasma drug concentrations are high, secretary transport may approach an upper limit (transport maximum). Substances with similar physicochemical properties may compete for excretion. Drugs with a molecular weight of >300 g/mol and with both polar and lipophilic groups are more likely to be excreted in bile; smaller molecules are generally excreted only in negligible amounts. In enterohepatic cycle, a drug secreted in bile is reabsorbed into the circulation from the intestine. Biliary excretion eliminates substances from the body only to the extent that enterohepatic cycling is incomplete—when some of the secreted drug is not reabsorbed from the intestine.

12.4.3 PHARMACODYNAMIC DRUG–DRUG INTERACTION

Drug interactions can occur through several mechanisms. One or more mechanisms may be involved in the expression of a clinically significant drug interaction. The primary mechanisms of drug interactions include effects of drugs on hepatic metabolism of pharmaceuticals including effects on cytochrome P450 (CYP) enzymes or effects on glucuronidation, medication effects on the function of the drug transporter, P-glycoprotein, and effects on absorption of drugs. Pharmacodynamic interaction occurs at the site of drug action. They usually lead to an altered sensitivity of the affected organ. Mostly drug effects are the result of binding of the drug to specialized areas on or within the cells, known as receptor site. The magnitude of the effect depends on the concentration of the free drug at its receptor site. The free drug or an active metabolite concentration at the receptor site depends on the amount of drug in the body, which turn is regulated by (i) physical and chemical properties of drug; (ii) altered gastrointestinal absorption or competition for protein binding site or receptor site; (iii) altered drug metabolism; (iv) change in acid base equilibrium; and (v) alteration of hemodynamic or renal tubular function influencing renal drug clearance.

12.4.3.1 Drugs Having Opposing Pharmacological Effect

Drugs with opposing or antagonistic pharmacodynamic effects reduce response to either drug. NSAIDs, especially the COX-2 inhibitors, increase the blood pressure by inhibiting the hypertensive action of diuretics, ACEIs and β-blockers. The effects of benzodiazepines get inhibited with the concurrent administration of theophylline.

However, a few antagonistic reactions can actually be beneficial such as the reversal of the effects of opium alkaloids with naloxone. Certain pharmacodynamic interactions occur indirectly wherein the toxic or therapeutic effects of either drug are not related directly and seem to act on separate parts of a common process; for example, warfarin could be involved in an indirect interaction with aspirin when other drugs such as dipyridamole, salicylates or phenylbutazone reduce platelet aggregation or in cases of thrombocytopenia.

NSAID's can cause gastric ulcer and patients having concomitant warfarin therapy run a risk of greatly increased bleeding. The aminoglycoside antibiotics (streptomycin, kanamycin, gentamicin) and the potent loop diuretic frusemide when used concurrently may causes severe ototoxicity.

12.4.3.2   Receptor Blockade

The development of drug that selectively block receptor, particularly those of the autonomic nervous system, has led to several important interaction, for example, α-adrenoreceptor blockade with phenoxybenzamine prevent the action of noradrenaline and other α-sympathomimetics: β-blockade with propranolol reduces or abolishes the cardiac stimulating activity of adrenaline and isoprenenaline. Anticholinergic action of tricyclic antidepressant and antihistaminic reduces the effect of chlonergic drug.

12.4.3.3   Drugs Having Similar Pharmacological Effect

The increased CNS depressant effect is experienced by individuals being treated with sedatives when they taken with alcoholic beverages.

Antipsychotic drugs, anti-Parkinson drugs and antidepressants differ in their pharmacological action but exhibit the same secondary effect may be administered concurrently. May patient being treated with chlorpromazine are also given trihexphenidyl (anti-Parkinson drug) to control the extrapyrimidal effect of chlorpromazine.

12.4.3.4   Alteration At Electrolyte Level

Therapy with diuretics usually results in depletion of electrolytes and ions like potassium, sodium, calcium, magnesium. Conditions like hypokalemia modify effect of drug like digitalis and lithium. Digitalis glycoside and thiazide diuretics, loop diuretics cause hypokalemia, result in arrhythmias. Lithium carbonate and diuretics concurrent administration leads to sodium depletion and increase risk of lithium toxicity.

12.5   RELATIVE IMPORTANCE OF DRUG INTERACTION

The overall safety of a given drug is determined by its toxicity, side effects, and drug-drug interactions. The World Health Organization reports that drug interactions are a leading cause of morbidity and mortality. This finding extends to medications used in the treatment of substance use disorders and pharmacotherapies used for treatment of medical or mental illnesses, as well as for abused substances—including alcohol, licit, and illicit substances. Following are the importance of study of drug interaction (Table 11.2):

- **To avoid harmful effect or ADR and determine appropriate precautions:** Many of the predisposing factors that determine whether an interaction will occur are known, in practice, it is still very difficult to predict exactly what will happen when an individual patient is administered two drugs that have the potential to interact. The solution to this practical problem is to choose noninteracting drugs. However, if such an alternative is not available, it is frequently possible to give interacting drugs together if appropriate precautions are taken. Every time a drug is administered with any other prescription medicine, OTC products, herbs or even food subject expose themselves to the risk of a potentially dangerous interaction. Understanding these potential reactions and their mechanisms help us to navigate the hazardous waters of combining drugs with other medicines, food, herbs and vitamins with confidence. Interactions should always be considered in the differential diagnosis of any unusual response occurring during drug therapy. Clinicians need to be aware of the fact that patients often see multiple physicians and come to them with a legacy of drugs acquired during their previous visits and therefore are not always aware of all the patient's medication; necessitating the need to obtain a thorough and meticulous drug history that should include the use of OTC products and health foods.
- **Early detection of interaction:** Certain drugs consistently run the risk of generating certain interactions through well understood

TABLE 12.2　Reported Interaction of Some Common Drugs

S. No.	Name of the Drug	Reported Drug Interaction
1	**5-Fluorouracil**	It is contraindicated in patients that are severely debilitated or in patients with myelosuppression due to either radiotherapy or chemotherapy. It is likewise contraindicated in pregnant or breastfeeding women. It should also be avoided in patients that do not have malignant illnesses.
2	**Acarbose**	This drug should not be used with the following medications because very serious interactions may occur: activated charcoal, digestive enzyme products (e.g., amylase, pancreatin).
3	**Acetazolamide**	It might interact with Amphetamines, Other carbonic anhydrase inhibitors, Ciclosporin Antifolates such as trimethoprim, methotrexate, pemeterxed and raltitrexed, Hypoglycemics, acetazolamide Lithium, Methenamine Phenytoin, Primidone, Quinidine, Salicylates, Sodium bicarbonate, Anticoagulants, cardiac glycosides.
4	Acyclovir	• Probenecid: Reports of increased half-life of acyclovir, as well as decreased urinary excretion and renal clearance have been shown in studies where probenecid is given simultaneously with acyclovir.
		• Interferon: Synergistic effects when administered with acyclovir and caution should be taken when administering acyclovir to patients' receiving IV interferon.
		• Zidovudine: Although administered often with acyclovir in HIV patients, neurotoxicity has been reported in at least one patient who presented with extreme drowsiness and lethargy 30–60 days after receiving IV acyclovir; symptoms resolved when acyclovir was discontinued.
5	Albendazole	The drugscarbamazepine, phenytoin, andphenobarbitallower the plasmatic concentration and thehalf lifeof albendazole. The drugcimetidineheightensserumalbendazole concentrations, and increases the half-life of albendazole. This might be a helpful interaction on more severe cases, because it boosts thepotencyof albendazole.
6	**Alcohol**	Alcohol often has harmful interactions with prescription medications, over the counter drugs, and even some herbal remedies. Alcohol interactions with medications may cause problems such as:
		• Nausea and vomiting.
		• Headaches.

S. No.	Name of the Drug	Reported Drug Interaction
		• Drowsiness.
		• Dizziness.
		• Fainting.
		• Changes in blood pressure.
		• Abnormal behavior.
		• Loss of coordination.
		Mixing alcohol and medications also may increase the risk of complications such as:
		• Liver damage.
		• Heart problems.
		• Internal bleeding.
		• Impaired breathing.
		• Depression.
7	Allopurinol	Following medications may interact with allopurinol:
		• Azathioprine.
		• Chlorpropamide.
		• Cyclosporine.
		• Mercaptopurine.
		• Antibiotics such as ampicillin or amoxicillin
		• Blood thinners such as warfarin.
		• Diuretics

TABLE 12.2 (Continued)

S. No.	Name of the Drug	Reported Drug Interaction
8	Alprazolam	Alprazolam is primarily metabolized via CYP3A4. Combining CYP3A4 inhibitors such as cimetidine, erythromycin, fluoxetine, fluvoxamine, itraconazole, ketoconazole, nefazodone, propoxyphene, and ritonavir delay the hepatic clearance of alprazolam, which may result in excessive accumulation of alprazolam. This may result in exacerbation of its adverse effect profile.
9	Amifostine	Drugs that cause drowsiness such as: certainantihistamines (e.g., diphenhydramine), antiseizure drugs (e.g., carbamazepine), medicine orsleeporanxiety (e.g., alprazolam, diazepam, zolpidem), muscle relaxants, narcotic pain relievers (e.g., codeine), psychiatric medicines (e.g., chlorpromazine, risperidone, trazodone). Check the labels on all your medicines (e.g., cough-and-cold products) because they may contain ingredients that cause drowsiness.
10	Aminoglutethimide	• Anticoagulants (e.g., warfarin), corticosteroids (e.g., dexamethasone), or digitoxin because the effectiveness of these medicines may be decreased.
11	Aminophylline	Alcohol and benzodiazepines such as alprazolam taken in combination have a synergistic effect on one another, which can cause severe sedation, behavioral changes, and intoxication. The more alcohol and alprazolam taken the worse the interaction. Combination of alprazolam with the herbkavacan result in the development of a semicomatose state. Hypericumconversely can lower the plasma levels of alprazolam and reduce its therapeutic effect. • Monoamine oxidase inhibitors as it can potentially induce a serotonin syndrome. • CYP2D6 inhibitors and substrates such as fluoxetine due to the potential for an increase in plasma concentrations of the drug to be seen. • Guanethidine as it can reduce the antihypertensive effects of this drug. • Anticholinergic agents such as benztropine, hyoscine (scopolamine) and atropine, because the two might exacerbate each other's anticholinergic effects, including paralytic ileus and tachycardia. • Antipsychotics due to the potential for them to exacerbate the sedative, anticholinergic, epileptogenic and pyrexic (fever-promoting) effects. Also increases the risk of neuroleptic malignant syndrome.

S. No.	Name of the Drug	Reported Drug Interaction
		Cimetidine due to the potential for it to interfere with hepatic metabolism of amitriptyline and hence increasing steady-state concentrations of the drug.
		• Disulfiram due to the potential for the development of delirium.
		• ECT may increase the risks associated with this treatment.
		• Antithyroid medications may increase the risk of agranulocytosis
		• Thyroid hormones have a potential for increased adverse effects such as CNS stimulation and arrhythmias.
		• Analgesics, such as tramadol, due to the potential for an increase in seizure risk.
		• Medications subject to gastric inactivation (e.g., levodopa) due to the potential for amitriptyline to delay gastric emptying and reduce intestinal motility.
		• Medications subject to increased absorption given more time in the small intestine (e.g., anticoagulants).
		• Serotoninergic agents such as the SSRIs and triptans due to the potential for serotonin syndrome.
12	Amiodarone	• In patients with severe coronary artery disease, amlodipine can increase the frequency and severity of angina or actually cause a heart attack on rare occasions.
		• Excessive lowering of blood pressure during initiation of amlodipine treatment can occur, especially in patients already taking another medication for lowering blood pressure. This includes medications for erectile dysfunction, such as sildenafil, which can also lower blood pressure. In rare instances, congestive heart failure has been associated with amlodipine, usually in patients already on a beta blocker.
		• Amlodipine is primarily metabolized by the liver, via the cytochrome P450 isoenzyme CYP3A4. As a result, serum levels can potentially be affected by drugs which inhibit oractivate CYP3A4. Grapefruit juice can inhibit the cytochrome P450 system, but the predicted interaction risk with amlodipine is low. Eating pomegranate or drinking pomegranate juice might cause similar side effects.

TABLE 12.2 (Continued)

S. No.	Name of the Drug	Reported Drug Interaction
13	Amitriptyline	• Monoamine oxidase inhibitors as it can potentially induce a serotonin syndrome.
		• CYP2D6 inhibitors and substrates such as fluoxetine due to the potential for an increase in plasma concentrations of the drug to be seen.
		• Guanethidine as it can reduce the antihypertensive effects of this drug.
		• Anticholinergic agents such as benztropine, hyoscine (scopolamine) and atropine, because the two might exacerbate each other's anticholinergic effects, including paralytic ileus and tachycardia.
		• Antipsychotics due to the potential for them to exacerbate the sedative, anticholinergic, epileptogenic and pyrexic (fever-promoting) effects. Also increases the risk of neuroleptic malignant syndrome.
		• Cimetidine due to the potential for it to interfere with hepatic metabolism of amitriptyline and hence increasing steady-state concentrations of the drug.
		• Disulfiram due to the potential for the development of delirium.
		• ECT may increase the risks associated with this treatment.
		• Antithyroid medications may increase the risk of agranulocytosis.
		• Thyroid hormones have a potential for increased adverse effects such as CNS stimulation and arrhythmias.
		• Analgesics, such as tramadol, due to the potential for an increase in seizure risk.
		• Medications subject to gastric inactivation (e.g., levodopa) due to the potential for amitriptyline to delay gastric emptying and reduce intestinal motility.
		• Medications subject to increased absorption given more time in the small intestine (e.g., anticoagulants).
		• Serotoninergic agents such as the SSRIs and triptans due to the potential for serotonin syndrome.

S. No.	Name of the Drug	Reported Drug Interaction
14	Amlodipine	• In patients with severe coronary artery disease, amlodipine can increase the frequency and severity of angina or actually cause a heart attack on rare occasions.
		• Excessive lowering of blood pressure during initiation of amlodipine treatment can occur, especially in patients already taking another medication for lowering blood pressure. This includes medications for erectile dysfunction, such as sildenafil, which can also lower blood pressure. In rare instances, congestive heart failure has been associated with amlodipine, usually in patients already on a beta blocker.
		• Amlodipine is primarily metabolized by the liver, via the cytochrome P450 isoenzyme CYP3A4. As a result, serum levels can potentially be affected by drugs which inhibit or activate CYP3A4. Grapefruit juice can inhibit the cytochrome P450 system (Sloan, 1983), but the predicted interaction risk with amlodipine is low (Malone, 2005). Eating pomegranate or drinking pomegranate juice might cause similar side effects.
15	Amodiaquine	No data regarding the interactions of Amodiaquine was found.
16	Amoxapine	• Potential additive CNS effects.
		• Potential for decreased amoxapine metabolism.
		• Possible additive anticholinergic effects; hyperthermia, particularly during hot weather, and paralytic ileus also possible.
		• Potential for tricyclic toxicity, particularly anticholinergic adverse effects.
		• Potentiates the effects of CNS depressants.
		• Possible antagonism of the antihypertensive effects of guanethidine and related compounds.
		• May interfere with levodopa absorption.
		• Potentially life-threatening serotonin syndrome.
		• Potential for decreased metabolism and increased therapeutic efficacy and toxicity of TCAs.
		• Possible serotonin syndrome.

TABLE 12.2 (Continued)

S. No.	Name of the Drug	Reported Drug Interaction
17	Amphotericin	• Increased vasopressor, cardiac effects.
		• Possible cardiac arrhythmias.
		• May enhance potential for renal toxicity, bronchospasm, or hypotension.
		• *In vitro* evidence of antagonism against *Candida* or *Aspergillus fumigatus*.
		• Amphotericin B-induced hypokalemia may potentiate toxicity of cardiac glycosides.
		• Possible enhanced potassium depletion.
		• *In vitro* evidence of synergistic antifungal activity against *Candida* or *Cryptococcus neoformans*.
		• Possibility of increased risk of flucytosine toxicity with conventional amphotericin B; may occur as the result of increased cellular uptake and/or decreased renal excretion of flucytosine.
		• Administration of conventional amphotericin B during or shortly after leukocyte transfusions has been associated with acute pulmonary reactions.
18	Ampicillin	• Possible increased incidence of rash.
		• *In vitro* evidence of synergistic antibacterial effects against enterococci; used to therapeutic advantage in treatment of endocarditis and other severe enterococcal infection Potential in vitro and in vivo inactivation of aminoglycosides.
		• *In vitro* evidence of antagonism. Possible decreased efficacy of estrogen-containing oral contraceptives and increased incidence of breakthrough bleeding.
		• Possible decreased renal clearance of methotrexate with penicillins; possible increased methotrexate concentrations and hematologic and GI toxicity.
		• Decreased renal tubular secretion of ampicillin; increased and prolonged ampicillin concentrations may occur. Synergistic bactericidal effect against many strains of β-lactamase-producing bacteria.
		• *In vitro* evidence of antagonism. Possible false-positive reactions in urine glucose tests using Clinitest, Benedict's solution, or Fehling's solution.

S. No.	Name of the Drug	Reported Drug Interaction
19	Artemether	• Possible falsely increased serum uric acid concentrations when copper-chelate method is used; phosphotungstate and uricase methods appear to be unaffected by the ampicillin.
		• Possible additive effects on prolongation of QT interval. Possible additive effects on prolongation of QT interval. Possible additive effects on prolongation of QT interval.
		• Increased plasma concentrations of artemether, the active metabolite of artemether, and lumefantrine. Antimalarial agents: Safety data on concomitant use limited.
		• Decreased plasma concentrations of lumefantrine; no change in artemether or mefloquine concentrations when mefloquine is administered immediately before artemether and lumefantrine.
		• Antiretroviral agents: Possible pharmacokinetic interaction; altered plasma concentrations of the antiretroviral agent, artemether, or lumefantrine; possible additive effects on prolongation of QT interval.
		• Lopinavir/ritonavir: Increased plasma concentrations of lumefantrine; slightly decreased concentrations of artemether and the active metabolite of artemether. Possible decreased plasma concentrations of the components of the contraceptive.
20	Artesunate	• Antimalarial potentiating action seen with mefloquine, primaquire and tetracycline. Additive effect with chloroquine. Antagonistic effect with pyrimethamine and sulphonamides.

TABLE 12.2 (Continued)

S. No.	Name of the Drug	Reported Drug Interaction
21	Aspirin	• Reduced BP response to ACE inhibitors Possible attenuation of hemodynamic actions of ACE inhibitors in patients with CHF.
		• Reduced hyponatremic effect of ACE inhibitors.
		• Drugs that decrease urine pH may decrease salicylate excretion.
		• Increased risk of bleeding.
		• Drugs that increase urine pH may increase salicylate excretion.
		• Increased risk of bleeding.
		• May displace warfarin from protein-binding sites, leading to prolongation of PT and bleeding time.
		• May displace phenytoin from binding sites; possible decrease intotal plasma phenytoin concentrations, with increased free fraction.
		• May displace valproic acid from binding sites; possible increase in free plasma valproic acid concentrations possible increased risk of bleeding.[h]
		• Potential for increased hypoglycemic effect.
		• Increased risk of salicylate toxicity.
		• Increased plasma acetazolamide concentrations; increased risk of acetazolamide toxicity.
		• Increased plasma methotrexate concentrations.
		• Inhibition of renal clearance of methotrexate leading to bone marrow toxicity, especially in geriatric patients or patients with renal impairment.
22	Atenolol	• Potential additive effect.
		• Increased risk of hypotension and heart failure.
		• Potential for additive effects (increased hypotension and marked bradycardia).

S. No.	Name of the Drug	Reported Drug Interaction
		• May exacerbate rebound hypertension following discontinuance of clonidine.
		• Additive hypotensive effect; may be used to therapeutic advantage.
		• Additive or potentiated hypotensive effect; may be used to therapeutic advantage.
		• Potential for decreased atenolol antihypertensive effect.
23	Atropine	• Increased anticholinergic effects.
		• Decreased GI absorption of atropine.
		• Increased anticholinergic effects.
		• Increased IOP.
		• Increased serum digoxin.
		• Increased gastric pH decreases ketoconazole absorption.
		• Increased GI metabolism of levodopa and decreased systemic concentrations.
		• Decreased GI absorption rate of mexiletine; no effect on bioavailability.
24	Azathioprine	• Potential for increased toxicity (anemia, severe leukopenia).
		• Allopurinol inhibits metabolic pathway catalyzed by xanthine oxidase; may increase risk of azathioprine toxicity.
		• Aminosalicylates inhibit metabolic pathway catalyzed by TPMT; may increase risk of azathioprine toxicity. Possible increased leukopenia, especially in renal transplant recipients.
		• Ribavirin inhibits metabolic pathway catalyzed by inosine monophosphate dehydrogenase, resulting in accumulation of myelotoxic metabolite of azathioprine; severe pancytopenia reported.
25	Azithromycin	• Increased peak plasma concentration and AUC of azithromycin when azithromycin, albendazole, and ivermectin given concomitantly.
		• Conventional tablets or oral suspension: Decreased azithromycin plasma concentrations; AUC unaffected.

TABLE 12.2 (Continued)

S. No.	Name of the Drug	Reported Drug Interaction
		• Extended-release oral suspension: Rate and extent of azithromycin absorption not affected.
		• Does not appear to affect PT response to a single warfarin dose; increased anticoagulant effects reported rarely with azithromycin and also reported with other macrolides.
		• Modest effect on pharmacokinetics of midazolam or triazolam; decreased clearance of triazolam and increased pharmacologic effect reported with other macrolides.
		• No clinically important pharmacokinetic interactions between chloroquine and azithromycin.
		• *In vitro* evidence of additive to synergistic effects against *P. falciparum*, including multidrug-resistant strains.
		• Potential increased pimozide plasma concentrations and risk of prolonged QT interval and serious cardiac arrhythmias reported with other macrolides.
26	Baclofen	• Additive CNS depression.
		• Hypotension and dyspnea may occur when administered with intrathecal baclofen.
27	Barbiturates	• Possible decreased plasma warfarin concentrations.
		• Antidepressant may precipitate seizures, resulting in decreased seizure control.
		• Potentiation of respiratory depression following toxic doses of tricyclic antidepressants.
		• Possible additive depressant effects.
		• Possible enhanced metabolism of estrogenic and progestinic components; potential for decreased oral contraceptive effectiveness and increased risk of pregnancy with phenobarbital pretreatment or concurrent therapy.
		• Possible decreased half-life of doxycycline; effect may persist up to 2 weeks after discontinuance of Phenobarbital.
		• Possible decreased griseofulvin absorption, resulting in decreased blood concentrations.

S. No.	Name of the Drug	Reported Drug Interaction
28	Beclomethasone	• Increased, decreased, or no change in plasma phenytoin concentrations reported.
		• May increase blood glucose concentrations in patients with diabetes mellitus.
		• Possible increased risk of GI ulceration.
		• Decreased serum salicylate concentrations.
		• When corticosteroids are discontinued, serum salicylate concentration may increase possibly resulting in salicylate intoxication.
		• May cause a diminished response to toxoids and live or inactivated vaccines.
		• May potentiate replication of some organisms contained in live, attenuated vaccines.
		• Can aggravate neurologic reactions to some vaccines (supraphysiologic dosages).
29	Betaxolol	• Although ophthalmic betaxolol used alone has little or no effect on pupil size, mydriasis resulting from concomitant therapy with epinephrine has been reported occasionally.
		• Close observation of the patient is recommended when a beta-blocker is administered to patients receiving oral beta-adrenergic blocking drugs, or catecholamine-depleting drugs such as reserpine, because of possible additive effects. Caution should be exercised in patients using concomitant adrenergic psychotropic drugs.
		• Pregnancy: There have been no adequate and well-controlled studies in pregnant women. Because animal reproduction studies are not always predictive of human response, this drug should be used during pregnancy only if clearly indicated.
		• Lactation: It is not known whether betaxolol is excreted in human milk. Because many drugs are excreted in human milk, caution should be exercised when ophthalmic betaxolol is administered to nursing women.
		• Children: Clinical studies to establish the safety and efficacy in children have not been performed.

TABLE 12.2   (Continued)

S. No.	Name of the Drug	Reported Drug Interaction
30	Bisacodyl	• Administration of delayed-release (enteric-coated) tablets within 1 h of antacids results in rapid erosion of the coating.
		• Administration of delayed-release (enteric-coated) tablets within 1 h of cimetidine results in rapid erosion of the coating.
		• Administration of delayed-release (enteric-coated) tablets within 1 h of milk results in rapid erosion of the coating.
		• Administration of delayed-release (enteric-coated) tablets within 1 h of famotidine results in rapid erosion of the coating.
		• Increased gastric pH results in rapid erosion of the coating of delayed-release (enteric-coated) tablets.
		• Administration of delayed-release (enteric-coated) tablets within 1 h of ranitidine results in rapid erosion of the enteric coating.
31	Bisphosphonates	• Co-administration of FOSAMAX and calcium, antacids, or oral medications containing multivalent cations will interfere withabsorptionof FOSAMAX. Therefore, instruct patients to wait at least one-half hour after taking FOSAMAX before taking any other oral medications.
		• In clinical studies, the incidence of upper gastrointestinal adverse events was increased in patients receiving concomitant therapy with daily doses of FOSAMAX greater than 10 mg and aspirin-containing products.
		• FOSAMAX may be administered to patients taking nonsteroidal antiinflammatory drugs (NSAIDs). In a 3-year, controlled, clinical study ($n = 2027$) during which a majority of patients received concomitant NSAIDs, the incidence of upper gastrointestinal adverse events was similar in patients taking FOSAMAX 5 or 10 mg/day compared to those taking placebo. However, sinceNSAIDuse is associated with gastrointestinal irritation, caution should be used during concomitant use with FOSAMAX.

S. No.	Name of the Drug	Reported Drug Interaction
32	Bleomycin	• Increased risk of bleomycin-induced pulmonary toxicity.
		• Bleomycin shown to be inactivated in vitro by ascorbic acid and riboflavin.
33	Bromocriptine	• May potentiate adverse effects of bromocriptine.
		• Possible decreased alcohol tolerance.
		• Potential reduced efficacy of bromocriptine.
		• Possible interference with bromocriptine metabolism.
		• Potential additive hypotensive effects.
		• Possible reduced efficacy of bromocriptine.
		• Potential reduced bromocriptine concentrations.
		• Potential for severe adverse effects (e.g., hypertension, myocardial infarction).
		• Possible interference with bromocriptine metabolism.
34	Bupivacaine	• The administration of local anesthetic solutions containingepinephrineor norepinephrineto patients receiving monoamine oxidase inhibitors ortricyclic antidepressantsmay produce severe, and prolonged hypertension. Concurrent use of these agents should generally be avoided. In situations in which concurrenttherapyis necessary, careful patient monitoring isessential.
		• Concurrent administration of vasopressor drugs and of ergot-type oxytocic drugs may cause severe, persistent hypertension orcerebrovascular accidents.
		• Phenothiazines and butyrophenones may reduce or reverse thepresser effect of epinephrine.
35	Buprenorphine	• Possible potentiation of anesthetic effectand more rapid onset and prolonged duration of analgesia.
		• Increased risk of QT-interval prolongation reported with transdermal buprenorphine.
		• Possible purpuric response.

TABLE 12.2 (Continued)

S. No.	Name of the Drug	Reported Drug Interaction
		• Reports of death or coma when buprenorphine was misused (e.g., self-injection of crushed tablets) via IV injection with benzodiazepines by drug abusers.
		• May alter usual ceiling on buprenorphine-induced respiratory depression, making buprenorphine's respiratory-depressant effects appear similar to those of full opiate agonists.
		• Respiratory and cardiovascular collapse reported in several patients receiving usual doses of IV buprenorphine and oral diazepam concomitantly.
		• Bradycardia, respiratory depression, and prolonged drowsiness reported following IV buprenorphine administration during surgery in a patient who had received oral lorazepam preoperatively.
36	Buspirone	• *Amitriptyline:* After addition of buspirone to the amitriptyline dose regimen, no statistically significant differences in the steady-state pharmacokinetic parameters (C_{max}, AUC, and C_{min}) of amitriptyline or its metabolite nortriptyline were observed.
		• *Diazepam:* After addition of buspirone to the diazepam dose regimen, no statistically significant differences in the steady-state pharmacokinetic parameters (C_{max}, AUC, and C_{min}) were observed for diazepam, but increases of about 15% were seen for nordiazepam, and minor adverse clinical effects (dizziness, headache, and nausea) were observed.
		• *Haloperidol:* In a study in normal volunteers, concomitant administration of buspirone and haloperidol resulted in increased serum haloperidol concentrations. The clinical significance of this finding is not clear.
37	Busulphan	• Itraconazole decreases busulfan clearance by up to 25%, and may produce an AUC greater than 1500 μM.min in some patients. Fluconazole (200 mg) has been used with BUSULFEX.
		• Phenytoin increases the clearance of busulfan by 15% or more, possibly due to the induction of glutathione-S-transferase. Since the pharmacokinetics of BUSULFEX were studied in patients treated with phenytoin, the clearance of BUSULFEX at the recommended dose may be lower and exposure (AUC) higher in patients not treated with phenytoin.

S. No.	Name of the Drug	Reported Drug Interaction
		• Because busulfan is eliminated from the body via conjugation with glutathione, use ofacetaminophenprior to (less than 72 h) or concurrent with BUSULFEX may result in reduced busulfan clearance based upon the known property of acetaminophen to decrease glutathione levels in the blood and tissues.
38	Calcitrol	• Increased risk of hypercalcemia. • Concurrent use of vitamin D analogs and cardiac glycosides may result in cardiac arrhythmias. • Intestinal absorption of calcitriol may be decreased. • Corticosteroids may counteract effects of vitamin D analogs. • Endogenous serum calcitriol may be reduced.
39	Capreomycin	• Increased risk of *nephrotoxicity* if given with: aminoglycosides. colistimethate sodium. • Increased risk of *ototoxicity* if given with: aminoglycosides.
40	Captopril	• Dual blockade of the RAS with angiotensin receptor blockers, ACE inhibitors, or aliskiren is associated with increased risks of hypotension, hyperkalemia, and changes in renal function (includingacute renal failure) compared to monotherapy. Closely monitor blood pressure, renal function and electrolytes in patients on Capoten and other agents that affect the RAS. • Do not coadminister aliskiren with Capoten in patients withdiabetes. Avoid use of aliskiren with Capoten in patients with renal impairment (GFR <60 mL/min). • In patients who are elderly, volume-depleted (including those on diuretic therapy), or with compromised renal function, coadministration of NSAIDs, including selective COX-2 inhibitors, with ACE inhibitors, including captopril, may result in deterioration of renal function, including possible acute renal failure. These effects are usually reversible. Monitor renal function periodically in patients receiving captopril and NSAID therapy. The antihypertensive effect of ACE inhibitors, including captopril, may be attenuated by NSAIDs.

TABLE 12.2 (Continued)

S. No.	Name of the Drug	Reported Drug Interaction
41	Carbamazepine	• There has been a report of a patient who passed an orange rubbery precipitate in his stool the day after ingesting Tegretol suspension immediately followed by Thorazine®* solution. Subsequent testing has shown that mixing Tegretol suspension and chlorpromazine solution (both generic and brand name) as well as Tegretol suspension and liquid Mellaril® resulted in the occurrence of this precipitate. Because the extent to which this occurs with other liquid medications is not known, Tegretol suspension should not be administered simultaneously with other liquid medicinal agents or diluents.
		• When carbamazepine is given with drugs that can increase or decrease carbamazepine levels, close monitoring of carbamazepine levels is indicated and dosage adjustment may be required.
		• CYP 3A4 inhibitors inhibit Tegretol metabolism and can thus increase plasma carbamazepine levels. Drugs that have been shown, or would be expected, to increase plasma carbamazepine levels include: ibuprofen, fluconazole, trazodone, dantrolene, cimetidine, acetazolamide, fluoxetine, ciprofloxacin, danazol, diltiazem, voriconazole, erythromycin, troleandomycin, clarithromycin, propoxyphene, fluvoxamine, olanzapine, loratadine, ketaconazole, terfenadine, omeprazole, oxybutynin, isoniazid, niacinamide, nicotinamide, aprepitant, itraconazole, verapamil, and ticlopidine.
42	Carbidopa	• Symptomatic postural hypotension occurred when SINEMET (carbidopa-levodopa) was added to the treatment of a patient receivingantihypertensive drugs. Therefore, when therapy with SINEMET (carbidopa-levodopa) is started, dosage adjustment of the antihypertensive drug may be required.
		• For patients receiving MAO inhibitors (Type A or B). Concomitant therapy with selegiline and carbidopa-levodopa may be associated with severeorthostatic hypotensionnot attributable to carbidopa-levodopa alone.
		• There have been rare reports of adverse reactions, including hypertension anddyskinesia, resulting from the concomitant use of tricyclic antidepressantsand SINEMET.

S. No.	Name of the Drug	Reported Drug Interaction
		• Dopamine D_2 receptor antagonists (e.g., phenothiazines, butyrophenones, risperidone) and isoniazid may reduce the therapeutic effects of levodopa. In addition, the beneficial effects of levodopa inParkinson's diseasehave been reported to be reversed by phenytoin and papaverine. Patients taking these drugs with SINEMET (carbidopa-levodopa) should be carefully observed for loss of therapeutic response.
43	Carbimazole	• Potentially Fatal: Increased sensitivity to warfarin in hyperthyroidism and careful control required as patient is rendered euthyroid.
44	Carboplatin	• The renal effects of nephrotoxic compounds may be potentiated by PARAPLATIN (carboplatin).
45	Carboprost	• HEMABATE may augment the activity of other oxytocic agents. Concomitant use with other oxytocic agents is not recommended.
46	Carisoprodol	• The sedative effects of SOMA and other CNS depressants (e.g., alcohol, benzodiazepines, opioids, tricyclic antidepressants) may be additive. Therefore, caution should be exercised with patients who take more than one of these CNS depressants simultaneously. Concomitant use of SOMA and meprobamate, a metabolite of SOMA, is not recommended.
		• Carisoprodol is metabolized in the liver by CYP2C19 to form meprobamate. Co-administration of CYP2C19 inhibitors, such as omeprazole or fluvoxamine, with SOMA could result in increased exposure of carisoprodol and decreased exposure of meprobamate. Co-administration of CYP2C19 inducers, such as rifampin or St. John's Wort, with SOMA could result in decreased exposure of carisoprodol and increased exposure of meprobamate. Low dose aspirin also showed an induction effect on CYP2C19. The full pharmacological impact of these potential alterations of exposures in terms of either efficacy or safety of SOMA is unknown.
		• Soma contains carisoprodol, a Schedule IV controlled substance. Carisoprodol has been subject to abuse, misuse, and criminal diversion for nontherapeutic use.

TABLE 12.2 (Continued)

S. No.	Name of the Drug	Reported Drug Interaction
47	Cefaclor	• Patients receiving Ceclor (cefaclor) may show a false-positive reaction for glucose in the urine with tests that use Benedict's and Fehling's solutions and also with Clinitesr tablets.
		• There have been reports of increased anticoagulant effect when Ceclor (cefaclor) and oral anticoagulants were administered concomitantly.
48	Cefadroxil	• Positive direct Coombs' tests have been reported during treatment with the cephalosporin antibiotics. In hematologic studies or intransfusioncross-matching procedures when antiglobulin tests are performed on the minor side or in Coombs' testing of newborns whose mothers have received cephalosporin antibiotics beforeparturition, it should be recognized that a positive Coombs' test may be due to the drug.
49	Cefazolin	• Probenecid may decrease renal tubular secretion of cephalosporins when used concurrently, resulting in increased and more prolonged cephalosporin blood levels.
		• Afalse positivereaction for glucose in the urine may occur with Benedict's solution, Fehling's solution or with Clinitest® tablets, but not with enzyme-based tests such as Clinistix®.
		• Positive direct and indirect antiglobulin (Coombs) tests have occurred; these may also occur in neonates whose mothers received cephalosporins before delivery.
50	Cefixime	• Elevated carbamazepine levels have been reported in postmarketing experience when cefixime is administered concomitantly. Drug monitoring may be of assistance in detecting alterations in carbamazepine plasma concentrations.
		• Increasedprothrombintime, with or without clinical bleeding, has been reported when cefixime is administered concomitantly.
		• A false-positive reaction for ketones in the urine may occur with tests using nitroprusside but not with those using nitroferricyanide.

S. No.	Name of the Drug	Reported Drug Interaction
		The administration of cefixime may result in a false-positive reaction for glucose in the urine using Clinitest®, Benedict's solution, or Fehling's solution. It is recommended that glucose tests based on enzymatic glucose oxidase reactions be used. A false-positive direct Coombs test has been reported during treatment with other cephalosporins; therefore, it should be recognized that a positive Coombs test may be due to the drug.
51	Cefoperazone	• A false-positive reaction for glucose in the urine may occur with Benedict's or Fehling's solution.
52	Cefotaxime	• Increased nephrotoxicity has been reported following concomitant administration of cephalosporins and aminoglycoside antibiotics.
		• Probenecid interferes with the renal tubular transfer of cefotaxime, decreasing the total clearance of cefotaxime by approximately 50% and increasing the plasma concentrations of cefotaxime. Administration of cefotaxime in excess of 6 grams/day should be avoided in patients receiving probenecid.
		• Cephalosporins, including cefotaxime sodium, are known to occasionally induce a positive direct Coombs' test.
		• A false-positive reaction for glucose in the urine may occur with copper reduction tests (Benedict's or Fehling's solution or with CLINITEST® tablets), but not with enzyme-based tests for glycosuria. There are no reports in published literature that link elevations of plasma glucose levels to the use of cefotaxime.
53	Cefoxitin	• Increased nephrotoxicity has been reported following concomitant administration of cephalosporins and aminoglycoside antibiotics.
		• Laboratory Tests: In patients treated with cefoxitin a false-positive reaction to glucose in the urine may occur with Benedict's or Fehling's solutions but not with the use of specific glucose oxidase methods.
		• Using the Jaffe Method falsely high creatinine values in serum may occur if serum concentrations of cefoxitin exceed 100 µg/mL. Serum samples from patients treated with cefoxitin should not be analyzed for creatinine if withdrawn within 2 h of drug administration.

TABLE 12.2 (Continued)

S. No.	Name of the Drug	Reported Drug Interaction
54	Ceftizoxime	• High concentrations of cefoxitin in the urine may interfere with measurement of urinary 17-hydroxy-corticosteroids by the Porter-Silber reaction and produce false increases of modest degree in the levels reported.
		• The concomitant administration of some cephalosporins and aminoglycosides has caused nephrotoxicity. The effect of administering ceftizoxime concomitantly with aminoglycosides is not known.
		• Pregnancy: The safety of ceftizoxime in pregnancy has not been established. Its use in pregnant women requires that the likely benefit from the drug be weighed against the possible risk to the mother and fetus. The pharmacokinetics in pregnant patients have not been investigated.
		• Reproduction studies performed in rats and rabbits have revealed no evidence of impaired fertility or harm to the fetus caused by ceftizoxime. Animal reproduction studies, however, are not always predictive of human response.
		• Labor and Delivery: The safety and efficacy of ceftizoxime use during labor and delivery has not been investigated.
		• Lactation: Ceftizoxime is excreted in human milk in low concentrations (less than 4% of serum concentrations at 1 h after dosing). The clinical significance of this is unknown; therefore caution should be exercised if ceftizoxime is to be administered to a nursing woman.
		• Infants and Children: The safety in infants less than 6 months of age has not been established. In children 6 months of age and older, treatment with ceftizoxime has been associated with transient elevated levels of eosinophils, AST, ALT, and CPK (creatine phosphokinase). The CPK elevation may be related to i.m. administration.
		• Geriatrics: The elimination of ceftizoxime may be reduced due to an age-dependent reduction in renal function.

S. No.	Name of the Drug	Reported Drug Interaction
55	Cefuroxime	• The concomitant administration of some cephalosporins and aminoglycosides has caused nephrotoxicity. The effect of administering ceftizoxime concomitantly with aminoglycosides is not known.
		• Pregnancy: The safety of ceftizoxime in pregnancy has not been established. Its use in pregnant women requires that the likely benefit from the drug be weighed against the possible risk to the mother and fetus. The pharmacokinetics in pregnant patients have not been investigated. Reproduction studies performed in rats and rabbits have revealed no evidence of impaired fertility or harm to the fetus caused by ceftizoxime. Animal reproduction studies, however, are not always predictive of human response.
		• Labor and Delivery: The safety and efficacy of ceftizoxime use during labor and delivery has not been investigated.
		• Lactation: Ceftizoxime is excreted in human milk in low concentrations (less than 4% of serum concentrations at 1 h after dosing). The clinical significance of this is unknown; therefore caution should be exercised if ceftizoxime is to be administered to a nursing woman.
		• Infants and Children: The safety in infants less than 6 months of age has not been established. In children 6 months of age and older, treatment with ceftizoxime has been associated with transient elevated levels of eosinophils, AST, ALT, and CPK (creatine phosphokinase). The CPK elevation may be related to i.m. administration.
		• Geriatrics: The elimination of ceftizoxime may be reduced due to an age-dependent reduction in renal function.
56	Cetrizine	• There was a small decrease in the clearance of cetirizine caused by a 400-mg dose of theophylline; it is possible that larger theophylline doses could have a greater effect.
57	Chlorambucil	• Vaccinations with live organism vaccines are not recommended in immunocompromised individuals (see Precautions). Animal studies indicate that patients who receive phenylbutazone may require a reduction of the standard chlorambucil doses because of the possibility of enhanced chlorambucil toxicity.

TABLE 12.2 (Continued)

S. No.	Name of the Drug	Reported Drug Interaction
58	Chloramphenicol	• Possible delayed response to iron preparations, vitamin B_{12}, or folic acid.
		• Possible increased anticoagulant effect;[c] chloramphenicol may impair utilization of prothrombin or decrease vitamin K production by intestinal bacteria.
		• *In vitro* evidence of antagonism with penicillins or cephalosporins[c].
		• Potential additive bone marrow depression.
59	Chlormezanone	• Interaction with acetaminophen, aspirin, buprenorphine, codeine, chlorpheniramine, ibuprofen, morphine, and tramadol.
60	Chloroquine	• Antacids and kaolin: Antacids and kaolin can reduce absorption of chloroquine; an interval of at least 4 h between intake of these agents and chloroquine should be observed.
		• Cimetidine can inhibit the metabolism of chloroquine, increasing its plasma level. Concomitant use of cimetidine should be avoided.
		• In a study of healthy volunteers, chloroquine significantly reduced the bioavailability of ampicillin. An interval of at least two hours between intake of this agent and chloroquine should be observed.
		• After introduction of chloroquine (oral form), a sudden increase in serum cyclosporine level has been reported. Therefore, close monitoring of serum cyclosporine level is recommended and, if necessary, chloroquine should be discontinued.
61	Chlorothiazide	• Increased risk of postural hypotension. Thiazides may cause slightly more alkaline urinary pH; may decrease urinary excretion of some amines (e.g., amphetamine) with concurrent use. Additive/potentiated potassium loss.
		• Postulated that may antagonize oral anticoagulant effects.
		• Thiazide hyperglycemic effect may exacerbate diabetes mellitus, increase antidiabetic agent requirements, and/or cause temporary loss of diabetic control or secondary failure to antidiabetic agent.

S. No.	Name of the Drug	Reported Drug Interaction
		• May potentiate diazoxide hyperglycemic, hypotensive, and hyperuricemic effects.
		• Addition of thiazide to stabilized regimen with potent hypotensive agent (e.g., guanethidine sulfate, methyldopa, ganglionic blocking agent) may cause severe postural hypotension.
		• May exacerbate diabetes mellitus, increase insulin requirements, cause temporary loss of diabetic control, or secondary failure to insulin[b] Thiazides (sometimes used with lithium to reduce lithium-induced polyuria), Reduced renal lithium clearance within several days.
		• Can increase serum lithium concentrations and the risk of lithium intoxication. Urinary alkalinization may decrease the effectiveness of methenamine compounds which require a urinary pH of ≤ 5.5 for optimal activity.
		• Increased risk of NSAIA-induced renal failure secondary to prostaglandin inhibition and decreased renal blood flow.
		• Also blocks renal tubular secretion of thiazide, but effect on thiazide duration of action apparently not studied.
		• Apparently enhances excretion of calcium, magnesium, and citrate during thiazide therapy, but urinary calcium concentrations remain below normal.
		• Sodium, potassium, ammonia, chloride, bicarbonate, phosphate, and titratable acid excretion apparently not affected by concomitant probenecid and thiazide.
		• Decreased values by interfering in vitro with the absorbance in the modified Glenn-Nelson technique for urinary 17-hydroxycorticosteroids; may also decrease urinary cortisol.
62	Chlorpheniramine	• MAO inhibitors prolong and intensify anticholinergic effects of antihistamines.
63	Chlorpromazine	• Potential additive CNS effects; concomitant use with alcohol potentiates hypotension observed with chlorpromazine. Potential decreased effect of oral anticoagulants. Chlorpromazine may lower seizure threshold; CNS depressant effects do not potentiate anticonvulsant activity of anticonvulsants.

TABLE 12.2 (Continued)

S. No.	Name of the Drug	Reported Drug Interaction
		• Chlorpromazine may interfere with phenytoin metabolism and precipitate phenytointoxicity. Phenobarbital may decrease plasma chlorpromazine concentrations.
		• Possible potentiated anticholinergic effects. Possible disruption of body temperature regulation. Possible additive effects or potentiated action of other CNS depressants.
		• An acute encephalopathic syndrome reported occasionally, especially when high serum lithium concentrations presentPossible increased plasma concentrations of chlorpromazine and propranolol.
		• Potential false-positive test results may occur during phenothiazine use.
		• False-positive results reported in some patients receiving phenothiazines; less likely to occur when serum test is used.
		• Potential for increased orthostatic hypotension.
64	Chlorpropamide	• Disulfiram-like reactions reported. Moderate-to-large amounts of alcohol may increase the risk of hypoglycemia.
		• Increased plasma concentrations of sulfonylureas and hypoglycemic effect. Possible potentiation of hypoglycemic effects. May prolong action of barbiturates. Possible potentiation of hypoglycemic effects.
		• Signs of hypoglycemia may be masked by β-adrenergic blocking agents. Possible potentiation of hypoglycemic effects. Potential for decreased hypoglycemic effect.
		• Possible displacement of chlorpropamide from plasma proteins and potentiation of hypoglycemic effects.

S. No.	Name of the Drug	Reported Drug Interaction
65	Chlorthalidone	• Increased risk of postural hypotension.
		• Thiazides may cause slightly more alkaline urinary pH; may decrease urinary excretion of some amines (e.g., amphetamine) with concurrent use.
		• Additive/potentiated potassium loss.
		• Postulated that may antagonize oral anticoagulant effects.
		• Thiazide hyperglycemic effect may exacerbate diabetes mellitus, increase antidiabetic agent requirements, and/or cause temporary loss of diabetic control or secondary failure to antidiabetic agent.
		• Increased risk of postural hypotension with thiazides.
66	Cholestyramine	• Cholestyramine for Oral Suspension USP may delay or reduce the absorption of concomitant oral medication such as phenylbutazone, warfarin, thiazide diuretics (acidic), or propranolol (basic), as well as tetracycline, penicillin G, phenobarbital, thyroid and thyroxine preparations, estrogens and progestins, and digitalis. Interference with the absorption of oral phosphate supplements has been observed with another positively charged bile acid sequestrant. Cholestyramine may interfere with the pharmacokinetics of drugs that undergo entero-hepatic circulation. The discontinuance of Cholestyramine could pose a hazard to health if a potentially toxic drug such as digitalis has been titrated to a maintenance level while the patient was taking Cholestyramine.
		• Because Cholestyramine binds bile acids, Cholestyramine may interfere with normal fat digestion and absorption and thus may prevent absorption of fat-soluble vitamins such as A, D, E and K. When Cholestyramine is given for long periods of time, concomitant supplementation with water-miscible (or parenteral) forms of fat-soluble vitamins should be considered.

TABLE 12.2 (Continued)

S. No.	Name of the Drug	Reported Drug Interaction
67	Cimetidine	• *Tagamet (cimetidine)*, apparently through an effect on certain microsomalenzymesystems, has been reported to reduce the hepaticmetabolismof warfarin-type anticoagulants, phenytoin, propranolol, nifedipine, chlordiazepoxide, diazepam, certain tricyclic antidepressants, lidocaine, theophylline and metronidazole, thereby delaying elimination and increasing bloodlevels of these drugs. • Clinically significant effects have been reported with thewarfarin anticoagulants; therefore, close monitoring ofprothrombin timeis recommended, and adjustment of theanticoagulantdose may be necessary when*Tagamet (cimetidine)* is administered concomitantly. Interaction with phenytoin, lidocaine and theophylline has also been reported to produce adverse clinical effects.
68	Cinnarizine	• CNS depressant effect enhanced with alcohol. Action potentiated by domperidone.
69	Ciprofloxacin	• In a pharmacokinetic study, systemic exposure of tizanidine (4 mg single dose) was significantly increased (Cmax 7-fold, AUC 10-fold) when the drug was given concomitantly with ciprofloxacin (500 mg BID for 3 days). Thehypotensiveandsedativeeffects of tizanidine were also potentiated. Concomitant administration of tizanidine and ciprofloxacin is contraindicated. • As with some other quinolones, concurrent administration of ciprofloxacin with theophylline may lead to elevated serum concentrations of theophylline and prolongation of its elimination half-life. This may result in increased risk of theophylline-related adverse reactions. If concomitant use cannot be avoided, serum levels of theophylline should be monitored and dosage adjustments made as appropriate. • Some quinolones, including ciprofloxacin, have also been shown to interfere with themetabolismofcaffeine. This may lead to reduced clearance of caffeine and a prolongation of its serum half-life. On concurrent administration of ciprofloxacin and caffeine or pentoxifylline containing products, elevated serum concentrations of thesexanthinederivatives were reported.

S. No.	Name of the Drug	Reported Drug Interaction
70	Cisapride	• Cisapride is metabolized mainly via the cytochrome P450 3A4 enzyme. In some cases where seriousventricular arrhythmias, QT prolongation, and torsades de pointes have occurred when cisapride was taken in conjunction with one of the cytochrome P450 3A4 inhibitors, elevated blood cisapride levels were noted at the time of the QT prolongation.
		• *In vitro* and/or in vivo data show that clarithromycin, erythromycin, and troleandomycin markedly inhibit themetabolismof cisapride, which can result in an increase in plasma cisapride levels and prolongation of the QT interval on theECG.
		• Concurrent administration of certain anticholinergic compounds, such as belladonna alkaloids and dicyclomine, would be expected to compromise the beneficial effects of cisapride.
71	Cisplatin	• Increased risk of nephrotoxicity. Increased risk of ototoxicity Increased risk of nephrotoxicity.
		• Possible decreased phenytoin concentrations (due to decreased absorption and/or increased metabolism).
		• Possible altered renal elimination of bleomycin Increased risk of ototoxicity. Possible decreased etoposide elimination.
		• Possible altered renal elimination of methotrexate.
		• Response duration adversely affected when used concomitantly with cisplatin and altretamine (hexamethylmelamine).
72	Clarithromycin	• Decreased clarithromycin AUC and peak plasma concentrations; increased 14-hydroxyclarithromycin AUC and peak plasma concentrations;no effect on AUC of efavirenz; rash reported with concomitant administration.
		• Potential decreased clearance of midazolam or triazolam and increased pharmacologic effects of the benzodiazepine.
		• Somnolence and confusion reported with clarithromycin and triazolam.
		• Increased atazanavir plasma concentrations; increased clarithromycin plasma concentrations and decreased 14-hydroxyclarithromycin plasma concentrations; increased clarithromycin concentrations may cause QTc prolongation.

TABLE 12.2 (Continued)

S. No.	Name of the Drug	Reported Drug Interaction
73	**Clofibrate**	• Potential inhibition of rifabutin metabolism and induction of clarithromycin metabolism.
		• Clofibrate is known to interact with several classes of drugs like warfarin, probenecid, chlorpropamide, rifampin, furosemide, rifapentine, tolazamide, ciprofibrate, and all HMG CoA inhibitor class.
74	Clonazepam	• Possible increase in plasma clonazepam concentrations.
		• Decreased plasma clonazepam concentrations; carbamazepine pharmacokinetics not affected. Additive CNS effect.
		• Decreased plasma clonazepam concentrations; phenobarbital pharmacokinetics not affected.
		• Possible decrease in plasma clonazepam concentrations.
75	Clonidine	• Epidural clonidine may prolong the duration of the pharmacologic effects, including both sensory and motor blockade of epidural local anesthetics. May inhibit the hypotensive effect of clonidine.
		• The increase in BP usually occurs during the second week of tricyclic antidepressant therapy, but occasionally may occur during the first several days of concomitant therapy.
		• Clonidine withdrawal may result in an excess of circulating catecholamines; therefore, caution should be exercised in concomitant use of drugs that affect the tissue uptake of these amines. Additive/potentiated hypotensive effect. Possible additive bradycardia AV block when clonidine is used with drugs that affect sinus nodal function or AV nodal conduction.
		• β-Adrenergic blocking agents may exacerbate rebound hypertension that may occur following discontinuance of clonidine therapy. Possible additive bradycardia, AV block when clonidine is used with drugs that affect sinus nodal function or AV nodal conduction.
76	Clotrimazole	• Clotrimazole top will decrease the level or effect of everolimus oral by affects how the drug is eliminated from the body.

S. No.	Name of the Drug	Reported Drug Interaction
77	Clozapine	• Risk of concomitant use with other drugs not systematically evaluated, but clinical experience and/or theoretical considerations indicate certain potential drug interactions exist.
		• Metabolized by many CYP isoenzymes, particularly CYP1A2, CYP2D6, and CYP3A4. 320 May inhibit CYP2D6; may make normal CYP2D6 metabolizers resemble "poor metabolizers" with regard to concomitant therapy with other drugs metabolized by CYP2D6.
		• Inhibitors or inducers of CYP1A2, CYP2D6, or CYP3A4: potential pharmacokinetic interaction (altered clozapine metabolism). Risk of metabolic interaction caused by an effect of an individual isoform minimized, but use concomitantly with caution; dosage adjustment of clozapine and/or other drug may be necessary.
		• Substrates of CYP2D6: potential pharmacokinetic interaction (decreased metabolism of substrate); reduced dosage of either clozapine or substrate may be required.
78	Codeine	• Possible paralytic ileus.
		• Potentiation of antidepressant effect.
		• Additive CNS effects.
79	Cotrimoxazole	• Toxic delirium reported in an individual who received amantadine and cotrimoxazole concomitantly.
		• Possible decreased efficacy of the tricyclic antidepressant.
		• Reversible nephrotoxicity reported in renal transplant recipients receiving cyclosporine and cotrimoxazole concomitantly.
		• Possible increased digoxin concentrations, especially in geriatric patients.
		• Possible increased incidence of thrombocytopenia and purpura if certain diuretics (principally thiazides) are used concomitantly, especially in geriatric patients.
		• Possible increased sulfamethoxazole concentrations.

TABLE 12.2 (Continued)

S. No.	Name of the Drug	Reported Drug Interaction
		• Co-trimoxazole can displace methotrexate from plasma protein-binding sites resulting in increased free methotrexate concentrations.
		• Possible interference with serum methotrexate assays if competitive protein binding technique is used with a bacterial dihydrofolate reductase as the binding protein; interference does not occur if methotrexate is measured using radioimmunoassay.
80	Colchicine	• Increased plasma concentrations of colchicines.
		• Decreased metabolism and increased plasma concentrations of colchicine; fatal colchicine toxicity reported.
		• Possible additive nephrotoxic effects; increased concentrations of cyclosporine in biological fluid.
		• Increased colchicine concentrations; fatal colchicine toxicity reported.
		• Rhabdomyolysis reported.
		• Increased plasma concentrations of colchicine; neuromuscular toxicity reported.
		• Oral contraceptives: No change in plasma concentrations of ethinyl estradiol or norethindrone.
		• Addition of a fibrate to long-term colchicine therapy or addition of colchicine to long-term fibrate therapy has resulted in myopathy and rhabdomyolysis.
		• Minimal change in plasma colchicine concentration reported, though increased colchicine concentrations reported with other moderate CYP3A4 inhibitors; increased colchicine concentrations possible.
		• Addition of a statinto long-term colchicine therapy or addition of colchicine to long-term statin therapy has resulted in myopathy and rhabdomyolysis.
		• Increased plasma concentrations of colchicine; neuromuscular toxicity reported.

S. No.	Name of the Drug	Reported Drug Interaction
81	Cyclophosphamide	• Concomitant administration may increase the incidence of bone marrow depression.
		• Possible potentiation of cardiotoxic effects.
		• May inhibit microsomal enzyme activity and decrease cyclophosphamide decrease cyclophosphamide metabolism.
		• Discontinuance or reduction in steroid dosage may increase cyclophosphamide toxicity.
		• Interacts chemically with urotoxic cyclophosphamide metabolites (and/or their precursors) to prevent or decrease incidence and severity of bladder toxicity (e.g., hemorrhagic cystitis).
		• Cyclophosphamide may reduce serum pseudocholinesterase concentrations and prolong neuromuscular blocking activity of succinylcholine (especially in very ill patients receiving large IV cyclophosphamide doses).
82	Cycloserine	• Possible increased risk of seizures in chronic alcoholics and other frequent users of alcohol, especially those receiving high cycloserine dosage.
		• Possible increased risk of adverse effects; seizures reported.
		• Adverse CNS effects (e.g., dizziness, drowsiness) may be additive.
		• Cycloserine inhibits hepatic metabolism of phenytoin.
83	Cyclosporine	• Possible hyperkalemia.
		• Increased plasma or blood concentrations of cyclosporine.
		• Decreased plasma or blood concentrations of cyclosporine.
		• Possible additive nephrotoxic effects.
		• Possible increased risk of acute tubular necrosis in renal allograft recipients.
		• Diltiazem, nicardipine, verapamil: Increased plasma or blood concentrations of cyclosporine.
		• Nifedipine: Frequent gingival hyperplasia Possible additive nephrotoxic effects; increased plasma or blood concentrations of cyclosporine.

TABLE 12.2 (Continued)

S. No.	Name of the Drug	Reported Drug Interaction
		• Possible decreased clearance of colchicine increasing the potential for enhanced colchicine toxicity (myopathy, neuropathy). Increased plasma or blood concentrations of cyclosporine; possible decreased clearance of prednisolone; seizures reported with combined cyclosporine and high-dose corticosteroid therapy.
		• Increased blood sirolimus concentrations with concomitant administration.
		• Possible decreased immune response to vaccination.
84	Cyproheptidine	• Possible additive CNS depression.
		• Reversal of fluoxetine's antidepressant effects reported in limited number of patients, possibly due to inhibition of fluoxetine's serotonergic effects.
		• MAO inhibitors prolong and intensify anticholinergic effects of antihistamines.
		• Inhalation-challenge testing with histamine or antigen: Possible suppression of test response.
		• Antigen skin testing: Possible suppression of wheal and flare reactions.
85	Cytarabine	• Possible enhanced neurotoxicity when intrathecal conventional cytarabine used concomitantly with other intrathecal cytotoxic agents.
		• GI absorption of oral digoxin tablets may be substantially reduced when used concomitantly with conventional cytarabine.
		• Possible inhibition of antiinfective activity by competitive inhibition of uptake by fungi when flucytosine was used concomitantly with conventional cytarabine.
		• *In vitro* evidence of inhibition of antibacterial activity against *Klebsiella pneumoniae* with conventional cytarabine.
		• Liposomal cytarabine vesicles are similar in size and appearance to WBC.
86	Dacarbazine	• Possible increased dacarbazine metabolism.

S. No.	Name of the Drug	Reported Drug Interaction
87	Dapsone	• Dapsone may interfere with some antiinflammatory effects of clofazimine in patients with erythema nodosumleprosum (ENL) reactions.
		• Possible increased dapsone concentrations. Studies using buffered didanosine indicate no clinically important effect on pharmacokinetics of a single dose of dapsone.
		• Possible decreased GI absorption of dapsone and decreased dapsone efficacy for PCP prophylaxis (greater relapse rate) reported in some HIV-infected patients receiving didanosine.
		• Additive adverse hematologic effects; increased risk of agranulocytosis.
		• Rifabutin: Decreased dapsone AUC.[j]
		• Rifampin or rifapentine: Potential increased metabolism and decreased plasma concentrations of dapsone. Possible increased plasma dapsone concentrations.
		• Possible improved efficacy for treatment of PCP compared with dapsone alone; possible increased risk of adverse effects (e.g., methemoglobinemia, rash, abnormal liver function tests).
		• Possible increased plasma trimethoprim concentrations.
88	Dantrolene	• Cardiovascular collapse reported rarely.
		• Decreased binding of dantrolene to plasma proteins.
		• Possible additive CNS effects (e.g., dizziness). Additive sedative effect; dantrolene metabolism and protein binding unaffected.
		• Possible increased frequency of hepatotoxicity in women >35 years of age.
		• Pharmacokinetic interaction unlikely; dantrolene metabolism unaffected.
		• No change in binding of dantrolene to plasma proteins.
		• Increased binding of dantrolene to plasma proteins.
		• Potentiation of vecuronium-induced neuromuscular blockade.
		• Decreased binding of dantrolene to plasma proteins.

TABLE 12.2 (Continued)

S. No.	Name of the Drug	Reported Drug Interaction
89	Desmopressin	• Possible decreased antidiuretic response to desmopressin.
		• Concomitant therapy has been used without adverse effects.
		• Hyponatremic seizures reported rarely in patients receiving desmopressin and imipramine during postmarketing surveillance.
		• Possible increased risk of water intoxication with hyponatremia.
		• Prior administration of carbamazepine decreased duration of action of desmopressin.
		• Possible increased risk of water intoxication with hyponatremia.
		• Potentiation and prolongation of antidiuretic effect of desmopressin.
		• Large doses of epinephrine may decrease the antidiuretic response to desmopressin.
		• Hyponatremic seizures reported rarely in patients receiving desmopressin and oxybutynin during postmarketing surveillance.
90	Dexamethasone	• Conflicting reports of alterations in the anticoagulant response.
		• Increased blood glucose concentrations in diabetes mellitus.
		• Decreased blood concentrations of dexamethasone.
		• Enhance the potassium-wasting effects of glucocorticoids.
		• May interfere with dexamethasone suppression tests.
		• Decreased plasma concentrations of indinavir.
		• False-negative results in the dexamethasone suppression test.
		• Inhibits adrenal corticosteroid synthesis, causing adrenal insufficiency during corticosteroid withdrawal.
		• Increased plasma dexamethasone concentrations. Increases the risk of GI ulceration.

S. No.	Name of the Drug	Reported Drug Interaction
		• Decreased serum salicylate concentrations.
		• When corticosteroids are discontinued, serum salicylate concentration may increase, possibly resulting in salicylate intoxication.
		• Decreased blood concentrations of dexamethasone.
		• Conflicting reports of increased and decreased blood phenytoin concentrations leading to alterations in seizure control.
91	Diazepam	• Possible increased plasma amitriptyline concentrations.
		• Possible decreased rate of diazepam absorption.
		• Possible decreased sedative effect.
		• Increased plasma diazepam concentrations Possible additive CNS effect.
		• Possible decreased plasma diazepam concentrations. Possible decreased renal excretion and increased plasma concentrations of digoxin.
		• Potential for increased plasma diazepam concentrations.
		• Decreased clearance of diazepam Possible increased plasma diazepam concentrations.
		• Possible decreased GI absorption of diazepam. Possible decreased plasma diazepam concentrations.
		• Possible false positive reactions for glucose with Clinistix and Diastix.
		• Possible increased plasma diazepam concentrations.

TABLE 12.2 (Continued)

S. No.	Name of the Drug	Reported Drug Interaction
92	Diazoxide	• May potentiate hypotensive effect.
		• May increase risk of hyperglycemia.
		• Potentiation of the hyperglycemic, hyperuricemic, or hypotensive effects of diazoxide.
		• May stimulate metabolism (e.g., decrease serum concentration and half-life) of diazoxide. With concomitant phenytoin, conflicting reports of phenytointoxicity or decreased serum phenytoin concentrations. May potentiate hypotensive effect.
		• Inhibits glucagon-stimulated insulin release and will cause a false-negative insulin response to glucagon.
		• Displaces warfarin from its protein binding sites in vitro.
93	Diclofenac	• Reduced BP response to ACE inhibitor.
		• Reduced BP response to angiotensin II receptor antagonist.
		• Possible deterioration of renal function in individuals with renal impairment.
94	Dicyclomine	• Antimuscarinics may antagonize the effects of drugs that alter GI motility.
		• Decreased GI absorption of dicyclomine.
		• Dicyclomine may antagonize the effects of antiglaucoma agents.
		• Increased anticholinergic effects.
		• Increased IOP.
		• Increased serum digoxin.
		• Inhibitory effects of anticholinergic drugs on gastric hydrochloric acid secretion are antagonized by agents used to treat achlorhydria.
		• May increase certain actions or adverse effects of dicyclomine.

S. No.	Name of the Drug	Reported Drug Interaction
95	Didanosine	• Aluminum- and magnesium-containing antacids increase oral bioavailability of didanosine.
		• Buffered didanosine: Decreased atazanavir concentrations and AUC; decreased didanosine concentrations and AUC.
		• Didanosine delayed-release capsules: Decreased didanosine concentrations and AUC if given with atazanavir and food; no change in atazanavir concentrations.
		• Itraconazole: Decreased itraconazole concentrations with buffered didanosine.
		• Ketoconazole: Decreased ketoconazole peak plasma concentrations and AUC with buffered didanosine; no changes in ketoconazole concentrations with didanosine delayed-release capsules.
		• No in vitro evidence of antagonistic antiretroviral effects.
		• Possible increased antacid adverse effects if additional antacids are used in patients receiving didanosine pediatric oral solution admixed with antacid.
		• Possible increased risk of didanosine toxicity (Seth, 2004).
		• Rifabutin: Slightly increased didanosine concentrations no clinically important pharmacokinetic interactions.
		• Substantially increased didanosine concentrations and AUC.
96	Digoxin	• Increased serum digoxin concentrations and toxicity.
		• Magnitude of the increase may be much greater in children.
		• Reduce GI absorption of digoxin (resulting in low plasma digoxin concentrations), especially when administered at the same time as digoxin.
		• Quinidine, procainamide, disopyramide, phenytoin, propranolol, and lidocaine may have negative inotropic effects with larger than usual doses, especially in patients with cardiac glycoside toxicity (propranolol has negative inotropic effects with usual doses).

TABLE 12.2 (Continued)

S. No.	Name of the Drug	Reported Drug Interaction
97	Diltiazem	• Concomitant use of digoxin and β-adrenergic blocking agents can have additive negative effects on AV conduction, which can result in complete heart block.
		• Reduce GI absorption of digoxin (resulting in low plasma digoxin concentrations), especially when administered at the same time as digoxin.
		• Increased plasma concentration and AUC of buspirone.
		• Increased plasma concentrations of propranolol and metoprolol.
		• Possible increased benzodiazepine plasma concentrations and AUC resulting in increased adverse effects (e.g., prolonged sedation, respiratory depression).
		• Possible increased blood cyclosporine concentration and consequent nephrotoxicity.
		• Possible increased depression of cardiac contractility, conductivity, and automaticity as well as vascular dilation.
		• Possible increased plasma diltiazem concentrations and additive effect on PR interval prolongation.
		• Possible increased serum or plasma carbamazepine concentrations and associated neurologic and sensory manifestations of carbamazepine toxicity.
		• Potential for additive negative effects on myocardial contractility, heart rate, and prolonging AV conduction.
		• Potential for increased effects and toxicity of buspirone.
98	Dimenhydrinate	• Anticholinergic effects may be potentiated.
		• May enhance the effects of other CNS depressants, including alcohol.
		• May mask the early manifestations of ototoxicity.
99	Diphenhydramine	• MAO inhibitors prolong and intensify anticholinergic effects of antihistamines.

S. No.	Name of the Drug	Reported Drug Interaction
100	Dipyridamole	• Antagonizes anticholinesterase effects.
		• Increased plasma adenosine concentration.
		• Inhibits dipyridamole vasodilatory and bronchospastic effects.
		• Possible increased risk of bleeding complications.
		• Possible increased risk of bleeding, particularly during or after surgery; however, concomitant use does not appear to increase frequency or severity of bleeding compared with use of warfarin alone.
		• Potentiation of adenosine vasoactive effects.
101	Disulfiram	• Acute psychoses and confusion reported.
		• Decreased total blood clearance and increased half-life of caffeine, possibly resulting from inhibition of hepatic metabolism of caffeine.
		• Enhanced alcohol-disulfiram reaction.
		• Increased sensitivity to alcohol, resulting in disulfiram-alcohol reaction.
		• Inhibition of hepatic metabolism of phenytoin, possibly resulting in phenytoin intoxication.
		• Interference with hepatic metabolism of barbiturate, possibly resulting in increased blood concentrations and toxicity of barbiturate.
		• Possible behavioral changes, marked changes in mental status, psychotic reactions, incoordination, or unsteady gait.
		• Possible increased plasma concentrations of anticoagulant, resulting in prolonged PT.
102	Dobutamine	• Animal studiesindicatethat dobutamine may be ineffective if the patient has recently received a β-blocking drug. In such a case, theperipheralvascular resistancemay increase.
		• Preliminary studies indicate that the concomitant use of dobutamine and nitroprusside results in a highercardiac outputand, usually, a lower pulmonarywedge pressure than when either drug is used alone.

TABLE 12.2 (Continued)

S. No.	Name of the Drug	Reported Drug Interaction
103	Domperidone	• There was no evidence of drug interactions in clinical studies in which dobutamine was administered concurrently with other drugs, including digitalis preparations, furosemide, spironolactone, lidocaine, glyceryl trinitrate, isosorbidedinitrate, morphine, atropine, heparin, protamine, potassiumchloride, folic acid, andacetaminophen.
		• *In vivo* interaction studies have shown that ketoconazole strongly inhibits the CYP3A4-dependent metabolism of domperidone.
		• Pharmacokinetic studies showed 3–10 fold increase in the area under ratio-Domperidone Page 7 of 25 curve (AUC) and the peak concentration (Cmax) of domperidone when ketoconazole was coadministered.
		• This coadministration resulted also in a prolongation of the QT interval (maximum of 10–20 msec) which was greater than the prolongation observed with ketoconazole alone.
		• QT prolongation was not observed at oral doses of domperidone of up to 160 mg/day, i.e., twice the maximum recommended daily therapeutic dose cardiac arrhythmia and death were reported following high parenteral doses of domperidone.
		• Results of the interaction study should be considered when domperidone is prescribed with CYP3A4 inhibitors (which may increase plasma levels of domperidone) or with drugs that can cause QT prolongation or torsade de pointes, especially in patients at risk for torsade de pointes (see Contraindications, Warnings and Precautions, Cardiovascular sections).
104	Dopamine	• Adverse cardiovascular effects may be potentiated by tricyclic antidepressants.
		• Can suppress the dopaminergic renal and mesenteric vasodilation induced with low-dose dopamine.
		• Cardiac effects of dopamine are antagonized.
		• Concomitant use may result in severe hypertension.

S. No.	Name of the Drug	Reported Drug Interaction
		• Diuretic effects (on urine flow) of low dopamine dosages may be additive with or potentiated by diuretics.
		• Inhibits dopamine metabolism; dopamine effects are prolonged and intensified by MAO inhibiters.
		• May increase cardiac irritability, resulting in ventricular arrhythmias and hypertension with usual dopamine doses during halogenated hydrocarbon (e.g., halothane) or cyclopropane anesthesia.
		• Peripheral vasoconstriction of high dopamine doses is antagonized.
		• Suppresses pituitary secretion.
105	Dothiepin	• Barbiturates reduce antidepressant effect of dosulepin. Raised serum concentrations with methylphenidate. Cigarette smoking enhances metabolism.
		• Increased risk of ventricular arrhythmias when used with drugs that prolong QT interval. Potentially Fatal: Should not be given concurrently or within 14 days of stopping MAOIs. Potentiates catecholamines, sympathomimetics, narcotics and alcohol.
		• Reduces hypotensive activity of bethanidine, guanethidine. See below for more dosulepin drug interactions.
106	Doxazosin	• Increased AUC (10%) of doxazosin (Seth, 2004).
		• Adverse effects with concomitant use generally reflect combined toxicity profile of each drug alone.
107	Doxepin	• May potentiate sedative effects.
		• Possible increased plasma doxepin concentrations.
		• Potentially life-threatening serotonin syndrome.
		• Severe hypoglycemia reported in a diabetic patient receiving oral doxepin.

TABLE 12.2 (Continued)

S. No.	Name of the Drug	Reported Drug Interaction
108	Doxorubicin	• Additive toxicity may occur especially with regard to bone marrow/hematologic and gastrointestinal effects (*see* Warnings and Precautions). The use of doxorubicin in combination chemotherapy with other potentially cardiotoxic drugs, as well as the concomitant use of other cardioactive compounds (e.g., calcium channel blockers), requires monitoring of cardiac function throughout treatment. Changes in hepatic function induced by concomitant therapies may affect doxorubicin metabolism, pharmacokinetics, therapeutic efficacy and/or toxicity.
		• Literature reports have also described the following drug interactions: Paclitaxel can cause increased plasma-concentrations of doxorubicin and/or its metabolites when given prior to doxorubicin.
		• Phenytoin levels may be decreased by doxorubicin.
		• Streptozocin may inhibit hepatic metabolism of doxorubicin. Exacerbation of cyclophosphamide induced hemorrhagic cystitis. Enhancement of the hepatotoxicity of 6-mercaptopurine.
109	Doxycycline	• Additive adverse CNS effect of pseudotumorcerebri (benign intracranial hypertension).
		• Decreased absorption of doxycycline.
		• Decreased effectiveness of oral contraceptive.
		• Decreased efficacy of penicillins.
		• Decreased plasma prothrombin activity.
		• Fatal renal toxicity.
		• Possible decreased doxycycline half-life.
		• Possible false elevation secondary to interference with fluorescence test.

S. No.	Name of the Drug	Reported Drug Interaction
110	Enalapril	• Enalapril may increase potassium levels (hyperkalemia) in blood. Therefore, there is an increased risk of hyperkalemia when enalapril is given with potassium supplements or drugs that increase potassium levels (e.g., spironolactone [Aldactone]).
		• There have been reports of increased lithium (Eskalith, Lithobid) levels when lithium is used in combination with ACE inhibitors. The reason for this interaction is not known, but the increased levels may lead to toxicity from lithium. There have been reports that aspirin and other nonsteroidal antiinflammatoy drugs (NSAIDs) such as ibuprofen (Advil, Children's Advil/Motrin, Medipren, Motrin, Nuprin, PediaCare Fever, etc.), indomethacin (Indocin, Indocin-SR), and naproxen (Anaprox, Naprelan, Naprosyn, Aleve) may reduce the effects of ACE inhibitors.
		• Nitritoid reactions (symptoms include facial flushing, nausea, vomiting and low blood pressure) may occur when injectable gold (sodium aurothiomalate [Myochrysine]), used in the treatment of rheumatoid arthritis, is combined with ACE inhibitors, including enalapril.
111	Ergometrine	• Increase the effect of oxytocin and vice versa. Syntometrine may enhance the blood pressure raising effect of vasoconstrictors (medicines given to constrict the blood vessels).
		• Some inhaled anesthetics used for general anesthesia, such as cyclopropane and halothane, may reduce the effect of oxytocin and ergometrine. There may also be an increased risk of a drop in blood pressure and abnormal heart beats if oxytocin is given with these general anesthetics.
112	Ergotamine	• Inhibition of ergotamine metabolism; increased risk of potentially fatal cerebral ischemia and/or ischemia of the extremities.
		• Increased plasma ergotamine concentrations.
		• Inhibition of ergotamine metabolism.
		• increased risk of potentially fatal cerebral ischemia and/or ischemia of the extremities.
		• Potential for excessive vasoconstriction.

TABLE 12.2 (Continued)

S. No.	Name of the Drug	Reported Drug Interaction
113	Erythromycin	• Diltiazem and Verapamil: Possible increased erythromycin concentrations and possible increased risk of sudden death from cardiac causes.
		• Increased carbamazepine concentrations and risk of carbamazepine toxicity.
		• Increased concentrations of the antiarrhythmic agent and increased risk of serious adverse cardiovascular effects.
		• Increased plasma concentrations of benzodiazepines; possible prolonged sedative and hypnotic effects of the drugs.
		• Nefedipine: No evidence of increased risk of sudden death from cardiac causes.
		• Pharmacokinetic interaction and potential for serious or life-threatening reactions (e.g., cardiac arrhythmias) with astemizole or terfenadine (drugs no longer commercially available in the US).
		• Possible increased erythromycin concentrations and increased risk of sudden death from cardiac causes.
		• Possible prolonged PT.
114	Erythropoietin	• Aluminum-containing antacids: Decreased ethambutol serum concentrations and urinary excretion; possible decreased oral absorption of the antimycobacterial.
		• Pharmacokinetic interaction are most unlikely to occur.
115	Ethambutol	• The effects of some drugs can change if you take other drugs or herbal products at the same time. This can increase your risk for serious side effects or may cause your medications not to work correctly. These drug interactions are possible, but do not always occur. Your doctor or pharmacist can often prevent or manage interactions by changing how you use your medications or by close monitoring.

S. No.	Name of the Drug	Reported Drug Interaction
116	Ethinyloestradiol	• Drugs that may interact with BREVICON include: Anticonvulsants (carbamazepine, ethosuximide, felbamate, oxcarbazepine, phenobarbital, phenytoin, primidone, topiramate, lamotrigine) Antibiotics (ampicillin, cotrimoxazole, penicillin, rifampin, chloramphenicol, metronidazole, neomycin, nitrofurantoin, sulfonamides, tetracyclines, troleandomycin, rifabutin) Antifungals (griseofulvin, fluconazole).
		• Cholesterol Lowering Agents (clofibrate, atorvastatin), Sedatives and Hypnotics (benzodiazepines, barbiturates, chloral hydrate, glutethimide, meprobamate, chlordiazepoxide, lorazepam, oxazepam, diazepam), Antacids Alpha-II Adrenoreceptor Agents (clonidine), Antidiabetic Drugs (oral hypoglycemics and insulin), Antihypertensive Agents (guanethidine, methyldopa and beta blockers), Antipyretics (acetaminophen, antipyrine, ASA), Betamimetic Agents (isoproterenol).
117	Ethosuximide	• The effects of some drugs can change if you take other drugs or herbal products at the same time.
		• This can increase your risk for serious side effects or may cause your medications not to work correctly. These drug interactions are possible, but do not always occur. Your doctor or pharmacist can often prevent or manage interactions by changing how you use your medications or by close monitoring.
		• To help your doctor and pharmacist give you the best care, be sure to tell your doctor and pharmacist about all the products you use (including prescription drugs, nonprescription drugs, and herbal products) before starting treatment with this product.
		• While using this product, do not start, stop, or change the dosage of any other medicines you are using without your doctor's approval.
		• Tell your doctor or pharmacist if you are taking other products that cause drowsiness, including alcohol, antihistamines (such ascetirizine, diphenhydramine), drugs for sleep or anxiety (such asalprazolam, diazepam, zolpidem), muscle relaxants, and narcotic pain relievers (such ascodeine). Check the labels on all your medicines (such asallergyorcough-and-cold products) because they may contain ingredients that cause drowsiness.

TABLE 12.2 (Continued)

S. No.	Name of the Drug	Reported Drug Interaction
		• Ask your pharmacist about using those products safely.
		• This document does not contain all possible interactions.
		• Keep a list of all the products you use. Share this list with your doctor and pharmacist to lessen your risk for serious medication problems.
118	Famotidine	• Does not appear to inhibit hepatic metabolism of drugs by hepatic CYP isoenzymes.
		• Antacids appear to cause slight but clinically unimportant decrease in bioavailability.
		• May concomitantly administer with antacids.
119	Fentanyl	• Additive Effects of other CNS depressants: The concomitant use of FENTORA (fentanyl buccal/sublingual effervescent tablets) with other CNS depressants, including other opioids, sedatives or hypnotics, general anesthetics, phenothiazines, tranquilizers, skeletal muscle relaxants, sedating antihistamines, and alcoholic beverages may produce increased depressant effects (e.g., respiratory depression, hypotension, and profound sedation).
		• Patients on concomitant CNS depressants must be monitored for a change in opioid effects.
		• Consideration should be given to adjusting the dose of FENTORA if warranted. CYP3A4 Inhibitors Fentanyl is metabolized mainly via the human CYP3A4 isoenzyme system therefore potential interactions may occur when FENTORA is given concurrently with agents that affect CYP3A4 activity.
120	Finasteride	• Finasteride does not appear to affect significantly the cytochrome P450-linked drug metabolizing enzyme system.
		• Drug-laboratory interactions in clinical studies with PROPECIA® in men 18–41 years of age, the mean value of serum prostate-specific antigen (PSA) decreased from 0.7 ng/mL at baseline to 0.5 ng/mL at month 12.
		• Finasteride does not appear to affect significantly the cytochrome P450-linked drug metabolizing enzyme system.

S. No.	Name of the Drug	Reported Drug Interaction
		• Compounds which have been tested in man have included antipyrine, digoxin, glyburide, propranolol, theophylline, and warfarin and no interactions were found.
		• However, patients on medication with narrow therapeutic indices, such as phenytoin, should be carefully monitored when treatment with PROPECIA®.
		• When finasteride is used in older men who have benign prostatic hyperplasia (BPH), PSA levels are decreased by approximately 50%. Until further information is gathered in men >41 years of age without BPH, consideration should be given to doubling the PSA level in men undergoing this test while taking PROPECIA®.
121	Flecainide	• Additive effects of other CNS depressants. The concomitant use of FENTORA (fentanyl buccal/ sublingual effervescent tablets) with other CNS depressants, including other opioids, sedatives or hypnotics, general anesthetics, phenothiazines, tranquilizers, skeletal muscle relaxants, sedating antihistamines, and alcoholic beverages may produce increased depressant effects (e.g., respiratory depression, hypotension, and profound sedation).
		• Patients on concomitant CNS depressants must be monitored for a change in opioid effects.
		• CYP3A4 inhibitors fentanyl is metabolized mainly via the human CYP3A4 isoenzyme system; therefore potential interactions may occur when FENTORA is given concurrently with agents that affect CYP3A4 activity.
		• The concomitant use of FENTORA with CYP3A4 inhibitors (e.g., indinavir, nelfinavir, ritonavir, clarithromycin, itraconazole, ketoconazole, nefazodone, saquinavir, telithromycin, arprepitant, diltiazem, erythromycin, fluconazole, grapefruit juice, verapamil, or cimetidine) may result in a potentially dangerous increase in fentanyl plasma concentrations, which could increase or prolong adverse drug effects and may cause potentially fatal respiratory depression.
122	Fluconazole	• Fluconazole exhibit one of the highest drug interactions.
		• The drug primarily interacts with alprazolam, atorvastatin, clopiogrel, erythromycin, guaifenesin, haloperidol, ibuprofen, quinidine, saquinavir, warfarin, ziprasidone.

TABLE 12.2 (Continued)

S. No.	Name of the Drug	Reported Drug Interaction
123	Fluoxetine	• Serotonin release by platelets plays an important role in hemostasis. Altered anticoagulant effects, including increased bleeding, have been reported when SNRIs or SSRIs are coadministered with warfarin.
		• Patients receiving warfarin therapy should be carefully monitored when fluoxetine is initiated or discontinued.
		• Concomitant use in patients taking pimozide is contraindicated. Pimozide can prolong the QT interval.
		• Fluoxetine can increase the level of pimozide through inhibition of CYP2D6. Fluoxetine can also prolong the QT interval. Clinical studies of pimozide with other antidepressants demonstrate an increase in drug interaction or QT prolongation.
		• Coadministration of fluoxetine with other drugs that are metabolized by CYP2D6, including certain antidepressants (e.g., TCAs), antipsychotics (e.g., phenothiazines and most atypicals), and antiarrhythmics (e.g., propafenone, flecainide, and others) should be approached with caution.
		• If fluoxetine is added to the treatment regimen of a patient already receiving a drug metabolized by CYP2D6, the need for decreased dose of the original medication should be considered.
124	Fluphenazine	• Phenothiazines may lower seizure threshold, but CNS depressant effects do not potentiate anticonvulsant activity of anticonvulsants.
		• Possible potentiated effects of atropine in some patients receiving fluphenazine because of added anticholinergic effects.
		• Possible additive effects or potentiated action of other CNS depressants.
		• Reversal of epinephrine action.
		• An acute encephalopathic syndrome reported occasionally, especially when high serum lithium concentrations present.

S. No.	Name of the Drug	Reported Drug Interaction
125	Flutamide	• False-positive test results may occur during phenothiazine use.
		• Increased risk of facial flushing.
		• Pharmacokinetic interaction unlikely increased PT reported (Seth, 2004).
126	Framycetin	• Although certain medicines should not be used together at all, in other cases two different medicines may be used together even if an interaction might occur. In these cases, your doctor may want to change the dose, or other precautions may be necessary.
127	Frusemide	• Additive antihypertensive effect; orthostatic hypotension may occur.
		• Additive diuretic effect.
		• Additive hypokalemic effects.
		• May aggravate orthostatic hypotension.
		• Possible reduced diuretic effect.
		• Possible antagonism of hypoglycemic effect as result of hypokalemia.
		• Possible reaction characterized by diaphoresis, flushes, hypertension, and uneasiness in patients with acute MI and CHF.
		• Share similar diuretic mechanisms.
128	Furazolidine	• Disulfiram-like reaction with alcohol. Potential hypertensive crisis with sympathomimetic, tyramine-containing foods, levodopa. **Potentially Fatal:** Potentiation of MAOIs. Toxic psychosis with amitriptyline.
129	Gabapentin	• Reduced bioavailability of gabapentin (Seth, 2004).
		• Plasma concentrations of carbamazepine, phenytoin, valproic acid, phenobarbital, and diazepam in existing treatment regimens not affected by gabapentin.
		• Pharmacokinetics of gabapentin not affected by these drugs.
		• Possible dose-dependent decrease in plasma concentrations of hydrocodone; possible increase in plasma concentrations of gabapentin.

TABLE 12.2 (Continued)

S. No.	Name of the Drug	Reported Drug Interaction
130	Ganciclovir	• Increased bioavailability of gabapentin at subtherapeutic dosages of both drugs.
		• Increases in serum creatinine were observed in patients treated with CYTOVENE (ganciclovir)-IV plus either cyclosporine.
		• Amphotericin B, drugs with known potential for nephrotoxicity.
131	Gemifibrozil	• **Anticoagulants:** caution should be exercised when warfarin is given in conjunction with lopid. The dosage of warfarin should be reduced to maintain theprothrombin time at the desired level to prevent bleeding complications. Frequent prothrombin determinations are advisable until it has been definitely determined that the prothrombin level has stabilized.
		• **Repaglinide:** In healthy volunteers, coadministration with gemfibrozil (600 mg twice daily for 3 days) resulted in a 8.1-fold (range 5.5- to 15.0-fold) higher repaglinide AUC and a 28.6-fold (range 18.5- to 80.1-fold) higher repaglinide plasma concentration 7 h after the dose. In the same study, gemfibrozil (600 mg twice daily for 3 days) + itraconazole (200 mg in the morning and 100 mg in the evening at Day 1, then 100 mg twice daily at Day 2–3) resulted in a 19.4-(range 12.9- to 24.7-fold) higher repaglinide AUC and a 70.4-fold (range 42.9- to 119.2-fold) higher repaglinide plasma concentration 7 h after the dose. In addition, gemfibrozil alone or gemfibrozil + itraconazole prolonged the hypoglycemic effects of repaglinide. Co-administration of gemfibrozil and repaglinide increases the risk of severe hypoglycemia and is contraindicated.
		• **Bile Acid-Binding Resins:** Gemfibrozil AUC was reduced by 30% when gemfibrozil was given (600 mg) simultaneously with resin-granule drugs such as colestipol (5 g). Administration of the drugs two hours or more apart is recommended because gemfibrozil exposure was not significantly affected when it was administered two hours apart from colestipol.
		• **Colchicine:** Myopathy, including rhabdomyolysis, has been reported with chronic administration of colchicine at therapeutic doses. Concomitant use of LOPID may potentiate the development of myopathy. Patients with renal dysfunction and elderly patients are at increased risk. Caution should be exercised when prescribing LOPID with colchicine, especially in elderly patients or patients with renal dysfunction.

S. No.	Name of the Drug	Reported Drug Interaction
132	Gentamicin	• Adequate hydration is crucial to minimize the risk of ototoxicity and nephrotoxicity associated with the use of aminoglycosides. Dehydration should preferably be corrected prior to initiation of therapy. In patients who may be at risk for dehydration, such as those with severe and/or prolonged diarrhea or vomiting, fluid status should be monitored closely. If signs of renal irritation develop during therapy, hydration should be increased as indicated, accompanied by a reduction in dosage if necessary. Therapy should be withdrawn if urinary output decreases progressively or azotemia increases.
133	Gliclazide	There may be inerction of Gliclazide with following drugs: • mifepristone. • milk thistle. • monoamine oxidase inhibitors (MAOIs; e.g., moclobemide, phenelzine, rasagiline, selegiline, tranylcypromine). • nicotinic acid. • nonsteroidal antiinflammatory drugs (e.g., ibuprofen, naproxen). • omeprazole. • oral contraceptives (birth control pills)
134	Glipizide	• Thehypoglycemicaction of sulfonylureas may be potentiated by certain drugs including nonsteroidal antiinflammatory agents, some azoles, and other drugs that are highlyproteinbound, salicylates, sulfonamides, chloramphenicol, probenecid, coumarins, monoamine oxidase inhibitors, andbeta adrenergic blocking agents.

TABLE 12.2 (Continued)

S. No.	Name of the Drug	Reported Drug Interaction
		• When such drugs are administered to a patient receiving GLUCOTROL (glipizide), the patient should be observed closely for hypoglycemia. When such drugs are withdrawn from a patient receiving GLUCOTROL (glipizide), the patient should be observed closely for loss of control. *In vitro*binding studies with humanserumproteinsindicatethat GLUCOTROL (glipizide) binds differently than tolbutamide and does not interact with salicylate or dicumarol. However, caution must be exercised in extrapolating these findings to the clinical situation and in the use of GLUCOTROL (glipizide) with these drugs.
		• Certain drugs tend to producehyperglycemiaand may lead to loss of control. These drugs include the thiazides and other diuretics, corticosteroids, phenothiazines, thyroidproducts, estrogens, oral contraceptives, phenytoin, nicotinic acid, sympathomimetics, calciumchannel blocking drugs, and isoniazid. When such drugs are administered to a patient receiving GLUCOTROL (glipizide), the patient should be closely observed for loss of control. When such drugs are withdrawn from a patient receiving GLUCOTROL (glipizide), the patient should be observed closely for hypoglycemia.
		• A potential interaction between oral miconazole and oral hypoglycemic agents leading to severe hypoglycemia has been reported. Whether this interaction also occurs with theintravenous, topical, or vaginal preparations of miconazole is not known. The effect of concomitant administration of DIFLUCAN® (fluconazole) and GLUCOTROL (glipizide) has been demonstrated in aplacebo-controlledcrossover studyin normal volunteers. All subjects received GLUCOTROL (glipizide) alone and following treatment with 100 mg of DIFLUCAN as a single daily oral dose for 7 days. The mean percentage increase in the GLUCOTROL (glipizide) AUC after fluconazole administration was 56.9% (range: 35 to 81).

S. No.	Name of the Drug	Reported Drug Interaction
135	Glyburide	• Thehypoglycemicaction of sulfonylureas may be potentiated by certain drugs including nonsteroidal antiinflammatory agents and other drugs that are highly protein bound, salicylates, sulfonamides, chloramphenicol, probenecid, coumarins, monoamine oxidase inhibitors, and beta adrenergic blocking agents. When such drugs are administered to a patient receiving MICRONASE, the patient should be observed closely forhypoglycemia. When such drugs are withdrawn from a patient receiving MICRONASE, the patient should be observed closely for loss of control.
		• An increased risk of liver enzyme elevations was observed in patients receiving glyburide concomitantly with bosentan. Therefore, concomitant administration of MICRONASE and bosentan is contraindicated.
		• Certain drugs tend to producehyperglycemiaand may lead to loss of control. These drugs include the thiazides and other diuretics, corticosteroids, phenothiazines, thyroidproducts, estrogens, oral contraceptives, phenytoin, nicotinic acid, sympathomimetics, calcium channel blocking drugs, and isoniazid. When such drugs are administered to a patient receiving MICRONASE, the patient should be closely observed for loss of control. When such drugs are withdrawn from a patient receiving MICRONASE, the patient should be observed closely for hypoglycemia.
		• A possible interaction between glyburide and ciprofloxacin, a fluoroquinoloneantibiotic, has been reported, resulting in a potentiation of the hypoglycemic action of glyburide. The mechanism for this interaction is not known.
		• A potential interaction between oral miconazole and oral hypoglycemic agents leading to severe hypoglycemia has been reported. Whether this interaction also occurs with the intravenous, topical or vaginal preparations of miconazole is not know.

TABLE 12.2 (Continued)

S. No.	Name of the Drug	Reported Drug Interaction
136	Glyceryl Trinitrate	• Nitric oxide (NO) donor medicines, such as other long-acting glyceryl trinitrate products, isosorbide dinitrate, amyl nitrite and butyl nitrate used to treat angina and/or heart failure.
		• Sildenafil, tadalafil and vardenafil used to treat impotence.
		• If you are currently using any of these medications, tell your doctor or pharmacist before starting glyceryl trinitrate ointment.
		• Before using this medication, tell your doctor or pharmacist of all prescription and nonprescription/ herbal products you may use, especially of:
		– Aspirin.
		– Non-steroidal antiinflammatory drugs (some types of pain killer).
		– Any medicines taken to lower blood pressure.
		– Tricyclic antidepressants to treat depression such as imipramine.
		– Antipsychotics used to treat mental disorders such as chlorpromazine.
137	Griseofulvin	• Patients on warfarin-type anticoagulant therapy may require dosage adjustment of the anticoagulant during and after griseofulvin therapy. Concomitant use of barbiturates usually depresses griseofulvin activity and may necessitate raising the dosage.
		• The concomitant administration of griseofulvin has been reported to reduce the efficacy of oral contraceptives and to increase the incidence of breakthrough bleeding.

S. No.	Name of the Drug	Reported Drug Interaction
138	Halofantrine	• Pregnancy: The use of halofantrine in pregnant women is not recommended, unless the benefits are considered to outweigh the potential risk. No teratogenicity was seen in rat studies at doses up to 120 mg/kg/day (4 times the recommended human dose based on mg/kg). Studies in rabbits revealed that doses up to 60 mg/kg/day (2 times the recommended human dose based on mg/kg) did not produce harmful effects to the mother or fetus. Severe maternal toxicity was evident in a second rabbit teratology study where a slight increase in the incidence of skeletal malformations was observed in the high dose group only (120 mg/kg which is 4 times the recommended human dose based on mg/kg).
		• Lactation: The use of halofantrine is not recommended in breast-feeding mothers. Animal data suggest that halofantrine may be secreted in maternal milk, resulting in reduced rate of weight gain of offspring.
139	Haloperidol	• Since QT-prolongation has been observed during Haldol treatment, caution is advised when prescribing to patient with QT-prolongation conditions (long QT-syndrome, hypokalemia, electrolyteimbalance) or to patients receiving medications known to prolong the QT-interval or known to cause electrolyte imbalance.
		• If concomitant antiparkinson medication is required, it may have to be continued after HALDOL is discontinued because of the difference in excretion rates. If both are discontinued simultaneously, extrapyramidal symptoms may occur. The physician should keep in mind the possible increase inintraocular pressurewhen anticholinergic drugs, including antiparkinson agents, are administered concomitantly with HALDOL.
		• As with other antipsychotic agents, it should be noted that HALDOL may be capable of potentiating CNS depressants such as anesthetics, opiates and alcohol.
		• Ketoconazole is a potent inhibitor of CYP3A4. Increases in QTc have been observed when haloperidol was given in combination with themetabolicinhibitors ketoconazole (400 mg/day) and paroxetine (20 mg/day). It may be necessary to reduce the haloperidol dosage.

TABLE 12.2 (Continued)

S. No.	Name of the Drug	Reported Drug Interaction
140	Halothane	• FLUOTHANE (halothane) augments the action of nondepolarising muscle relaxants and the muscle relaxant effects of aminoglycosides.
		• FLUOTHANE (halothane) may augment the hypotension caused by the ganglionic-blocking effect of tubocurarine.
		• Caution should be exercised during the administration of adrenaline to patients anesthetized with FLUOTHANE (halothane) as arrhythmias may be precipitated. For this reason the dose of adrenaline should be restricted and an antiarrhythmic agent administered as appropriate. Caution should also be applied for other sympathomimetics, and for aminophylline and theophylline and tricyclic antidepressants, which may also precipitate arrhythmias.
141	Heparin	*Oral anticoagulants.*
		• Heparin sodium may prolong the one-stage prothrombin time. Therefore, when heparin sodium is given with dicumarol or warfarin sodium, a period of at least 5 h after the last intravenous dose should elapse before blood is drawn if a valid PROTHROMBIN time is to be obtained.
		• *Platelet inhibitors.*
		• Drugs such as acetylsalicylic acid, dextran, phenylbutazone, ibuprofen, indomethacin, dipyridamole, hydroxychloroquine and others that interfere with platelet aggregation reactions (the main hemostatic defense of heparinized patients) may induce bleeding and should be used with caution in patients receiving heparin sodium.
		• *Other interactions.*
		• Digitalis, tetracyclines, nicotine, antihistamines or i.v. nitroglycerin may partially counteract the anticoagulant action of heparin sodium.
		• *Hyperaminotransferasemia.*

S. No.	Name of the Drug	Reported Drug Interaction
142	Hydralazine	• Significant elevations of aminotransferase (SGOT [S-AST] and SGPT [SALT]) levels have occurred in a high percentage of patients (and healthy subjects) who have received heparin. Since aminotransferase determinations are important in the differential diagnosis of myocardial infarction, liver disease, and pulmonary emboli, rises that might be caused by drugs (like heparin) should be interpreted with caution
		• MAO inhibitors should be used with caution in patients receiving hydralazine.
		• When other potent parenteral antihypertensive drugs, such as diazoxide, are used in combination with hydralazine, patients should be continuously observed for several hours for any excessive fall in blood pressure. Profound hypotensive episodes may occur when diazoxide injection and Apresoline (hydralazine) are used concomitantly.
143	Hydrocortisone	• The pharmacokinetic interactions listed below are potentially clinically important. Drugs that induce hepatic enzymes such as phenobarbital, phenytoin and rifampin may increase the clearance of corticosteroids and may require increases in corticosteroid dose to achieve the desired response. Drugs such as troleandomycin and ketoconazole may inhibit the metabolism of corticosteroids and thus decrease their clearance. Therefore, the dose of corticosteroid should be titrated to avoid steroid toxicity. Corticosteroids may increase the clearance of chronic high dose aspirin. This could lead to decreased salicylate serum levels or increase the risk of salicylate toxicity when corticosteroid is withdrawn. Aspirin should be used cautiously in conjunction with corticosteroids in patients suffering from hypoprothrombinemia. The effect of corticosteroids on oral anticoagulants is variable. There are reports of enhanced as well as diminished effects of anticoagulants when given concurrently with corticosteroids. Therefore, coagulation indices should be monitored to maintain the desired anticoagulant effect.
144	Hydroxyurea	• Prospective studies on the potential for hydroxyurea to interact with other drugs have not been performed.
		• Concurrent use of hydroxyurea and other myelosuppressive agents or radiation therapy may increase the likelihood of bone marrow depression or other adverse events.

TABLE 12.2 (Continued)

S. No.	Name of the Drug	Reported Drug Interaction
145	Ibuprofen	• *ACE-inhibitors*
		• Reports suggest that NSAIDs may diminish the antihypertensive effect of ACE-inhibitors. This interaction should be given consideration in patients taking NSAIDs concomitantly with ACE-inhibitors.
		• *Aspirin*
		• When MOTRIN (ibuprofen) tablets are administered with aspirin, its protein binding is reduced, although the clearance of free MOTRIN (ibuprofen) tablets is not altered. The clinical significance of this interaction is not known; however, as with other NSAIDs, concomitant administration ofibuprofenand aspirin is not generally recommended because of the potential for increased adverse effects.
		• *Diuretics*
		• Clinical studies, as well as post marketing observations, have shown that MORTIN tablets can reduce the natriuretic effect-of furosemide and thiazides in some patients. This response has been attributed to inhibition of renal prostaglandin synthesis. During concomitant therapy with NSAIDs, the patient should be observed closely for signs of renal failure as well as to assure diuretic efficacy.
		• *Lithium*
		• Ibuprofen produced an elevation of plasma lithium levels and a reduction in renal lithium clearance in a study of 11 normal volunteers. The mean minimum lithium concentration increased 15% and the renal clearance of lithium was decreased by 19% during this period of concomitant drug administration. This effect has been attributed to inhibition of renal prostaglandin synthesis by ibuprofen. Thus, when ibuprofen and lithium are administered concurrently, subjects should be observed carefully for signs of lithium toxicity. (Read circulars for lithium preparation before use of such concurrent therapy.).
		• *Methotrexate*

S. No.	Name of the Drug	Reported Drug Interaction
		• NSAIDs have been reported to competitively inhibit methotrexate accumulation in rabbit kidney slices. This may indicate that they could enhance the toxicity of methotrexate. Caution should be used when NSAIDs are administered concomitantly with methotrexate.
		Warfarin-type anticoagulants
		• Several short-term controlled studies failed to show that MOTRIN (ibuprofen) tablets significantly affectedprothrombintimes or a variety of other clotting factors when administered to individuals on coumarin-type anticoagulants. However, because bleeding has been reported when MOTRIN (ibuprofen) tablets and other NSAIDs have been administered to patients on coumarin-type anticoagulants, the physician should be cautious when administering MOTRIN (ibuprofen) tablets to patients on anticoagulants. The effects ofwarfarinand NSAIDs on GI bleeding are synergistic, such that the users of both drugs together have a risk of serious GI bleeding higher than users of either drug alone.
		H-2 Antagonists
		• In studies with human volunteers, coadministration of cimetidine or ranitidine with ibuprofen had no substantive effect on ibuprofen serum concentration.
146	Imipramine	• Drug interactions may change how your medications work or increase your risk for serious side effects. This document does not contain all possible drug interactions. Keep a list of all the products you use (including prescription/nonprescription drugs and herbal products) and share it with your doctor and pharmacist. Do not start, stop, or change the dosage of any medicines without your doctor's approval.
		• Some products that may interact with this drug include: anticholinergics (e.g., atropine, belladonna alkaloids, scopolamine, drugs for Parkinson's disease such as benztropine), certain drugs for high blood pressure (e.g., clonidine, guanadrel, guanethidine, reserpine), digoxin, disopyramide, levodopa, thyroid supplements, valproic acid.

TABLE 12.2 (Continued)

S. No.	Name of the Drug	Reported Drug Interaction
		• Taking MAO inhibitors with this medication may cause a serious (possibly fatal) drug interaction. Avoid taking MAO inhibitors (isocarboxazid, linezolid, methylene blue, moclobemide, phenelzine, procarbazine, rasagiline, selegiline, tranylcypromine) during treatment with this medication. Most MAO inhibitors should also not be taken for two weeks before and after treatment with this medication. Ask your doctor when to start or stop taking this medication.
		• The risk of serotonin syndrome/toxicity increases if you are also taking other drugs that increase serotonin. Examples include street drugs such as MDMA/"ecstasy," St. John's wort, certain antidepressants (including SSRIs such as fluoxetine/paroxetine, SNRIs such as duloxetine/venlafaxine), among others. The risk of serotonin syndrome/toxicity may be more likely when you start or increase the dose of these drugs.
		• Other medications can affect the removal of imipramine from your body, which may affect how imipramine works. Examples include alcohol, barbiturates (such as phenobarbital), cimetidine, cisapride, haloperidol, certain drugs for heart rhythm (such as flecainide, propafenone), halofantrine, certain HIV protease inhibitors (such as fosamprenavir), phenothiazines (such as thioridazine), pimozide, certain antiseizure drugs (such as carbamazepine, phenytoin), terbinafine, trazodone, among others.
		• Many drugs besides imipramine may affect the heart rhythm (QT prolongation), including amiodarone, dofetilide, pimozide, procainamide, quinidine, sotalol, macrolide antibiotics (such as erythromycin), sparfloxacin, among others. Therefore, before using imipramine, report all medications you are currently using to your doctor or pharmacist.
		• Also report the use of drugs which might increase seizure risk (decrease seizure threshold) when combined with this medication such as isoniazid (INH), theophylline, or tramadol, among others. Consult your doctor or pharmacist for details.

S. No.	Name of the Drug	Reported Drug Interaction
147	Indomethacin	• *ACE-Inhibitors and Angiotensin II Antagonists.*
		• Reports suggest that NSAIDs may diminish theantihypertensiveeffect of ACE-inhibitors and angiotensin II antagonists. INDOCIN (indomethacin) can reduce the antihypertensive effects of captopril and losartan. These interactions should be given consideration in patients taking NSAIDs concomitantly with ACE-inhibitors or angiotensin II antagonists. In some patients with compromised renal function, the coadministration of anNSAID and an ACE-inhibitor or an angiotensin II antagonist may result in further deterioration of renal function, including possibleacute renal failure, which is usually reversible.
		• *Aspirin*
		• When INDOCIN (indomethacin) is administered with aspirin, itsproteinbinding is reduced, although the clearance of free INDOCIN (indomethacin) is not altered. The clinical significance of this interaction is not known.
		• The use of INDOCIN (indomethacin) in conjunction with aspirin or other salicylates is not recommended. Controlled clinical studies have shown that the combined use of INDOCIN (indomethacin) and aspirin does not produce any greatertherapeuticeffect than the use of INDOCIN (indomethacin) alone. In a clinical study of the combined use of INDOCIN (indomethacin) and aspirin, the incidence of gastrointestinal side effects was significantly increased with combined therapy.
		• In a study in normal volunteers, it was found thatchronicconcurrent administration of 3.6 g of aspirin per day decreases indomethacinbloodlevels approximately 20%.
		• *Beta-adrenoceptor blocking agents*
		• Blunting of the antihypertensive effect of beta-adrenoceptor blocking agents by nonsteroidal antiinflammatory drugs including INDOCIN (indomethacin) has been reported. Therefore, when using these blocking agents to treat hypertension, patients should be observed carefully in order to confirm that the desired therapeutic effect has been obtained.
		• *Cyclosporine*

TABLE 12.2 (Continued)

S. No.	Name of the Drug	Reported Drug Interaction
		• Administration of nonsteroidal antiinflammatory drugs concomitantly with cyclosporine has been associated with an increase in cyclosporine-inducedtoxicity, possibly due to decreasedsynthesisof renal prostacyclin. NSAIDs should be used with caution in patients taking cyclosporine, and renal function should be carefully monitored.
		Diflunisal
		• In normal volunteers receiving indomethacin, the administration of diflunisal decreased the renal clearance and significantly increased theplasmalevels of indomethacin. In some patients, combined use of INDOCIN (indomethacin) and diflunisal has been associated with fatal gastrointestinal hemorrhage. Therefore, diflunisal and INDOCIN (indomethacin) should not be used concomitantly.
		Digoxin
		• INDOCIN (indomethacin) given concomitantly with digoxin has been reported to increase theserumconcentration and prolong the half-life of digoxin. Therefore, when INDOCIN (indomethacin) and digoxin are used concomitantly, serum digoxin levels should be closely monitored.
		Diuretics
		• In some patients, the administration of INDOCIN (indomethacin) can reduce thediuretic, natriuretic, and antihypertensive effects of loop, potassium-sparing, and thiazide diuretics. This response has been attributed to inhibition of renalprostaglandinsynthesis.
		• INDOCIN (indomethacin) reduces basal plasma renin activity (PRA), as well as those elevations of PRA induced by furosemide administration, or salt or volume depletion. These facts should be considered when evaluating plasma renin activity inhypertensivepatients.
		• It has been reported that the addition of triamterene to a maintenance schedule of INDOCIN (indomethacin) resulted in reversible acute renal failure in two of four healthy volunteers. INDOCIN (indomethacin) and triamterene should not be administered together.

S. No.	Name of the Drug	Reported Drug Interaction
		• INDOCIN (indomethacin) and potassium-sparing diuretics each may be associated with increased serumpotassiumlevels. The potential effects of INDOCIN (indomethacin) and potassium-sparing diuretics on potassiumkineticsand renal function should be considered when these agents are administered concurrently.
		• Most of the above effects concerning diuretics have been attributed, at least in part, to mechanisms involving inhibition of prostaglandin synthesis by INDOCIN (indomethacin).
		• During concomitant therapy with NSAIDs, the patient should be observed closely for signs of renal failure (seeWARNINGS, Renal Effects), as well as to assure diuretic efficacy.
		• *Lithium.*
		• Capsules INDOCIN (indomethacin) 50 mgt.i.d.produced a clinically relevant elevation of plasma lithium and reduction in renal lithium clearance inpsychiatricpatients and normal subjects with steady state plasma lithium concentrations. This effect has been attributed to inhibition of prostaglandin synthesis. As a consequence, when NSAIDs and lithium are given concomitantly, the patient should be carefully observed for signs of lithium toxicity. (Read circulars for lithium preparations before use of such concomitant therapy.) In addition, the frequency of monitoring serum lithium concentration should be increased at the outset of such combination drug treatment.
		• *Methotrexate*
		• NSAIDs have been reported to competitively inhibit methotrexate accumulation in rabbitkidneyslices. This mayindicatethat they could enhance the toxicity of methotrexate. Caution should be used when NSAIDs are administered concomitantly with methotrexate.
		• *NSAIDs*
		• The concomitant use of INDOCIN (indomethacin) with other NSAIDs is not recommended due to the increased possibility of gastrointestinal toxicity, with little or no increase in efficacy.
		• *Oral anticoagulants*

TABLE 12.2 (Continued)

S. No.	Name of the Drug	Reported Drug Interaction
		• Clinical studies have shown that INDOCIN (indomethacin) does not influence the hypoprothrombinemia produced by anticoagulants. However, when any additional drug, including INDOCIN (indomethacin), is added to the treatment of patients onanticoagulanttherapy, the patients should be observed for alterations of theprothrombin time. In postmarketing experience, bleeding has been reported in patients on concomitant treatment with anticoagulants and INDOCIN (indomethacin). Caution should be exercised when INDOCIN (indomethacin) and anticoagulants are administered concomitantly. The effects ofwarfarinand NSAIDs on GI bleeding are synergistic, such that users of both drugs together have a risk of serious GI bleeding higher than users of either drug alone.
		• *Probenecid*
		• When INDOCIN (indomethacin) is given to patients receiving probenecid, the plasma levels of indomethacin are likely to be increased. Therefore, a lower total daily dosage of INDOCIN (indomethacin) may produce a satisfactory therapeutic effect. When increases in the dose of INDOCIN (indomethacin) are made, they should be made carefully and in small increments.
148	Isoniazid	• **Food:**Isoniazid should not be administered with food. Studies have shown that the bioavailability of isoniazid is reduced significantly when administered with food. Tyramine- and histamine-containing foods should be avoided in patients receiving isoniazid. Because isoniazid has some monoamine oxidase inhibiting activity, an interaction with tyramine-containing foods (cheese, red wine) may occur. Diamine oxidase may also be inhibited, causing exaggerated response (e.g., headache, sweating, palpitations, flushing, hypotension) to foods containing histamine (e.g., skipjack, tuna, other tropical fish).

S. No.	Name of the Drug	Reported Drug Interaction
		• **Acetaminophen:** a report of severe acetaminophen toxicity was reported in a patient receiving Isoniazid It is believed that the toxicity may have resulted from a previously unrecognized interaction between isoniazid and acetaminophen and a molecular basis for this interaction has been proposed. However, current evidence suggests that isoniazid does induce P-450IIE1, a mixed-function oxidase enzyme that appears to generate the toxic metabolites, in the liver. Furthermore it has been proposed that isoniazid resulted in induction of P-450IIE1 in the patients liver which, in turn, resulted in a greater proportion of the ingested acetaminophen being converted to the toxic metabolites. Studies have demonstrated that pretreatment with isoniazid potentiates acetaminophen hepatotoxicity in rats.
		• **Carbamazepine:**Isoniazid is known to slow the metabolism of carbamazepine and increase its serum levels. Carbamazepine levels should be determined prior to concurrent administration with isoniazid signs and symptoms of carbamazepine toxicity should be monitored closely, and appropriate dosage adjustment of theanticonvulsantshould be made.
		• **Ketoconazole:**Potential interaction of Ketoconazole and Isoniazid may exist. When Ketoconazole is given in combination with isoniazid and rifampin the AUC of ketoconazole is decreased by as much as 88% after 5 months of concurrent Isoniazid and Rifampin therapy.
		• **Phenytoin:**Isoniazid may increase serum levels of phenytoin. To avoid phenytoin intoxication appropriate adjustment of the anticonvulsant should be made.
		• **Theophylline:**A recent study has shown that concomitant administration of isoniazid and theophylline may cause elevated plasma levels of theophylline, and in some instances a slight decrease in the elimination of isoniazid. Since the therapeutic range of theophylline is narrow, theophylline serum levels should be monitored closely, and appropriate dosage adjustments of theophylline should be made (Blaschke et al., 1981).
		• **Valproate:** A recent case study has shown a possible increase in the plasma level of valproate when coadministered with isoniazid. Plasma valproate concentration should be monitored when isoniazid and valproate are coadministered, and appropriate dosage adjustments of valproate should be made (Das, 2004).

TABLE 12.2 (Continued)

S. No.	Name of the Drug	Reported Drug Interaction
149	Isosorbidedinitrate	• The vasodilating effects of isosorbide dinitrate may be additive with those of othervasodilators. Alcohol, in particular, has been found to exhibit additive effects of this variety.
		• Concomitant use of Isordil Titradose with phosphodiesterase inhibitors in any form is contraindicated Concomitant use of Isordil Titradose with riociguat, a soluble guanylate cyclase stimulator, is contraindicated.
150	Itraconazole	• Itraconazole and its major metabolite, hydroxyitraconazole, are inhibitors of CYP3A4. Therefore, the following drug interactions may occur.
		• SPORANOX® (itraconazole capsules) may decrease the elimination of drugs metabolized by CYP3A4, resulting in increased plasma concentrations of these drugs when they are administered with SPORANOX® (itraconazole capsules). These elevated plasma concentrations may increase or prolong both therapeutic and adverse effects of these drugs. Whenever possible, plasma concentrations of these drugs should be monitored, and dosage adjustments made after concomitant SPORANOX® (itraconazole capsules) therapy is initiated. When appropriate, clinical monitoring for signs or symptoms of increased or prolonged pharmacologic effects is advised. Upon discontinuation, depending on the dose and duration of treatment, itraconazole plasma concentrations decline gradually (especially in patients with hepatic cirrhosis or in those receiving CYP3A4 inhibitors). This is particularly important when initiating therapy with drugs whose metabolism is affected by itraconazole.
		• Inducers of CYP3A4 may decrease the plasma concentrations of itraconazole. SPORANOX® (itraconazole capsules) may not be effective in patients concomitantly taking SPORANOX® (itraconazole capsules) and one of these drugs. Therefore, administration of these drugs with SPORANOX® (itraconazole capsules) is not recommended.
		• Other inhibitors of CYP3A4 may increase the plasma concentrations of itraconazole. Patients who must take SPORANOX® (itraconazole capsules) concomitantly with one of these drugs should be monitored closely for signs or symptoms of increased or prolonged pharmacologic effects of SPORANOX® (itraconazole capsules).

S. No.	Name of the Drug	Reported Drug Interaction
151	Kanamycin	• Possible increased incidence of nephrotoxicity and/or neurotoxicity.
		• Anti-emetics that suppress nausea and vomiting of vestibular origin and vertigo may mask symptoms of vestibular ototoxicity.
		• Possible increased incidence of nephrotoxicity and/or neurotoxicity.
		• *In vitro* evidence of additive or synergistic antibacterial effects between penicillins and aminoglycosides against some enterococci, Enterobacteriaceae, or *Ps. aeruginosa*; used to therapeutic advantage (e.g., treatment of endocarditis).
		• Possible increased incidence of nephrotoxicity reported with some cephalosporins; cephalosporins may spuriously elevate creatinine concentrations.
		• Potential in vitro and in vivo inactivation of aminoglycosides **(Seth, 2004)**.
		• *In vitro* evidence of additive or synergistic antibacterial effects with aminoglycosides against some gram-positive bacteria (*Enterococcus fecalis, S. aureus, Listeria monocytogenes*).
152	Ketamine	• Prolonged recovery time may occur if barbiturates and/or narcotics are used concurrently with ketamine.
		• Ketamine is clinically compatible with the commonly used general and localanestheticagents when an adequaterespiratoryexchange is maintained.

TABLE 12.2 (Continued)

S. No.	Name of the Drug	Reported Drug Interaction
153	Ketorolac	• Increased risk of renal impairment (Seth, 2004).
		• Reduced BP response to ACE inhibitor.
		• No alteration in the protein binding of ketorolac.
		• Reduced BP response to angiotensin II receptor antagonist.
		• Possible deterioration of renal function in individuals with renal impairment.
		• No effect on the extent of oral ketorolac absorption.
		• Seizures reported in patients receiving carbamazepine or phenytoin.
		• Phenytoin does not alter the protein binding of ketorolac.
		• Hallucinations reported in patients receiving fluoxetine, thiothixene, or alprazolam (Seth, 2004).
		• Possible increased risk of bleeding.
		• No alteration in the protein binding of either drug.
		• Reduced natriuretic effect.
		• Increased risk of bleeding complications.
		• Increased bleeding time when administered with heparin 5000 units; concurrent use with heparin 2500–5000 units sub-Q every 12 h not studied extensively.
154	Labetalol	• Labetalol and ethanol may have additive effects in lowering your blood pressure. You may experience headache, dizziness, lightheadedness, fainting, and/or changes in pulse or heart rate. These side effects are most likely to be seen at the beginning of treatment, following a dose increase, or when treatment is restarted after an interruption.

S. No.	Name of the Drug	Reported Drug Interaction
		• Beta-adrenergic receptor blocking agents (aka beta-blockers) may alter serum lipid profiles. Increases in serum VLDL and LDL cholesterol and triglycerides, as well as decreases in HDL cholesterol, have been reported with some beta-blockers. Patients with preexisting hyperlipidemia may require closer monitoring during beta-blocker therapy, and adjustments made accordingly in their lipid-lowering regimen.
		• Drug interactions between Aspirin Low Strength and labetalol.
155	Lamotrigine	• Using Lamotrigine together with ethanol can increase nervous system side effects such as dizziness, drowsiness, and difficulty concentrating.
		• Rash remained the prime adverse symptom of drug use.
		• Severe Potential Hazard, High plausibility.
156	Lansoprazole	• Moderate potential hazard, moderate plausibility.
		• Lansoprazole has only rarely been associated with hepatic injury.
157	Levamisole	• You should avoid drinking alcohol, it can increase some of the side effects of levamisole. This can cause nausea, dizziness, vomiting, and upset stomach. It is important to tell your doctor about all other medications you use, including vitamins and herbs.
158	Levodopa	• Using propoxyphene together with levodopa may increase side effects such as dizziness, drowsiness, confusion, difficulty concentrating, and other nervous system or mental effects.
		• Some people, especially the elderly, may also experience impairment in thinking, judgment, and coordination.
159	Levonorgestrol	• The central nervous system effects and blood levels of ethanol may be increased in patients taking oral contraceptives, although data are lacking and reports are contradictory. The mechanism may be due to enzyme inhibition. Consider counseling women about this interaction which is unpredictable.

TABLE 12.2 (Continued)

S. No.	Name of the Drug	Reported Drug Interaction
		• An increased risk of benign hepatic adenomas and hepatocellular carcinomas has been associated with long-term, oral estrogen–progestin contraceptive use of at least 4 years and 8 years of levonorgestrel contraceptive implants is contraindicated in patients with a current or past history of idiopathic intracranial hypertension (pseudotumorcerebri, benign intracranial hypertension).
160	Lignocaine	• Common medications checked in combination with lidocaine.
		• Advair Diskus (fluticasone/salmeterol) Ambien (zolpidem) Ativan (lorazepam) Benadryl (diphenhydramine) Colace (docusate) Cymbalta (duloxetine) Fish Oil (omega-3 polyunsaturated fatty acids) Lasix (furosemide) Lyrica (pregabalin) MiraLax (polyethylene glycol 3350) Neurontin (gabapentin) Nexium (esomeprazole) ProAir HFA (albuterol) Tylenol (acetaminophen) Vicodin (acetaminophen/hydrocodone) Vitamin B12 (cyanocobalamin) Vitamin C (ascorbic acid) Vitamin D3 (cholecalciferol) Xanax (alprazolam) Zofran (ondansetron).
161	Lithium	• Lithium increases a brain chemical called serotonin. Some medications for depression also increase the brain chemical serotonin.
		• Taking lithium along with these medications for depression might increase serotonintoo much and cause serious side effects including heart problems, shivering, and anxiety.
		• Lithium increases a chemical in the brain. This chemical is called serotonin. Some medications used for depression also increase serotonin. This could cause serious side effects including heart problems, shivering, and anxiety. Some of these medications used for depression include phenelzine (Nardil), tranylcypromine (Parnate).
162	Lorazepam	• Using Lorazepam together with ethanol can increase nervous system side effects such as dizziness, drowsiness, and difficulty concentrating.
		• Some people may also experience impairment in thinking and judgment.
		• Tolerance may develop with chronic ethanol use. The mechanism may be decreased clearance of the benzodiazepines because of CYP450 hepatic enzyme inhibition.

S. No.	Name of the Drug	Reported Drug Interaction
163	Losartan	• If taking losartan then avoid potassium-containing salt substitutes. This can cause high levels of potassium in your blood. High levels of potassium can cause weakness, irregular heartbeat, confusion, tingling of the extremities, or feelings of heaviness in the legs.
		• Some patients grapefruits and grapefruit juice may decrease the efficacy of losartan.
		• Grapefruit juice may modestly decrease and delay the conversion of losartan to its active metabolite, E3174.
		• The proposed mechanism is inhibition of CYP450 3A4-mediated first-pass metabolism in the gut wall by certain compounds present in grapefruits.
164	Lovastatin	• Increased risk of myopathy and/or rhabdomyolysis when used with another statin.
		• Chlorpropamide or glipizide: Pharmacokinetic interactions not reported during concomitant use.
		• Inhibition of lovastatin metabolism via CYP3A4.
		• Itraconazole, ketoconazole, or posaconazole: Inhibition of CYP3A4-dependent metabolism of lovastatin, resulting in increased lovastatin plasma concentrations and AUC and increased risk of myopathy and/or rhabdomyolysis. Voriconazole: Possible inhibition of lovastatin metabolism, resulting in increased risk of myopathy and/or rhabdomyolysis.
		• Increased lovastatin AUC and increased risk of myopathy and/or rhabdomyolysis.
		• Increased risk of myopathy and/or rhabdomyolysis, particularly with higher lovastatin dosages.
165	Mebendazole	• Decreased plasma mebendazole concentrations.
		• Increased plasma mebendazole concentrations
166	Meclozine	• Additive CNS depression

TABLE 12.2 (Continued)

S. No.	Name of the Drug	Reported Drug Interaction
167	Mefenamic acid	• Reduced BP response to ACE inhibitor possible.
		• Possible deterioration of renal function in individuals with renal impairment.
		• Increased peak plasma concentrations and AUC of mefenamic acid.
		• Possible bleeding complications.
		• Increased risk of GI ulceration or other complications.
		• No consistent evidence that low-dose aspirin mitigates the increased risk of serious cardiovascular events associated with NSAIAs.
		• Reduced natriuretic effects possible
168	Mefloquine	• Possible decreased concentrations of the anticonvulsant and loss of seizure control.
		• Possibility of ECG abnormalities and increased risk of seizures with quinine, quinidine, or chloroquine.
		• Use of halofantrine after mefloquine has resulted in potentially fatal prolongation of the QT_c interval; data not available regarding use of mefloquine after halofantrine.
		• Possibility of interference with immune response to typhoid vaccine live oral since mefloquine has in vitro activity against *Salmonella typhi*
169	Meprobamate	• Additive CNS effects
170	Mercaptopurine	• Possible decreased mercaptopurine metabolism and increased risk of myelotoxicity.
		• Possible inhibition of anticoagulant effect.

S. No.	Name of the Drug	Reported Drug Interaction
171	Mestranol	• Common medications checked in combination with mestranol/norethindrone.
		• Actos (pioglitazone), Advair Diskus (fluticasone/salmeterol), Crestor (rosuvastatin), Cymbalta (duloxetine), Diovan (valsartan), Diovan HCT (hydrochlorothiazide/valsartan), Effexor XR (venlafaxine), Klor-Con (potassium chloride), Levaquin (levofloxacin), Levoxyl (levothyroxine), Lexapro (escitalopram), Lipitor (atorvastatin), Nexium (esomeprazole), Plavix (clopidogrel), Prevacid (lansoprazole), ProAir HFA (albuterol), Proventil HFA (albuterol), Singulair (montelukast), Synthroid (levothyroxine), Vytorin (ezetimibe/simvastatin).
172	Methadone	• Potential for increased respiratory depressant effects and elevation of CSF pressure in patients with increased intracranial pressure, head trauma, or other intracranial lesions. Monitor susceptible patients for sedation and respiratory depression, particularly during initiation of therapy
173	Methotrexate	• Decreased adalimumab clearance.
		• Possible increased methotrexate toxicity because of displacement from protein binding sites.
		• Possible inhibition of methotrexate metabolism with prolonged administration (2 weeks) of oral amiodarone.
		• Administered concurrently in clinical studies, but specific drug interactions not evaluated in humans.
		• No effect on clearance or toxicologic profile of either drug when administered concurrently in rats.
		• Decreased effectiveness of methotrexate during period of asparagine suppression.
		• Possible increased methotrexate toxicity because of displacement from protein binding sites.
		• Possible decreased intestinal absorption of methotrexate or interference with enterohepatic circulation.

TABLE 12.2 (Continued)

S. No.	Name of the Drug	Reported Drug Interaction
174	Methyldopa	• Potential for hypotension.
		• Possible decreased hypotensive effect.
		• Additive/potentiated hypotensive effect.
		• Possible psychomotor retardation, memory impairment, and inability to concentrate in nonschizophrenic patients.
		• Concomitant administration may decrease oral absorption and alter the metabolism of methyldopa.
		• Possible increased BP.
		• May interfere with measurement of creatinine by the alkaline picrate method.
175	Methylphenidate	• Possible inhibition of anticonvulsant metabolism.
		• Rare cases of serious cardiovascular effects, including death; causality not established.
		• Potential for increased gastric pH to alter release characteristics of extended-release capsules (RitalinLA); clinical importance not established.
		• Antagonism of hypotensive effect.
176	Metoclopramide	• It is often associated with cholinergic effects, CNS depression, and strong hyperprolactinemic effects.
177	Metoprolol	• Catecholaminedepleting drugs (e.g., reserpine, monoamine oxidase (MAO) inhibitors) may have an additive effect when given with beta-blocking agents. Observe patients treated with TOPROL-XL plus a catecholamine depletor for evidence of hypotension or marked bradycardia, which may producevertigo, syncope, orpostural hypotension.

S. No.	Name of the Drug	Reported Drug Interaction
		• Drugs that inhibit CYP2D6 such as quinidine, fluoxetine, paroxetine, and propafenone are likely to increase metoprolol concentration. In healthy subjects with CYP2D6 extensive metabolizer phenotype, coadministration of quinidine 100 mg and immediate-release metoprolol 200 mg tripled the concentration of S-metoprolol and doubled the metoprolol elimination half-life. In four patients withcardiovascular disease, coadministration of propafenone 150 mg t.i.d. with immediate-release metoprolol 50 mg t.i.d. resulted in two- to five-fold increases in the steady-state concentration of metoprolol. These increases in plasma concentration would decrease the cardioselectivity of metoprolol.
178	Metronidazole	• Alcohol: Patients taking metronidazole should be warned against consuming alcohol (during therapy and for 24 h posttreatment) because of a possible disulfiram-like reaction.
		• Anticoagulants: Metronidazole has been reported to potentiate the anticoagulant effect of warfarin resulting in a prolongation of prothrombin time. This possible drug interaction should be considered when metronidazole is prescribed for patients on this type of anticoagulant therapy.
		• Barbiturates: metronidazole metabolism may be enhanced causing reduced serum concentrations.
		• Disulfiram: Administering disulfiram and metronidazole together may result in confusion and psychotic reactions because of combined toxicity.
		• Lithium: Initiation of metronidazole therapy has been associated with increased serum lithium levels and, in a few cases, signs of lithium toxicity.
		• Pregnancy: Metronidazole crosses the placental barrier. Metronidazole should be withheld during the first trimester. In addition, it is advisable that administration be avoided during the second and third trimesters; however, if metronidazole treatment is considered necessary, its use requires that the potential benefits be weighed against the possible risks.
		• Lactation: Metronidazole is distributed into milk. Any unnecessary exposure to metronidazole should be avoided. If a nursing mother is treated with metronidazole, the breast milk should be expressed and discarded during treatment. Breast-feeding can be resumed 24 to 48 h after treatment.
		• Children: Controlled studies in children are limited.

TABLE 12.2 (Continued)

S. No.	Name of the Drug	Reported Drug Interaction
179	Miconazole	• Potential for increased plasma warfarin concentrations with intravaginal miconazole.
		• Potential for interaction with miconazole applied topically to skin is unknown.
180	Midazolam	• CYP3A4 substrate.
		• CNS depression.
		• GI absorption enhanced by GLP-2 receptor agonist.
		• Possible antagonism of sedative effect during anesthesia.
		• Minimum alveolar concentration of halothane required for general anesthesia appears to be decreased in a linear, dose-related manner with concomitant administration of IV midazolam.
		• Decreased peak plasma concentration and AUC of midazolam reported with concomitant use of oral midazolam and phenytoin or carbamazepinesimilar effects expected with phenobarbita.
		• Decreased plasma clearance and increased peak plasma concentration of midazolam reported with oral midazolam;potential for intense and prolonged sedation and respiratory depression
181	Mifepristone	• Inhibitors of CYP3A4: Potential pharmacokinetic interaction (increased plasma mifepristone concentrations); however, specific drugs and food not studied to date.
		• Inducers of CYP3A4: Potential pharmacokinetic interaction (decreased plasma mifepristone concentrations); however, specific drugs and food not studied to date.
		• Possible decreased serum mifepristone concentrations
182	Minocycline	• Tetracyclines have been shown to depress plasma prothrombin activity, patients who are on anticoagulant therapy may require downward adjustment of their anticoagulant dosage.
		• Sincebacteriostaticdrugs may interfere with the bactericidal action of penicillin, it is advisable to avoid giving tetracycline-class drugs in conjunction with penicillin.

S. No.	Name of the Drug	Reported Drug Interaction
		• Absorption of tetracyclines is impaired by antacids containingaluminum, calcium, ormagnesium, and iron-containing preparations.
		• The concurrent use oftetracyclineand methoxyflurane has been reported to result in fatal renal toxicity.
		• Concurrent use of tetracyclines with oral contraceptives may render oral contraceptives less effective.
		• Administration of isotretinoin should be avoided shortly before, during, and shortly after minocycline therapy. Each drug alone has been associated with pseudotumor cerebri.
		• Increased risk ofergotismwhen ergot alkaloids or their derivatives are given with tetracyclines.
		• False elevations of urinary catecholamine levels may occur due to interference with the fluorescence test.
183	Minoxidil	• Additive hypotensive effect.[b] Concomitant use may prevent sodium retention and increased plasma volume that may occur with minoxidil therapy.
		• Possibly profound orthostatic hypotensive effects.
184	Morphine	• Concurrent use of MS CONTIN and othercentral nervous system(CNS) depressants including sedatives or hypnotics, general anesthetics, phenothiazines, tranquilizers, and alcohol can increase the risk of respiratory depression, hypotension, profound sedation or coma. Monitor patients receiving CNS depressants and MS CONTIN for signs of respiratory depression and hypotension. When such combined therapy is contemplated, reduce the initial dose of one or both agents.
		• Mixedagonist/antagonistanalgesics (i.e., pentazocine, nalbuphine, butorphanol) may reduce theanalgesiceffect of MS CONTIN or may precipitatewithdrawal symptomsin these patients. Avoid the use of agonist/antagonist analgesics in patients receiving MS CONTIN.
		• Morphinemay enhance theneuromuscularblocking action ofskeletal musclerelaxants and produce an increased degree of respiratory depression. Monitor patients receiving muscle relaxants and MS CONTIN for signs of respiratory depression that may be greater than otherwise expected.

TABLE 12.2 (Continued)

S. No.	Name of the Drug	Reported Drug Interaction
185	Mustine	• No any drug interaction has been found.
186	Mycophenolate Mofetil	• Coadministration of mycophenolate mofetil (1 g) and acyclovir (800 mg) to 12 healthy volunteers resulted in no significant change in MPA AUC and Cmax. However, MPAG and acyclovir plasma AUCs were increased 10.6% and 21.9%, respectively. Because MPAG plasma concentrations are increased in the presence of renal impairment, as are acyclovir concentrations, the potential exists for mycophenolate and acyclovir or itsprodrug(e.g., valacyclovir) to compete for tubular secretion, further increasing the concentrations of both drugs.
		• Increased risk of bone marrow suppression.
		• Decreased mycophenolic acid AUC.
187	Naloxone	• Serious adverse cardiovascular effects (e.g., ventricular tachycardia and fibrillation, pulmonary edema, cardiac arrest) resulting in death, coma, and encephalopathy reported in postoperative patients.
		• Methohexital appears to block the acute onset of withdrawal symptoms induced by naloxone in opiate addicts
188	Nandrolone	• Anticoagulants.
		• Anabolic steroids may increase sensitivity to oral anticoagulants.
		• Dosage of the anticoagulant may have to be decreased in order to maintain the prothrombin time at the desired therapeutic level.
		• Patients receiving oral anticoagulant therapy require close monitoring, especially when anabolic steroids are started or stopped.

S. No.	Name of the Drug	Reported Drug Interaction
189	Naproxen	• NSAIDs may diminish theantihypertensiveeffect of ACE-inhibitors, ARBs, or beta-blockers (including propanolol).
		• Monitor patients taking NSAIDs concomitantly with ACE-inhibitors, ARBs, or beta blockers for changes in blood pressure.
		• In addition, in patients who are elderly, volume-depleted (including those ondiuretic therapy), or have compromised renal function, coadministration of NSAIDs with ACE inhibitors or ARBs may result in deterioration of renal function, including possibleacute renal failure. Monitor these patients closely for signs of worsening renal function.
		• Antacids and Sucralfate.
		• Concomitant administration of some antacids (magnesium oxide or aluminum hydroxide) and sucralfate can delay theabsorptionof naproxen.
		• Aspirin.
		• When naproxen as NAPROSYN, EC-NAPROSYN, ANAPROX, ANAPROX DS or NAPROSYN Suspension is administered withaspirin, its protein binding is reduced, although the clearance of free NAPROSYN, ECNAPROSYN, ANAPROX, ANAPROX DS or NAPROSYN Suspension is not altered. The clinical significance of this interaction is not known; however, as with other NSAIDs, concomitant administration of naproxen and naproxen sodium and aspirin is not generally recommended because of the potential of increased adverse effects.
		• Cholestyramine.
		• As with other NSAIDs, concomitant administration of cholestyramine can delay the absorption of naproxen.

TABLE 12.2 (Continued)

S. No.	Name of the Drug	Reported Drug Interaction
190	Neomycin	• Concurrent or serial use may enhance nephrotoxicity, ototoxicity, and/or potentiate neuromuscular blockade.
		• Possible increased incidence of nephrotoxicity and/or neurotoxicity.
		• Neomycin may enhance warfarin effects by decreasing vitamin K availability.
		• Anti-emetics that suppress nausea and vomiting of vestibular origin and vertigo may mask symptoms of vestibular ototoxicity.
		• Neomycin inhibits GI absorption of oral vitamin B_{12}.
191	Neostigmine	• Certain antibiotics, especially neomycin, streptomycin and kanamycin, have a mild but definite nondepolarising blocking action which may accentuate neuromuscularblock. These antibiotics should be used in the myasthenic patient only where definitely indicated, and then careful adjustment should be made of adjunctive anticholinesterase dosage.
		• Local and some general anesthetics, antiarrhythmic agents and other drugs that interfere with neuromuscular transmission should be used cautiously, if at all, in patients withmyasthenia gravis; the dose of Prostigmin (neostigmine) may have to be increased accordingly.
192	Nicorandil	• Smoking: The effect of smoking on the pharmacokinetics of nicorandil has not been studied.
		• Cimetidine: The effects of cimetidine (400 mg twice daily for 7 days) on the pharmacokinetics of nicorandil (20 mg twice daily given for 7 days alone and then another 7 days with cimetidine) were assessed in 12 healthy volunteers. The coadministration of cimetidine with nicorandil did not significantly modify the rate of absorption of nicorandil or other pharmacokinetic parameters (such as C_{max}, t_{max}, and urinary excretion parameters). Thus, cimetidine does not significantly inhibit the liver enzymes involved in the metabolism of nicorandil.
		• Phosphodiesterase 5 inhibitors As hypotensive effects of nitrates or nitric oxide donors are potentiated by phosphodiesterase 5 inhibitors (e.g., sildenafil, tadalafil, vardenafil), the concomitant use of nicorandil and phosphodiesterase 5 inhibitors is contraindicated

S. No.	Name of the Drug	Reported Drug Interaction
193	Nicotine	• Smoking Cessation.
		• Cessation of smoking may alter the response to concomitant administration of various drugs in patients who previously smoked.
		• Potential pharmacokinetic interaction (decreased metabolism and increased blood concentrations of certain drugs) with cessation of smoking.
		• Consider the effect of smoking cessation in patients receiving nicotine replacement therapy when patient is receiving other drugs concomitantly.
		• Sympathomimetic and Sympatholytic Drugs.
		• Potential pharmacodynamic interaction (increased circulating plasma concentrations of cortisol and catecholamines); may require dosage adjustment of sympathomimetic (adrenergic) or sympatholytic (adrenergic blocking) drugs.
		• Possible altered absorption of transdermal nicotine from drugs producing cutaneous vasoconstriction (e.g., sympathomimetic agents) or vasodilation (e.g., antihypertensive agents).
		• Foods Affecting Salivary Acidity.
		• Transient decrease of salivary pH may inhibit buccal absorption of nicotine from gum, lozenge, or oral inhaler.
194	Nicotinic acid	• Potential additive anticoagulant effect (increased PT and decreased platelet count reported).
		• Possible potentiation of hypotensive effects.
		• Decreased niacin metabolic clearance.
		• Decreased niacin bioavailability due to binding of niacinto bile acid sequestrant.
		• Increased risk of myopathy and rhabdomyolysis with antilipemic dosages (>1 g daily) of niacin.
		• Simvastatin: Increased risk of myopathy observed in Chinese versus non-Chinese patients receiving simvastatin 40 mg daily with antilipemic dosages of niacin.

TABLE 12.2 (Continued)

S. No.	Name of the Drug	Reported Drug Interaction
195	Nifedipine	• Nifedipine is mainly eliminated bymetabolismand is a substrate of CYP3A. Inhibitors and inducers of CYP3A can impact the exposure to nifedipine and consequently its desirable and undesirable effects. *In vitro*and *in vivo*data indicate that nifedipine can inhibit the metabolism of drugs that are substrates of CYP3A, thereby increasing the exposure to other drugs. Nifedipine is a vasodilator, and coadministration of other drugs affectingblood pressuremay result in pharmacodynamic interactions.
		• CYP3A inhibitors.
		• CYP3A inhibitors such as ketoconazole, fluconazole, itraconazole, clarithromycin, erythromycin (Azithromycin, although structurally related to the class of macrolide antibiotic is void of clinically relevant CYP3A4 inhibition), grapefruit, nefazodone, fluoxetine, saquinavir, indinavir, nelfinavir, and ritonavir may result in increased exposure to nifedipine when coadministered. Careful monitoring and dose adjustment may be necessary; consider initiating nifedipine at the lowest dose available if given concomitantly with these medications.
		• Strong CYP3A inducers.
		• Strong CYP3A inducers, such as rifampin, rifabutin, phenobarbital, phenytoin, carbamazepine, and St. John's Wort reduce the bioavailability and efficacy of nifedipine; therefore nifedipine should not be used in combination with strong CYP3A inducers such as rifampin.
196	Nimesulide	• Additive hepatotoxic effects with known hepatotoxins: anticonvulsants (e.g., valproic acid), antifungals (e.g., ketoconazole), antituberculous drugs (e.g., isoniazid), tacrine, pemoline, amiodarone, methotrexate, methyldopa, amoxicillin/clavulanic acid. May decrease the oral bioavailability of furosemide and the natriuretic and diuretic response to furosemide. Increased risks of GI and hepatic adverse effects with other NSAIDs, including aspirin. May increase anticoagulant effect of warfarin. Potentiates the action of phenytoin. May be displaced from binding sites with fenofibrate, salicylic acid, and tolbutamide. The drug often interacts wirth NSAIDs, lithium, probenecid and ciclosporin.

S. No.	Name of the Drug	Reported Drug Interaction
197	Nimodipine	• It is possible that the cardiovascular action of other calcium channel blockers could be enhanced by the addition of Nimotop® (nimodipine).
		• In Europe, Nimotop® (nimodipine) was observed to occasionally intensify the effect ofantihypertensivecompounds taken concomitantly by patients suffering from hypertension; this phenomenon was not observed in North American clinical trials.
		• A study in eight healthy volunteers has shown a 50% increase in mean peak nimodipine plasma concentrations and a 90% increase in mean area under the curve, after a one-week course of cimetidine at 1,000 mg/day and nimodipine at 90 mg/day. This effect may be mediated by the known inhibition of hepatic cytochrome P-450 by cimetidine, which could decrease first-pass metabolism of nimodipine.
198	Nitrazepam	• Nitrazepam may produce additive CNS depressant effects when coadministered with alcohol, sedative antihistamines, narcotic analgesics, anticonvulsants, or psychotropic medications which themselves can produce CNS depression.
		• Compounds which inhibit certain hepatic enzymes (particularly cytochrome P450) may enhance the activity of benzodiazepines. Examples include cimetidine or erythromycin.
199	Nitrofurantoin	• Decreased rate and extent of absorption of nitrofurantoin.
		• Possible inhibition of renal excretion of nitrofurantoin resulting in increased plasma nitrofurantoin concentrations (with increased risk of adverse effects) and decreased urine nitrofurantoin concentrations (with decreased efficacy in treatment of UTIs).
		• Some in vitro evidence of antagonism between quinolones and nitrofurantoin.
		• Possible false-positive reactions in urine glucose tests using Benedict's solution or Fehling's solution
200	Nitroprusside	• The hypotensive effects are additive when used with ganglionic blocking agents, negative inotropic agents, general anesthetics (e.g., halothane, enflurane), and most other circulatory depressants

TABLE 12.2 (Continued)

S. No.	Name of the Drug	Reported Drug Interaction
201	Noradrenaline	• Cyclopropane and halothane anesthetics increase cardiac autonomic irritability and therefore seem to sensitize the myocardium to the action of intravenously administered epinephrine or norepinephrine. Hence, the use of LEVOPHED (norepinephrine bitartrate) during cyclopropane and halothaneanesthesiais generally considered contraindicated because of the risk of producingventricular tachycardiaorfibrillation. The same type of cardiac arrhythmias may result from the use of LEVOPHED (norepinephrine bitartrate) in patients with profound hypoxia orhypercarbia. LEVOPHED (norepinephrine bitartrate) should be used with extreme caution in patients receiving monoamine oxidase inhibitors (MAOI) or antidepressants of the triptyline or imipramine types, because severe, prolonged hypertension may result.
202	Morfloxacin	• Quinolones, including norfloxacin, have been shown in vitro to inhibit CYP1A2. Concomitant use with drugs metabolized by CYP1A2 (e.g., caffeine, clozapine, ropinirole, tacrine, theophylline, tizanidine) may result in increased substrate drug concentrations when given in usual doses. Patients taking any of these drugs concomitantly with norfloxacin should be carefully monitored.
		• Elevated plasma levels of theophylline have been reported with concomitant quinolone use. There have been reports of theophylline-related side effects in patients on concomitant therapy with norfloxacin and theophylline. Therefore, monitoring of theophylline plasma levels should be considered and dosage of theophylline adjusted as required.
		• Elevated serum levels of cyclosporine have been reported with concomitant use of cyclosporine with norfloxacin. Therefore, cyclosporine serum levels should be monitored and appropriate cyclosporine dosage adjustments made when these drugs are used concomitantly.
		• Quinolones, including norfloxacin, may enhance the effects of oral anticoagulants, including warfarin or its derivatives or similar agents. When these products are administered concomitantly, prothrombin time or other suitablecoagulationtests should be closely monitored.

S. No.	Name of the Drug	Reported Drug Interaction
203	Norgestrel	• Contraceptive effectiveness may be reduced when hormonal contraceptives are coadministered with antibiotics, anticonvulsants, and other drugs that increase the metabolism of contraceptive steroids. This could result in unintended pregnancy or breakthrough bleeding. Examples include rifampin, rifabutin, barbiturates, primidone, phenylbutazone, phenytoin, dexamethasone, carbamazepine, felbamate, oxcarbazepine, topiramate, griseofulvin, and modafinil.
		• Several cases of contraceptive failure and breakthrough bleeding have been reported in the literature with concomitant administration of antibiotics such as ampicillin and other penicillins, and tetracyclines, possibly due to a decrease of enterohepatic recirculation of estrogens. However, clinical pharmacology studies investigating drug interactions between combined oral contraceptives and these antibiotics have reported inconsistent results. Enterohepatic recirculation of estrogens may also be decreased by substances that reduce gut transit time.
		• Several of the anti-HIV protease inhibitors have been studied with coadministration of oral combination hormonal contraceptives; significant changes (increase and decrease) in the plasma levels of the estrogen and progestin have been noted in some cases. The safety and efficacy of oral contraceptive products may be affected with coadministration of anti-HIV protease inhibitors. Health-care professionals should refer to the label of the individual anti-HIV protease inhibitors for further drug-drug interaction information.
204	Nortriptyline	• May block hypotensive actions of guanethidine and similar agents.
		• May enhance effects of alcohol use with caution in patients with a history of excessive alcohol consumption.
		• Possible pharmacokinetic (increased systemic exposure to nortriptyline) interaction with quinidine.

TABLE 12.2 (Continued)

S. No.	Name of the Drug	Reported Drug Interaction
205	Ofloxacin	• Quinolones form chelates with alkaline earth and transition metal cations. Administration of quinolones with antacids containing calcium, magnesium, or aluminum, with sucralfate, with divalent or trivalent cations such as iron, or with multivitamins containing zinc or with Videx® (didanosine) may substantially interfere with the absorption of quinolones resulting in systemic levels considerably lower than desired. These agents should not be taken within the two-hour period before or within the two-hour period after ofloxacin administration.
		• Interactions between ofloxacin and caffeine have not been detected.
		• Cimetidine has demonstrated interference with the elimination of some quinolones. This interference has resulted in significant increases in half-life and AUC of some quinolones. The potential for interaction between ofloxacin and cimetidine has not been studied
206	Omeprazole	• Potential to prolong elimination of drugs metabolized by oxidation in the liver. Interaction reported with drugs metabolized by CYP isoenzymes; monitor and adjust dosage of these drugs if necessary.
		• Potential pharmacologic interaction (possible increased risk of hypomagnesemia). Consider monitoring magnesium concentrations prior to initiation of prescription proton-pump inhibitor therapy and periodically thereafter
207	Ondansetron	• Inhibitors or inducers of CYP1A2, CYP2D6, or CYP3A4; potential pharmacokinetic interaction (altered ondansetron metabolism). Based on available data, no dosage adjustment recommended for patients on these drugs.
		• Potential additive effect on QT-interval prolongation.
		• May induce electrolyte disorders and increase risk of cardiac arrhythmias (e.g., QT-interval prolongation, torsades de pointes).
		• Substantial increase in ondansetron clearance (decreased plasma concentrations and half-life)
208	Oxazepam	• Additive CNS effect.
		• Apparent increase in oxazepam metabolism

S. No.	Name of the Drug	Reported Drug Interaction
209	Oxymetazoline	• Interactions occur with amitriptyline, clomipramine, desipramine, doxepin, imipramine, nortriptyline, ergotamine, dihydroergotamine, ergonovine, methylergonovine, isocarboxazid, linezolid, phenelzine, rasagiline, selegiline, and tranylcypromine.
210	Paclitaxel	• In a Phase 1 trial using escalating doses of TAXOL (110–200 mg/m²) and cisplatin (50 or 75 mg/m²) given as sequential infusions, myelosuppression was more profound when TAXOL was given after cisplatin than with the alternate sequence (ie, TAXOL before cisplatin). Pharmacokinetic data from these patients demonstrated a decrease in paclitaxel clearance of approximately 33% when TAXOL was administered following cisplatin.
		• The metabolism of TAXOL is catalyzed by cytochrome P450 isoenzymes CYP2C8 and CYP3A4. Caution should be exercised when TAXOL is concomitantly administered with known substrates (e.g., midazolam, buspirone, felodipine, lovastatin, eletriptan, sildenafil, simvastatin, and triazolam), inhibitors (e.g., atazanavir, clarithromycin, indinavir, itraconazole, ketoconazole, nefazodone, nelfinavir, ritonavir, saquinavir, and telithromycin), and inducers (e.g., rifampin and carbamazepine) of CYP3A4.
		• Potential interactions between TAXOL, a substrate of CYP3A4, and protease inhibitors (ritonavir, saquinavir, indinavir, and nelfinavir), which are substrates and/or inhibitors of CYP3A4, have not been evaluated in clinical trials.
211	Pancuronium	• Increased potency of neuromuscular blockade.
		• Possible ventricular arrhythmias in patients receiving tricyclic antiderpessants concomitantly with pancuronium and halothane.
		• Possible prolonged duration of neuromuscular blockade.
		• Prior administration of succinylcholine may increase potency and prolong duration of neuromuscular blockade.

TABLE 12.2 (Continued)

S. No.	Name of the Drug	Reported Drug Interaction
212	Paracetamol	• Increased risk of acetaminophen-induced hepatotoxicity.
		• Increased conversion of acetaminophen to hepatotoxic metabolites; increased risk of hepatotoxicity.
		• Possible increased PT.
		• Possible increased risk of hepatotoxicity.
213	Paroxetine	• Does not potentiate cognitive and motor effects of alcohol possible serotonergically mediated pharmacodynamic interaction in CNS.
		• Pharmacokinetic interactions unlikely.
		• Possible inhibition of metabolism by paroxetine.
		• Increased peak plasma concentrations, AUC, and elimination half-life of TCA.
		• Increased peak plasma concentrations and AUCs of atomoxetine.
		• Pharmacokinetics of paroxetine not affected.
		• Pharmacokinetic or pharmacologic interactions unlikely.
214	Pemoline	• Patients who are receiving CYLERT (pemoline) concurrently with other drugs, especially drugs with CNS activity, should be monitored carefully.
		• Decreased seizure threshold has been reported in patients receiving CYLERT (pemoline) concomitantly with *antiepileptic medications*.

S. No.	Name of the Drug	Reported Drug Interaction
215	Penicillin G	• *In vitro* evidence of synergistic antibacterial effects against enterococci or viridans streptococci;[1] used to therapeutic advantage in treatment of endocarditis and other severe enterococcal infections.
		• Potential in vitro and in vivo inactivation of aminoglycosides.
		• Possible decreased efficacy of estrogen-containing oral contraceptives and increased incidence of breakthrough bleeding reported with some penicillins (penicillin V, ampicillin).
		• Possible decreased renal clearance of methotrexate with penicillins: possible increased methotrexate concentrations and hematologic and GI toxicity.
		• Possible false-positive reactions in urine glucose tests using Clinitest, Benedict's solution, or Fehling's solution.
		• Possible falsely increased serum uric acid concentrations when copper-chelate method is used; phosphotungstate and uricase methods appear to be unaffected.
216	Pentazocine	• Possible additive effects.
		• Possible transient symptoms (e.g., diaphoresis, ataxia, flushing, tremor) suggestive of serotonin syndrome.
		• Possible decrease in urinary 17-hydroxycorticosteroid determinations (Porter-Silber reaction).
217	Perphenazine	• Potential additive CNS effects; concomitant use with alcohol potentiates hypotension observed with perphenazine.
		• Additive anticholinergic effect.
		• Perphenazine may lower seizure threshold.
		• Possible increased plasma concentrations of perphenazine.

TABLE 12.2 (Continued)

S. No.	Name of the Drug	Reported Drug Interaction
218	Phenindione	• Reduced effect with haloperidol. Increased effect with clofibrate.
		• Increased prothrombin levels with cotrimoxazole.
		• May increase toxicity of chlorpropamide, phenytoin, sulfonamides and tolbutamide.
		• Potential increase in anticoagulant effects:
		• Alcohol, anesthetics, antithyroids, aspirin and salicylates, aztreonam, bretylium, cathartics, cimetidine, ciprofloxacin, dextran, disulfiram, ethacrynic acid, fibrinolysin, gemfibrozil, glucagon, guanethidine, heparin, iodine, methaqualone, methotrexate, methyldopa, methylphenidate, nalidixic acid, nicotinic acid, NSAIDs, paracetamol, phenothiazines, phenytoin, probenecid, quinine, quinidine, sulfinpyrazone, thyroid hormones, tolbutamide, vitamin B complex, X-ray contrast media.
		• Potential decrease in anticoagulant effects: Antacids, antihistamines, barbiturates, bioflavonoids, digitalis glycosides, diuretics, ethchlorvynol, fibrinogen, glutethimide, griseofulvin, haloperidol, meprobamate, oral contraceptives, thrombin, trifluperidol, vitamin K, xanthines.
219	Pheniramine	• MAO inhibitors.
220	Phentolamine	• Possible false-positive test for pheochromocytoma.
		• Possible paradoxical fall in blood pressure.
221	Phenylbutazone	• Avoid combining with other antiinflammatory drugs that tend to cause GI ulcers, such as corticosteroids and other NSAIDs. Avoid combining with anticoagulant drugs particularly coumarin derivatives. Avoid combining with other hepatotoxic drugs.
		• Phenylbutazone may affect blood levels and duration of action of phentoin, valproic acid, sulfonamides, sulfonylurea antidiabetic agents, barbiturates, promethazine, rifampin, chlorpheniramine, diphenhydramine, penicillin G.
		• May reduce phenytoin or warfarin metabolism and methotrexate excretion.

S. No.	Name of the Drug	Reported Drug Interaction
222	Phenylephrine	• Vasopressor response to phenylephrine is decreased by prior administration of an α-adrenergic blocking agent.
		• Cardiostimulating effects of phenylephrine are blocked by prior administration of β-adrenergic blocking drug.
		• Cyclopropane or halogenated hydrocarbon general anesthetics increase cardiac irritability, may sensitize the myocardium to phenylephrine, and may result in arrhythmia.
		• May potentiate the vasopressor effects of phenylephrine.
		• Blocks the reflex bradycardia caused by phenylephrine and enhances the pressor response to phenylephrine.
		• Cardiac and pressor effects of phenylephrine are potentiated by prior administration of MAO inhibitors because metabolism of phenylephrine is reduced[b].
		• The potentiation is greater following oral administration of phenylephrine than after parenteral administration of the drug because reduction of the metabolism of phenylephrine in the intestine results in increased absorption of the drug
223	Phenytoin	• There are many drugs which may increase or decrease phenytoin levels or which phenytoin may affect. The most commonly occurring drug interactions are listed below (see also Drug Interactions table in the Clin-Info section).
		• When adding or deleting phenytoin from a patient's therapeutic regimen, pharmacotherapy must be monitored closely as dosage adjustment may be necessary. Serum level determinations of each drug are especially helpful when possible drug interactions are suspected.
		• Drugs which may increase phenytoin serum levels include: amiodarone, chloramphenicol, cimetidine, disulfiram, erythromycin, fluconazole, fluoxetine, isoniazid, ketoconazole, methylphenidate, omeprazole, phenylbutazone, salicylates, sulfonamides, trazodone, warfarin and acute alcohol ingestion.

TABLE 12.2 (Continued)

S. No.	Name of the Drug	Reported Drug Interaction
		• Drugs which may decrease phenytoin levels include: carbamazepine, chronic alcohol abuse, diazoxide, rifampin and theophylline.
		• Drugs which may either increase or decrease phenytoin serum levels include: phenobarbital, valproic acid, and sodium valproate. Similarly, the effect of phenytoin on phenobarbital, valproic acid and sodium valproate serum levels are unpredictable.
		• Drugs whose efficacy is impaired by phenytoin include: corticosteroids, diazoxide, digitalis glycosides, doxycycline, estrogens, furosemide, levodopa, methadone, oral contraceptives, quinidine, theophylline, vitamin D and warfarin.
		• Administration of phenytoin with sulcralfate, enteral feeds, antacids or calcium preparations should be separated by at least 3 h to prevent a decrease in phenytoin absorption.
224	Physostigmine	• Aminoglycosides, clindamycin, colistin, cyclopropane, halogenated inhalational anesthetics and atropine may antagonize the effects of physotigmine. Quinine, chloroquine, hydroxychloroquine, quinidine, procainamide, propafenone, lithium and β-blockers may reduce treatment efficacy. Prolonged bradycardia may occur when taken with β-blockers. May prolong the action of suxamethonium when used together.
225	Pilocarpine	• Additive IOP lowering effect.
		• Competitive inhibition of miotic effect and presumably IOP-lowering effect.
226	Pindolol	• Pharmacokinetic interaction unlikely.
		• Possible decreases in serum digoxin concentrations.
		• Possible increased hypotensive effects.
		• Increased serum concentrations of thioridazine (Seth, 2004) and metabolites; higher than expected serum concentrations of pindolol.
		• Increased thioridazine concentrations may cause prolongation of the QT_c interval and a possible increase in the risk of serious, potentially fatal cardiac arrhythmia (e.g., torsades de pointes).

S. No.	Name of the Drug	Reported Drug Interaction
227	Pipecuronium	• Actions antagonized by cholinesterases and long-term carbamazepine, phenytoin or corticosteroids usage. Enhanced block when used with drugs that have neuromuscular blocking activity such as lidocaine, quinidine, verapamil and aminoglycosides. • **Potentially Fatal:** Effects enhanced by volatile inhalational anesthetics, ketamine (IV), antiarrhythmics, antibacterials, K depleting diuretics, parenteral Mg salts.
228	Piperazine	• Antagonism if pyrantel and piperazine are used together. Piperazine may potentiate extrapyramidal effects of chlorpromazine and other phenothiazines.
229	Piroxicam	• Reduced BP response to ACE inhibitor. • Reduced BP response to angiotensin II receptor antagonist. • Possible bleeding complications. • Reduced natriuretic effects. • NSAIAs including aspirin: Increased risk of GI ulceration and other complications. • Aspirin: No consistent evidence that use of low-dose aspirin mitigates the increased risk of serious cardiovascular events associated with NSAIAs. • Decreased plasma piroxicam concentrations with concomitant use of 20 mg piroxicam and 3.9 g aspirin daily.
230	Pralidoxime	• **Potentially Fatal:** Potentiates toxicity by carbamate pesticides. When atropine and pralidoxime chloride are used together, the signs of atropinization (flushing, mydriasis, tachycardia, dryness of the mouth and nose) may occur earlier than might be expected when atropine is used alone. This is especially true if the total dose of atropine has been large and the administration of pralidoxime chloride has been delayed.

TABLE 12.2 (Continued)

S. No.	Name of the Drug	Reported Drug Interaction
		• The following precautions should be kept in mind in the treatment of anticholinesterase poisoning, although they do not bear directly on the use of pralidoxime chloride: since barbiturates are potentiated by the anticholinesterases, they should be used cautiously in the treatment of convulsions; morphine, theophylline, aminophylline, reserpine, and phenothiazine-type tranquilizers should be avoided in patients with organophosphate poisoning. Prolonged paralysis has been reported in patients when succinylcholine is given with drugs having anticholinesterase activity; therefore, it should be used with caution.
231	Praziquantel	• Plasma praziquantel concentrations were undetectable in 7 out of 10 subjects. When a single 40 mg/kg dose of praziquantel was administered to these healthy subjects two weeks after discontinuation of rifampin, the mean praziquantel AUC and Cmax were 23% and 35% lower, respectively, than when praziquantel was given alone. In patients receiving rifampin, for example, as part of a combination regimen for the treatment oftuberculosis, alternative agents forschistosomiasisshould be considered.
232	Prednisolone	• Inhibitors of CYP3A4: Potential pharmacokinetic interaction (decreased prednisolone metabolism). • May enhance the potassium-wasting effect of glucocorticoids^c. • Conflicting reports of diminished as well as enhanced response to anticoagulants. • Severe weakness with concomitant use of anticholinesterase agents and corticosteroids in patients with myasthenia gravis. • Glucocorticoids may increase blood glucose concentration.
233	Prednisone	• May enhance the potassium-wasting effect of glucocorticoids. • Conflicting reports of diminished as well as enhanced response to anticoagulants. • Severe weakness with concomitant use of anticholinesterase agents and corticosteroids in patients with myasthenia gravis.

S. No.	Name of the Drug	Reported Drug Interaction
		• Glucocorticoids may increase blood glucose concentration.
		• Concomitant administration of prednisone and cyclosporine may result in inhibition of metabolism of either agent. Seizures reportedly have occurred in adult and pediatric patients receiving high-dose glucocorticoid therapy concurrently with cyclosporine.
234	Primaquine	• Drug Interactions Caution is advised if primaquine is used concomitantly with other drugs that prolong the QT interval.
235	Primidone	• Possible increased metabolism of both the estrogenic and progestinic components of oral contraceptives.
		• Possible increased sedation.
		• Possible increase in amount of primidone converted to phenobarbital and increased sedation.
		• Increased plasma phenobarbital concentrations and excessive somnolence.
236	Probenecid	• Possible increased peak plasma concentrations of acetaminophen.
		• Potential for increased serum urate concentrations.
		• Probenecid increases excretion of allopurinol's active metabolite.
		• Possible increased plasma concentrations of oral sulfonylurea antidiabetic agents resulting in hypoglycemia.
		• Increased serum urate concentrations; possible uric acid nephropathy.
		• Decreased renal excretion of β-lactam antibiotics and increased concentrations of β-lactam antibiotics[a].
		• Psychic disturbances reported with concomitant use.

TABLE 12.2 (Continued)

S. No.	Name of the Drug	Reported Drug Interaction
237	Procainamide	• Possibility that potentially serious cardiac arrhythmias, including torsades de pointes, could occur if procainamide were used concomitantly with other drugs that prolong the QT_c interval. Use with caution, if at all, in combination with other drugs that prolong the QT interval; consider expert consultation.
		• Concurrent use with class IA antiarrhythmics (e.g., disopyramide, quinidine) can enhance conduction prolongation, contractility depression, and hypotension, especially in cardiac decompensation. Reserve combined use for serious arrhythmias unresponsive to monotherapy; monitor closely.
		• Enhances acetylation of procainamide to NAPA; alcohol consumption may reduce half-life.
		• Possible increased plasma procainamide and NAPA concentrations and subsequent toxicity
238	Procaine	• Possible antagonism of aminosalicylic acid activity.
		• Possible severe, prolonged hypertension or disturbances of cardiac rhythm due to epinephrine component.
		• Possible reduction or reversal of pressor effect of epinephrine.
		• Possible severe, persistent hypertension or cerebrovascular accidents (e.g., rupture of cerebral blood vessel) due to epinephrine component.
		• Possible antagonism of sulfonamide activity
239	Procarbazine	• Possible disulfiram-like reactions.
		• Potential serious and life-threatening serotonin syndrome.
		• Possible additive CNS depression.
		• Potential for hypertensive reactions.

S. No.	Name of the Drug	Reported Drug Interaction
240	Proguanil	• Decreased plasma atovaquone concentrations.
		• Decreased trough concentrations of indinavir; no change in peak plasma concentrations or AUC of indinavir.
		• Decreased bioavailability of atovaquone.
		• Possible potentiation of the anticoagulant effects of warfarin
241	Promethazine	• Additive effects.
		• Reversal of vasopressor effect of epinephrine.
		• Additive anticholinergic effects.
		• Increased extrapyramidal effects.
		• Possible test interference: false-positive Gravindextest and false-negative Prepurexand Daptest.
242	Propantheline	• Decrease rate but not extent of acetaminophen absorption; may delay onset of acetaminophen therapeutic effects.
		• Possible decreased absorption of antimuscarinic.
		• Possible additive adverse anticholinergic effect.
		• Possible additive adverse anticholinergic effects.
		• Increased serum digoxin concentration with slowly dissolving digoxin tablets.
243	Prothionamide	• Cycloserine: possible increased risk of neurotoxicity.
		• Isoniazid: increased serum concentrations.
		• P-aminosalicylic acid: increased risk of hypothyroidism, possible increased risk of hepatoxicity.
		• Rifampicin: increased risk of hepatoxicity.

TABLE 12.2 (Continued)

S. No.	Name of the Drug	Reported Drug Interaction
244	Pyrazinamide	• Severe liver injuries, including some fatalities, reported in patients receiving a 2-month daily regimen of rifampin and pyrazinamide for treatment of LTBI.
		• Pyrazinamide produces a pink-brown color that may interfere with interpretation of test.
245	Pyridostigmine	• Possible delayed onset of IV pyridostigmine.
		• Possible prolongation of neuromuscular blockade or resistance to reversal if used in conjunction with nondepolarising neuromuscular blocking agents.
		• Interfere with neuromuscular transmission.
		• Antagonizes muscarinic effects of pyridostigmine.
		• Converted to pantothenic acid in vivo; possible additive effects due to increased acetylcholine production.
246	Pyrimethamine	• Possibility that sulfonamide component of Fansidar may potentiate the effects of coumarin anticoagulants by displacing them from their protein-binding sites or by impairing anticoagulant metabolism.
		• Increased incidence and severity of adverse effects reported when Fansidar used with chloroquine.
		• Risk of adverse hematologic effects.
		• Additive adverse hematologic effects; increased risk of agranulocytosis.
		• Because pyrimethamine and sulfonamides interfere with folic acid synthesis in susceptible organisms, this synergism is used to therapeutic advantage in the treatment of toxoplasmosisand has been used to therapeutic advantage in the prevention or treatment of malaria.
		• Increased risk of bone marrow suppression if pyrimethamine or Fansidaris used with other folic acid antagonists.

S. No.	Name of the Drug	Reported Drug Interaction
247	Quinidine	• Appears to be metabolized principally by CYP3A4 Not metabolized by CYP2D6.
		• Inhibits CYP2D6; therapeutic serum quinidine concentrations may effectively convert CYP2D6 extensive metabolizers into CYP2D6 poor metabolizers.
		• Drugs that increase urine pH may reduce renal excretion of quinidine; toxicity may occur.
		• Increased serum quinidine concentrations.
		• Concomitant administration may delay oral absorption of quinidine.
		• Concomitant use of aluminum hydroxide antacid and oral quinidine gluconate does not have a clinically important effect on quinidine absorption.
		• Quinidine potentiates anticoagulant effect of warfarin.
248	Quinine	• Metabolized principally by CYP3A4; may also be metabolized by other CYP enzymes, including 1A2, 2C8, 2C9, 2C19, 2D6, and 2E1.
		• quinine prolongs the QT interval, an additive effect on the QT interval might occur if the drug is administered with other drugs that prolong the QT interval. Concomitant use of quinine in patients receiving other drugs known to cause QT prolongation, including class IA antiarrhythmic agents (e.g., quinidine, procainamide, disopyramide) and class III antiarrhythmic agents (e.g., amiodarone, sotalol, dofetilide) not recommended.
		• Delayed or decreased absorption of quinine.
		• Possible interference with the anticoagulant effects of heparin
249	Ramipril	• Pharmacokinetic interaction unlikely.
		• Possible hypoglycemia in diabetic patients.
		• Increased hypotensive effect.
		• Increased serum lithium concentrations; possible toxicity.

TABLE 12.2 (Continued)

S. No.	Name of the Drug	Reported Drug Interaction
250	Ranitidine	• Potential for reduction of renal function and increase in serum potassium.
		• No interaction observed with indomethacin.
		• Dose-dependent inhibition of acetaminophen metabolism in vitro.
		• Moderate alcohol consumption by individuals receiving concurrent ranitidine unlikely to result in clinically important alterations of blood alcohol concentration and/or alcohol metabolism.
		• Low doses (10–15 mEq HCl neutralizing capacity/10 mL) do not appear to decrease absorption or plasma concentrations of ranitidine (Barar, 2000; Das, 2004; Katzung, 2007; Mozayani, 2011; Seth, 2004). Higher doses (e.g., 150 mEq HCl neutralizing capacity/30 mL) decrease absorption by 33%, decrease plasma concentrations, and AUC.
		• Lorazepam elimination half-life, volume of distribution, clearance unaffected.
		• Midazolam oral bioavailability may be increased by ranitidine.
		• Triazolam oral bioavailability may be increased by elevated gastric pH (Seth, 2004) clinical importance unknown.
251	Reserpine	• MAO inhibitors (isocarboxazid, linezolid, methylene blue, moclobemide, phenelzine, procarbazine, rasagiline, selegiline, tranylcypromine).\
		• Check the labels on all your medicines (such as cough-and-cold products, diet aids, nonsteroidal antiinflammatory drugs-NSAIDs such as ibuprofen for pain/fever reduction) because they may contain ingredients that cause drowsiness or could increase your blood pressure or heart rate.
		• This medication may interfere with certain urine laboratory tests (including 17-OHCS and 17-ketosteroids test), possibly causing false test results.

S. No.	Name of the Drug	Reported Drug Interaction
252	Ribavirin	• The use of ribavirin for the treatment of chronic hepatitis C in patients receiving azathioprine has been reported to induce severe pancytopenia and may increase the risk of azathioprine-related myelotoxicity. Inosine monophosphate dehydrogenase (IMDH) is required for one of the metabolic pathways of azathioprine.
		• Ribavirin is known to inhibit IMDH, thereby leading to accumulation of an azathioprine metabolite, 6-methylthioinosine monophosphate (6-MTITP), which is associated with myelotoxicity (neutropenia, thrombocytopenia, and anemia).
		• Patients receiving azathioprine with ribavirin should have complete blood counts, including platelet counts, monitored weekly for the first month, twice monthly for the second and third months of treatment, then monthly or more frequently if dosage or other therapy changes are necessary.
253	Rifampicin	• Rifampin is known to induce certain cytochrome P-450 enzymes. Administration of rifampin with drugs that undergo biotransformation through these metabolic pathways may accelerate elimination of coadministered drugs. To maintain optimum therapeutic blood levels, dosages of drugs metabolized by these enzymes may require adjustment when starting or stopping concomitantly administered rifampin.
		• Rifampin has been reported to substantially decrease the plasma concentrations of the following antiviral drugs: atazanavir, darunavir, fosamprenavir, saquinavir, and tipranavir. These antiviral drugs must not be coadministered with rifampin.
		• Rifampin has been reported to accelerate the metabolism of the following drugs: anticonvulsants (e.g., phenytoin), digitoxin, antiarrhythmics (e.g., disopyramide, mexiletine, quinidine, tocainide), oral anticoagulants, antifungals (e.g., fluconazole, itraconazole, ketoconazole), barbiturates, beta-blockers, calcium channel blockers (e.g., diltiazem, nifedipine, verapamil), chloramphenicol, clarithromycin, corticosteroids, cyclosporine, cardiac glycoside preparations, clofibrate, oral or other systemic hormonal contraceptives, dapsone, diazepam, doxycycline, fluoroquinolones (e.g., ciprofloxacin), haloperidol, oral hypoglycemic agents (sulfonylureas), levothyroxine, methadone, narcotic analgesics, progestins, quinine, tacrolimus, theophylline, tricyclic antidepressants (e.g., amitriptyline, nortriptyline) and zidovudine. It may be necessary to adjust the dosages of these drugs if they are given concurrently with rifampin.

TABLE 12.2 (Continued)

S. No.	Name of the Drug	Reported Drug Interaction
254	Risperidone	• Patients using oral or other systemic hormonal contraceptives should be advised to change to nonhormonal methods of birth control during rifampin therapy.
		• The dose of RISPERDAL® should be adjusted when used in combination with CYP2D6 enzyme inhibitors (e.g., fluoxetine, and paroxetine) and enzyme inducers (e.g., carbamazepine).
		• Repeated oral doses of RISPERDAL® (3 mg twice daily) did not affect the exposure (AUC) or peak plasma concentrations (Cmax) of lithium (n=13). Dose adjustment for lithium is not recommended.
		• Repeated oral doses of RISPERDAL® (4 mg once daily) did not affect the predose or average plasma concentrations and exposure (AUC) of valproate (1000 mg/day in three divided doses) compared to placebo ($n = 21$). However, there was a 20% increase in valproate peak plasma concentration (Cmax) after concomitant administration of RISPERDAL®. Dose adjustment for valproate is not recommended.
		• RISPERDAL® (0.25 mg twice daily) did not show a clinically relevant effect on the pharmacokinetics of digoxin. Dose adjustment for digoxin is not recommended.
		• Given the primary CNS effects of risperidone, caution should be used when RISPERDAL® is taken in combination with other centrally acting drugs and alcohol.
255	Ritodrine	• Streptococcal infection, recent (antistreptococcal antibodies are likely to be present in the circulation; these antibodies may cause a temporary resistance to the therapeutic effects of anistreplase or streptokinase and/or an increased risk of severe allergic reactions to the medication; although resistance may be overcome by increasing the dosage, use of an alternate thrombolytic agent [alteplase or urokinase] is advisable if thrombolytic therapy is needed within 1 year after anistreplase or streptokinase therapyor streptococcal infection).
		• Trauma, severe, recent, other than to the CNS. Tuberculosis, active, with cavitation of recent onset. Infection at or near site of thrombus, obstructed intravenous catheter, or occluded arteriovenous cannula (risk of spreading the infection into and via the circulation).

S. No.	Name of the Drug	Reported Drug Interaction
256	Salbutamide	• Beta-adrenergic-receptor blocking agents not only block the pulmonary effect of beta-agonists, such as PROVENTIL HFA Inhalation Aerosol, but may produce severe bronchospasm in asthmatic patients. Therefore, patients with asthma should not normally be treated with beta-blockers. However, under certain circumstances, e.g., as prophylaxis after myocardial infarction, there may be no acceptable alternatives to the use of beta-adrenergic blocking agents in patients with asthma. In this setting, cardioselective beta-blockers should be considered, although they should be administered with caution.
		• The ECG changes and/or hypokalemia which may result from the administration of nonpotassium-sparing diuretics (such as loop or thiazide diuretics) can be acutely worsened by beta-agonists, especially when the recommended dose of the beta-agonist is exceeded. Although the clinical significance of these effects is not known, caution is advised in the coadministration of beta-agonists with nonpotassium-sparing diuretics.
		• PROVENTIL HFA Inhalation Aerosol should be administered with extreme caution to patients being treated with monoamine oxidase inhibitors or tricyclic antidepressants, or within 2 weeks of discontinuation of such agents, because the action of albuterol on the cardiovascular system may be potentiated.
257	Salmetrol	• Salmeterol is a substrate of CYP3A4. The use of strong CYP3A4 inhibitors (e.g., ritonavir, atazanavir, clarithromycin, indinavir, itraconazole, nefazodone, nelfinavir, saquinavir, ketoconazole, telithromycin) with SEREVENT DISKUS is not recommended because increasedcardiovascular adverse effects may occur.
		• SEREVENT DISKUS should be administered with extreme caution to patients being treated with monoamine oxidase inhibitors or tricyclic antidepressants, or within 2 weeks of discontinuation of such agents, because the action of salmeterol on the vascular system may be potentiated by these agents.
		• In one study, concomitant administration of digoxin with simvastatin resulted in a slight elevation in digoxin concentrations in plasma. Patients taking digoxin should be monitored appropriately when simvastatin is initiated.

TABLE 12.2 (Continued)

S. No.	Name of the Drug	Reported Drug Interaction
258	Selegilline	• The occurrence of stupor, muscular rigidity, severe agitation, and elevated temperature has been reported in some patients receiving the combination of selegiline and meperidine. Symptoms usually resolve over days when the combination is discontinued. This is typical of the interaction of meperidine and MAOIs.
		• Other serious reactions (including severe agitation, hallucinations, and death) have been reported in patients receiving this combination. Severe toxicity has also been reported in patients receiving the combination of tricyclic antidepressants and ELDEPRYL (selegilinehcl) and selective serotoninreuptake inhibitors and ELDEPRYL.
		• One case of hypertensive crisis has been reported in a patient taking the recommended doses of selegiline and a sympathomimetic medication (ephedrine).
259	Simvastatin	• Strong CYP3A4 inhibitors: Simvastatin, like several other inhibitors of HMG-CoA reductase, is a substrate of CYP3A4. Simvastatin is metabolized by CYP3A4 but has no CYP3A4 inhibitory activity; therefore it is not expected to affect the plasma concentrations of other drugs metabolized by CYP3A4.
		• Elevated plasma levels of HMG-CoA reductase inhibitory activity increases the risk of myopathy and rhabdomyolysis, particularly with higher doses of simvastatin.
		• Cyclosporine or Danazol: The risk of myopathy, including rhabdomyolysis is increased by concomitant administration of cyclosporine or danazol. Therefore, concomitant use of these drugs is contraindicated.
260	Sparfloxacin	• Sparfloxacin has no effect on the pharmacokinetics of digoxin.
		• Sparfloxacin does not increase the anticoagulant effect of warfarin.

S. No.	Name of the Drug	Reported Drug Interaction
		• Aluminum and magnesium cations in antacids and sucralfate form chelation complexes with sparfloxacin. The oral bioavailability of sparfloxacin is reduced when an aluminum–magnesium suspension is administered between 2 h before and 2 h after sparfloxacin administration. Similarly, the oral bioavailability of sparfloxacin may be reduced when Videx® (Didanosine), chewable/buffered tablets or the pediatric powder for oral solution is administered between 2 h before and 2 h after sparfloxacin administration. The oral bioavailability of sparfloxacin is not reduced when the aluminum–magnesium suspension is administered 4 h following sparfloxacin administration.
		• Absorption of quinolones is reduced significantly by these preparations. These products may be taken 4 h after sparfloxacin administration.
261	Strptokinase	• Anticoagulants, agents that alter platelet function (e.g., aspirin, other NSAIDs, dipyridamole), other thrombolytic agents, agents that alter coagulation.
		• May increase the risk of bleeding.
		• Do not add other medication to the streptokinase container.
		• Will cause marked decreases in plasminogen and fibrinogen levels and increases in thrombin time, activated partial thromboplastin time, and prothrombin time, which usually normalize within 12 to 24 h.
262	Succinylcholine	• Some in vitro evidence of antagonism with aminoglycosides;in vivo antagonism has not been demonstrated and the drugs have been administered concomitantly with no apparent decrease in activity.
		• Possible increased incidence of nephrotoxicity and/or neurotoxicity.
		• In vitro evidence of additive or synergistic antibacterial effects between penicillins and aminoglycosides against some enterococci, viridans streptococci, Enterobacteriaceae, or Ps. aeruginosa; used to therapeutic advantage (e.g., treatment of endocarditis).
		• Possible increased incidence of nephrotoxicity reported with some cephalosporins (Seth, 2004).
		• Potential in vitro and in vivo inactivation of aminoglycosides.

TABLE 12.2 (Continued)

S. No.	Name of the Drug	Reported Drug Interaction
263	Strptomycin	• Possible increased incidence of nephrotoxicity and/or neurotoxicity.
		• *In vitro* evidence of additive or synergistic antibacterial effects between penicillins and aminoglycosides against some enterococci, viridans streptococci, Enterobacteriaceae, or *Ps. aeruginosa*; used to therapeutic advantage (e.g., treatment of endocarditis).
		• Possible increased incidence of nephrotoxicity reported with some cephalosporins (Seth, 2004).
		• Potential in vitro and in vivo inactivation of aminoglycosides.
		• *In vitro* evidence of additive or synergistic antibacterial effects with aminoglycosides against some gram-positive bacteria (*E. fecalis, S. aureus, L. monocytogenes*).
264	Sucralfate	• Some studies have shown that simultaneous sucralfate administration in healthy volunteers reduced the extent of absorption (bioavailability) of single doses of the following: cimetidine, digoxin, fluoroquinolone antibiotics, ketoconazole, l-thyroxine, phenytoin, quinidine, ranitidine, tetracycline, and theophylline. Subtherapeutic prothrombin times with concomitant warfarin and sucralfate therapy have been reported in spontaneous and published case reports. However, two clinical studies have demonstrated no change in either serum warfarin concentration or prothrombin time with the addition of sucralfate to chronic warfarin therapy.
265	Sulfonamide	• Possibility that sulfonamides may potentiate the effects of coumarin anticoagulants by displacing them from their protein-binding sites or by impairing anticoagulant metabolism.
		• Sulfonamides may potentiate the hypoglycemic effects by displacing the antidiabetic agents from their protein-binding sites.
		• Possible decreased GI absorption of digoxin.
		• May displace sulfonamides from plasma albumin and increase concentrations of free drug in plasma.
		• May displace sulfonamides from plasma albumin and increase concentrations of free drug in plasma.

S. No.	Name of the Drug	Reported Drug Interaction
266	Sumatriptan	• Pretreatment with oral sumatriptan followed by acetaminophen affected rate, but not extent of acetaminophen absorption over 8 h.
		• Administration of alcohol 30 min prior to oral sumatriptan did not affect sumatriptan pharmacokinetics.
		• Concomitant use did not affect sumatriptan efficacy.
		• Potentially life-threatening serotonin syndrome.
		• MAO-A inhibitors decrease sumatriptan clearance, resulting in substantially increased systemic exposure; no substantial effect on sumatriptan metabolism seen with an MAO-B inhibitor.
267	Suxamethonium	• Concurrent use with anticholinesterases, cyclophosphamide, antiarrhythmics, aminoglycosides, lincosamides (clindamycin and lincomycin), anticonvulsants, phenelzine, magnesium, metoclopramide, inhalation anesthetics, exposure to organophosphate insecticides may enhance neuromuscular block of suxamethonium. Increased risk of arrhythmias with cardiac glycosides.
268	Tacrine	• Tacrine is primarily eliminated byhepaticmetabolismvia cytochrome P450 drug metabolizingenzymes. Drug-drug interactions may occur when Cognex® (tacrine) is given concurrently with agents such as theophylline that undergo extensive metabolism via cytochrome P450 IA2.
		• Coadministration of tacrine with theophylline increased theophylline elimination half-life and averageplasmatheophylline concentrations by approx-imately 2-fold. Therefore, monitoring of plasma theophylline concentrations and appropriate reduction of theophylline dose are recommended in patients receiving tacrine and theophylline concurrently. The effect of theophylline on tacrine pharmacokinetics has not been assessed.
		• Cimetidine increased the Cmax and AUC of tacrine by approximately 54% and 64%, respectively.
		• Because of its mechanism of action, Cognex® (tacrine) has the potential to interfere with the activity ofanticholinergicmedications.

TABLE 12.2   (Continued)

S. No.	Name of the Drug	Reported Drug Interaction
269	Tacrolimus	• With a given dose of mycophenolic acid (MPA) products, exposure to MPA is higher with Prograf coadministration than with cyclosporine coadministration because cyclosporine interrupts the enterohepatic recirculation of MPA while tacrolimus does not. Clinicians should be aware that there is also a potential for increased MPA exposure after crossover from cyclosporine to Prograf in patients concomitantly receiving MPA-containing products.
		• Grapefruit juice inhibits CYP3A-enzymes resulting in increased tacrolimus whole blood trough concentrations, and patients should avoid eating grapefruit or drinking grapefruit juice with tacrolimus.
		• Most protease inhibitors inhibit CYP3A enzymes and may increase tacrolimus whole blood concentrations. It is recommended to avoid concomitant use of tacrolimus with nelfinavir unless the benefits outweigh the risks.
		• Whole blood concentrations of tacrolimus are markedly increased when coadministered with telaprevir or with boceprevir.
		• Monitoring of tacrolimus whole blood concentrations and tacrolimus-associated adverse reactions, and appropriate adjustments in the dosing regimen of tacrolimus are recommended when tacrolimus and protease inhibitors (e.g., ritonavir; telaprevir, boceprevir) are used concomitantly.
270	Tamoxifen	• When NOLVADEX (tamoxifen citrate) is used in combination with coumarin-type anticoagulants, a significant increase in anticoagulant effect may occur. Where such coadministration exists, careful monitoring of the patient's prothrombin time is recommended.
		• In the NSABP P-1 trial, women who required coumarin-type anticoagulants for any reason were ineligible for participation in the trial.
		• There is an increased risk of thromboembolic events occurring when cytotoxic agents are used in combination with NOLVADEX (tamoxifen citrate).

S. No.	Name of the Drug	Reported Drug Interaction
		• Tamoxifen reduced letrozole plasma concentrations by 37%. The effect of tamoxifen on metabolism and excretion of other antineoplastic drugs, such as cyclophosphamide and other drugs that require mixed function oxidases for activation, is not known. Tamoxifen and N-desmethyltamoxifen plasma concentrations have been shown to be reduced when coadministered with rifampin or aminoglutethimide. Induction of CYP3A4-mediated metabolism is considered to be the mechanism by which these reductions occur; other CYP3A4 inducing agents have not been studied to confirm this effect.
		• Concomitant bromocriptine therapy has been shown to elevate serum tamoxifen and N-desmethyltamoxifen.
271	Tetrazosin	• Concomitant administration of HYTRIN (terazosin hcl) with a phosphodiesterase-5 (PDE-5) inhibitor can result in additive blood pressure lowering effects and symptomatic hypotension.
		• In controlled trials, HYTRIN (terazosin hcl) tablets have been added to diuretics, and several beta-adrenergic blockers; no unexpected interactions were observed. HYTRIN (terazosin hcl) tablets have also been used in patients on a variety of concomitant therapies; while these were not formal interaction studies, no interactions were observed. HYTRIN (terazosin hcl) tablets have been used concomitantly in at least 50 patients on the following drugs or drug classes: 1) analgesic/antiinflammatory (e.g., acetaminophen, aspirin, codeine, ibuprofen, indomethacin); 2) antibiotics (e.g., erythromycin, trimethoprim and sulfamethoxazole)
272	Terbinafine	• Co-administration of a single dose of fluconazole (100 mg) with a single dose of terbinafine resulted in a 52% and 69% increase in terbinafine Cmax and AUC, respectively. Fluconazole is an inhibitor of CYP2C9 and CYP3A enzymes. Based on this finding, it is likely that other inhibitors of both CYP2C9 and CYP3A4 (e.g., ketoconazole, amiodarone) may also lead to a substantial increase in the systemic exposure (Cmax and AUC) of terbinafine when concomitantly administered.
		• There have been spontaneous reports of increase or decrease in prothrombin times in patients concomitantly taking oral terbinafine and warfarin, however, a causal relationship between Lamisil Tablets and these changes has not been established.

TABLE 12.2 (Continued)

S. No.	Name of the Drug	Reported Drug Interaction
273	Terbutaline	• The influence of terbinafine on the pharmacokinetics of fluconazole, cotrimoxazole (trimethoprim and sulfamethoxazole), zidovudine or theophylline was not considered to be clinically significant.
		• Terbutaline should be administered with extreme caution to patients being treated with monoamine oxidase inhibitors or tricyclic antidepressants, or within 2 weeks of discontinuation of such agents, since the action of terbutaline on the vascular system may be potentiated.
		• Beta-adrenergic receptor blocking agents not only block the pulmonary effect of beta-agonists, such as terbutaline, but may produce severe bronchospasm in asthmatic patients. Therefore, patients with asthma should not normally be treated with beta-blockers. However, under certain circumstances, e.g., as prophylaxis after myocardial infarction, there may be no acceptable alternatives to the use of beta-adrenergic blocking agents in patients with asthma. In this setting, cardioselective beta-blockers could be considered, although they should be administered with caution.
		• The ECG changes and/or hypokalemia that may result from the administration of nonpotassium-sparing diuretics (such as loop or thiazide diuretics) can be acutely worsened by beta-agonists, especially when the recommended dose of the beta-agonist is exceeded. Although the clinical significance of these effects is not known, caution is advised in the coadministration of beta-agonists with nonpotassium-sparing diuretics.
274	Tetracycline	• Bacteriostatic drugs like tetracycline may interfere with the bactericidal action of penicillin, it is advisable to avoid giving tetracycline in conjunction with penicillin.
		• Tetracyclines have been shown to depress plasma prothrombin activity, patients who are on anticoagulant therapy may require downward adjustment of their anticoagulant dosage.
		• Concurrent use of tetracycline may render oral contraceptives less effective.
		• The concurrent use of tetracycline and methoxyflurane has been reported to result in fatal renal toxicity.

S. No.	Name of the Drug	Reported Drug Interaction
275	Theophylline	• Theophylline interacts with a wide variety of drugs. The interaction may be pharmacodynamic, i.e., alterations in the therapeutic response to theophylline or another drug or occurrence of adverse effects without a change in serum theophylline concentration. More frequently, however, the interaction is pharmacokinetic, i.e., the rate of theophylline clearance is altered by another drug resulting in increased or decreased serum theophylline concentrations. Theophylline only rarely alters the pharmacokinetics of other drugs.
		• Theophylline blocks adenosine receptors.
		• A single large dose of alcohol (3 mL/kg of whiskey) decreases theophylline clearance for up to 24 h.
		• Decreases theophylline clearance at allopurinol doses $\geq$600 mg/day.
		• Increases theophylline clearance by induction of microsomal enzyme activity.
276	Thiabendazole	• Theophylline, for sites of metabolism in the liver, thus elevating the serum levels of such compounds to potentially toxic levels. Therefore, when concomitant use of thiabendazole and xanthine derivatives is anticipated, it may be necessary to monitor blood levels and/or reduce the dosage of such compounds. Such concomitant use should be administered under careful medical supervision.
277	Thioridazine	• Reduced cytochrome P450 2D6 isozyme activity, drugs which inhibit this isozyme (e.g., fluoxetine and paroxetine), and certain other drugs (e.g., fluvoxamine, propranolol, and pindolol) appear to appreciably inhibit the metabolism of thioridazine. The resulting elevated levels of thioridazine would be expected to augment the prolongation of the QTc interval associated with thioridazine and may increase the risk of serious, potentially fatal. cardiac arrhythmias, such as torsade de pointes-type arrhythmias. Such an increased risk may result also from the additive effect of coadministering thioridazine with other agents that prolong the QTc interval. Therefore, thioridazine is contraindicated with these drugs as well as in patients, comprising about 7% of the normal population, who are known to have a genetic defect leading to reduced levels of activity of P450 2D6.

TABLE 12.2 (Continued)

S. No.	Name of the Drug	Reported Drug Interaction
278	ThioTEPA	• Thiotepa injection combined with other alkylating agents such as nitrogen mustard or cyclophosphamide or thiotepa injection combined with irradiation would serve to intensify toxicity rather than to enhance therapeutic response. If these agents must follow each other, it is important that recovery from the first agent, as indicated by white blood cell count, be complete before therapy with the second agent is instituted.
279	Ticlopidine	• Therapeutic doses of TICLID (ticlopidine hcl) caused a 30% increase in the plasma half-life of antipyrine and may cause analogous effects on similarly metabolized drugs. Therefore, the dose of drugs metabolized by hepatic microsomal enzymes with low therapeutic ratios or being given to patients with hepatic impairment may require adjustment to maintain optimal therapeutic blood levels when starting or stopping concomitant therapy with ticlopidine.
		• Administration of TICLID (ticlopidine hcl) after antacids resulted in an 18% decrease in plasma levels of ticlopidine.
		• Coadministration of TICLID (ticlopidine hcl) with digoxin resulted in a slight decrease (approximately 15%) in digoxin plasma levels. Little or no change in therapeutic efficacy of digoxin would be expected.
280	Tinidazole	• Alcoholic beverages or preparations containing alcohol or propylene glycol may cause abdominal cramps, nausea, vomiting, headaches, and flushing.
		• Possible increased PT and enhanced anticoagulant effects.
		• Possible prolonged half-life, decreased clearance, and increased plasma concentrations of tinidazole.
		• Studies using metronidazole indicate cholestyramine decreases oral bioavailability of the nitroimidazole by 21%.
		• Experience with metronidazole and disulfiram indicates psychotic reactions can occur; such reactions not reported to date with tinidazole.

S. No.	Name of the Drug	Reported Drug Interaction
281	Tocainide	• Beta-blocking agents, other antiarrhythmic agents, anticoagulants, and diuretics, without evidence of clinically significant interactions. Nevertheless, caution should be exercised in the use of multiple drug therapy.
		• Effective in digitalized and nondigitalized patients. In 17 patients withrefractoryventricular arrhythmiason concomitant therapy, serumdigoxin levels (1.1 ±0.4 ng/mL) remained in the expectednormal range(0.5–2.5 ng/mL) during tocainide administration.
282	Tolbutamide	• The hypoglycemia action of sulfonylurea may be potentiated by certain drugs including nonsteroidal antiinflammatory agents and other drugs that are highly protein bound, salicylates, sulfonamides, chloramphenicol, probenecid, coumarins, monoamine oxidase inhibitors, and beta-adrenergic blocking agents. When such drugs are administered to a patient receiving Tolbutamide, the patient should be observed closely for hypoglycemia. When such drugs are withdrawn from a patient receiving Tolbutamide, the patient should be observed closely for loss of control.
		• Certain drugs tend to produce hyperglycemia and may lead to loss of control. These drugs include the thiazides and other diuretics, corticosteroids, phenothiazines, thyroid products, estrogens, oral contraceptives, phenytoin, nicotinic acid, sympathomimetics, calcium channel blocking drugs, and isoniazid. When such drugs are administered to a patient receiving Tolbutamide, the patient should be closely observed for loss of control. When such drugs are withdrawn from a patient receiving Tolbutamide, the patient should be observed closely for hypoglycemia.
		• A potential interaction between oral miconazole and oral hypoglycemic agents leading to severe hypoglycemia has been reported. Whether this interaction also occurs with the intravenous, topical or vaginal preparations of miconazole is not known.
283	Topiramate	• Increased plasma amitriptyline concentrations.
		• Decreased plasma concentrations of topiramate.
		• Possible increased risk of kidney stone formation; possible increased risk of hyperthermia.
		• Enhanced CNS depression.

TABLE 12.2 (Continued)

S. No.	Name of the Drug	Reported Drug Interaction
284	Tramadol	• Possible decrease in serum digoxin concentrations.
		• Increased plasma topiramate concentrations.
		• Interaction With Central Nervous System (CNS) Depressants ULTRAM® should be used with caution and in reduced dosages when administered to patients receiving CNS depressants such as alcohol, opioids, anesthetic agents, narcotics, phenothiazines, tranquilizers or sedative hypnotics. ULTRAM® increases the risk of CNS and respiratory depression in these patients. Interactions with Alcohol and Drugs of Abuse Tramadol may be expected to have additive effects when used in conjunction with alcohol, other opioids, or illicit drugs that cause central nervous system depression. Increased Intracranial Pressure or Head Trauma ULTRAM® should be used with caution in patients with increased intracranial pressure or head injury.
		• The respiratory depressant effects of opioids include carbon dioxide retention and secondary elevation of cerebrospinal fluid pressure, and may be markedly exaggerated in these patients. Additionally, pupillary changes (miosis) from tramadol may obscure the existence, extent, or course of intracranial pathology.
285	Trazodone	• Potential for drug interactions when trazodone is given with CYP3A4 inhibitors. Ritonavir, a potent CYP3A4 inhibitor, increased the Cmax, AUC, and elimination half-life, and decreased clearance of trazodone after administration of ritonavir twice daily for 2 days.
		• Carbamazepine reduced plasma concentrations of trazodone when coadministered. Patients should be closely monitored to see if there is a need for an increased dose of trazodone when taking both drugs.
		• Increased serum digoxin or phenytoin levels have been reported to occur in patients receiving.
		• Concurrent administration withelectroshock therapyshould be avoided because of the absence of experience in this area.
		• There have been reports of increased and decreased prothrombin time occurring in warfarinized patients who take DESYREL (trazodone hydrochloride

S. No.	Name of the Drug	Reported Drug Interaction
286	Triazolam	• Possible increased plasma concentrations of triazolam.
		• Increased plasma concentrations and decreased clearance of triazolam.
		• Possible increased plasma concentrations of triazolam.
		• Additive CNS depressant effects.
287	Trimethoprim	• Increased dapsone concentrations and increased risk of dapsone-associated adverse effects (e.g., methemoglobinemia);possible increased trimethoprim concentrations, but no evidence of increased risk of trimethoprim-associated adverse effects.
		• May inhibit phenytoin metabolism resulting in increased half-life and decreased clearance of phenytoin.
		• Possible interference with Jaffe alkaline picrate assay resulting in falsely elevated creatinine concentrations.
		• Possible interference with serum methotrexate assays if competitive binding protein technique (CBPA) is used with a bacterial dihydrofolate reductase as the binding protein;interference does not occur if methotrexate is measured using radioimmunoassay (RIA).
288	Tropicamide	• Tropicamide may interfere with ocular antihypertensive action of carbachol.
		• Tropicamide may interfere with ocular antihypertensive action of ophthalmic cholinesterase inhibitors.
		• Tropicamide may interfere with ocular antihypertensive action of pilocarpine
289	Urokinase	• Increased risk of hemorrhage.
290	Valciclovir	• No effect on acyclovir pharmacokinetics.
		• Potential increased peak plasma concentrations and AUC of acyclovir.
		• No effect on pharmacokinetics of acyclovir or digoxin

TABLE 12.2 (Continued)

S. No.	Name of the Drug	Reported Drug Interaction
291	Valproate	• Limited pharmacokinetic studies reveal little to no interaction following concomitant administration.
		• May reduce plasma anticonvulsant concentrations to subtherapeutic levels; an increase in seizure frequency and a worsening in the EEG may be observed.
		• Additive CNS depression may occur.
		• Decreased plasma clearance of amitriptyline and nortriptyline (the pharmacologically active metabolite of amitriptyline).
		• May increase unbound fraction of warfarin.
		• Depression may occur (particularly with phenobarbital and primidone).
		• Concomitant administration of valproic acid and phenobarbital (or primidone which is metabolized to phenobarbital) can result in increased plasma phenobarbital concentrations and excessive somnolence.
		• May displace phenytoin from protein binding and inhibit its metabolism.
292	Vancomycin	• *In vitro* evidence of synergistic antibacterial activity against *S. aureus*, nonenterococcal group D streptococci (*S. bovis*), enterococci, and viridians streptococciIncreased risk of ototoxicity and/or nephrotoxicity.
		• Possible increased risk of anaphylactoid reactions and increased risk of vancomycin infusion reactions in patients receiving anesthetic agents; erythema and histamine-like flushing reported.
293	Venlafaxine	• Metabolized by CYP isoenzymes, principally by CYP2D6 to O-desmethylvenlafaxine (ODV), its major active metabolite. Also metabolized by CYP3A4 (Seth, 2004). Relatively weak inhibitor of CYP2D6. Does not inhibit CYP1A2, CYP2C9, CYP2C19, or CYP3A4.

S. No.	Name of the Drug	Reported Drug Interaction
		• Inhibitors of CYP2D6 or 3A4: Potential pharmacokinetic interaction (increased plasma venlafaxine concentrations). Use caution if administered concomitantly with drugs that inhibit both CYP2D6 and 3A4.
		• Potential pharmacokinetic interaction (increased substrate plasma concentrations) with concomitant use of drugs that are metabolized by CYP2D6.
		• Increased plasma venlafaxine concentrations, but no effect on ODV pharmacokinetics.
294	Verapamil	• Additive hypotensive effects.
		• Increased hypotensive effect, possibly excessive in some patients.
		• Additive negative effects on myocardial contractility, heart rate, and AV conductionExcessive bradycardia and AV block, including complete heart block, reported in hypertensive patients.
		• Increased serum concentrations and efficacy of doxorubicinDecreased verapamil absorption when used with COPP (cyclophosphamide, vincristine, procarbazine, prednisone) or VAC (vindesine, doxorubicin, cisplatin) regimen. Decreased paclitaxel clearance (interaction with R-verapamil).
		• Pharmacokinetic interaction not observed; no evidence of additive CNS depressant effects.
295	Vigabatrin	• Both increased and decreased plasma carbamazepine concentrations reported; no apparent effect on plasma vigabatrin concentrations.
		• Increased peak plasma concentrations and decreased time to peak concentration of clonazepam; no substantial change in plasma concentrations of vigabatrin. No evidence of additive CNS effects.
		• Slightly increased AUC of vigabatrin; no change in plasma concentrations or AUC of felbamateClinically important pharmacokinetic interaction unlikely.
		• No substantial effect on CYP3A4-mediated metabolism of ethinyl estradiol/levonorgestrel; unlikely to affect efficacy of steroid oral contraceptivesPharmacokinetics of vigabatrin not substantially affected.

TABLE 12.2 (Continued)

S. No.	Name of the Drug	Reported Drug Interaction
296	Vinblastine	• Inhibitors of CYP3A: potential pharmacokinetic interaction (inhibition of vinblastine metabolism); earlier onset and/or increased severity of adverse effects of vinblastine may occur. Use concomitantly with caution.
		• Itraconazole: Earlier onset and/or increased severity of neuromuscular effects reported with another vinca alkaloid (vincristine). Voriconazole: Possible neurotoxicity.
		• Decreased serum concentrations of phenytoin and increased seizure activity reported.
		• Possible increased tolterodine concentrations.
297	Vincristine	• Possible reduction in hepatic clearance of vincristine.
		• Earlier onset and/or increased severity of neuromuscular adverse effects reported with concomitant use, probably related to inhibition of vincristine metabolism.
		• Potential increased risk of serious adverse respiratory effects with concomitant use, particularly in preexisting pulmonary dysfunction.
		• Decreased serum concentrations of phenytoin and increased seizure activity reported with combination chemotherapy regimens that included vincristine, possibly as a result of decreased absorption and/or increased metabolism of phenytoin.
298	Warfarin	• warfarin and other coumarin derivatives can occur via pharmacodynamic interactions (e.g., impaired hemostasis; increased or decreased intestinal synthesis or absorption of vitamin K; altered distribution or metabolism of vitamin K; increased warfarin affinity for receptor sites; decreased synthesis and/or increased catabolism of functional blood coagulation factors II, VII, IX, and X; interference with platelet function or fibrinolysis; ulcerogenic effects) or pharmacokinetic interactions (e.g., increased or decreased rate of warfarin metabolism; increased or decreased protein binding Such interactions may increase or decrease response to coumarin derivatives.

S. No.	Name of the Drug	Reported Drug Interaction
		• Potential pharmacokinetic interaction with inhibitors or inducers of CYP2C9, 1A2, or 3A4 (*increased* warfarin exposure with concomitant inhibitors, *decreased* warfarin exposure with concomitant inducers). Closely monitor INR in patients who initiate, discontinue, or change dosages of these concomitant drugs. List of drugs is not all-inclusive.
		• Potential for increased anticoagulant effects however, conflicting data exist regarding clinical importance.
299	Xylometazoline	• May increase the risk of hypertensive crisis. Maprotiline, tricyclic antidepressants: May increase effects of xylometazoline. Herbal None known. Food None known.
300	Zidovudine	• No clinically important pharmacokinetic interactions. *In vitro* evidence of synergistic antiretroviral effects.
		• Pharmacokinetic interactions unlikely.
		• Increased toxicity reportedhas been used concomitantly without increased toxicity.
		• Rifabutin: Pharmacokinetic interactions unlikely.
		• Rifampin: Decreased zidovudine AUC.
		• No change in zidovudine AUC; possible decreased trough zidovudine concentrations.
		• No *in vitro* evidence of antagonistic antiretroviral effects.
		• No pharmacokinetic interaction with cidofovir; however, cidofovir must be given concomitantly with probenecid and probenecid can reduce zidovudine clearance.
		• Decreased zidovudine concentrations and AUC; no effect on didanosine concentrations or AUC.
		• *In vitro* evidence of synergistic antiretroviral effects.

mechanisms so when such drugs are started or stopped the prescriber must be alert to the possibility of drug interactions. Their early detection could enable reconsideration of the culprit treatment regimen and prudent management if they do lead to adverse events.

- **Recognizing potentially harmful interactions:** Drug interactions present a health risk to patients and a great challenge to the physician as monitoring the patient's therapy is a standard of care expected by the patients and the liability of interactions rests squarely on the physician who fails to recognize potentially harmful interactions to avoid extra costs of healthcare.

- **Adjustment of the dosage:** Many interactions are dose related; therefore, by monitoring the patient closely, the effects of the interaction can often be allowed for by adjusting the dosage. Some interactions can be avoided by using another member of the same group of drugs.

12.6   ROLE OF DRUG INTERACTION CENTER

Concomitant use different drugs may yield excessive risk for adverse drug reactions and it is a challenging task to do surveillance on the safety profile of the interaction between different drugs.

Following are the some significant role of drug interaction detection centers:

- **Detection of drug–drug interactionsin the post marketing period:** Generally, the detection of possible interactions is based on the following concept: when a suspected ADR is reported more frequently in the combination of two drugs compared with the situation where they are used alone, this association might indicate the existence of a drug–drug interaction.

- **Surveillance** schemes based on spontaneous reporting systems (SRS) are a cornerstone of the early detection of drug hazards that are novel by virtue of their clinical nature, severity and/or frequency.

- **Looking for new safety concerns**: The FDA Adverse Event Reporting System (FAERS) is a useful tool for FDA for activities such as looking for new safety concerns that might be related to a marketed product, evaluating a manufacturer's compliance to reporting regulations and responding to outside requests for information.

The reports in FAERS are evaluated by clinical reviewers in the Center for Drug Evaluation and Research (CDER) and the Center for Biologics Evaluation and Research (CBER) to monitor the safety of products after they are approved by FDA. If a potential safety concern is identified in FAERS, further evaluation is performed.

- **Regulatory action:** Based on an evaluation of the potential safety concern, FDA may take regulatory action(s) to improve product safety and protect the public health, such as updating a product's labeling information, restricting the use of the drug, communicating new safety information to the public, or, in rare cases, removing a product from the market. All reports are individually reviewed and detection depends upon the skills and memory of the professionals involved. Since this procedure is time consuming and the number of reports is increasing, there is a growing need for an automated system that could facilitate the process of data arrangement and detection of ADR, especially in some special cases such as drug–drug interactions.

KEYWORDS

- **drug**
- **food**
- **interaction**
- **mechanism**
- **pharmacodynamic**
- **pharmacokinetic**

REFERENCES

Barar, F. S. K. (2000). *Essentials of Pharmacotherapeutics*, S. Chand and Company Ltd. New Delhi.

Blaschke, T. F., Cohen S. N., Tatro D. S., & Rubin P. C. (1981). Drug-Drug-Interactions and aging. In: Jarvi, K. L. F., Greenbatt, D. J., Harman, D. (eds.). *Clinical Pharmacology in the Aged Patient*. New York, Raven.

Das, P. K. (2004). *Pharmacology*, 2nd ed., Elsevier: New Delhi.

Golan, D. E., Tashjian, A. H. Jr., & Armstrong, E. J. (2008). *Principle of Pharmacology. The Pathophysiologic Basis of Drug Therapy*, Wolter Kluwer Pvt. Ltd. New Delhi.

Katzung, B. G. (2007). *Basic and Clinical Pharmacology*, 10th ed.: McGraw-Hill: Boston.

Linnarson, R. (1993). Drug interactions in primary health care. A retrospective data base study and its implications for the design of a computerized decision support system. *Scand J Prim Health Care, 11*, 181–186.

Malone, D. C. (2005). Assessment of potential drug–drug interactions with a prescription claims database. *Am J Health-System Pharm, 62*, 1983–1991.

Mozayani, A. (2011). *Handbook of Drug Interaction: A Clinical and Forensic Guide*, 2nd ed., Humana Press: New Jersey.

Seth, S. D. (2004). *Textbook of Pharmacology*, 2nd ed., Elsevier: New Delhi.

Sloan, R. W. (1983). Drug Interactions. *Am Fam Physician, 27*, 229.

INDEX